T0211839

Lecture Notes in Artificial Intelligence   12274

Subseries of Lecture Notes in Computer Science

More information about this series at http://www.springer.com/series/1244

Gang Li · Heng Tao Shen ·
Ye Yuan · Xiaoyang Wang ·
Huawen Liu · Xiang Zhao (Eds.)

Knowledge Science, Engineering and Management

13th International Conference, KSEM 2020
Hangzhou, China, August 28–30, 2020
Proceedings, Part I

 Springer

Editors
Gang Li (ID)
Deakin University
Geelong, VIC, Australia

Ye Yuan
Beijing Institute of Technology
Beijing, China

Huawen Liu (ID)
Zhejiang Normal University
Jinhua, China

Heng Tao Shen (ID)
University of Electronic Science
and Technology of China
Chengdu, China

Xiaoyang Wang (ID)
Zhejiang Gongshang University
Hangzhou, China

Xiang Zhao (ID)
National University of Defense Technology
Changsha, China

ISSN 0302-9743   ISSN 1611-3349 (electronic)
Lecture Notes in Artificial Intelligence
ISBN 978-3-030-55129-2   ISBN 978-3-030-55130-8 (eBook)
https://doi.org/10.1007/978-3-030-55130-8

LNCS Sublibrary: SL7 – Artificial Intelligence

This Springer imprint is published by the registered company Springer Nature Switzerland AG
The registered company address is: Gewerbestrasse 11, 6330 Cham, Switzerland

Preface

The International Conference on Knowledge Science, Engineering and Management (KSEM) provides a forum for researchers in the broad areas of knowledge science, knowledge engineering, and knowledge management to exchange ideas and to report state-of-the-art research results. KSEM 2020 is the 13th in this series, which builds on the success of 12 previous events in Guilin, China (KSEM 2006); Melbourne, Australia (KSEM 2007); Vienna, Austria (KSEM 2009); Belfast, UK (KSEM 2010); Irvine, USA (KSEM 2011); Dalian, China (KSEM 2013); Sibiu, Romania (KSEM 2014); Chongqing, China (KSEM 2015); Passau, Germany (KSEM 2016); Melbourne, Australia (KSEM 2017); Changchun, China (KSEM 2018); and Athens, Greece (KSEM 2019).

The selection process this year was, as always, competitive. We received received 291 submissions, and each submitted paper was reviewed by at least three members of the Program Committee (PC) (including thorough evaluations by the PC co-chairs). Following this independent review, there were discussions between reviewers and PC chairs. A total of 58 papers were selected as full papers (19.9%), and 27 papers as short papers (9.3%), yielding a combined acceptance rate of 29.2%.

We were honoured to have three prestigious scholars giving keynote speeches at the conference: Prof. Zhi Jin (Peking University, China), Prof. Fei Wu (Zhejiang University, China), and Prof. Feifei Li (Alibaba Group, China). The abstracts of Prof. Jin's and Prof Wu's talks are included in this volume.

We would like to thank everyone who participated in the development of the KSEM 2020 program. In particular, we would give special thanks to the PC for their diligence and concern for the quality of the program, and also for their detailed feedback to the authors. The general organization of the conference also relies on the efforts of KSEM 2020 Organizing Committee.

Moreover, we would like to express our gratitude to the KSEM Steering Committee honorary chair, Prof. Ruqian Lu (Chinese Academy of Sciences, China), the KSEM Steering Committee chair, Prof. Dimitris Karagiannis (University of Vienna, Austria), Prof. Chengqi Zhang (University of Technology Sydney, Australia), who provided insight and support during all the stages of this effort, and the members of the Steering Committee, who followed the progress of the conference very closely with sharp comments and helpful suggestions. We also really appreciate the KSEM 2020 general co-chairs, Prof. Hai Jin (Huazhong University of Science and Technology, China), Prof. Xuemin Lin (University of New South Wales, Australia), and Prof. Xun Wang (Zhejiang Gongshang University, China), who were extremely supportive in our efforts and in the general success of the conference.

We would like to thank the members of all the other committees and, in particular, those of the Local Organizing Committee, who worked diligently for more than a year to provide a wonderful experience to the KSEM participants. We are also grateful to Springer for the publication of this volume, who worked very efficiently and effectively.

Finally and most importantly, we thank all the authors, who are the primary reason why KSEM 2020 is so exciting, and why it will be the premier forum for presentation and discussion of innovative ideas, research results, and experience from around the world as well as highlight activities in the related areas.

June 2020

Gang Li
Heng Tao Shen
Ye Yuan

Organization

Steering Committee

Ruqian Lu (Honorary Chair)	Chinese Academy of Sciences, China
Dimitris Karagiannis (Chair)	University of Vienna, Austria
Yaxin Bi	Ulster University, UK
Christos Douligeris	University of Piraeus, Greece
Zhi Jin	Peking University, China
Claudiu Kifor	University of Sibiu, Romania
Gang Li	Deakin University, Australia
Yoshiteru Nakamori	Japan Advanced Institute of Science and Technology, Japan
Jorg Siekmann	German Research Centre of Artificial Intelligence, Germany
Martin Wirsing	Ludwig-Maximilians-Universität München, Germany
Hui Xiong	Rutgers University, USA
Bo Yang	Jilin University, China
Chengqi Zhang	University of Technology Sydney, Australia
Zili Zhang	Southwest University, China

Organizing Committee

Honorary Co-chairs

Ruqian Lu	Chinese Academy of Sciences, China
Chengqi Zhang	University of Technology Sydney, Australia

General Co-chairs

Hai Jin	Huazhong University of Science and Technology, China
Xuemin Lin	University of New South Wales, Australia
Xun Wang	Zhejiang Gongshang University, China

Program Committee Co-chairs

Gang Li	Deakin University, Australia
Hengtao Shen	University of Electronic Science and Technology of China, China
Ye Yuan	Beijing Institute of Technology, China

Keynote, Special Sessions, and Tutorial Chair

Zili Zhang   Southwest University, China

Publication Committee Co-chairs

Huawen Liu   Zhejiang Normal University, China
Xiang Zhao   National University of Defense Technology, China

Publicity Chair

Xiaoqin Zhang   Wenzhou University, China

Local Organizing Committee Co-chairs

Xiaoyang Wang   Zhejiang Gongshang University, China
Zhenguang Liu   Zhejiang Gongshang University, China
Zhihai Wang   Zhejiang Gongshang University, China
Xijuan Liu   Zhejiang Gongshang University, China

Program Committee

Klaus-Dieter Althoff   DFKI and University of Hildesheim, Germany
Serge Autexier   DFKI, Germany
Massimo Benerecetti   Università di Napoli Federico II, Italy
Salem Benferhat   Université d'Artois, France
Xin Bi   Northeastern University, China
Robert Andrei Buchmann   Babes-Bolyai University of Cluj Napoca, Romania
Chen Chen   Zhejiang Gongshang University, China
Hechang Chen   Jilin University, China
Lifei Chen   Fujian Normal Univeristy, China
Dawei Cheng   Shanghai Jiao Tong University, China
Yurong Cheng   Beijing Institute of Technology, China
Yong Deng   Southwest University, China
Linlin Ding   Liaoning University, China
Shuai Ding   Hefei University of Technology, China
Christos Douligeris   University of Piraeus, Greece
Xiaoliang Fan   Xiamen University, China
Knut Hinkelmann   FHNW University of Applied Sciences and Arts
                                        Northwestern Switzerland, Switzerland
Guangyan Huang   Deakin University, Australia
Hong Huang   UGOE, Germany
Zhisheng Huang   Vrije Universiteit Amsterdam, The Netherlands
Frank Jiang   Deakin University, Australia
Jiaojiao Jiang   RMIT University, Australia
Wang Jinlong   Qingdao University of Technology, China
Mouna Kamel   IRIT, Université Toulouse III - Paul Sabatier, France
Krzysztof Kluza   AGH University of Science and Technology, Poland

Longbin Lai	Alibaba Group, China
Yong Lai	Jilin University, China
Qiujun Lan	Hunan University, China
Cheng Li	National University of Singapore, Singapore
Ge Li	Peking University, China
Jianxin Li	Deakin University, Australia
Li Li	Southwest University, China
Qian Li	Chinese Academy of Sciences, China
Shu Li	Chinese Academy of Sciences, China
Ximing Li	Jilin University, China
Xinyi Li	National University of Defense Technology, China
Yanhui Li	Northeastern University, China
Yuan Li	North China University of Technology, China
Shizhong Liao	Tianjin University, China
Huawen Liu	Zhejiang Normal University, China
Shaowu Liu	University of Technology Sydney, Australia
Zhenguang Liu	Zhejiang Gongshang University, China
Wei Luo	Deakin University, Australia
Xudong Luo	Guangxi Normal University, China
Bo Ma	Chinese Academy of Sciences, China
Yuliang Ma	Northeastern University, China
Stewart Massie	Robert Gordon University, UK
Maheswari N	VIT University, India
Myunghwan Na	Chonnam National University, South Korea
Bo Ning	Dalian Maritime University, China
Oleg Okun	Cognizant Technology Solutions GmbH, China
Jun-Jie Peng	Shanghai University, China
Guilin Qi	Southeast University, China
Ulrich Reimer	University of Applied Sciences St. Gallen, Switzerland
Wei Ren	Southwest University, China
Zhitao Shen	Ant Financial Services Group, China
Leilei Sun	Beihang University, China
Jianlong Tan	Chinese Academy of Sciences, China
Zhen Tan	National University of Defense Technology, China
Yongxin Tong	Beihang University, China
Daniel Volovici	ULB Sibiu, Romania
Quan Vu	Deakin University, Australia
Hongtao Wang	North China Electric Power University, China
Jing Wang	The University of Tokyo, Japan
Kewen Wang	Griffith University, Australia
Xiaoyang Wang	Zhejiang Gongshang University, China
Zhichao Wang	Tsinghua University, China
Le Wu	Hefei University of Technology, China
Jia Xu	Guangxi University, China
Tong Xu	University of Science and Technology of China, China
Ziqi Yan	Beijing Jiaotong University, China

Bo Yang	Jilin University, China
Jianye Yang	Hunan University, China
Shiyu Yang	East China Normal University, China
Shuiqiao Yang	University of Technology Sydney, Australia
Yating Yang	Chinese Academy of Sciences, China
Feng Yi	UESTC: Zhongshan College, China
Min Yu	Chinese Academy of Sciences, China
Long Yuan	Nanjing University of Science and Technology, China
Qingtian Zeng	Shandong University of Science and Technology, China
Chengyuan Zhang	Central South University, China
Chris Zhang	Chinese Science Academy, China
Chunxia Zhang	Beijing Institute of Technology, China
Fan Zhang	Guangzhou University, China
Songmao Zhang	Chinese Academy of Sciences, China
Zili Zhang	Deakin University, Australia
Xiang Zhao	National University of Defense Technology, China
Ye Zhu	Monash University, Australia
Yi Zhuang	Zhejiang Gongshang University, China
Jiali Zuo	Jiangxi Normal University, China

Additional Reviewers

Weronika T. Adrian	Piotr Wiśniewski
Taotao Cai	Yanping Wu
Xiaojuan Cheng	Zhiwei Yang
Viktor Eisenstadt	Xuan Zang
Glenn Forbes	Yunke Zhang
Nur Haldar	Qianru Zhou
Kongzhang Hao	Borui Cai
Sili Huang	Hui Chen
Francesco Isgro	Shangfu Duan
Gongjin Lan	Uno Fang
Enhui Li	Huan Gao
Shuxia Lin	Xin Han
Patryk Orzechowski	Xin He
Roberto Prevete	Xuqian Huang
Najmeh Samadiani	Krzysztof Kutt
Bi Sheng	Boyang Li
Beat Tödtli	Jiwen Lin
Beibei Wang	Yuxin Liu
Yixuan Wang	Ning Pang

Pascal Reuss
Jakob Michael Schoenborn
Atsushi Suzuki
Ashish Upadhyay
Shuai Wang
Yuhan Wang

Haiyan Wu
Zhengyi Yang
Guoxian Yu
Roozbeh Zarei
Qing Zhao
Xianglin Zuo

Abstracts of Invited Talks

Learning from Source Code

Zhi Jin

Key Laboratory of High-Confidence of Software Technologies (MoE),
Peking University, China
zhijin@pku.edu.cn

Abstract. Human beings communicate and exchange knowledge with each other. The system of communication and knowledge exchanging among human beings is natural language, which is an ordinary, instinctive part of everyday life. Although natural languages have complex forms of expressive, it is most often simple, expedient and repetitive with everyday human communication evolved. This naturalness together with rich resources and advanced techniques has led to a revolution in natural language processing that help to automatically extract knowledge from natural language documents, i.e. learning from text documents.

Although program languages are clearly artificial and highly restricted languages, programming is of course for telling computers what to do but is also as much an act of communication, for explaining to human beings what we want a computer to do[1]. In this sense, we may think of applying machine learning techniques to source code, despite its strange syntax and awash with punctuation, etc., to extract knowledge from it. The good thing is the very large publicly available corpora of open-source code is enabling a new, rigorous, statistical approach to wide range of applications, in program analysis, software mining and program summarization.

This talk will demonstrate the long, ongoing and fruitful journey on exploiting the potential power of deep learning techniques in the area of software engineering. It will show how to model the code[2,3]. It will also show how such models can be leveraged to support software engineers to perform different tasks that require proficient programming knowledge, such as code prediction

---

[1] A. Hindle, E. T. Barr, M. Gabel, Z. Su and P. Devanbu, On the Naturalness of Software, Communication of the ACM, 59(5): 122–131, 2016.

[2] L. Mou, G. Li, L. Zhang, T. Wang and Z. Jin, Convolutional Neural Networks over Tree Structures for Programming Language Processing, AAAI 2016: 1287–1293.

[3] F. Liu, L. Zhang and Z. Jin, Modeling Programs Hierarchically with Stack-Augmented LSTM, The Journal of Systems and Software, https://doi.org/10.1016/j.jss.2020.110547.

and completion[4], code clone detection[5], code comments[6,7] and summarization[8], etc. The exploratory work show that code implies the learnable knowledge, more precisely the learnable tacit knowledge. Although such knowledge is difficult to transfer among human beings, it is able to transfer among the automatically programming tasks. A vision for future research in this area will be laid out as the conclusion.

Keywords: Software · Source code · Program languages · Programming knowledge

[4] B. Wei, G. Li, X. Xia, Z. Fu and Z. Jin, Code Generation as a Dual Task of Code Summarization, NeurIPS 2019.

[5] W. Wang, G. Li, B. Ma, X. Xia and Z. Jin, Detecting Code Clones with Graph Neural Network and Flow-Augmented Abstract Syntax Tree, SANER 2020: 261–271.

[6] X. Hu, G. Li, X. Xia, D. Lo, S. Lu and Z. Jin, Deep Code Comment Generation, ICPC 2018: 200–210.

[7] X. Hu, G. Li, X. Xia, D. Lo, S. Lu and Z. Jin, Deep Code Comment Generation with Hybrid Lexical and Syntactical Information, Empirical Software Engineering (2020) 25: 2179–2217.

[8] X. Hu, G. Li, X. Xia, D. Lo, S. Lu and Z. Jin, Summarizing Source Code with Transferred API Knowledge, IJCAI 2018: 2269–2275.

Memory-Augmented Sequence2equence Learning

Fei Wu

Zhejiang University, China
wufei@cs.zju.edu.cn

Abstract. Neural networks with a memory capacity provide a promising approach to media understanding (e.g., Q-A and visual classification). In this talk, I will present how to utilize the information in external memory to boost media understanding. In general, the relevant information (e.g., knowledge instance and exemplar data) w.r.t the input data is sparked from external memory in the manner of memory-augmented learning. Memory-augmented learning is an appropriate method to integrate data-driven learning, knowledge-guided inference and experience exploration.

Keywords: Media understanding · Memory-augmented learning

Contents – Part I

Knowledge Graph

Knowledge Representation

Knowledge-Based Systems

Data Processing and Mining

Contents – Part II

Machine Learning

Recommendation Algorithms and Systems

Social Knowledge Analysis and Management

Text Mining and Document Analysis

Deep Learning

Knowledge Graph

Event-centric Tourism Knowledge Graph—A Case Study of Hainan

Jie Wu[1], Xinning Zhu[2($\boxtimes$)], Chunhong Zhang[2], and Zheng Hu[1]

[1] State Key Laboratory of Networking and Switching Technology,
Beijing University of Posts and Telecommunications, Beijing, China
[2] School of Information and Communication Engineering,
Beijing University of Posts and Telecommunications, Beijing, China
zhuxn@bupt.edu.cn

Abstract. Knowledge graphs have become increasingly important in tourism industry recently for their capability to power insights for applications like recommendations, question answering and so on. However, traditional tourism knowledge graph is a knowledge base which focuses on the static facts about entities, such as hotels, attractions, while ignoring events or activities of tourists' trips and temporal relations.

In this paper, we first propose an Event-centric Tourism Knowledge Graph (ETKG) in order to model the temporal and spatial dynamics of tourists trips. ETKG is centered on activities during the trip and regards tourists' trajectories as carriers. We extract valuable information from over 18 thousand travel notes crawled from Internet, and define an ETKG schema to model tourism-related events and their key properties. An ETKG based on touristic data in Hainan is presented which incorporates 86977 events (50.61% of them have complete time, activity, location information) and 7132 journeys. To demonstrate the benefits of ETKG, we propose an Event-centric Tourism Knowledge Graph Convolutional Network (ETKGCN) for POI recommendations, which facilitates incorporating tourists behavior patterns obtained from ETKG, so as to capture the relations between users and POIs more efficiently. The offline experiment results show that our approach outperforms strong recommender baselines, so that it validates the effectiveness of ETKG.

Keywords: Knowledge graph · Event evolutionary graph · Intelligent tourism · Recommendation system

1 Introduction

With the uprising of large scale knowledge graphs (KGs) like Wikidata, DBpedia, and YAGO, various knowledge-based applications, such as semantic search, question answering [12], recommendation system [14] and so on, can benefit from the substantial and valuable knowledge in the KGs, and achieve significant performance improvement. Recently, in the field of tourism, knowledge graphs are

© Springer Nature Switzerland AG 2020
G. Li et al. (Eds.): KSEM 2020, LNAI 12274, pp. 3–15, 2020.
https://doi.org/10.1007/978-3-030-55130-8_1

leveraged to recommend travel attractions or tour plans [2,6]. However, existing tourism knowledge graphs mainly focus on modeling static knowledge about tourists trips, such as opening hours of attractions, nearby restaurants, addresses of hotels and so on. This static knowledge may not provide enough information to meet the needs of tourists in some cases. We analyze these issues.

- Tourist activities have not received enough attention in the existing tourism knowledge graphs. In fact, various activities in travel have an important impact on the experience of tourists.
- In the current knowledge graphs, tourist attractions, which make up the entire trip, are described as separate entities, ignoring the transfer relationship between attractions. Therefore, it's difficult to recommend or answer questions about travel routes.
- It is difficult to answer questions concerned with most" and best" based on existing knowledge graphs since there is no relation strength or entity popularity provided. For example, when it comes to recommending the most suitable places to surf for a tourist, it's hard to get the best results with traditional knowledge graphs.

Inspired by the event-centric knowledge graph proposed by Rospocher [13] and event evolutionary graph by Liu [10], we turn the focus of knowledge graph into events. In the literature, an event refers to a specific occurrence at a certain time and location, involving one or more participants. In this study, we consider tourist activities as events in tourism scenario, such as rain forest rafting, water-skiing, and whale watching, etc., which happens sometime at certain places. Efficiently incorporating these events and spatial-temporal information into a variety of tourism applications will be very beneficial. So we propose an Event-centric Tourism Knowledge Graph (ETKG) that interconnects tourist activities using temporal relations in chronological order, and with spatial information at that time. In addition, features about tourist trips like trip type, trip consumption are integrated into ETKG to enrich it. In this work, we construct an ETKG based on a large number of travel notes about tourism in Hainan province, which is a popular tourist destination in China. The travel notes record where the tourists visited and what they did during the trips for the purpose of sharing their travel experiences, mostly in chronological order. The information about tourism activities included in travel notes is of high quality and suitable for generating the initial ETKG.

To verify the effectiveness of ETKG, we present an application of POI recommendation which can make use of information generated from ETKG so as to improve the accuracy of recommendation. We propose Event-centric Tourism Knowledge Graph Convolutional Networks (ETKGCN), which can better learn user and item representations by incorporating information inferred from ETKG. Through experiments on real-world datasets in the tourism, we demonstrate that ETKGCN achieves substantial gains over several state-of-the-art baselines.

In summary, our key contributions are presented as follows:

– We present an Event-centric Tourism Knowledge Graph (ETKG) which integrates tourism events and their temporal relations to represent tourists' activities in an efficient way. The schema of ETKG built upon the Simple Event Model [5] with additional properties and classes is also presented.
– We propose an efficient pipeline to construct ETKG. Event information can be extracted from travel notes accurately and organized in an appropriate form. Based on the pipeline, an ETKG about Hainan tourism is presented, which incorporates 86977 events, 680 locations, 7132 journey and 79845 temporal relations.
– A new recommendation system framework is proposed based on ETKG— ETKGCN, which demonstrates the effectiveness of ETKG.

The paper is organized as follows. In Sect. 2, we describe the related work. In Sect. 3, we present the pipeline to construct the graph. An application of ETKG for POI recommendation is shown in Sect. 4. Finally in Sect. 5, we provide a conclusion.

2 Related Work

Knowledge Graphs (KGs) are used extensively to enhance the results provided by popular search engines at first. These knowledge graphs are typically powered by structured data repositories such as Freebase, DBpedia, Yago, and Wikidata.

With the development of NLP technology, more and more knowledge graphs are constructed based on unstructured corpus [13], using named entity recognition [9] and relation extraction [8,11] technology. The most commonly used model of named entity recognition(NER) is BiLSTM+CRF [9]. As for relation extraction, there are also many new approaches in academia, such as distant supervision and transformer [3].

In the last few years, knowledge graphs have been used in many fields. In the tourism scene, some experts also begin to build knowledge graphs to solve tourists' problems during the travel [2,6]. DBtravel [2] is a tourism-oriented knowledge graph generated from the collaborative travel site Wikitravel. Jorro-Aragoneses and his team proposed an adaptation process to generate accessible plans based on the retrieved plans by the recommender system in 2018 [6]. A KG is a type of directed heterogeneous graph in which nodes correspond to entities and edges correspond to relations. The introduction of Graph Convolutional Networks [17] accelerates the application of knowledge graph in industry, especially in the field of recommendation [14].

Event-centric knowledge graphs (ECKG) [13] was proposed in 2016. Compared with traditional KG, the focus of ECKG turns into the events in real world. In 2018, Simon Gottschalk built EventKG, a multilingual knowledge graph that integrates and harmonizes event-centric and temporal information regarding historical and contemporary events [4]. Both ECKG and EventKG were constructed based on SEM [5] and supported W3C standard [15]. In the same year, Professor Liu Ting's team from Harbin Institute of Technology put forward the concept of event evolutionary graph [10]. Event evolutionary graph puts more emphasis on the logical relationship between events and achieved excellent result in script event prediction task. However, it didn't support W3C standard. Our goal is to define a KG with the features of EventKG and Event evolutionary graph.

3   Construction of ETKG

In contrast with existing tourism knowledge graphs, Event-centric Tourism Knowledge Graph (ETKG) is centered on events, i.e. tourism activities in this study, and interconnects events using temporal relations in chronological order. Moreover, some additional features such as spatial information, attributes of journeys are incorporated into ETKG. In this way, we can obtain not only static information of tourism, but also some hidden patterns of various tourist behaviors inferred from inter-event relationship, which can be utilized as prior knowledge for tourism applications. We design an ETKG schema based on the Simple Event Model [5].

3.1   Generation Pipeline of ETKG

The construction process of ETKG is shown in Fig. 1 which generally consists of the following three parts. Firstly, we crawl travel notes from the web page, for the reason that travel notes not only record the activities of tourists in various scenic spots, but also contain some important tags, such as trip consumption, travel companions, types of travel and so on. These tags are made by tourists themselves when their travel notes are published, which has a high credibility. Secondly, we extract valuable information from travel notes, including trajectories, activities and tags mentioned above. Finally, the information extracted from travel notes was organized based on ETKG schema we proposed in order to model events in RDF. The ETKG schema is given in Fig. 2.

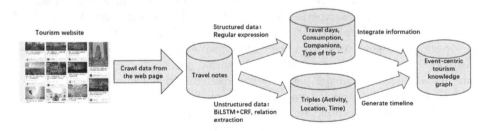

Fig. 1. ETKG generation pipeline.

In this work, we use a case study of Hainan tourism to illustrate the construction of ETKG and its application of POI recommendation. We crawl users' travel notes about Hainan from Ctrip, one of China's largest travel websites (https://www.ctrip.com). About 18 thousand travel notes related to Hainan are crawled. However, some of them just mention Hainan instead of recording the travel experiences in Hainan. After filtering these out, we get 7132 travel notes and then construct an Event-centric Tourism Knowledge Graph based on them.

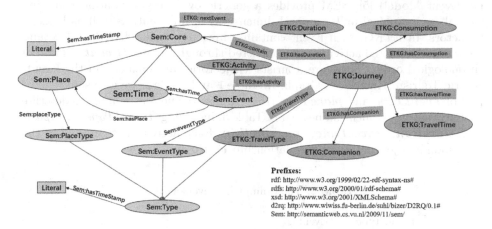

Fig. 2. Schema of Event-centric Tourism Knowledge Graph (*ETKG*). Our schema is based on *SEM* and the classes are colored gray. The blue arrows denote *rdfs:subClassOf* properties, while the black arrows visualize *rdfs:domain* and *rdfs:range* restrictions on properties. Classes and properties introduced in *ETKG* are colored green. (Color figure online)

Table 1. The examples of place type

Place type	Place
Shopping	Shopping mall, supermarket, store
Transportation	Airport, railway station, long-distance bus station, customs ports
Recreation	Park, gym, beach, soccer field, zoo, museum, scenic spots
Entertainment	Tea room, coffee shop, nightclub, pub, theater, beauty salon

Travel notes contain both structured and unstructured information. Structured information includes: travel days, trip consumption, travel companions, type of trip (self-driving travel, travel on a budget and so on) and month of travel. The structured information can be extracted by regular expression. As for unstructured information, this paper uses BiLSTM+CRF to extract events and entities from the body of travel notes, including: scenic spots visited by tourist, tourist's activities which correspond to events in this study, time for tourists to participate in activities.

The schema of the Event-centric Tourism Knowledge Graph is shown in Fig. 2. As we can see, the description of activity is inspired by SEM (Simple Event Model) [5]. SEM provides a generic event representation including topical, geographical and temporal dimensions of an event, as well as links to its actors. In ETKG, each Event contains three properties:*Activity*, *Place* and *Time*. The extracted activities are connected through the **nextEvent** relation in chronological order. Therefore, tourists' trajectories are included in the Event-centric Tourism Knowledge Graph. In Table 1 and Table 2, we give some examples of **eventType** and **placeType** [16]. We also define 5 attributes to describe tourists' journey, which can be seen in Table 3, including: **travelType**, **companion**, **duration**, **travelTime**, **consumption**. Each activity is connected to the corresponding journey through relationship **contain**.

Table 2. The examples of event type

Event type	Event
Dining	Eat BBQ, have a buffet, enjoy seafood, walk down the food court...
Sightseeing	See the sunrise, watch the sunset, go surfing, take a motorboat...
Rest	Sleep, have a rest, sunbathe
Entertainment	Go shopping, go to the spa, go to the coffee shop, go to nightclub...

Table 3. The definition of five attributes

Attribute	Meaning	Examples
TravelTime	The month when tourists went to Hainan	$1 \sim 12$
Duration	The number of days tourists visited Hainan	$2 \sim 20$
Consumption	Per capita consumption on the trip	$1000 \sim 10000$
TravelType	A brief summary of the journey	Self-driving, gourmet tour
Companion	The people who the user travel with	Parents, friends

3.2   Information Extraction and Organization

In this subsection, we give a detailed description how to extract events and entities from unstructured data. We have to extract three types of elements: activity, location, and time from the texts of travel notes. We turn the problem into a named entity identification task [9]. Our purpose is to tag every words of travel notes with labels: activity, location, time, None. BiLSTM+CRF [9] is selected to accomplish this task due to its high performance in this area.

To annotate travel notes and construct training and validation set, we built the dictionary of Hainan tourism after discussing with local tourism experts. The dictionary contains 79 kinds of activities, 680 locations and 5 types of time. The 79 kinds of activities are the most concerned by tourists and they are determined by experts. As the carrier of activity information, in the tourism scene, trajectories are often made up of places that tourists pay attention to during the trip, such as attractions, hotels and so on. The 680 locations almost cover all the scenic spots and hotels in Hainan. As for time, we divide a day into early morning, morning, noon, evening and late night after reading some travel notes to differentiate happening time for various activities.

After the processing of BiLSTM+CRF model, we get corresponding tags (activity, location, time, None) for each word in the travel notes. Then we need to organize them according to the ETKG schema we designed as shown in Fig. 2. There are two tasks to construct ETKG. First, an event is represented by a triple (activity, location, time) in the schema, so we have to match the time, location with the associated activity. Second, we need to get the temporal relations of events to interlink events within each specific journey. We propose an approach

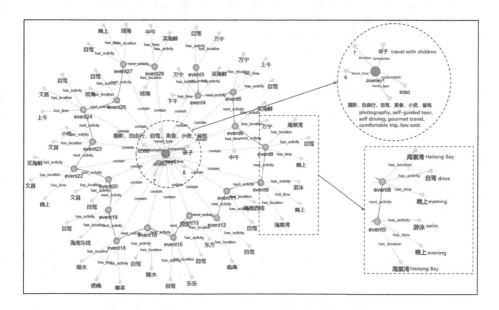

Fig. 3. Event-centric tourism knowledge graph of one journey.

to associate location and time with each activity and get relations between activities, which is shown in Algorithm 1. Based on an assumption that most tourists prefer to record their journey in chronological order, we get tourists' trajectories by the order of locations appearing in travel notes. Algorithm 1 generate a chain of events for each travel note tagged by BiLSTM+CRF model. And then combined with the attributes extracted from the structured data to construct an ETKG according to ETKG schema, so that it clearly describes the tourist's journey.

Through the above information extraction and organization methods, the Event-centric Tourism Knowledge Graph of 7132 journeys in Hainan is constructed which incorporates 86977 events, 140726 entities and 227703 relations (79845 temporal relations in it). 50.61% of these events contain time and location information. The ETKG of Hainan covers most of tourist attractions (680 sites) and most kinds of travel (26 types of journeys: self-driving tours, package tour and so on).

Algorithm 1. Approach to generate the chain of events

Input:
>   A travel note tagged by BiLSTM+CRF, T;
>   Trajectory of the tourist, $L(l_1, l_2, \ldots)$;

Output:
>   The chain of events in the travel note, $E(e_1, e_2, \ldots)$;

1: Initialize E with an empty list.
2: **for all** $l \in L$ **do**
3:     Find the paragraph P containing location l in T.
4:     Generate a sequence $A(a_1, a_2, \ldots)$ containing all the activity in P.
5:     **if** $len(A) = 0$ **then**
6:         Append triple $(None, l, None)$ to E
7:         continue
8:     **end if**
9:     **for all** $a \in A$ **do**
10:        Find the sentence S containing activity a in P
11:        **if** there is time t in S **then**
12:            Append triple (a, l, t) to E
13:        **else**
14:            Append triple $(a, l, None)$ to E
15:        **end if**
16:    **end for**
17: **end for**
18: **return** E

Figure 3 shows the ETKG of one journey. Obviously, we can learn most key information about this journey. The tourist came to Hainan with their children and have a seven-day road trip. Each person spent 6500 yuan during the trip and took many pictures. We can also know where the user had been and what activities they participated in at each location. With ETKG, the activities and travel routes for each journey can be obtain easily. After aggregations of each kind of journeys in ETKG, we can get event popularity and relation strength between events or entities, so as to answer questions about "most" or "best" by using SPARQL. Here is a case when tourists ask for best places for diving.

Question: Could you recommend suitable places for diving?
SPARQL: SELECT ?location WHERE {?e rdf:type:Event. ?e :hasActivity "diving". ?e :hasLocation ?location.}

We can get a list of scenic spots where other tourists choose to dive, and we would return the answer that appears most frequently to users.

4   An Application of ETKG for POI Recommendation

As mentioned above, from ETKG, some hidden patterns of tourist behaviors can be inferred, such as transfer relationship between different POIs, popularity indications of various tourist activities, etc. With the information obtained by ETKG, it offers an opportunity to improve the accuracy of POI recommendation. In this section, we propose a new knowledge-aware POI recommendation framework, ETKGCN (Event-centric Tourism Knowledge Graph Convolutional Network), which makes full use of the information in ETKG. Experimental results on a real world dataset show that when applied to the knowledge-aware POI recommendation, ETKG is superior to the traditional tourism KG.

We formulate our knowledge-aware recommendation problem as follows. A KG $\mathcal{G}$ is comprised of entity-relation-entity triples (h, r, t). Here $h, t \in \mathcal{E}$ denote the head entity and tail entity of a knowledge triple respectively. $r \in \mathcal{R}$ denotes the relation between h and t. $\mathcal{E}$ and $\mathcal{R}$ are the set of entities and relations in the knowledge graph. We have a set of M users $\mathcal{U} = \{u_1, u_2, \ldots, u_M\}$, a set of N POIs $\mathcal{V} = \{v_1, v_2, \ldots, v_N\}$. The user-poi interaction matrix $\mathbf{Y} \in \mathbb{R}^{M \times N}$ is defined according to users' historical trajectories, where $y_{uv} = 1$ indicates that user u has been to POI v, otherwise $y_{uv} = 0$. The POI set $\mathcal{V}$ is a subset of entity set $\mathcal{E}$. Our goal is to learn a prediction function $\hat{y}_{uv} = \mathcal{F}(u, v | \Theta, \mathbf{Y}, \mathcal{G})$, where $\hat{y}_{uv}$ $(0 \leq \hat{y}_{uv} \leq 1)$ denotes the probability that user u would go to POI v, and Θ denotes the model parameters of function $\mathcal{F}$.

4.1   Framework of ETKGCN

The framework is proposed to capture high-order structural proximity among entities in ETKG. In real life, tourists choose whether to go to a POI not only considering their personal preferences, but also referring to the opinions of most other tourists. We will take these two factors into account in ETKGCN framework.

Here we define π_r^u which characterizes the importance of relation r to user u in ETKG.

$$\pi_r^u = t_r g(\mathbf{u}, \mathbf{r}) \tag{1}$$

where $\mathbf{u} \in \mathbb{R}^d$ and $\mathbf{r} \in \mathbb{R}^d$ are the representations of user u and relation r. Function g is to compute the inner product of $\mathbf{u}$ and $\mathbf{r}$, which is used to model the tourist's travel preferences. As for the experiences from most other tourists, our framework introduces knowledge matrix $T \in \mathbb{R}^{|\mathcal{E}| \times |\mathcal{E}|}$ that is calculated from ETKG. t_r is the value of the element r in T which indicates the relation strength of its head and tail entities. As shown in Fig. 4, we compute t_{r1} as follows (with the help of ETKG):

$$t_{r1} = \frac{|E_{v,e1}|}{|E_v|} \tag{2}$$

Here, E_v is the event set that happened in Yalong Bay (v) while $E_{v,e1}$ is event set that represents tourists swam $(e1)$ in Yalong Bay (v). So for a user loves swimming, if a large proportion of tourists go to Yalong Bay for swimming, π_{r1}^u will be high and it means the user would pay more attention to $r1$ in this subgraph. We consider both their own preferences and experience of others.

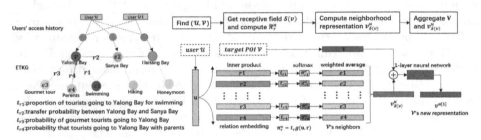

Fig. 4. The detail of an ETKGCN-layer.

As shown in Fig. 4, inspired by Graph Attention Network (GAT), consider a candidate pair of user u and POI v (an entity in ETKG). We first find the target entity v in ETKG and use $\mathcal{N}(v)$ denote the set of entities directly connected to v. In ETKG, the size of $\mathcal{N}(v)$ may vary significantly over all entities. We uniformly sample K neighbors and define the receptive field of v as $\mathcal{S}(v) \triangleq \{e|e \in \mathcal{N}(v)\}, |\mathcal{S}(v)| = K$. If we use $\mathbf{e}$ to denote the representation of an entity e, in order to characterize the topological proximity structure of POI v, we compute the linear combination of v's neighborhood through user-relation score π_r^u:

$$\tilde{\pi}_{r_{v,e}}^u = \frac{\exp\left(\pi_{r_{v,e}}^u\right)}{\sum_{e \in \mathcal{S}(v)} \exp\left(\pi_{r_{v,e}}^u\right)} \tag{3}$$

$$\mathbf{v}_{\mathcal{S}(v)}^u = \sum_{e \in \mathcal{S}(v)} \tilde{\pi}_{r_{v,e}}^u \mathbf{e} \tag{4}$$

where $\tilde{\pi}_{r_{v,e}}^u$ denotes the normalized user-relation score, and it acts as personalized filters when computing vs neighborhood representation $\mathbf{v}_{\mathcal{S}(v)}^u$.

Then we would aggregate the target entity representation $\mathbf{v} \in \mathbb{R}^d$ and its neighborhood representation $\mathbf{v}_{\mathcal{S}(v)}^u \in \mathbb{R}^d$ to get the new representation of POI v.

$$\mathbf{v}^{u[1]} = \sigma\left(\mathbf{W} \cdot \left(\mathbf{v} + \mathbf{v}_{\mathcal{S}(v)}^u\right) + \mathbf{b}\right) \tag{5}$$

where $\mathbf{W}$ and $\mathbf{b}$ are transformation weight and bias, and σ is the activation function.

Here, we use $\mathbf{v}^{u[1]}$ to denote the 1-order representation of POI v for user u. If we want to get h-order representation, we should sample v's neighbors up to h hops away(in our framework, $h = 2$). After obtaining the final representation $\mathbf{v}^u$ of v, we can predict the probability of user u going to POI v:

$$\hat{y}_{uv} = f\left(\mathbf{u}, \mathbf{v}^u\right) \tag{6}$$

Function f can be a multilayer perceptron. If y_{uv} denotes whether user u intends to go to POI v, the loss function can be described as:

$$\mathcal{L} = \sum_{u \in \mathcal{U}} \left(\sum_{v \in h} \mathcal{J}\left(y_{uv}, \hat{y}_{uv}\right) \right) + \lambda \|\mathcal{F}\|_2^2 \tag{7}$$

$\mathcal{J}$ is cross-entropy loss and h denotes the training set coming from users' access history. The last term is the L2-regularizer.

4.2   Experiments

We obtain POI visit records of 6166 anonymous tourists of Hainan in 2015 which are generated from a call detail records (CDRs) data set, and carried out experiments on these data. To verify the effectiveness of ETKG, we compare the proposed ETKGCN with some classic recommendation algorithms, as well as those traditional entity-centric tourism KG-based models. The entity-centric tourism KG contain 10000 entities and nearly 87000 relations in Hainan, which covers the same tourist attractions as ETKG.

Baselines. We compare the proposed ETKGCN with the following baselines:

- **SVD** is a classic CF-based model using inner product to model user-item interactions.
- **LightFM** [7] is a feature-based factorization model in recommendation scenarios.
- **LightFM+KG** extends LightFM [7] by attaching an entity representation learned by TransE [1] using traditional tourism KG.
- **LightFM+ETKG** extends LightFM [7] by attaching an entity representation learned by TransE [1] using ETKG.
- **GCN+KG** [14] is a Graph Convolutional Neural Network method cooperating with traditional tourism KG.
- **GCN+ETKG** [14] is a Graph Convolutional Neural Network method cooperating with ETKG.

Results. The experiment results are shown in Table 4. We use AUC, F1score and Top-N precision to evaluate the performance of POI recommendation models. We have the following observations:

- ETKGCN performs best among all the methods. Specifically, ETKGCN outperforms baselines by 4.6% to 21.7% on AUC and 2.7% to 17.2% on F1score. ETKGCN also achieves outstanding performance in top-n recommendation.
- KG-aware models perform much better than KG-free baselines (like SVD, LightFM). It means that the information in KGs plays a positive role in building tourists' preferences.
- GCN-based models get better result than traditional graph representation learning method (such as transE). This demonstrates that GCN can make full use of information in KGs.
- The models cooperating with ETKG (GCN+ETKG and LightFM+ETKG) perform better than those with traditional tourism KG (GCN+KG and LightFM+KG). This demonstrates that ETKG may be more suitable for POI recommendation task, for the reason that ETKG integrates more information than the traditional tourism KG, especially transfer relationship between POIs in this case.

Table 4. The result of experiments.

Model	AUC	F1score	Top-n precision				
			1	2	5	10	15
SVD	0.754 (−21.7%)	0.746 (−17.2%)	0.280	0.232	0.241	0.212	0.234
LightFM	0.758 (−21.3%)	0.734 (−18.4%)	0.480	0.475	0.458	0.490	0.479
LightFM+KG	0.809 (−16.2%)	0.822 (−9.6%)	0.400	0.415	0.434	0.478	0.500
LightFM+ETKG	0.830 (−14.1%)	0.862 (−5.6%)	0.560	0.505	0.486	0.491	0.520
GCN+KG	0.903 (−6.8%)	0.889 (−2.9%)	0.602	0.602	0.572	0.577	0.592
GCN+ETKG	0.925 (−4.6%)	0.891 (−2.7%)	**0.607**	**0.613**	0.601	0.595	0.602
ETKGCN	**0.971**	**0.918**	0.592	0.610	**0.611**	**0.613**	**0.624**

5   Conclusion

This paper presented an Event-centric Tourism Knowledge Graph (ETKG) to interconnect events using temporal relations. We built an ETKG of Hainan and realized an application of POI recommendation based on it. Our evaluations show that ETKG performs very well in solving the problems related to routes and activities of tourists during the trip. The code and data of ETKGCN are available at: https://github.com/xcwujie123/Hainan_KG and we will constantly update the work on it.

Acknowledgements. This work was supported by the National Key Research and Development Project, 2018YFE0205503 and 2019YFF0302601.

References

1. Bordes, A., Usunier, N., Garcia-Duran, A., Weston, J., Yakhnenko, O.: Translating embeddings for modeling multi-relational data. In: Advances in Neural Information Processing Systems, pp. 2787–2795 (2013)
2. Calleja, P., Priyatna, F., Mihindukulasooriya, N., Rico, M.: DBtravel: a tourism-oriented semantic graph. In: Pautasso, C., Sánchez-Figueroa, F., Systä, K., Murillo Rodríguez, J.M. (eds.) ICWE 2018. LNCS, vol. 11153, pp. 206–212. Springer, Cham (2018). https://doi.org/10.1007/978-3-030-03056-8_19
3. Du, J., Han, J.: Multi-level structured self-attentions for distantly supervised relation extraction (2018)
4. Gottschalk, S., Demidova, E.: EventKG: a multilingual event-centric temporal knowledge graph. In: Gangemi, A., Navigli, R., Vidal, M.-E., Hitzler, P., Troncy, R., Hollink, L., Tordai, A., Alam, M. (eds.) ESWC 2018. LNCS, vol. 10843, pp. 272–287. Springer, Cham (2018). https://doi.org/10.1007/978-3-319-93417-4_18
5. Hage, W.R.V., Malaisé, V., Segers, R., Hollink, L., Schreiber, G.: Design and use of the simple event model (SEM). Social Science Electronic Publishing (2011)
6. Jorro-Aragoneses, J.L., Bautista-Blasco, S.: Adaptation process in context-aware recommender system of accessible tourism plan. In: Pautasso, C., Sánchez-Figueroa, F., Systä, K., Murillo Rodríguez, J.M. (eds.) ICWE 2018. LNCS, vol. 11153, pp. 292–295. Springer, Cham (2018). https://doi.org/10.1007/978-3-030-03056-8_29
7. Kula, M.: Metadata embeddings for user and item cold-start recommendations (2015)
8. Kumar, S.: A survey of deep learning methods for relation extraction (2017)
9. Lample, G., Ballesteros, M., Subramanian, S., Kawakami, K., Dyer, C.: Neural architectures for named entity recognition (2016)
10. Li, Z., Xiao, D., Liu, T.: Constructing narrative event evolutionary graph for script event prediction (2018)
11. Pawar, S., Palshikar, G.K.: Relation extraction : a survey (2017)
12. Qu, Y., Liu, J., Kang, L., Shi, Q., Ye, D.: Question answering over freebase via attentive RNN with similarity matrix based CNN (2018)
13. Rospocher, M., et al.: Building event-centric knowledge graphs from news. Web Semant. Sci. Serv. Agents World Wide Web (2016)
14. Wang, H.: Knowledge graph convolutional networks for recommender systems (2019)
15. Web, S.: World wide web consortium (W3C) (2010)
16. Wei, T., et al.: Coupling mobile phone and social media data: a new approach to understanding urban functions and diurnal patterns. Int. J. Geog. Inf. Sci. **31**(12), 2331–2358 (2017)
17. Ying, R., He, R., Chen, K., Eksombatchai, P., Hamilton, W.L., Leskovec, J.: Graph convolutional neural networks for web-scale recommender systems (2018)

Extracting Short Entity Descriptions for Open-World Extension to Knowledge Graph Completion Models

Wangpeng Zhu[1], Xiaoli Zhi[1,2](✉), and Weiqin Tong[1,2]

[1] School of Computer Engineering and Science,
Shanghai University, Shanghai, China
{shucs_zwp,xlzhi,wqtong}@shu.edu.cn
[2] Shanghai Institute for Advanced Communication and Data Science,
Shanghai, China

Abstract. Great advances have been made in closed-world Knowledge Graph Completion (KGC). But it still remains a challenge for open-world KGC. A recently proposed open-world KGC model called OWE found a method to map the text space embedding obtained from the entity name and description to a pre-trained graph embedding space, by which OWE can extend the embedding-based KGC models to the open world. However, OWE uses average aggregation to obtain the text representation, no matter the entity description is long or short. It uses much unnecessary textual information and may become unstable. In this paper, we propose an extension to OWE, which is named OWE-MRC, to extract short expressions for entities from long descriptions by using a Machine Reading Comprehension (MRC) model. After obtaining short descriptions for entities, OWE-MRC uses the extension method of OWE to extend the embedding-based KGC models to the open world. We have applied OWE-MRC to extend common KGC models, such as ComplEx and Graph Neural Networks (GNNs) based models, to perform open-world link prediction. Our experiments on two datasets FB20k and DBPedia50k indicate that (1) the MRC model can effectively extract meaningful short descriptions; (2) our OWE-MRC uses much less textual information than OWE, but achieves competitive performance on open-world link prediction. In addition, we have used OWE to extend the GNN-based model to the open world. And our extended GNN model has achieved significant improvements on open-world link prediction comparing to the state-of-the-art open-world KGC models.

Keywords: Open-world knowledge graph completion · Machine Reading Comprehension · Graph Neural Networks

1 Introduction

Knowledge graphs (KGs) are usually composed of a large number of structured triples to store facts in the form of relations between different entities. For example, the triple *(Yao Ming, birthplace, Shanghai)* can be used to represent the fact

© Springer Nature Switzerland AG 2020
G. Li et al. (Eds.): KSEM 2020, LNAI 12274, pp. 16–27, 2020.
https://doi.org/10.1007/978-3-030-55130-8_2

'Yao Ming was born in Shanghai'. Formally, a KG $\mathcal{G} \subset E \times R \times E$ consists of a series of triples (h, r, t), where E and R are the set of entities and relations, h and t represent head and tail entities, r is the relation between entities, $h, t \in E$ and $r \in R$ [12]. KGs can provide high-quality data and rich semantic information, so they are widely used in many fields of artificial intelligence such as information extraction [4], question answering [5] and search [14]. Despite their usefulness and popularity, KGs are often noisy and incomplete. For example, 71% of people in Freebase have no place of birth, 75% no nationality [4].

In order to solve the problem of incomplete KG, Knowledge Graph Completion (KGC) has been developed to fill missing links in KG. KGC aims to discover the implicit information that exists in the Knowledge Graph and improve the KG by evaluating the possibility of triples that do not exist in the graph. Most existing works in KGC assume that KGs are fixed. They can only process entities and relationships that already exist in the KG, and cannot handle new entities. This type of completion is known as closed-world KGC. However, in real applications, most KGs are not fixed. For example, in the six months from October 2015 to April 2016, DBPedia added 36,340 new English entities, which is equivalent to adding 200 new entities per day [13]. Obviously, the pure closed-world KGC cannot meet the actual needs. A new challenge is emerging in the field of KGC, which is known as open-world KGC. Regarding the formal definitions of close-world and open-world, readers can refer to the work of Shi et al. [13].

Recently, a model called OWE was proposed by Shah et al. [12] to map the text space embedding obtained from the entity name and description to a pre-trained graph embedding space. OWE can be used to extend the embedding-based KGC model to the open world. However, it uses average aggregation to obtain the text representation, no matter the entity description is long or short. Much unnecessary textual information is involved in the model. For example, on Wikipedia, there is a long introduction to *Yao Ming* with hundreds of words. However, when we are completing the triple *(Yao Ming, birthplace, ?)*, we only need to use a small snippet from the long description: Yao Ming was born in Shanghai. In addition to the OWE model, Shah et al. proposed a standard dataset FB15k-237-OWE [12]. In this dataset, the entity descriptions are short descriptions with an average of 5 words. Shah claimed that 'short description' is the major contributing factor to the excellent performance of OWE model on FB15k-237-OWE.

Motivated by the aforementioned observations, in this paper, we propose an extension to OWE which we named OWE-MRC. Different from OWE, given a new entity, we propose to use a MRC model to draw a short description from long description. After obtaining short description for entity, we extend the embedding-based KGC model to open-world by using the extension approach of OWE [12]. We have applied OWE-MRC to extend common KGC models, such as ComplEx and Graph Neural Networks (GNNs) based models, to perform open-world link prediction. Since the MRC model we used can extract meaningful short descriptions, we can draw precise text space embeddings for entities. The models extended to the open-world by OWE-MRC use much less textual information than OWE, but obtain competitive results on open-world link prediction. In addition, we have used OWE to extend the GNN-based model to the open world. And our extended GNN model has achieved excellent performance on open-world link prediction.

2   Related Work

Machine Reading Comprehension. Machine reading comprehension (MRC) which requires machines to answer certain given questions has been widely used in many areas of artificial intelligence. Chen et al. uses MRC to the open-domain questions and its question system is known as DRQA [2]. Das et al. uses MRC model to build dynamic Knowledge Graphs from text [3]. In this paper, we use MRC model to select short descriptions for entities from long descriptions.

Knowledge Graph Completion. The widespread popularity of KGs, coupled with its incompleteness, stimulated a lot of research on KGC. The most representative work is the TransE [1] model proposed by Bordes et al. This model is a translation model. A triple (h, r, t) is considered correct if it satisfies $h + r \approx t$, that is, the tail entity is obtained by the head entity through the relation translation. TransE is also complemented by many other models, such as DistMult [19] and ComplEx [16]. Another typical models are the compositional models, such as RESCAL [10] and NTN [15]. Both RESCAL and NTN use tensor products. These tensor products can capture rich interactions, but they also require more parameters to model the relationship. In addition to the above two types of models, there is another type of model that uses GNNs, such as GCN [7] and R-GCN [11]. R-GCN is an extension of GCN. It introduces a relationship-specific transformation that depends on the type and direction of edges. Recently, a work named *Learning Attention-based Embeddings for Relation Prediction in Knowledge Graphs* [8] used Graph Attention Network (GAT) in KG. It learns new graph attention based embeddings, which are specifically for KG's relation prediction. The above works has been successful in the KGC field. However, none of these works involve the open-world KGC that we will address in this paper.

Open-World Knowledge Graph Completion. Only few works involve KGC in the open world. Among them, the representative is OOKB [6], which is based on auxiliary triples. In this paper, we pay more attention to the approach of using entity's textual information (names and descriptions) to complete open-world KGC. DKRL [17] uses entity descriptions in the knowledge graph to predict entities and relationships. By studying the description content, this method can not only obtain the structural information in the triples, but also the keywords in the entity description and the textual information hidden in the word order. ConMask [13] learns the embedding of entities names and part of their textual descriptions to connect unseen entities to the KG. A recently proposed open-world KGC model called OWE [12] presents a method to map the text space embedding obtained from the entity name and description to a pre-trained graph embedding space, by which OWE can extend embedding-based KGC models to the open world.

3 Approach

Our model is mainly an extension of OWE. Same as OWE, we start by training a link prediction model (also referred to as a graph-based model). The link prediction model evaluates triples through a scoring function:

$$score\left(h, r, t\right) = \phi\left(u_h, u_r, u_t\right) \tag{1}$$

where u represents the embeddings of entities or relationships in the graph space and ϕ is the scoring function of the link prediction model.

Link Prediction. Link prediction aims to predict a triple whose h or t is missing. For tail prediction, only the head entity h and the relation r are given, the score function for the link prediction model is used to score each $t \in E$, and the one of the highest score is considered to be the target tail entity:

$$tail\_pre = arg\max_{t \in E} score\left(h, r, t\right) \tag{2}$$

Similarly, for the prediction of the head entity:

$$head\_pre = arg\max_{h \in E} score\left(h, r, t\right) \tag{3}$$

In the remaining sections, we only discuss tail prediction for brevity.

Open-World Extension with MRC (OWE-MRC). Open-world link prediction aims to predict the triples (h', r, t), where h' is a new entity. For the link prediction model, it also refers to the prediction of entity facts that have not been seen during training. Given a new head entity $h' \notin E$, we use an MRC model to draw a short description for the entity from long description:

$$long\ description \xrightarrow{MRC} short\ description \tag{4}$$

After obtaining short description for the entity, we concatenate its name and short description into a word sequence W. We then use pre-trained word embeddings (such as Wikipedia2Vec) to convert the word sequence W into an embedding sequence. Finally, we employ a simple semantic averaging function that combines word embeddings to represent entity name and short description by a single embedding $v_{h'}$:

$$v_{h'} = \Psi^{avg}\left(v_{w_1}, v_{w_2}, ..., v_{wn}\right) = \frac{1}{n}\sum_{1}^{n} v_{w_{[1:n]}} \tag{5}$$

The embeddings of the graph space u_x and the embedding of the text space $v_{h'}$ are trained independently, the text space embeddings $v_{h'}$ cannot be used directly in graph-based models. Therefore a transformation Ψ^{map} should be learned from text space embedding to graph space embedding. After training the transformation, we can get $\Psi^{map}(v_{h'}) \approx u_{h'}$. The converted embedding $\Psi^{map}(v_{h'})$ can be recognized by the graph-based model, and we can use Eq. 1 to score the triples with the unseen entities:

$$score\,(h', r, t) = \phi\left(\Psi^{map}(v_{h'}), u_r, u_t\right) \qquad (6)$$

We can easily extend the graph-based model to the open world using the textual information (name and description) of the entity. The extension process takes three steps: (1) A graph-based model is firstly trained to get graph space embeddings; (2) An MRC model is used to draw short descriptions for entities from long descriptions. After obtaining short descriptions for entities, we employ a simple semantic averaging function to aggregate word embeddings to a text-based embedding; (3) A transformation is learned to map the embeddings of text space to the graph-based embeddings. Figure 1 provides an architectural illustration of our approach and the individual steps are described below in more detail.

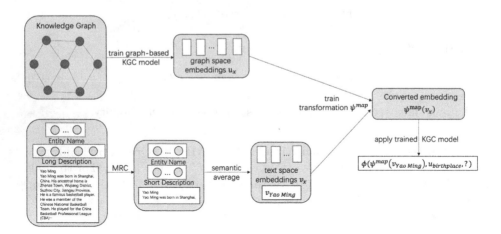

Fig. 1. The OWE-MRC in a nutshell.

3.1   Graph-Based Model

In this paper, we test two models: ComplEx [16] and *Learning Attention-based Embeddings for Relation Prediction in Knowledge Graphs* [8], which is a GNN-based link prediction model and will be referred as GNN-LAEKG in the following for the convenience of explanation. The scoring functions of these two models are as follows:

$$ComplEx : \phi\,(u_h, u_r, u_t\,) = Re\,(\langle u_h, u_r, \overline{u}_t \rangle) \qquad (7)$$

$$GNN - LAEKG : \phi\,(u_h, u_r, u_t\,) = concat\,(g\,([u_h, u_r, u_t\,] * \Omega\,)) \cdot w \qquad (8)$$

In Formula 7, $\overline{u}$ denotes the complex conjugate. Since GNN-LEAKG is an encoder-decoder model and its decoder is ConvKB [9], its scoring function is the same as ConvKB, as shown in Formula 8, where g denotes a non-linear function, $*$ denotes a convolution operator, $\cdot$ denotes a dot product and Ω denotes a set of filters.

3.2   Machine Reading Comprehension and Semantic Averaging

Rather than designing a sophisticate MRC model from the beginning, we will use a modified version of a widely used model DrQA [2] to draw short descriptions for entities from long description. For more details of the DrQA model, the reader is referred to the original publication [2]. In short, it uses a multi-layer recurrent neural network (RNN) architecture to encode paragraphs and question text, and uses self-attention to match the two encodings. For each token in the text, it outputs a score indicating the likelihood that it will become the beginning or end of the span of answering questions [3]. In order to enable this model to effectively extract short descriptions, we did the following modifications:

1. On dataset FB20k, we generate a simple natural language question for every entity e, such as "What is e ?" or "Where is e ?". We then take this simple problem as the question text for DrQA, treat the entity's long textual description as DrQA's paragraphs, and finally use DrQA to extract the short descriptions for entities.
2. On dataset DBPedia50k, we only take entity name as the question text for DrQA, treat the entity long textual description as DrQA's paragraphs, and finally use DrQA to extract the short descriptions for entities.

   After obtaining short descriptions for entities, we employ a simple semantic averaging function $\Psi^{avg}\,(v_{w_1}, v_{w_2}, ..., v_{wn}) = \frac{1}{n}\sum_1^n v_{w[1:n]}$ that combines word embeddings v_{w_x} to represent entity names and short descriptions. In this paper, we use 300-dimensional Wikipedia2Vec embeddings [18].

3.3   Transformation Function

The core idea of open-world extension is the transformation from text space embeddings to graph space embeddings[12]. By transforming the two spaces, the trained graph-based model can identify new entities by their descriptions and names. To achieve this transformation, we use an Affine transformation: $\Psi^{map}\,(v) = A \cdot v + b$

To train the transformation, first we need to train the graph-based model on the full graph and obtain the embeddings in the graph space $u_1, ...u_n$. Then we choose all entities $e_{i_1}, ..., e_{i_m}$ with textual information (description and name) and get text space embeddings $v_1, ... v_n$ for them through MRC model and semantic averaging aggregation. After obtaining graph space embeddings and text space embeddings an Affine transformation Ψ_θ^{map} is trained to map text space embeddings to graph space embeddings. Ψ_θ^{map} is trained by minimizing the loss function:

$$L\left(\theta\right) = \sum_m^{k=1} ||\Psi_{map}^\theta \left(v_{i_k}\right) - u_{i_k}||_2 \tag{9}$$

4   Experiment

4.1   Datasets and Evaluation Protocol

Datasets. To evaluate our proposed approach, we use two benchmark datasets: FB20k [17], DBPedia50k [13]. FB20k is built on the basis of FB15k by introducing unknown triples into the test set. In FB20k and DBPedia50k, the description of entities is biased towards long descriptions, which averages 454 words on DBPedia50k, and 147 words on FB20k. Statistics of the datasets can be seen in Table 1. And *EOpen* represent entities used for open-world link prediction.

Table 1. Datasets statistics.

Dataset	For closed-world KGC					For open-world KGC		
	Entities	Relations	Number of triples			EOpen	Tail pred number	
			Train	Valid	Test		Valid	Test
FB20k	14904	1341	472860	48991	57803	5019	–	11586
DBPedia50k	24624	351	32388	123	2095	3636	164	4320

Evaluation Protocol. We evaluate the performance of our approach by making tail predictions on the test set. For each test triple (h', r, t), where $h' \notin E$ is the head of the open-world, we use the extended trained graph-based model to get the score $\phi(h', r, t)$ of each $t \in E$, and sort the score in descending order. Finally, we use mean rank (MR), mean reciprocal rank (MRR) and Hits@1, Hits@3, Hits@10 to evaluate the target entity's rank. Like the previous works [12,13], we evaluate all the models in a filtered setting.

Naming Rules. ComplEx extended to open-world by using OWE-MRC is named ComplEx-OWE-MRC and GNN-LAEKG extended to open-world by using OWE-MRC is named GNN-LAEKG-MRC.

4.2   MRC Model Evaluation

A New Baseline. We evaluate the MRC model on the datasets DBPedia50k and FB20k. Before evaluation, we introduce a new baseline called ComplEx-OWE-Name: all description information of entities are discarded and only entity names are left as textual information; and the ComplEx is extended to open-world by using OWE.

Implementation Details. When we use the OWE model to extend the graph-based model to the open-world, we use the experimental parameters originally published by OWE. The Adam optimizer is used to train the Affine transformation in our OWE-MRC model. The learning rate is 0.001 and the batch size is 256. When employing a semantic averaging function that combines word embeddings to represent entity names and short descriptions, we use no dropout. (This strategy is carried out through all the following experiments). Since FB20k lacks an open-world validation set, 10% triples randomly selected from the test set are used as the validation data. We use ComplEx as our graph-based model and use the OpenKE *framework*[1] to train ComplEx. (For the following experiments on ComplEx, we all use this framework).

Table 2. Evaluation results on datasets DBPedia50k and FB20k.

Model	DBPedia50k				FB20k			
	H@1	H@3	H@10	MRR	H@1	H@3	H@10	MRR
DKRL	–	–	40.0	23.0	–	–	–	–
ComplEx-OWE-Name	32.2	44.8	54.5	40.0	33.4	43.6	53.9	40.7
ComplEx-OWE-MRC	**37.8**	**45.5**	**58.8**	**44.2**	**37.7**	**52.2**	**61.4**	**46.5**

Result and Analysis. Table 2 presents our evaluation results on datasets DBPedia50k and FB20k. On DBPedia50k and FB20k, compared with ComplEx-OWE-Name, our ComplEx-OWE-MRC performs better on all metrics. On DBPedia50k, compared with ComplEx-OWE-Name, our ComplEx-OWE-MRC model has the biggest improvement in Hit@1, which has increased by about 5.6%, and its H@10 has increased by 4.3%. On FB20k, the improvement of H@3 is the largest, which has increased by 8.6%, and its H@10 has increased by 7.5%. The only difference between ComplEx-OWE-Name and ComplEx-OWE-MRC is that ComplEx-OWE-MRC adds the short descriptions selected by MRC when extending to open-world. The textual information used by ComplEx-OWE-Name when expanding to open-world is only the entity name, while the textual information used by ComplEx-OWE-MRC is the entity name and the short description selected by MRC model. So, the excellent performances of our ComplEx-OWE-MRC on DBPedia50k and FB20k testify that the MRC model can extract effective short descriptions. On DBPedia50k, our ComplEx-OWE-MRC also performs far better than DKRL.

Short Descriptions Extracted by MRC. Table 3 shows some results of our short descriptions selected from long descriptions by using the MRC model. FB15k-237-OWE [12] is a standard dataset proposed by Shah et al. In this

---
[1] https://github.com/thunlp/OpenKE.

dataset, the entity descriptions are as short as 5 words on average. Comparing the short descriptions selected by MRC model with the descriptions of the standard dataset FB15k-237-OWE, we found that the short descriptions selected by MRC model have a high similarity with FB15k-237-OWE. Table 3 shows some results. Comparison with the standard dataset also confirms that the MRC we used can indeed screen out effective short descriptions. At the same time, OWE's great success on the short text dataset FB15k-237-OWE is largely due to the short description in FB15k-237-OWE which also proves that our work is meaningful.

Table 3. Some short descriptions extracted by MRC.

Entity	Original desc length	Short desc of FB-15k-237-OWE	Short desc selected by MRC
Hackensack, New Jersey	268 words	City and county seat of Bergen County, New Jersey, United States	A city in Bergen County, New Jersey, United State
Minority Report	198 words	2001 American science fiction drama film by Steven Spielberg	A 2002 American neo-noir science fiction thriller film
Adam Carolla	221 words	American radio personality, television host, comedian, and actor	An American comedian, radio personality, television host and actor

4.3 Extending ComplEx and GNN-LAEKG to Open-World Using OWE-MRC

After evaluating the MRC model on the datasets DBPedia50k and FB20k. We use our OWE-MRC model to extend ComplEx [16] and GNN-LAEKG [8] to open-world. We evaluate our extended model on the dataset FB20k.

Implementation Details. For training the model GNN-LAEKG, we use the *implementation*[2] provided by Nathani et al.

Note on Data. An asterisk (*) indicates that the result is different from the original published result. We used OWE's original parameters and methods, but were unable to reproduce the results of FB20k. Therefore, in this experiment, we use our measured data and also use the original published data. ComplEx-OWE-300* represents our measured data.

---

[2] https://github.com/deepakn97/relationPrediction.

Table 4. Comparison with other open-world KGC models on tail prediction on FB20k.

Model	FB20k			
	H@1	H@3	H@10	MRR
ConMask	42.3	57.3	**71.7**	53.3
ComplEx-OWE-300*	40.9*	53.7*	63.6*	49.2*
ComplEx-OWE-MRC	37.7	52.2	61.4	46.5
ComplEx-OWE-300	**44.8**	57.1	69.1	53.1
GNN-LAEKG-MRC	44.7	**58.5**	71.4	**54.0**

Result and Analysis. It can be seen from Table 4, our extended model GNN-LAEKG-MRC has a certain improvement compared with ConMask, ComplEx-OWE-300. The performance on H@3 and MRR is the best, and other metrics are also very close to state of the art. Compared to our measured model ComplEx-OWE-300*, the extended model ComplEx-OWE-MRC also performed well, and all indicators were only a little lower. Compared with OWE, our OWE-MRC model uses short descriptions. OWE uses long descriptions with an average length of 147 words, while the short descriptions used by our OWE-MRC model have an average length of 15 words or less. We use much less textual information, but have achieved very competitive results. This also shows that the MRC model we used can extract meaningful short descriptions, and that our work is meaningful.

4.4   Extending GNN-LAEKG to Open-World Using OWE

Due to their powerful capability of processing graph structure data, more and more studies are now introducing GNNs to KGC. Typical examples are R-GCN [11], OOKB [6] and GNN-LAEKG [8]. Although our OWE-MRC has achieved good results, the OWE's performance with long descriptions is still slightly better. In order to make better use of GNN-LAEKG, at the end of this section, we use OWE to extend GNN-LAEKG and the extended model is named GNN-LAEKG-OWE.

Table 5. Comparison with other open-world KGC models on tail prediction.

Model	DBPedia50k				FB20k			
	H@1	H@3	H@10	MRR	H@1	H@3	H@10	MRR
Target Filt. Base	4.5	9.7	23.0	11.0*	17.5	32.1	41.2	27.2
DKRL	–	–	40.0	23.0	–	–	–	–
ConMask	47.1	64.5	**81.0**	58.4*	42.3	57.3	71.7	53.3
ComplEx-OWE-300	51.9	**65.2**	76.0	60.3	44.8	57.1	69.1	53.1
GNN-LAEKG-OWE	**55.9**	62.2	71.3	**61.9**	**48.4**	**61.2**	**73.6**	**56.9**

Comparison with State of the Art. We compare our GNN-LAEKG-OWE with other open-world KGC models. Like the previous work [12,13], we evaluate all the models in a filtered setting. On the datasets DBPedia50 and FB20k, we use 300-dimensional Wikipedia2Vec embedding. The experimental results of Baseline, DKRL, ConMask and ComplEx-OWE use data published by Shah et al. [12]. It can be seen from Table 5, our GNN-LAEKG-OWE has achieved very good results. On DBPedia50k, compared with ConMask and ComplEx-OWE-300, our GNN-LAEKG-OWE has greatly improved on H@1, and the value of MRR has also increased. On FB20k, our GNN-LAEKG-OWE model greatly improves all metrics.

5   Conclusion

In this paper, we propose an extension to OWE called OWE-MRC. It integrates a MRC model to extract short entity descriptions for open-world extension. We have applied OWE-MRC to extend common KGC models, such as ComplEx and GNN-based models, to perform open-world link prediction. Experimental results on the datasets DBPedia50k and FB20k show that the MRC model can extract meaningful short descriptions from long descriptions. The long text used by OWE averages 454 words on DBPedia50k, and 147 words on FB20k. Our OWE-MRC uses short descriptions selected by MRC, only with an average length of 15 words or less. Using much less textual information than OWE, our OWE-MRC still achieves competitive results, which shows the significance of our MRC extension to OWE. In addition, we have used OWE to extend the GNN-based model to the open world. And our extended GNN model has achieved excellent performance on open-world link prediction.

For now, the MRC model only uses entity information to extract short descriptions and ignores the relation information in KG. In the near future, an endeavor will be made to design an MRC model that can exploit both KG's entity and relationship information, expecting more accurate short descriptions can be extracted for entities.

Acknowledgements. We sincerely acknowledge the High Performance Computing Center of Shanghai University and Shanghai Engineering Research Center of Intelligent Computing System (No. 19DZ2252600) for providing the computing resources and technical support to run our experiments. This work is supported in part by science and technology committee of shanghai municipality under grant No.19511121002.

References

1. Bordes, A., Usunier, N., Garcia-Duran, A., Weston, J., Yakhnenko, O.: Translating embeddings for modeling multi-relational data. In: Advances in Neural Information Processing Systems, pp. 2787–2795 (2013)
2. Chen, D., Fisch, A., Weston, J., Bordes, A.: Reading Wikipedia to answer open-domain questions. arXiv preprint arXiv:1704.00051 (2017)

3. Das, R., Munkhdalai, T., Yuan, X., Trischler, A., McCallum, A.: Building dynamic knowledge graphs from text using machine reading comprehension. arXiv preprint arXiv:1810.05682 (2018)

4. Dong, X., et al..: Knowledge vault: a web-scale approach to probabilistic knowledge fusion. In: Proceedings of the 20th ACM SIGKDD International Conference on Knowledge Discovery and Data Mining, pp. 601–610 (2014)

5. Ferrucci, D., et al.: Building Watson: an overview of the DeepQA project. AI Mag. **31**(3), 59–79 (2010)

6. Hamaguchi, T., Oiwa, H., Shimbo, M., Matsumoto, Y.: Knowledge transfer for out-of-knowledge-base entities: a graph neural network approach. arXiv preprint arXiv:1706.05674 (2017)

7. Kipf, T.N., Welling, M.: Semi-supervised classification with graph convolutional networks. arXiv preprint arXiv:1609.02907 (2016)

8. Nathani, D., Chauhan, J., Sharma, C., Kaul, M.: Learning attention-based embeddings for relation prediction in knowledge graphs. arXiv preprint arXiv:1906.01195 (2019)

9. Nguyen, D.Q., Nguyen, T.D., Nguyen, D.Q., Phung, D.: A novel embedding model for knowledge base completion based on convolutional neural network. arXiv preprint arXiv:1712.02121 (2017)

10. Nickel, M., Tresp, V., Kriegel, H.P.: A three-way model for collective learning on multi-relational data. In: ICML, vol. 11, pp. 809–816 (2011)

11. Schlichtkrull, M., Kipf, T.N., Bloem, P., van den Berg, R., Titov, I., Welling, M.: Modeling relational data with graph convolutional networks. In: Gangemi, A., et al. (eds.) ESWC 2018. LNCS, vol. 10843, pp. 593–607. Springer, Cham (2018). https://doi.org/10.1007/978-3-319-93417-4_38

12. Shah, H., Villmow, J., Ulges, A., Schwanecke, U., Shafait, F.: An open-world extension to knowledge graph completion models. In: Proceedings of the AAAI Conference on Artificial Intelligence, vol. 33, pp. 3044–3051 (2019)

13. Shi, B., Weninger, T.: Open-world knowledge graph completion. In: Thirty-Second AAAI Conference on Artificial Intelligence (2018)

14. Singhal, A.: Introducing the knowledge graph: things, not strings (2012). https://googleblog.blogspot.ie/2012/05/introducing-knowledge-graph-things-not.html

15. Socher, R., Chen, D., Manning, C.D., Ng, A.: Reasoning with neural tensor networks for knowledge base completion. In: Advances in Neural Information Processing Systems, pp. 926–934 (2013)

16. Trouillon, T., Welbl, J., Riedel, S., Gaussier, É., Bouchard, G.: Complex embeddings for simple link prediction. In: International Conference on Machine Learning (ICML) (2016)

17. Xie, R., Liu, Z., Jia, J., Luan, H., Sun, M.: Representation learning of knowledge graphs with entity descriptions. In: Thirtieth AAAI Conference on Artificial Intelligence (2016)

18. Yamada, I., Shindo, H., Takeda, H., Takefuji, Y.: Joint learning of the embedding of words and entities for named entity disambiguation. arXiv preprint arXiv:1601.01343 (2016)

19. Yang, B., Yih, W.T., He, X., Gao, J., Deng, L.: Embedding entities and relations for learning and inference in knowledge bases. arXiv preprint arXiv:1412.6575 (2014)

Graph Embedding Based on Characteristic of Rooted Subgraph Structure

Yan Liu[1,2]([⊠]) [iD], Xiaokun Zhang[1], Lian Liu[3], and Gaojian Li[1]

[1] PLA Strategic Support Force Information Engineering University, Zhengzhou 450001, China
ms.liuyan@foxmail.com
[2] State Key Laboratory of Mathematical Engineering and Advanced Computing,
Zhengzhou 450001, China
[3] Investigation Technology Center PLCMC, Beijing 100000, China

Abstract. Given the problem that currently distributed graph embedding models have not yet been effectively modeled of substructure similarity, biased-graph2vec, a graph embedding model based on structural characteristics of rooted subgraphs is proposed in this paper. This model, based on the distributed representation model of the graph, has modified its original random walk process and converted it to a random walk with weight bias based on structural similarity. The appropriate context is generated for all substructures. Based on preserving the tag features of the nodes and edges in the substructure, the representation of the substructure in the feature space depends more on the structural similarity itself. Biased-graph2vec calculates the graph representations with unsupervised algorithm and could build the model for both graphs and substructures via universal models, leaving complex feature engineering behind and has functional mobility. Meanwhile, this method models similar information among substructures, solving the problem that typical random walk strategies could not capture similarities of substructures with long distance. The experiments of graph classification are carried out on six open benchmark datasets. The comparison among our method, the graph kernel method, and the baseline method without considering the structural similarity of long-distance ions is made. Experiments show that the method this paper proposed has varying degrees inordinately improved the accuracy of classification tasks.

Keywords: Graph data · Network embedding · Graph embedding · Structural similarity · Graph classification

1 Introduction

Graph data is a data form widely exist in the field of biochemistry, social network & network security, in which tasks like the prediction of biochemical characteristics [1], community detection [2], malicious code detection [3], *etc.* often share a tight link with graph classifications and clusters. There is a need to put the original graph into representation as an eigenvector of fixed length to facilitate the application of mature classification and clustering algorithms in machine learning. Graph embedding, therefore, is the method that via studies upon how to maintain enough characteristics of the

G. Li et al. (Eds.): KSEM 2020, LNAI 12274, pp. 28–39, 2020.
https://doi.org/10.1007/978-3-030-55130-8_3

graph while fitting it into the characteristic space, making the original graph being presented in vectors as integrated as possible, and eigenvectors after representation could lead to a better outcome in the following tasks of graph-processing.

The work related to the graph representation can be broadly divided into two categories:

(1) **Graph Kernel Method.** The Graph kernel method is a widely used method to measure the similarities between different graph structures. For ordinary kernel methods, the primary thought is to map a low-dimension vector x to a higher dimension that reproduces kernel Hilbert space using a nonlinear mapping function $\emptyset$. This way, nonlinear tasks that are relatively hard to calculate in lower dimensions can be solved in higher dimension Hilbert feature space through linear algorithms. Early Graph embedding methods mainly include the graph kernel methods [4–6] and dimensionality-reducing methods (multidimensional scaling (MDS) [7], IsoMap [8], locally linear embedding – LLE [9]).

Though graph kernel methods stand their crucial role in multiple graph-related tasks and are now still widely used, the method has its restrictions: 1) The high-dimensional feature representation obtained by the graph kernel method has some information redundancy, which leads to the high cost of calculation and memory. 2) This method needs a predefined kernel function, which relies on feature engineering to get practical features. As a result, making the model insufficient mobility. 3) In the graph kernel method, we generally regard substructures as atomic structures, ignoring the structural similarities among substructures.

(2) **Graph Embedding Method**. Graph embedding could be regarded as a unique form of network embedding in particular circumstances. Graph embedding learning and network embedding learning both aim to learn the low dimension vector presentations in the feature space. However, there is some difference between graph embedding learning and network embedding learning. Network embedding learning faces networks with rich node properties information like social networks. Graph embedding learning, simultaneously, faces network with rich graph-structural data like biochemical structures and code flows, which do not contain much information of node properties but contain rich information like node labels, edge labels, and weights. DeepWalk [10] is the first article to apply the word vector model, word2vec, and random walk to the field of network embedding. Later embedding models like LINE [11], node2vec [12], *etc.* are based on the representation learning model in the framework of the random walk. As word vector models widely used, new research based on distributed word vector models has come out in graph embedding field, e.g., subgraph2vec [3], graph2vec [13], GE-FSG [14], *etc.* These methods share similarities in their general frames. They decompose the graphs into their atom substructures by considering the graph as the document, considering the atom substructures as words in the document. The graph-embedding models can use the word vector model to learn the low dimension embeddings of each graph.

However, the graph embedding model fundamentally differs from the word embedding model. In the word vector model, it is not easy to capture the similarities between words in the very beginning. The similarity can only be obtained via the model's embedding results. If there are two certain random words marked as $w1$, $w2$, the model will not be able to measure their similarity without information of their context properly. In graph embedding learning, the structural similarity between substructures is easy to measure. For example, given two substructures $g1$, $g2$, even if there is no information about their context, the model can still capture their similarity by measuring characteristics such as edges, nodes, degrees, graph-kernels, and more. Existing graph embedding models leave this similarity information behind.

Subgraph (substructure), as a significant characteristic of graphs and networks, obtains a higher level of abstraction information than characteristics like nodes, edges, degrees, *etc*. Many recent kinds of research regard subgraphs as the atom substructures of graphs and learn their representation via distributed learning models. However, those graph embedding models ignore the similarities between subgraphs that exist in the very beginning. This paper proposes **biased-graph2vec**, which is based on structural characteristics of rooted subgraphs. The model learns the vector representation of both graphs and rooted subgraphs in the same vector space. The classified task has been tested in six base data sets, and the result shows the accuracy of graph classifications, compare to the baseline method, which has been varying degrees improved.

2 Problem Definition

Given a graph set Γ, graphs in Γ represented as $\{G1, G2, \ldots\}$. The goal of graph embedding is to learn d-dimensional representation in the characteristic vector space of each graph G in Γ. In the learning process, it is vital that characteristic vectors reserve corresponding characteristics of labels and edges in substructures and context of the substructures. Moreover, the dimensionality d should be properly set to keep the memory and calculation cost of the representation matrix $\Phi \in R^{|\Gamma| \times d}$ low enough.

Graphs in the graph set Γ are defined as $G = (N, E, \lambda)$, in which N stands for the set of nodes, $E \subseteq (N \times N)$ represents the edge set. For that data used in graph embedding usually has relatively complete labels of edges and nodes, therefore in such a system, we define those graphs with label information as labeled graphs while those without defined as non-labeled graphs. In labeled graphs, there exist functions $\lambda : N \to \ell$, mapping each node to a corresponding character in the alphabet ℓ, in the same way, we define the edge-mapping function $\eta : E \to \varepsilon$.

For two given graphs, $G = (N, E, \lambda)$ and $G_{sg} = (N_{sg}, E_{sg}, \lambda_{sg})$, G_{sg} is the subgraph of G if and only if there exits an injective function $\mu : N_{sg} \to N$, which makes $(\mu(n_1), \mu(n_2)) \in E$ if and only if $(n_1, n_2) \in E_{sg}$.

3 Graph Embedding Model Based on Structural Characteristics of Rooted Graphs

According to descriptions in [15], the context of substructures only shows local characteristics of substructures. Therefore, if two substructures resemble each other but are far from each other in the context, it is almost impossible for existing distributed graph models to produce valid learning results. In a graph-related calculation, there is usually enough label information of nodes and edges, but apart from that, the similarities between substructures also mean they have similar characteristics. For example, in a malicious code detection task, the function call of code segment A is different from malicious code segment B, but they share a similar process of execution. Therefore code segment A is highly suspectable, which means the representation of A in feature space resembles B more than a random one. In the existing graph distributed representation model, A and B as independent atomic structures possess different labels and contexts, making the final results of representation differ a lot from each other. Currently, existing distributed graph embedding models are incapable of capturing structural similarities between distanced structures like these. As in Sect. 3, by bringing the substructure's similarity into consideration, this paper proposed biased-Graph2vec to solve the problem.

3.1 The Frame of Biased-Graph2vec

The d-dimensional subgraph of the node n in graph g is defined as a subgraph containing every single node that could be reached from node n in d hops and edges between these nodes.

The core of the model is to use the similarities information among substructures to create a hierarchical structure from where random walks are performed. This paper uses graph sets in experiments, to learn a graph's embedding the process is the same.

The biased-graph2vec mainly consists of two parts. The first part is to traverse all nodes to produce substructures related to every node, construct a 2-layer structure to capture similarity information among substructures, then use the biased random walk algorithm to perform a random walk in the two-layer graph structure to obtain contexts containing structural characteristics of the subgraphs. The second part is just like other graph embedding models, to fit all substructures into the word embedding model doc2vec. Treat substructures as a word in the word embedding model, the graph as the word sequence and substructures of the graph as words. Doc2vec model is applied to acquire the representation of the graph and substructures in the feature space of a lower dimension.

According to the specific task, the selection subgraphs is flexible choosing from subgraphs like frequent subgraph, rooted subgraph, ego graph, *etc.* Biased-graph2vec, in this paper, selects rooted subgraphs as its substructures. Advantages of rooted subgraphs over the other options are as followed:

(1) The computation cost is much less than the frequent subgraph mining algorithm.
(2) Rooted subgraphs, compared with other characteristics like nodes and edges, possess a higher level of abstraction, possibly containing more information on the graph. Once applied to the word vector model, for it is generated from node traversal, different rooted subgraphs share a similar order of magnitudes. The more valuable information the subgraph contains, the better the graph representation will be.
(3) The nonlinear substructure could better acquire the graph's characteristics in normal tasks than linear substructures. Weisfeiler-Lehan (WL) kernel [6] algorithm, for example, is the kind of algorithm based on characteristics of rooted graphs that appears to be more effective in both experiments and applications than other linear graph-kernel algorithms like random walk kernel algorithm and shortest path kernel algorithm.

The processing procedure of biased-graph2vec is shown in the Fig. 1. First, generate the set of rooted subgraphs, then calculate the structural similarities among substructures, build a model of the biased random walk from structural similarities, and the model of the document is built based on that model, finally output the low dimension representations vectors of substructures and graphs.

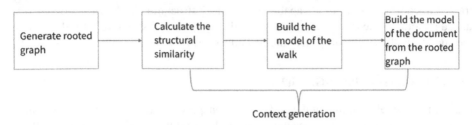

Fig. 1. A processing procedure of biased-graph2vec

3.2 Rooted Subgraphs Generation

Graph set denoted as $D = \{d_1, d_2, \ldots, d_N\}$, generate rooted subgraph $sg_v^{(h)}$ for every node $v \in d_i$ of every single graph $d_i \in D$, h denotes the depth of currently rooted subgraph. After h iterations, the result set contains all the root subgraphs of nodes v less than or equal to h-order neighbor nodes. The procedure of generating rooted subgraph refers to the labeling process of the Weisfeiler-Leman (WL) kernel [6] algorithm, the WL relabeling process, which is shown in Algorithm 1. The input of the algorithm is current node v, graph to be extracted G, hyper-parameter h depth of extraction. The process of extracting rooted subgraphs is recursive, in which h controls the of recursive depth the extraction, meaning that in the end, the rooted graph set of node v contains rooted graphs ordered from 0 to h (when h equals 0 the function returns to the current node). The larger h is, the more rooted subgraphs are extracted, and the more information about adjacent structures are contained, the more subsequent computations cost will be. The rooted subgraph generation algorithm is shown in Fig. 2.

3.3 Context Generation

After recursion of all nodes of each graph in graph set D, the set of all rooted subgraphs in the graph set is generated and denoted as $\mathbb{G}_{sg}$. Also, the neighboring rooted graphs of the targeted subgraph are obtained in the subgraph generating process. For example, if h is 3, the subgraph $sg_v^{(1)}$ has the context of $sg_v^{(0)}$, $sg_v^{(1)}$, $sg_v^{(2)}$ that represent local information of $sg_v^{(1)}$. According to WL relabeling process, it is able to represent every rooted subgraph in the form of unambiguous unique strings, which make up the vocabulary of rooted subgraphs, $Vocab_{sg}$.

The context obtained above only represents the local feature of the rooted subgraph. To capture the similarity between subgraphs far away and expand the range of the random walk, the context generating process should meet the requirements below:

Algorithm 1: GetWLSubgraph(v; G; h)

Input: v: *current node.*

   $G = (V, E, \lambda)$: *graph to perform subgraph extraction.*

   h: *extraction depth of rooted graph.*

Output: $sg_v^{(h)}$: rooted subgraph set of the current node with extraction depth h.

1: **function** $GetWLSubgraph(v; G; h)$

2:    $sg_v^{(h)} = \{\}$

3:    **if** $h = 0$ **then**

4:       $sg_v^{(h)} \leftarrow \lambda(v)$

5:    **else**

6:       $Neighbor_v := \{v' | (v, v') \in E\}$

7:       $M_v^{(h)} := \{GetWLSubgraph(v', G, h-1) | v' \in Neighbor_v\}$

8:       $sg_v^{(h)} := sg_v^{(h)} \cup GetWLSubgraph(v, G, h-1) \oplus sort(M_v^{(h)})$

9:    **end if**

10:   return $sg_v^{(h)}$

11: **end function**

Fig. 2. Process of GetWLSubgraph algorithm

For rooted subgraphs $sg_1, sg_2 \in \mathbb{G}_{sg}$, representations in feature space are correspondingly marked as $e(sg_1), e(sg_2)$. The distance of $e(sg_1)$, $e(sg_2)$ embedding in vector space should reflect not only the local context's similarity but also the structure similarity of sg_1, sg_2 themselves.

Structural Similarity Calculation. Depending on the specific situation, a standard similarity measure can be used to measure structural similarity such as node similarity, edge similarity, graph kernel, *etc.* Considering the calculation cost, for that the degree of the node reveals a structural similarity to some degree, this section calculates the similarities among each node's degree sequence of rooted subgraphs to infer the structural similarity of the rooted subgraphs.

For a given subgraph $sg \in \Gamma_{sg}$, its ordered sequence of nodes' degree is marked as $s(V_{sg})$, and V_{sg} stands for the node-set of subgraph sg. Due to the sequence length could be inequality, this section applies the Dynamic Time Warping (DTW) method to the calculation. In this method, all elements in two sequences will be correspondingly lined up individually, making the sum of aligned sequences' distance reduced to the least. Let sequences A, B represent the sequences of degrees of nodes, this section uses the formula below to calculate the distance between $a \in A, b \in B$:

$$d(a, b) = \frac{\max(a, b)}{\min(a, b)} - 1 \tag{1}$$

in which $\max(a, b)$ is the maximum of two node degrees and $\min(a, b)$ is the opposite. Formula (1) makes the distance clear to zero when the sequences are identical and the d in the formula becomes more sensitive towards the difference between a, b when a, b are smaller.

Calculation of structural similarities between two subgraphs sg_1 and sg_2 can be converted to the issue of calculating the distance of degree sequence of nodes from those two corresponding subgraphs. As defined in the DTW algorithm, the distance can be turned into an optimization problem. The sequence distance obtained by the DTW algorithm is the distance of structural similarity of rooted subgraphs sg_1, sg_2, which are denoted as $f(sg_1, sg_2)$.

Random Walk Process. Compared to structural similarities among rooted subgraphs, the similarity between two nodes is not meaningful, so in context generation, biased-graph2vec chooses a hyperparameter to avoid the calculation.

To generate enough context for single nodes and capture long distanced rooted subgraph's structural similarities, biased-graph2vec uses a two-layer network structure where context is captured through a cross-layer random walk. The two-layer walk model uses the similarity between the degree sequences of nodes to measure the structural similarity of nodes, and controls the walk jump probability through the similarity. The random walk process is shown in Fig. 3.

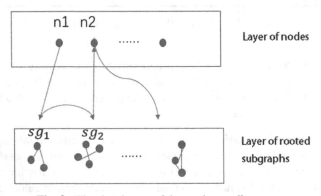

Fig. 3. The sketch map of the random walk process

The generated sequence of the random walk process acquired in Fig. 3 is denoted as $(n1, sg_1, sg_2, n2 \ldots)$. The first layer contains all nodes from each graph in the graph set $\mathbb{G}$, while the second layer includes all rooted subgraphs. The structural similarities of nodes are represented via the calculation of similarities of nodes' rooted subgraphs. In the random walk process, each node in the node layer and each rooted subgraph in the subgraph layer will be start point and do the random walk process. The random walk process is performed to generate a fixed-length sequence and repeat the process multiple times. The number of walks performed from every start point and the length of each step will be taken as hyper-parameters to control the scale of context generated. The random walk process can be described in three conditions:

1. Jump from the node layer to rooted subgraph layer

The transition probability of skipping from node layer $layer_n$ to rooted subgraph layer $layer_{sg}$ is 1. For node v of the node layer as the current node, the destination of skipping will be chosen from all rooted graph sets that include node v. As defined in formula (2), the likelihood of skipping is inversely proportional to the sum of the node's degree sequence of the currently rooted subgraph. If the number of nodes in a subgraph is smaller, or the degree of nodes is smaller, the structure of the root subgraph is simpler, that is, the sum of the sequence elements of the sequence composed of the degree of each node is smaller, then the probability of skipping to a rooted subgraph that contains node v and relatively simple in structure is more significant and vice versa. The probability of skipping from the node v of the node layer to rooted subgraph sg of the rooted subgraph layer can be calculated from the formula:

$$P_{layer_n \to layer_{sg}}(v, sg) = \frac{e^{-sum(s(V_{sg}))}}{M} \tag{2}$$

The $sum(s(V_{sg}))$ denotes the sum of the elements in the degree sequence of the current node. The normalizing factor M is defined as:

$$M = \sum_{sg' \in \Gamma_{sg}, sg' \neq sg, v \in sg'} e^{-sum(s(V_{sg'}))} \tag{3}$$

2. Jump from rooted subgraph layer to rooted subgraph layer

The structure of the rooted subgraph layer could be regarded as an undirected weighted graph, in which weight represents the transition probability in the random walk process. It is defined by structural similarities. In rooted subgraph layer, the transition probability is defined as followed:

$$p(sg_1, sg_2) = \frac{e^{-f(sg_1, sg_2)}}{Z} \tag{4}$$

Z is the normalizing factor defined as:

$$Z = \sum_{sg \in \Gamma_{sg}, sg \neq sg_1} e^{-f(sg_1, sg)} \tag{5}$$

The more similarity sg_1, sg_2 share, the smaller the similarity distance $f(sg_1, sg_2)$ will be, and synchronously, the probability of jump probability $p(sg_1, sg_2)$ grows.

3. Jump from rooted subgraph layer to node layer

The likelihood of skipping from rooted subgraph layer to the node layer is the hyperparameter q. The skipping destination is a random node from all nodes contained in the current subgraph. For each node in the current subgraph, the probability is equal.

The context acquired from the random walk above and context from the process of rooted subgraph generation are merged as the context of biased-graph2vec.

Model Construction Based on Rooted Subgraphs. Every rooted subgraph in the rooted subgraph set is correspondingly fitted into word2vec as a word while graph set $D = \{d_1, d_2, \ldots, d_N\}$ as the document. Applying the word2vec model and representation in a lower dimension of a graph and rooted subgraph in the same vector space is obtained. For that, the vocabulary size of the rooted subgraph is often relatively large. Applying negative sampling technology could effectively reduce the amount of calculation. Biased-graph2vec uses SGD (Stochastic Gradient Descent) to optimize the parameters of the model.

The low dimension representation acquired could be applied to the following tasks. Moreover, it is convenient to use the traditional machine learning algorithm. For example, in graph classification tasks, we can directly feed the acquired vectors into classifying algorithms like SVM. In graph clustering tasks, the vectors obtained can be used as the input of clustering algorithms like K-means.

4 Experiments

4.1 Experimental Settings

Experimental Data. The six open benchmark data sets used in the experiment from the field of biochemistry are Enzymes, Mutag, Nci1, Nci109, Proteins, and Ptc. Table 1 below shows the statistics of the data sets. The six data sets are all multi classification data, and their classification standards range from different protein structures to whether they cause cancer to experimental mice, including multiple classification standards in structure and function. In this experiment, the graph representation ability of the model is proved by the classification task experiment under different classification standards on six datasets.

Table 1. Statistics of the graph sets

Data sets	Enzymes	Mutag	Nci1	Nci109	Proteins	Ptc
Number of samples	600	188	4110	4127	1113	344
Average degree	33.5	17.9	29.8	29.6	39.1	25.5
Graph label	7	2	2	2	2	2
Node label	44	7	37	38	3	19

The Enzyme contains proteinic structures from enzymes of 6 kinds, moreover, 100 protein structures per kind. Mutag contains 188 structures of compounds classified and separately labeled according to their capability of inducing a certain kind of bacteria mutation. Ptc contains 344 compound structures classified and labeled by the fact of whether they are carcinogenic to mice. Protein has 1113 amino acids' second-level structures. Nci1 and Nci109 contain compounds related to cancer cell researching, which respectively have the sample numbers of 4110 and 4127.

Base Line Methods. The baseline methods apply the subgraph2vec, graph2vec algorithms, and kernel method WL, which have been mentioned in content-related sections of the paper.

Subgraph2vec [3] aims at the disadvantage that the substructure of the graph kernel method is entirely independent. It takes graph's substructures as words in the text while the graph as the document then applies a random walk and word embedding model to learn the representation of subgraphs. The algorithm applies the labeling process of the WL method to generate rooted subgraphs. Meanwhile, it modified the skip-gram model of the word2vec algorithm to make its inputting parameters able to suit vectors of indeterminate length.

Graph2vec [13] universally learns the representation of both graphs and substructures in the same characteristic space. The algorithm has two stages, first of which is to generate the rooted subgraphs, the second is to embed the vector of every rooted subgraph via doc2vec. The difference between the algorithm and subgraph2vec is that the model used in the second stage is different.

Weisfeiler-Lehman (WL) kernel [6] designs graph kernel algorithm from character- istics of the rooted subgraph. Compared with others, this is a better way to capture the characteristics of the graph itself.

In the experiment, the dimension of the vectors is uniformly set to 256. The length of the random walk sequence of biased-graph2vec and graph2vec is fixed to 10. The rooted subgraphs generated with the WL method have a max depth of 2, and the hyper-parameter of biased-graph2vec, q, is 0.3.

Evaluation Methodology. This section evaluates the effectiveness of the model with classification tasks. We use the accuracy rate as an evaluation indicator, which is to classify the nodes from their representations and evaluate the accuracy of the result. Process of calculation is as followed:

The vector representation of each rooted subgraph and node is obtained by modeling 90% of the data while verifying the representation in the remaining 10%. This paper applies SVM to classify the nodes of the remaining 10% to get accuracy. The formula of calculation of the accuracy is:

$$acc = \frac{TP + TN}{P + N} \times 100\% \tag{6}$$

P is the number of positive examples, while the N means the number of negative cases. Similarly, TP is the number of the positive examples that have been correctly classified and TN is the number of negative examples in their supposed places.

Graph Classification. The experiment was repeated ten times because some of the data sets were relatively small. In this paper, the average accuracy and standard deviation are taken as the evaluation index of the classification effect.

The parameters in the experiment are chosen via the grid-search method. The length of the random walk sequence is 15, and the walk has been performed five times. The probability of skipping in biased-graph2vec, q, is 0.1, and the depth of acquiring rooted subgraphs is 3.

Many results of the experiment are shown in Table 2.

In the multi-classification tasks, it is evident that the WL kernel method and the subgraph2vec have a rather poor performance. Probably for that, the lack of a unified model building of graphs and subgraphs leads to their vector representation emerging in different vector spaces.

Table 2. Accuracy of the experiment of graph classification

Data sets	Ptc	Proteins	Nci109	Nci1	Mutag	Enzymes
Biased-graph2vec	**70.86 ± 2.40**	**77.17 ± 4.40**	**71.82 ± 1.35**	**72.02 ± 2.33**	**88.42 ± 6.25**	**77.50 ± 6.90**
WL	58.78 ± 4.91	76.07 ± 1.16	69.93 ± 3.58	70.03 ± 1.35	67.11 ± 8.32	32.78 ± 8.34
Subgraph2vec	55.92 ± 9.01	75.18 ± 1.47	70.07 ± 3.21	71.14 ± 2.94	65.79 ± 8.89	44.00 ± 6.3
Graph2vec	68.10 ± 7.96	76.25 ± 4.75	67.19 ± 1.61	70.60 ± 1.60	86.47 ± 7.3	72.92 ± 4.52

In subgraph2vec, the output of the model is the representation of the substructure in the feature space; a similarity matrix of substructures is needed to get the representation of the graph at its final stage. Based on model doc2vec, biased-graph2vec uniformly does the model construction of both substructures and graphs. In this multi-classification experiment, doc2vec-based methods perform better.

In the experiment, biased-graph2vec is more effective than the other baseline methods in six data sets. It proves that compared with graph kernel method and method, ignoring the similarity among long distanced substructures, biased-graph2vec is more effective.

5 Summary

This paper proposed biased-graph2vec, the graph embedding model based on structural characteristics of rooted subgraphs, improved based on the vulnerability that existing projects have not done practical acquirement of the structural similarity of substructures. By building a suitable walking model for the substructures, biased random walks are performed to generate a moderate context. After that, the low dimensional representations of graphs and substructures are acquired via word embedding model. The experiment of graph classification proves the effectiveness of the model.

Directions of future research:

(1) For that compared with other graph embedding models, the biased-graph2vec model adds an additional part, that is, context generation, which leads to more computing work in the model. Next, we will further study how to reduce the computing cost.
(2) In the next step, we will also find out whether other effective fine-grained structures will not increase the calculation amount and can capture the similarity of substructures more precisely.

Acknowledgments. This work was supported by the National Natural Science Foundation of China (U1636219, 61602508, 61772549, U1736214, 61572052, U1804263, 61872448) and Plan for Scientific Innovation Talent of Henan Province (No. 2018JR0018).

References

1. Airola, A., Pyysalo, S., Björne, J., Pahikkala, T., Ginter, F., Salakoski, T.: All-paths graph kernel for protein-protein interaction extraction with evaluation of cross-corpus learning. BMC Bioinformatics **9**(S11), S2 (2008). https://doi.org/10.1186/1471-2105-9-S11-S2

2. Lancichinetti, A., Fortunato, S.: Community detection algorithms: a comparative analysis. Phys. Rev. E **80**(5), 056117 (2009)
3. Narayanan, A., Chandramohan, M., Chen, L., Liu, Y., Saminathan, S.: subgraph2vec: learning distributed representations of rooted sub-graphs from large graphs. arXiv preprint arXiv:1606. 08928 (2016)
4. Borgwardt, K.M., Kriegel, H.P.: Shortest-path kernels on graphs. In: Fifth IEEE International Conference on Data Mining, pp. 74–81. IEEE (2005)
5. Vishwanathan, S.V.N., Schraudolph, N.N., Kondor, R., Borgwardt, K.M.: Graph kernels. J. Mach. Learn. Res. **11**(Apr), 1201–1242 (2010)
6. Shervashidze, N., Schweitzer, P., Leeuwen, E.J.V., Mehlhorn, K., Borgwardt, K.M.: Weisfeiler-lehman graph kernels. J. Mach. Learn. Res. **12**(Sep), 2539–2561 (2011)
7. Kruskal, J.B.: Multidimensional scaling by optimizing goodness of fit to a nonmetric hypothesis. Psychometrika **29**(1), 1–27 (1964). https://doi.org/10.1007/BF02289565
8. Balasubramanian, M., Schwartz, E.L.: The isomap algorithm and topological stability. Science **295**(5552), 7 (2002)
9. Roweis, S.T., Saul, L.K.: Nonlinear dimensionality reduction by locally linear embedding. Science **290**(5500), 2323–2326 (2000)
10. Perozzi, B., Al-Rfou, R., Skiena, S.: DeepWalk: online learning of social representations. In: Proceedings of the 20th ACM SIGKDD International Conference on Knowledge Discovery and Data Mining, pp. 701–710. ACM, New York (2014)
11. Tang, J., Qu, M., Wang, M., Zhang, M., Yan, J. Mei, Q.: LINE: large-scale information network embedding. In: Proceedings of the 24th International Conference on World Wide Web, pp. 1067 1077. International World Wide Web Conferences Steering Committee, Florence (2015)
12. Grover, A., Leskovec, J.: node2vec: scalable feature learning for networks. In: Proceedings of the 22nd ACM SIGKDD International Conference on Knowledge Discovery and Data Mining, pp. 855–864. ACM, San Francisco (2016)
13. Prieto, L.P., Rodríguez-Triana, M.J., Kusmin, M., Laanpere, M.: graph2vec: learning distributed representations of graphs. arXiv preprint arXiv:1707.05005 (2017)
14. Nguyen, D., Luo, W., Nguyen, T.D., Venkatesh, S., Phung, D.: Learning graph representation via frequent subgraphs. In: Proceedings of the 2018 SIAM International Conference on Data Mining, pp. 306–314. Society for Industrial and Applied Mathematics, San Diego (2018)
15. Ribeiro, L.F., Savarese, P.H., Figueiredo, D.R.: struc2vec: learning node representations from structural identity. In: Proceedings of the 23rd ACM SIGKDD International Conference on Knowledge Discovery and Data Mining, pp. 385–394. ACM, Halifax (2017)

Knowledge Graphs Meet Geometry for Semi-supervised Monocular Depth Estimation

Yu Zhao⬤, Fusheng Jin$^{(\boxtimes)}$⬤, Mengyuan Wang⬤, and Shuliang Wang⬤

Beijing Institute of Technology,
5 South Zhongguancun Street, Haidian District, China
jfs21cn@bit.edu.cn

Abstract. Depth estimation from a single image plays an important role in computer vision. Using semantic information for depth estimation becomes a research hotspot. The traditional neural network-based semantic method only divides the image according to the features, and cannot understand the deep background knowledge about the real world. In recent years, the knowledge graph is proposed and used for model semantic knowledge. In this paper, we enhance the traditional depth prediction method by analyzing the semantic information of the image through the knowledge graph. Background knowledge from the knowledge graph is used to enhance the results of semantic segmentation, and further improve the depth estimation results. We conducted experiments on the KITTI driving dataset, and the results showed that our method outperformed the previous unsupervised learning methods and supervised learning methods. The result of the Apollo dataset demonstrates that our method can perform in the common case.

Keywords: Knowledge graph · Object detection · Depth prediction · Auto driving

1 Introduction

In the field of computer vision and robotics, estimating the depth structure of images has been extensively studied. The estimation of the depth structure is the foundation of motion planning and decision making. These technologies are widely used in autonomous moving platforms (AMP) [17].

The traditional monocular depth estimation methods obtain the depth structure of the scene by fusing information from different views, which facing the problems of mismatching and insufficient features [15]. In recent years, deep learning has been considered as a promising approach to overcome the limitations of traditional visual depth estimation methods [4]. Compared to traditional methods based on multi-view geometry, the learning-based methods predict the depth structure of a single view using prior knowledge learned from numerous training samples. Researchers also combine traditional methods with deep

© Springer Nature Switzerland AG 2020
G. Li et al. (Eds.): KSEM 2020, LNAI 12274, pp. 40–52, 2020.
https://doi.org/10.1007/978-3-030-55130-8_4

Fig. 1. Example predictions by our method on KITTI dataset of [7]. Compared against [23], our approach recovers more details in the scene.

learning [17], by predicting depth map with trained neural network and refine scene structure with multi-view geometry. The emergence of self-supervised and unsupervised methods makes the training of neural networks no longer require ground-truth depth data [6,13,23]. Meanwhile using semantic information for depth estimation also draws the attention of researchers [16].

State-of-the-art learning-based algorithms can effectively find some rules of training data, but they fail to analyze the background information of the scene like humans [5]. Given that images and videos are reflections of the world, exploiting the background knowledge can effectively improve the training effect of depth prediction.

In this paper, we introduce the knowledge graph into the depth estimation task, by boosting semantic segmentation using object detection information. Our key idea is first boosting object detection and semantic segmentation with the background knowledge given by the knowledge graph, then improving depth prediction result with semantic information. The detection result will help semantic segmentation by determining the area of the object [1]. Finally, the depth prediction task will benefit from semantic segmentation results. Some example predictions are shown in Fig. 1.

Our model is trained on the KITTI provided by [7] dataset and perform a comprehensive evaluation of the model. The evaluation includes semantic segmentation, depth prediction, and ego-motion prediction. The results of the assessment indicate that the knowledge graph can effectively improve the depth prediction results.

2   Related Work

Knowledge Graph Boosted Object Detection. Object detection is an essential task in the computer vision field. It is the backbone for many advanced applications, such as facial detection, autonomous driving, drone photography. Object detection can be divided into two main parts: location regression and image classification. Many new classification frameworks are trying to utilize extra information beyond the image itself.

Knowledge graph is an effective method to introduce background knowledge into the model [19], and it is widely used in the classification field. Lee *et al.* [11] proposed ML-ZSL framework, developing the relationship knowledge graph for zero-shot classification. By combining the location regression technique with the knowledge graph, object detection is boosted by knowing the background knowledge given by the graph. Fang *et al.* [5] propose a framework of knowledge-aware object detection, which enables the integration of external knowledge from knowledge graphs into any object detection algorithm. Liu et al. [22] integrate graph convolutional networks (GCN) into object detection framework to exploit the benefit of category relationship among objects. These methods help traditional object detection benefit from external knowledge.

Joint Detection and Segmentation. Object detection is an essential task in the computer vision field. It is the backbone for many advanced applications, such as facial detection, autonomous driving, drone photography. Object detection can be divided into two main parts: location regression and image classification. Many new classification frameworks are trying to utilize extra information beyond the image itself.

Detection and segmentation can reinforce each other, and improve the depth estimation. Yang *et al.* [21] use object detection to segment the image, they introduce a figure-ground masks and find relative depth ordering of it. Hariharan *et al.* [9] propose a learning-based model for simultaneous detection and segmentation, in which they refine pixel-level segment with boxes detection. Chen *et al.* [2] introduce a model for box detection, semantic segmentation, and direction prediction, and they refining the object detection result with semantic and direction features.

Learning Based Depth Prediction. With the advancement of convolutional neural networks (CNN), techniques for predicting scene depth structures from a single image using deep learning methods are receiving increasing attention. The neural networks is introduced to depth estimation [4,12], these methods require ground-truth depth information collected by depth sensors such as depth cameras or LIDAR for supervision. However, the noise and artifacts generated by the depth sensor severely limit their performance [8]. In order to solve these problems, self-supervised methods and unsupervised methods have emerged, which do not require depth information but calculate the loss function based on geometric constraints between different views to learn the spatial structure of scene from images or videos [6,23]. However, semantic information is no used in

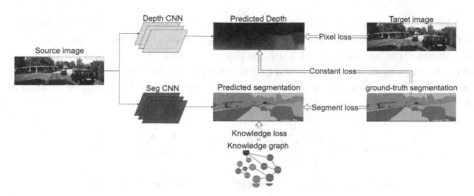

Fig. 2. The main process of our method. The depth and segmentation are predicted with the neural network. The depth domain is trained with multi-view geometric information, and the semantic segmentation is used to improve the result. The semantic domain is trained with ground-truth and the result is refined by the knowledge graph.

most unsupervised methods. Ramirez *et al.* [16] boost unsupervised method with semantic segmentation. Their overall learning framework is semi-supervised, as they deploy ground-truth data in the semantic domain. Their semantic segment is learned without external knowledge. With the help of knowledge graph for better detection and segmentation, depth prediction results can be further improved.

3   Method

Our method learns to predict the semantic segmentation and depth structure of the scene. The segmentation result is refined with external knowledge. The main process of our method is shown in Fig. 2. We first introduce the object detection and semantic segmentation with the knowledge graph. And then we introduce the semi-supervised learning of depth and segmentation with CNN. Finally other loss function for depth prediction and segmentation is introduced.

3.1   Detection with Knowledge Graph

Our key idea is utilizing information on objects' relationships for object detection tasks and improving the segmentation result with the information given by detection.

We utilize the objects' relationship of KITTI object detection dataset[7]. The category relationship knowledge graph is built following the way in [5]. For the established KITTI knowledge graph, each node represents each category and the graph edge from Node A to Node B is calculated by the conditional probability $P(B|A)$. For instance, if car and cyclist appear together 1000 times in dataset, and car appears 8000 times totally in the dataset, the edge from cyclist to car is defined as $P(Cyclist|car) = 1/8 = 0.125$.

After we get a matrix of the conditional probability, we need to further compare it with an existing model to enable knowledge-aware detection. The key idea is that two object with a high degree of semantic consistency are more likely to appear at the same time with comparable probability. For two bounding boxes of different objects b and b' in one image, the result of label will be punished if $P(b|b')$ or $P(b'|b)$ is small. The loss is designed as follows:

$$L_{knowledge} = \frac{1}{N} \sum_{b}^{B} \sum_{b'}^{B} (1 - P(b|b')) + \frac{1}{N} \sum_{b} \sum_{b'} (1 - P(b'|b)) \qquad (1)$$

in which B denotes all boxes in the image, N is the number of boxes.

Unlike object detection, semantic segmentation requires a label for each pixel in the image The initial set of detection is generated by classifying the fast edge boxes. Segmentation is guided by the initial localization and the corresponding appearance cues. We generate box map based on edge boxes, in which areas without any box is zero while other is label number. And we punish points of segmentation in the same category that are in the box. Let p_t be the semantic result given by the semantic network, $\bar{p}_t$ be the edge boxes map generated, the semantic-segmentation loss can be designed as:

$$L_{semseg} = \begin{cases} 1 & \bar{p}_t = 0 \\ H(p_t, \bar{p}_t) & other \end{cases} \qquad (2)$$

where H denotes the entropy. L_{semseg} is computed at full resolution only.

3.2   Depth Estimation with Multi-view and Semantic Information

In our method, the core of training the neural network is to simultaneously minimize the reprojection loss and semantic loss. The reprojection loss refers to the loss between the point of the projected target image, which is obtained by projecting the source image with a specific projection transformation. For the semantic loss, we smooth the image based on semantic segmentation, which makes the depth estimation result correspond to the semantic segmentation result.

We first introduce our reprojection loss. Given a pair of sequence frames I_{t-1} and I_t, in which I_{t-1} is the source view and I_t is the target view. For every point p_t on target view I_t, first warp them to the source view I_{t-1} using the predicted scene depth map and ego-motion, then calculate discrete pixel coordinates by bilinear interpolation. For each point p_t on the target image, its corresponding point p_{t-1} is

$$p_{t-1} \sim K T_{t \to t-1} D_t(p_t) K^{-1} p_t \qquad (3)$$

in which K denotes the camera intrinsic matrix and D_t denotes depth map predicted by neural networks and $T_{t \to t-1}$ is the rotation matrix from target view to source view.

Since p_{t-1} is continuous value, the discrete pixel coordinates can be calculated using the differentiable bilinear sampling method. That is, interpolation is performed according to four pixel points (upper left, lower left, upper right, lower right) adjacent to p_{t-1} to approximate. Finally, the reprojection loss can be expressed as follow:

$$L_{reproject} = \sum_p \left| I_t(p) - \hat{I}_t(p) \right| \tag{4}$$

in which $\hat{I}_t$ denotes warped target image, p denotes pixel index.

The image reconstruction loss is calculated using three pictures in the video sequence. We used the image mask structure proposed by [13]. The mask limits the valid range of image matching, preventing pixels projected outside the image from affecting the prediction result. The mask can be effectively calculated based on the depth of the scene and ego-motion. For image sequence I_{t-1}, I_t, I_{t+1}, we calculate two masks M_{t-1} and M_{t+1}, which represent the valid part of the projected image $\hat{I}_{t-1}$ and $\hat{I}_{t+1}$, so that we can avoid the invalid matching.

We perform Gaussian smoothing on the target image when calculating the loss, which makes pixel points easy to match with the source image. Gaussian smoothing is achieved by convolution, and the Gaussian convolution kernel can be calculated as follow:

$$G(u, v) = \frac{1}{2\pi\sigma^2} e^{-(u^2+v^2)/(2\sigma^2)} \tag{5}$$

in which u and v are the size of convolution kernel, σ denotes the smooth parameters for Gaussian. In our experiments, we use $u, v = 3$, $\sigma = 5$ to calculate the kernel. The image is smoothed by convolving the target image with the Gaussian kernel.

Finally, we normalize the depth image by a simple normalization method. The operator $\eta(\cdot)$ denotes dividing the input tensor by its mean:

$$\eta(d_i) = \frac{N d_i}{\sum_{j=1}^{N} d_j} \tag{6}$$

in which d_i and d_j donates the depth points in the depth map.

Combined with the above three methods, the final image reconstruction loss can be expressed as follow:

$$L_{rec} = \sum_p \left| (\zeta(I_t(p)) - \hat{I}_t^{t-1}(p)) M_{t-1}^p \right|$$
$$+ \sum_p \left| (\zeta(I_t(p)) - \hat{I}_t^{t+1}(p)) M_{t+1}^p \right| \tag{7}$$

in which ζ denotes the Gaussian smooth process and $\hat{I}_t^{t-1}$ denotes the projected image from source image I_{t-1}. And M is the masks calculated.

After the reprojection loss we introduce our semantic loss or this part, we calculate the standard cross-entropy between the predicted results and ground-truth semantic labels:

$$L_s = c(p_t, \bar{p}_t) = H(p_t, \bar{p}_t) + KL(p_t, \bar{p}_t) \tag{8}$$

where H is the entropy and KL is the KL-divergence. p_t is the points in predicted result and $\bar{p}_t$ is ground-truth label. The semantic term, L_s, is only computed at the full resolution.

The depth-semantic consist loss is designed as follow:

$$L_{consist} = \sum_{ij} ||\partial_x D^{ij}|| \, e^{-||\partial_x sem^{ij}||}$$
$$+ \sum_{ij} ||\partial_y D^{ij}|| \, e^{-||\partial_y sem^{ij}||} \tag{9}$$

where sem denotes the ground-truth semantic map and D is the predicted disparity map. Unlike smoothing the disparity domain directly, the novel $L_{consist}$ term detects discontinuities between semantic labels and the disparity domain, which keeps the gradient consistent.

3.3   Other Loss Functions

The structural similarity (SSIM) proposed in [20] is a common metric used to evaluate the quality of image predictions. It measures the similarity between two images. SSIM is calculated as follows:

$$SSIM(x, y) = \frac{(2\mu_x \mu_y + c_1)(2\sigma_{xy} + c_2)}{(\mu_x^2 + \mu_y^2 + c_1)(\sigma_x + \sigma_y + c_2)} \tag{10}$$

in which μ and σ denote the mean and the variance, respectively. We calculate μ and σ by pooling using the method proposed by [13]. Since SSIM needs to be maximized and the upper bound is 1, the SSIM loss is designed as:

$$L_{SSIM} = \sum_p \left[1 - SSIM(I_t^p, \hat{I}_t^p) \right] M_{t-1}^p$$
$$+ \sum_p \left[1 - SSIM(I_t^p, \hat{I}_t^p) \right] M_{t+1}^p \tag{11}$$

The depth output needs to be smoothed. We smoothed the depth map using a same smoothing, which ensures that the depth gradient are consistent with image gradient. The smooth loss can be designed as:

$$L_{sm} = \sum_{ij} ||\partial_x D^{ij}|| \, e^{-||\partial_x I^{ij}||}$$
$$+ \sum_{ij} ||\partial_y D^{ij}|| \, e^{-||\partial_y I^{ij}||} \tag{12}$$

in which I denotes the source image and D denotes the depth map predicted by the network.

4   Experiments

In this chapter, our network structure and training details is first introduced. Then we evaluated the performance of our method and compared it to previous work. We used the KITTI dataset for training. Using the data split method provided in the article [4], a total of 44540 image in the dataset were used, of which 40109 were used for training and 4431 were used for validation.

4.1   Network Structure

For depth estimation, our network structure includes two subnets: the depth estimation network and the ego-motion estimation network. The depth estimation network inputs a single 3-channel image and outputs depth values of four different scales. The network adopts the [14] network structure, that is also adopted by [23]. This is a network based on the encoder-decoder, adding a skip structure and adopting an output of four scales.

The ego-motion estimation network inputs three consecutive images and outputs 6-DOF of ego-motion. We used the same network structure used in [13], which inputs three consecutive images in the image sequence and outputs 6-DOF ego-motion.

For the semantic part, our network structure shares the same network with a depth estimation network. The detection network is fine-tuned from Faster-RCNN [10]. We fine-tune the network with a knowledge graph calculated from KITTI object detection dataset.

4.2   Metrics and Baseline for Depth Estimation

We evaluate depth estimation using the metrics the same as Eigen *et al.* [4]. 700 images are tested in the test data from the division of the KITTI dataset (this division excludes visually similar images). In our evaluation, we set the effective distance to 50 meters and 80 meters, respectively.

As this work aims at improving the depth estimation result by strengthening image correspondence, we mainly compare our approach to the learning-based method. We compared our method with the existing supervised and unsupervised methods. The actual depth value is obtained by projecting the LIDAR data to the image plane and interpolating.

4.3   Evaluation of Depth Estimation

Table 1 compares the results of our work with existing work in estimating the depth of the scene. As seen in Table 1, "Ours" and "Ours knowledge" indicate the results of using and not using the background knowledge loss, respectively. When trained only on the KITTI dataset with semantic knowledge as supervision, our model lowers the mean absolute relative depth prediction error from 0.136 to 0.133. Compare with [16], our method further enhance the segmentation

Table 1. Depth evaluation results for the KITTI test set, K indicates training on KITTI, and C indicates training on Cityscapes [3]. Ours indicates that the background knowledge is not used, and Ours knowledge indicates the result using the knowledge loss term.

Method	cap	Dataset	Supervised	Error				Accuracy metric		
				Abs Rel	Sq Rel	RMSE	RMSE log	$\delta < 1.25$	$\delta < 1.25^2$	$\delta < 1.25^3$
Eigen *et al.* [4] coarse	80	K	Depth	0.214	1.605	6.563	0.292	0.673	0.884	0.957
Eigen *et al.*[4] fine	80	K	Depth	0.203	1.548	6.307	0.282	0.702	0.890	0.958
Liu *et al.* [12]	80	K	Depth	0.202	1.614	6.307	0.282	0.678	0.895	0.965
Zhou *et al.* [23]	80	K		0.208	1.768	6.856	0.283	0.678	0.885	0.957
Zhou *et al.* [23]	80	K+C		0.183	1.595	6.720	0.270	0.733	0.901	0.959
Ramirez *et al.*[16]	80	K	Semantic	0.136	1.872	6.127	0.210	0.854	0.945	0.976
Ours	80	K	Semantic	0.134	1.557	6.131	0.219	0.847	0.936	0.970
Ours knowledge	80	K	Semantic	**0.133**	1.497	6.014	0.208	0.839	0.958	0.966
Garg *et al.* [6]	50	K	Pose	0.169	1.080	5.104	0.273	0.740	0.904	0.962
Zhou *et al.* [23]	50	K+C		0.173	1.151	4.990	0.250	0.751	0.915	0.969
Ramirez *et al.*[16]	50	K	Semantic	0.131	1.267	4.752	0.207	0.868	0.962	0.977
Ours	50	K	Semantic	0.131	1.301	4.600	0.213	0.845	0.953	0.972
Ours knowledge	50	K	Semantic	**0.130**	1.101	4.340	0.205	0.882	0.965	0.979

using background knowledge obtained from the knowledge graph. Figure 3 is a qualitative comparison of visualizations. However, our result in accuracy metric of $\delta < 1.25^1$ and $\delta < 1.25^3$, we believe this is due to some objects are incorrectly excluded in some pictures. Still, the results manifest the geometry understanding ability of our method, which successfully captures information from different images.

Input   Groundtruth   Eigen   Zhou   Ramirez   Ours

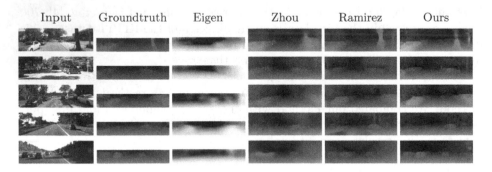

Fig. 3. Qualitative results on KITTI [7] test set. Our method captures details in thin structures and preserves consistently high-quality predictions both in close and distant regions.

The experimental results reflect our method's ability to understand 3D scenes, that is, the method successfully analyzes the 3D consistency of different scenes. For the using of background knowledge, our method further lowers the mean absolute relative depth prediction error from 0.134 to 0.133 in 80 m, and from 0.131 to 0.130 in 50 m. This reflects the effect of the using of knowledge graph.

To demonstrate that our proposed method can perform on common case, We directly apply our model trained on KITTI to the Apollo stereo test set[18]. We find that our method perform well on this dataset, even scenes structure is more complicated. As shown in Fig. 4, our method can recover more details.

Image	Zhou *et al.*	Ours

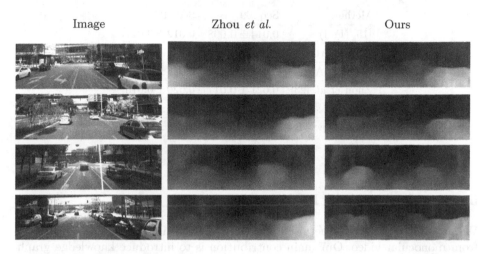

Fig. 4. Example predictions by our method on Apollo dataset of [18]. The model is only trained on KITTI dataset but also performs well in other cases. Compared with [23], our method recover more details.

4.4   Evaluation of Ego-Motion

During the training process, the depth of the scene and the accuracy of the ego-motion are closely related. To evaluate the performance of our camera position estimation network, we conducted an experiment on the official KITTI odometry split dataset, which included 11 actual history data obtained through IMU/GPS readings. We used the sequence 00–08 to train it, and the sequence 09–10 to evaluate. At the same time, we compared it with the traditional visual odometry method [15]. ORB-SLAM is a typical indirect sparse SLAM method and has a closed-loop detection based on graph optimization, which further constrains ego-motion by non-adjacent images. Therefore, we compare our approach to two different SLAM processes: (1) "ORB-SLAM (short)" containing only 5 frames as input, ie no closed-loop detection; (2) "ORB-SLAM (Full)" containing the entire process all frames. The scaling factor was optimized to be consistent with the actual data. Table 2 compares the results of our work with existing work on ego-motion estimation. Our approach goes beyond other unsupervised learning methods, approaching the traditional visual odometry method with global optimization.

Table 2. Absolute track error (ATE) tested on the KITTI odometry dataset [7]. Ours indicates that the background knowledge is not used, and Ours knowledge indicates the result using the knowledge loss term.

Method	Seq. 09	Seq. 10
[15] (Full)	0.014 ± 0.008	0.012 ± 0.011
[15] (Short)	0.064 ± 0.141	0.064 ± 0.130
Mean SLAM	0.032 ± 0.026	0.028 ± 0.023
Zhou *et al.* [23]	0.021 ± 0.017	0.020 ± 0.015
Ours	0.019 ± 0.017	0.019 ± 0.013
Ours knowledge	$\mathbf{0.017} \pm 0.015$	0.016 ± 0.012

5   Conclusions and Further Work

We proposed a novel unsupervised algorithm for learning depth and ego-motion from monocular video. Our main contribution is to introduce knowledge graph into geometric problems. We do so using a novel loss function which enhance the results of semantic segmentation, and further improve the result of depth and ego-motion estimation. Our algorithm needs only a single monocular video stream for training, and can produce depth from a single image at test time.

However, there are also some problems left. The main problem is the object categories in KITTI dataset are still not enough, the objects in the background like buildings and trees are not labeled in it. If the knowledge can be enlarged by adding more categories of objects, the result could be further improved.

References

1. Brazil, G., Yin, X., Liu, X.: Illuminating pedestrians via simultaneous detection & segmentation. In: Proceedings of the IEEE International Conference on Computer Vision, pp. 4950–4959 (2017)
2. Chen, L.C., Hermans, A., Papandreou, G., Schroff, F., Wang, P., Adam, H.: MaskLab: instance segmentation by refining object detection with semantic and direction features. In: Proceedings of the IEEE Conference on Computer Vision and Pattern Recognition, pp. 4013–4022 (2018)
3. Cordts, M., et al.: The cityscapes dataset for semantic urban scene understanding. In: Computer Vision and Pattern Recognition, pp. 3213–3223 (2016)
4. Eigen, D., Puhrsch, C., Fergus, R.: Depth map prediction from a single image using a multi-scale deep network. In: Advances in Neural Information Processing Systems, pp. 2366–2374 (2014)

5. Fang, Y., Kuan, K., Lin, J., Tan, C., Chandrasekhar, V.: Object detection meets knowledge graphs. In: International Joint Conference on Artificial Intelligence (2017)
6. Garg, R., B.G., V.K., Carneiro, G., Reid, I.: Unsupervised CNN for single view depth estimation: geometry to the rescue. In: Leibe, B., Matas, J., Sebe, N., Welling, M. (eds.) ECCV 2016. LNCS, vol. 9912, pp. 740–756. Springer, Cham (2016). https://doi.org/10.1007/978-3-319-46484-8_45
7. Geiger, A.: Are we ready for autonomous driving? The KITTI vision benchmark suite. In: Proceedings of the 2012 IEEE Conference on Computer Vision and Pattern Recognition (CVPR) (2012)
8. Godard, C., Aodha, O.M., Brostow, G.J.: Unsupervised monocular depth estimation with left-right consistency. In: Computer Vision and Pattern Recognition, pp. 6602–6611 (2017)
9. Hariharan, B., Arbeláez, P., Girshick, R., Malik, J.: Simultaneous detection and segmentation. In: Fleet, D., Pajdla, T., Schiele, B., Tuytelaars, T. (eds.) ECCV 2014. LNCS, vol. 8695, pp. 297–312. Springer, Cham (2014). https://doi.org/10.1007/978-3-319-10584-0_20
10. Huang, J., et al.: Speed/accuracy trade-offs for modern convolutional object detectors. In: Proceedings of the IEEE Conference on Computer Vision and Pattern Recognition, pp. 7310–7311 (2017)
11. Lee, C.W., Fang, W., Yeh, C.K., Frank Wang, Y.C.: Multi-label zero-shot learning with structured knowledge graphs. In: Proceedings of the IEEE Conference on Computer Vision and Pattern Recognition, pp. 1576–1585 (2018)
12. Liu, F., Shen, C., Lin, G., Reid, I.: Learning depth from single monocular images using deep convolutional neural fields. IEEE Trans. Pattern Anal. Mach. Intell. 38(10), 2024–2039 (2016)
13. Mahjourian, R., Wicke, M., Angelova, A.: Unsupervised learning of depth and ego-motion from monocular video using 3D geometric constraints. In: Computer Vision and Pattern Recognition, pp. 5667–5675 (2018)
14. Mayer, N., et al.: A large dataset to train convolutional networks for disparity, optical flow, and scene flow estimation. In: Computer Vision and Pattern Recognition, pp. 4040–4048 (2016)
15. Murartal, R., Montiel, J.M.M., Tardos, J.D.: ORB-SLAM: a versatile and accurate monocular SLAM system. IEEE Trans. Robot. 31(5), 1147–1163 (2015)
16. Ramirez, P.Z., Poggi, M., Tosi, F., Mattoccia, S., Di Stefano, L.: Geometry meets semantics for semi-supervised monocular depth estimation. In: Jawahar, C.V., Li, H., Mori, G., Schindler, K. (eds.) ACCV 2018. LNCS, vol. 11363, pp. 298–313. Springer, Cham (2019). https://doi.org/10.1007/978-3-030-20893-6_19
17. Tateno, K., Tombari, F., Laina, I., Navab, N.: CNN-SLAM: real-time dense monocular SLAM with learned depth prediction. In: Computer Vision and Pattern Recognition, pp. 6565–6574 (2017)
18. Wang, P., Huang, X., Cheng, X., Zhou, D., Geng, Q., Yang, R.: The ApolloScape open dataset for autonomous driving and its application. IEEE Trans. Pattern Anal. Mach. Intell. (2019)
19. Wang, X., Wang, S., Xin, Y., Yang, Y., Li, J., Wang, X.: Distributed Pregel-based provenance-aware regular path query processing on RDF knowledge graphs. World Wide Web, 1–32 (2019)
20. Wang, Z., Bovik, A.C., Sheikh, H.R., Simoncelli, E.P.: Image quality assessment: from error visibility to structural similarity. IEEE Trans. Image Process. 13(4), 600–612 (2004)

21. Yang, Y., Hallman, S., Ramanan, D., Fowlkes, C.C.: Layered object models for image segmentation. IEEE Trans. Pattern Anal. Mach. Intell. **34**(9), 1731–1743 (2011)
22. Liu, Z., Jiang, Z., Feng, W., Feng, H.: OD-GCN: object detection boosted by knowledge GCN. arXiv: Computer Vision and Pattern Recognition (2019)
23. Zhou, T., Brown, M., Snavely, N., Lowe, D.G.: Unsupervised learning of depth and ego-motion from video. In: Computer Vision and Pattern Recognition, pp. 6612–6619 (2017)

Topological Graph Representation Learning on Property Graph

Yishuo Zhang[1,7](✉) ⓘ, Daniel Gao[2] ⓘ, Aswani Kumar Cherukuri[3] ⓘ,
Lei Wang[1] ⓘ, Shaowei Pan[4] ⓘ, and Shu Li[5,6,7] ⓘ

[1] Xinjiang Technical Institute of Physics and Chemistry,
Chinese Academy of Sciences, Urumqi, China
yishuo.zhang@tulip.org.au
[2] Caulfield Grammar School,
Melbourne, VIC 3150, Australia
[3] School of Information Technology and Engineering,
Vellore Institute of Technology, Vellore, India
[4] Xi'an Shiyou University, Xi'an, Shannxi, China
[5] Institute of Information Engineering,
Chinese Academy of Sciences, Beijing, China
[6] School of Cyber Security,
University of Chinese Academy of Sciences, Beijing, China
[7] School of Information Technology,
Deakin University, Melbourne, Australia

Abstract. Property graph representation learning is using the property features from the graph to build the embeddings over the nodes and edges. There are many graph application tasks are using the property graph representation learning as part of the process. However, existing methods on Property graph representation learning ignore either the property features or the global topological structure information. We propose the TPGL, which utilizes the topological data analysis with a bias property graph representation learning strategy. The topological data analysis could augment the global topological information to the embedding and significantly improve the embedding performance on node classification experiments. Moreover, the designed bias strategy aggregated the property features into node embedding by using GNN. Particularly, the proposed TPGL outperformed the start of the art methods including PGE in node classification tasks on public datasets.

Keywords: Graph embedding · Topological data analysis · Property graph

1 Introduction

Graph data have been explored in many research fields such as the social network, sensor network and recommendation systems [8]. In simple terms, graph

© Springer Nature Switzerland AG 2020
G. Li et al. (Eds.): KSEM 2020, LNAI 12274, pp. 53–64, 2020.
https://doi.org/10.1007/978-3-030-55130-8_5

data is the data with structured items such as nodes and edges to represent the object information and their relationships. Applications such as graph classification and node classification are usually conducted directly on graph data [10]. Although many methods have been proposed, most are suffering from the complexity issue because the structural information is usually high dimensional and non-Euclidean. Graph kernel [7] or graph statistics only provides limited information and could not utilize the structural information sufficiently.

Graph representation learning aims to address the above-mentioned problem by learning the embedding for each node in the given graph [2]. Namely, the learning will map the graphical structural information to a node embedding like low-dimensional vectors which could use Euclidean metrics [2]. Many graph representation learning methods have shown satisfactory performance on graph classification and node classifications [5]. However, most graph representation learning methods consider the structural information but usually overlook other property features on both nodes and edges [5], and the traditional recommendation methods are highly relied on the property features [13,14] For example, in a buyer-product graph, the connected buyer nodes will have rich demographic information, and the same for the connected product nodes. Those property features enrich the graph's structural information (topological feature) by the label and the property information (property feature), and we call those graphs as the property graph [5]. When dealing with the property graph, many existing methods on graph representation learning are limited in utilizing both topological and property features efficiently [5].

Some existing methods could leverage the topological and property feature in the node embedding process, such as GCN [9] and GraphSAGE [4]. GCN could use the node property feature as the input and then process through multiple layers of neural network. GraphSAGE extends GCN by training the matrix before the embedding and then aggregates with property feature of the neighbour on each node. However, neither GCN [9] nor GraphSAGE [4] considered the dissimilarities between the nodes and its neighbours during the node embedding. The node and its neighbours are treated equally when the property features are aggregated, though it is not realistic in real-world scenarios. For instance, in the buyer-product graph, some products belong to the same category and with similar features, and some buyers are with a similar profile and from the same communities. Moreover, buyers coming from the same communities may have similar purchasing behaviour, and they constitute a similar neighbour structure in the graphs. Therefore, it is important to consider the dissimilar patterns in node embedding, though neither GCN or GraphSAGE considered this. Property Graph Embedding (PGE) [5], adopted the biased learning strategy when considering both topological and property features. However, PGE only considers the local neighbour topology and ignore the global graph topological information which is widely used in many graph application tasks by using the topological data analysis (TDA) [15,16].

In this paper, we propose a new framework, *Topological Property Graph Learning* (TPGL), to address the limitations of existing methods. In addition to a biased method to differentiate the neighbours of each node, TPGL incorporates the topological data analysis into the node embedding process. The added persistence feature provides the global graph topological information and enriches the node embedding with its property features comprehensively. We validated the performance of TPGL on node classification tasks by comparing with popular graph representation learning methods such as GCN, GraphSAGE, DeepWalk, Node2vec as well as the PGE. The performance confirms the importance of using topological and property feature as well as the global topological information in the graph representation learning. In summary, the contributions in this paper include: we designed a novel framework to incorporate the global topological features in graph representation learning by using topological data analysis; our novel *topological property graph learning* method (TPGL) is using a biased method that could largely differentiate the similar and dissimilar nodes in the aggregation process; the experiment results show that the *topological property graph learning* method (TPGL) outperforms all baseline methods in all node classification tasks on 3 public datasets.

2   Related Work

Existing graph representation learning methods could be categorized into the matrix factorization approach, the graph kernel approach and the deep learning approach.

Matrix factorization approach generates the graph embedding from the graph property by forming and factorizing the matrix. There are two types of matrix factorization [2]: the graph Laplacian Eigenmaps and the node proximity matrix. The work of GraRep [1] uses the adjacency matrix to measure the node similarity for learning. [11] constructs k nearest node neighbours from the graph to create the similarity matrix for obtaining the embedding. Another line of research is using the graph kernels to learn the embeddings. [12] proposed the *graphlet*, the small connected non-isomorphic induced sub-structures of a graph, to calculate the similarity between two graphs. [7] defined the fast random walk graph kernel with improved accuracy on many other methods. Moreover, with the success of deep learning in many fields [17], graph kernel methods can also utilize the powerful deep learning for the embeddings. [6] proposed the deep walk method for node-level embeddings. The method from [3] could perform both broad search and deep search on random walks and used the *word2vec* model on the walks to calculate the graph embedding. The graph neural networks have been used to learn the embeddings by encoding the node properties information into vectors from the neighbourhood. GCN [9] uses the graph convolutional networks to learn the node embeddings, and GraphSAGE [4] are able to capture the embedding for unseen nodes in the graph. Even though the graph-based neural network could consider the graph node structure and property information together, the neighbourhood aggregation treats the property information equally and fail to differentiate the dissimilarities. PGE [5] firstly proposed the bias strategy to make aggregation treat node property differently but failed to consider the global topology on the given graph. Considering the fact that graph global topology could provide more structural information on the node embedding, it is promising to overcome the limitations of existing methods by fully utilizing that information, as in the proposed method of this paper, the topological property graph learning (TPGL).

3   Topological Property Graph Learning

In this section, we propose a new framework, topological property graph learning (TPGL), which **TPGL** utilizes the topological data analytics with the current property graph learning method through a bias aggregation. The **TPGL** framework, not only employs the neighbour's property information and topological structure from local neighbours, but also utilizes the global topological structure information.

3.1   Problem Definition and Notations

We use the $\mathcal{G} = (\mathbf{V}_{\mathcal{G}}, \mathbf{E}_{\mathcal{G}}, \mathbf{P}_{\mathcal{G}}, \mathbf{L}_{\mathcal{G}})$ to represent the property graph. Particularly, $\mathcal{V}$ is the set of the vertex (nodes), $\mathcal{E}$ is the set of edges, $\mathcal{P}$ is the set of all property features and $\mathcal{L}$ is the set of labels on each vertex (nodes) in graph $\mathcal{G}$. Furthermore, we use $\mathcal{P} = \mathcal{P}_{\mathcal{V}} \cup \mathcal{P}_{\mathcal{E}}$ to denote the relationship between the node property feature and the edge property feature, and use $\mathcal{L} = \mathcal{L}_{\mathcal{V}} \cup \mathcal{L}_{\mathcal{E}}$ for the relationship between node labels and edge labels.

In addition to above notations, we use $\mathcal{N}_v$ to denote the set of neighbours of vertex $v \in \mathcal{V}$. TPGL attempts to integrate the global topological structure information and property information into the learning for improved performance. For a property graph $\mathcal{G} = (\mathbf{V}_{\mathcal{G}}, \mathbf{E}_{\mathcal{G}}, \mathbf{P}_{\mathcal{G}}, \mathbf{L}_{\mathcal{G}})$, we define the similarity between two nodes v_i and v_j as $s_G(v_i, v_j)$. Then, $s_G(v_i, v_j)$ could be further decomposed into the topological similarities $s_G^t(v_i, v_j)$ and the property similarities $s_G^p(v_i, v_j)$. Existing methods such as PGE, only use $s_G^t(v_i, v_j)$ to compare the local topological structure such as neighbours of each given node. In the TPGL, we further use the TDA to obtain the global graph topological structure to represent the similarity here.

To explain the TPGL, we further define the ideal embedding of vertex v_i and v_j as e_i and e_j by representation learning process. The learning of TPGL is to find the optimal encoding process which could minimize the gap $\sum_{v_i, v_j \in \mathcal{V}} (s_G(v_i, v_j) - e_i^\top e_j) = \sum_{v_i, v_j \in \mathcal{V}} (enc(s_G^t(v_i, v_j), s_G^p(v_i, v_j)) - e_i^\top e_j)$, where the $enc(\cdot)$ is the non-negative mapping, and TPGL is trying to find the representation learning process by minimizing the gap $\sum_{v_i, v_j \in \mathcal{V}} (s_G(v_i, v_j) - e_i^\top e_j)$.

The TPGL framework has four main steps as follows:

Property Node Clustering. In this step, we cluster the node $v \in \mathcal{G}$ by its property feature $\mathcal{P}_{v \in \mathcal{G}}$. The K clusters $C = C_1, C_2 ... C_k$ are obtained after the clustering process. The clustering algorithm is the standard Kmeans or Xmeans for completing this step.

Graph Topological Analytics. We will run the TDA on the entire graph in the node level by using its property features again. The edge weight will be calculated by using the Euclidean distance (dissimilarity) between two connected nodes. After running the TDA on the graph, each node v_i will have a persistence diagram set $[x_{v_i}, y_{v_i}]$. Then we could transfer those persistence diagram set into persistence value by using $Perv(v_i) = Perv(x_{v_i}, y_{v_i}) = |x_{v_i} - y_{v_i}|$. At the end of this step, each node $v_i \in \mathcal{G}$ will have its own persistence value $Perv(v_i)$.

Neighborhood Bias. Following the existing PGE method, we will assign the bias to the neighbourhood $\mathcal{N}_{v_i}$ of each node $v_i \in \mathcal{G}$. More specificity, for each neighbour $v' \in \mathcal{N}_{v_i}$, if the v' and v_i belongs to the same cluster, then we assign the bias b_s for neighbour node v' to express its similarity. However, if v' and v_i are not in the same cluster, then we assign a different bias b_d to indicate the dissimilarity. Here, the bias b_s is calculated by using the minimum distance to the cluster centre (the cluster of node v_i) for all the neighbour nodes $v' \in \mathcal{N}_{v_i}$. Also, the bias b_d is the max distance to the cluster centre for all the neighbour node $v' \in \mathcal{N}_{v_i}$.

Below Eqs. (1) and (2)

$$b_s = Min(D(v'_{(c^k,...,c^q)}, c^k(v_i))) \tag{1}$$

$$b_d = Max(D(v'_{(c^k,...,c^q)}, c^k(v_i))) \tag{2}$$

where D is the Euclidean distance. Obviously, it could be easily proofed that the $b_s < b_d$.

Neighborhood Aggregation. Based on the above steps, we will aggregate the property feature and the bias together to obtain the embedding $z_{v'_i}$ first. Then we will need to aggregate the persistence values for all the neighbour nodes into the persistence feature vector. Finally, we concatenate the $z_{v'_i}$ and the persistence feature vector together as the input into the graph neural networks.

3.2   Property Node Clustering

This step is required to distinguish the neighbourhood of each node v_i into two types based on their property features. That is, when the neighbour node $v' \in \mathcal{N}_{v_i}$ is similar to v_i according to its property features, then they will belong to the same cluster. Otherwise, we will regard $v' \in \mathcal{N}_{v_i}$ as a dissimilar neighbour of v_i.

The rationale is to use the property node clustering as a preparation step of bias assigning. In the property graph, without clustering the node by property feature, we will treat the property feature on each node equally and ignore the difference of information efficacy among the nodes in a property graph. The number of clusters could not be equal to the number of nodes in the property graph and also larger than 1 based on the algorithm of Kmeans or Xmeans.

3.3 Topological Analytics on Graph

Topological data analytics is the data analysis method based on techniques from topology theory [16]. TDA can extract the information from datasets that may be high-dimensional, incomplete and noisy. *Persistent homology* (PH) is one of the methods that could deliver the insights of the TDA. The essence of PH is the filtration process over the simplicial complex, which is a generation process for the series of topological spaces in the different scales (See Eqs. (1) and (2)).

PH is the process of filtration over a simplicial complex (clique complex) K, and the filtration is defined as below:

Definition 1. *Filtration over Simplicial Complex. A filtration of a simplicial complex K is a nested sequence of subcomplexes, $\emptyset \subseteq K^0 \subseteq K^1 \subseteq \cdots \subseteq K^{i-1} \subseteq K^n = K$. We call the simplicial complex K with a filtration a filtered complex.*

The intuitive explanation of the above filtration is as follows: given a graph $G = (V, E)$, the persistent homology is requiring the filtration on the graph G with a set of nested subgraphs (considering the graph is a simplicial complex). The node on each subgraph is formed by the same node set V as the original graph G, but the edges $E_i \in E$ on each subgraph are increasing by connecting more and more node pairs.

The filtration on a given graph G could be expressed as below:

$$\emptyset \subseteq G_0 \subseteq G_1 \subseteq \cdots \subseteq G_{i-1} \subseteq G_i \subseteq G \tag{3}$$

According to above filtration process, the newly added edge will merge two connected components. Therefore, the number of connected components will be reduced during the filtration process from G_{i-1} to G_i. Persistent homology could tract the changes during the filtration process from G_0 to G. Whenever a component disappears in graph G_i, there will be a new built tuple of $(0, i)$ stored as the persistent homology feature. The tuple is also called as the index persistence tuples. Under the above cardinality with filtration, each index persistence tuples could also be expressed as a pair of (x_{v_i}, y_{v_i}) by using the value of weights from graph. The set of pairs could form a persistence diagram or a persistence barcode for further calculation on persistence value $Perv(v_i)$ [15].

3.4 Neighbourhood Bias

As discussed in Sect. 3.1, according to the clustering results, we will give two different biases to each neighbour node $v' \in \mathcal{N}_{v_i}$ to distinguish the different information efficacy and the local topology concept (neighbourhood types). Intuitively, if all nodes are similar, then there is no difference in the embedding information across all nodes in the graph. So for a given node, the dissimilarity in neighbour nodes bring more information when compared with those similar neighbours. Thus, the bias of similar neighbours should have less weight than the dissimilar neighbour's node. We will properly proof above conclusion in later discussion on the efficacy of our TPGL. That is, in order to have better embedding results, we will need to make the b_s smaller than b_d.

3.5   Neighborhood Aggregation

This step involves two tasks: firstly, the property feature and the bias of node v_i need to be aggregated into embedding $z_{v_i'}$ via graph neural network. Then, the embedding $z_{v_i'}$ needs to be concatenated with the persistence value obtained in Sect. 3.1 to form the final embedding z_{v_i}.

For the first task, we use Eq. (4) to aggregate the embedding $z_{v_i'}$.

$$z_{v_i'} = \sigma(W' \cdot A(p_{v_i}, p_{v' \in C^{v_i}} \cdot b_s, p_{v' \notin C^{v_i}} \cdot b_d)) \tag{4}$$

For the second task, the persistence feature vector is obtained Eq. (5):

$$P(v_i) = [\sum Perv(v_{c_1}' \in \mathcal{N}_{v_i}), \sum Perv(v_{c_2}' \in \mathcal{N}_{v_i})... \sum_1^k Perv(v_{c_k}' \in \mathcal{N}_{v_i})] \tag{5}$$

Then we use the persistence feature vector $P(v_i)$ to aggregate the final embedding z_{v_i}:

$$z_{v_i} = \sigma(W^1 \cdot A(z_{v_i'}, P(v_i))) \tag{6}$$

where $A(\cdot)$ is the aggregation function which operates as concatenation, W' and W^1 are the weights to be learned via graph neural network, and $\sigma(\cdot)$ is the non-linear activation function. The weights will be optimized through the graph neural network optimizer, SGD optimizer, which will update the weights by minimizing the loss function as defined for specific applications such as node classification.

4   Theoretical Analysis of TPGL Bias

In this section, we will analyse the advantages of TPGL bias strategy and prove the validity of the bias value introduced in Sect. 3.1.

Efficacy on TPGL Biased Strategy. Existing research acknowledged that adding bias to similar or dissimilar neighbours of the graph node could improve the embedding results [5]. Particularly in PGE, a biased strategy statement was given as following [5].

Definition 2. *Biased Strategy. In graph embedding learning, there is an optimal strategy existing as B. For a biased strategy P and an unbiased strategy Q, if the inequality $\|B - P\|_1 < \|B - Q\|_1$ holds, where $B \neq Q$, then the biased strategy P could improve the embedding results.*

Therefore, to claim the efficacy of TPGL, we will need to prove that our bias strategy P holds the inequality $\|B - P\|_1 < \|B - Q\|_1$. Firstly, the optimal strategy B for a given node v_i and its neighbour could be expressed as: $B_{i,j} = \sum_{v_i \in \mathcal{V}} \sum_{v_j \in \mathcal{N}_{v_i}} *enc(b_{v_i,v_j})$ where b_{v_i,v_j} is the optimal bias for all associated nodes in $\mathbf{B}_{i,j}$, and $enc(\cdot)$ is the non-negative mapping function that assigns the bias to the nodes. Respectively, the unbiased strategy $\mathbf{Q}$ for a given node v_i and its neighbor could be denoted as: $\mathbf{Q}_{i,j} = \sum_{v_i \in \mathcal{V}} \sum_{v_j \in \mathcal{N}_{v_i}} *enc(\frac{1}{|\mathcal{N}_{v_i}|})$. Here, for unbiased strategy $\mathbf{Q}$, it means that no bias is assigned to the nodes in graph and to keep the consistency in optimal strategy $\mathbf{B}$, we give a constant value $\frac{1}{|\mathcal{N}_{v_i}|}$ in the $enc(\cdot)$ function. Therefore, we have Eq. (7):

$$\|\mathbf{B} - \mathbf{Q}\|_1 = \sum_{v_i \in \mathcal{V}} \sum_{v_j \in \mathcal{N}_{v_i}} |b_{v_i,v_j} - \frac{1}{|\mathcal{N}_{v_i}|}| \tag{7}$$

in which we omit the $enc(,)$ function as the mapping does not affect the calculation here. Similarly, we represent the biased strategy for a given node v_i and its neighbour as: $\mathbf{P}_{i,j} = \sum_{v_i \in \mathcal{V}} \sum_{v_j \in \mathcal{N}_{v_i}} *enc(b_s^{v_j \in \mathcal{N}_{v_i} \cap C^{v_i}} + b_d^{v_j \in \mathcal{N}_{v_i} \cap C_c^{v_i}})$. where $C_c^{v_i}$ means that the cluster is not the same as C^{v_i}. Then, we have Eq. (8) to denote the $\|\mathbf{B} - \mathbf{P}\|_1$:

$$\|\mathbf{B} - \mathbf{P}\|_1 = \sum_{v_i \in \mathcal{V}} \sum_{v_j \in \mathcal{N}_{v_i}} |b_{v_i,v_j} - (b_s^{v_j \in \mathcal{N}_{v_i} \cap C^{v_i}} + b_d^{v_j \in \mathcal{N}_{v_i} \cap C_c^{v_i}})| \tag{8}$$

Accordingly, to ensure the inequality $\|\mathbf{B} - \mathbf{P}\|_1 < \|\mathbf{B} - \mathbf{Q}\|_1$, the following needs to be satisfied:

$$\sum_{v_i \in \mathcal{V}} \sum_{v_j \in \mathcal{N}_{v_i}} |b_{v_i,v_j} - \frac{1}{|\mathcal{N}_{v_i}|}| > \sum_{v_i \in \mathcal{V}} \sum_{v_j \in \mathcal{N}_{v_i}} |b_{v_i,v_j} - (b_s^{v_j \in \mathcal{N}_{v_i} \cap C^{v_i}} + b_d^{v_j \in \mathcal{N}_{v_i} \cap C_c^{v_i}})| \tag{9}$$

Particularly, we will need to find out suitable b_s and b_d to ensure $b_s^{v_j \in \mathcal{N}_{v_i} \cap C^{v_i}} + b_d^{v_j \in \mathcal{N}_{v_i} \cap C_c^{v_i}}| > \frac{1}{|\mathcal{N}_{v_i}|}$, holds the inequality. As $\frac{1}{|\mathcal{N}_{v_i}|}$ is a constant that may be smaller than 1, we could easily find out a set of b_s and b_d to satisfy the above inequality. Thus, we could prove that the biased strategy in TPGL is better than the unbiased strategy when proper bias values are assigned to b_s and b_d.

The Discuss on Bias Values. Following the discussion in Sect. 4, we show that with proper bias value assigned to b_s and b_d the biased strategy in TPGL could improve the embedding results. In this subsection, we will deliberate our method on assigning bias values to b_s and b_d, and we will also verify the method with the corresponding proof.

Given a node v_i and its two neighbours $v_j, v_q \in \mathcal{N}_{v_i}$, we assume the local topological information of the two neighbors $v_j, v_q \in \mathcal{N}_{v_i}$ are the same. Then we will have the $\mathcal{N}_{v_j} = \mathcal{N}_{v_q}$ and also have the $s_G^t(v_i, v_j) = s_G^t(v_i, v_q)$. In the meantime, we assume that the neighbour node $v_j \in \mathcal{N}_{v_i}$ is more similar with node v_i, then we have $s_G(v_i, v_j) > s_G(v_i, v_q)$, $s_G^p(v_i, v_j) > s_G^p(v_i, v_q)$ and $|p_{v_i} - p_{v_j}| < |p_{v_i} - p_{v_q}|$. Where $|p_{v_i} - p_{v_j}|$ is the property feature difference between node v_i and v_j. According to the aggregation Eq. (4) and Eq. (6), we could have the following lemma:

Lemma 1. *if* $|p_{v_i} - p_{v_j}| < |p_{v_i} - p_{v_q}|$, *then* $|z_{v_i'} - z_{v_j'}| < |z_{v_i'} - z_{v_q'}|$ *and* $|z_{v_i} - z_{v_j}| < |z_{v_i} - z_{v_q}|$.

To prove above Lemma 1, we just need to check on Eq. (4) and Eq. (6). Because two neighbour nodes $v_j, v_q \in \mathcal{N}_{v_i}$ have the same local neighbour information ($\mathcal{N}_{v_j} = \mathcal{N}_{v_q}$), their global topological persistence value vectors are equal, and their neighbourhood bias values are equal as well. Hence, we conclude that the lemma is proper: if $|p_{v_i} - p_{v_j}| < |p_{v_i} - p_{v_q}|$ holds, then $|z_{v_i} - z_{v_j}| < |z_{v_i} - z_{v_q}|$ is appropriate.

According to the Lemma 1, when $|p_{v_i} - p_{v_j}| < |p_{v_i} - p_{v_q}|$ holds, we will have $s_G^p(v_i, v_j) > s_G^p(v_i, v_q)$. Therefore, we could argue that the larger dissimilarity between two nodes v_i, v_q will contribute larger changes regarding the embedding results $|z_{v_i} - z_{v_q}|$. As a consequence, b_d should be larger than b_s when assigning the bias in the TPGL bias strategy. From Eq. (1) and Eq. (2), we could easily see that the bias on b_d is larger than b_s because the max distance for all node to the cluster centre is always larger than the min distance. Therefore, our bias method is appropriate and properly proved.

5   Experiment on Node Classification

In this section, we evaluate the performance of TPGL on the task of node classification, using `MovieLens`, `PubMed` and `BlogCatalog` the three public real-world datasets, and analyse the effects of TGPGL parameters such as cluster number on the performance.

Particularly, in the experiment of node classification, we will compare the TPGL framework with five popular graph embedding methods: *DeepWalk*, *node2vec*, *GCN*, *GraphSAGE* and *PGE*.

5.1   Node Classification Performance

The nodes in the graph for all three datasets are divided into training nodes, validation nodes and test nodes. According to the TPGL learning steps, we firstly cluster all the nodes in the dataset. Particularly, we choose our number of cluster for `Movielens` as 10, `PubMed` as 5 and `BlogCatelog` as 15 for obtaining the best performance for TPGL. Then, we calculate the bias of b_s and b_d respectively, and run TDA over the graph by using the node property feature. Finally, we use the property feature and the bias to obtain the aggregation in graph neural network to complete the TPGL on node classification tasks. Notably, we use the F1-score to evaluate the performance of node classification over five methods on all three datasets. As the node classification tasks here is a multiple label classification, F1-score is a preferred metric in multiple-label classification evaluation.

Table 1 detailed the node classification results. It is clear that both PGE and TPGL have achieved higher F1-score when compared to the unbiased method such as DeepWalk, GCN and GraphSAGE. Particularly, the proposed TPGL achieves the best F1-score among all other baseline methods on all three datasets

Table 1: Node classification for all 3 data sets

Method	MovieLens	PubMed	BlogCatelog
DeepWalk	78.8 ± 0.54	76.2 ± 0.86	47.3 ± 3.22
Node2vec	80.3 ± 0.91	79.3 ± 1.39	45.8 ± 2.10
GCN	81.6 ± 0.61	79.7 ± 1.11	51.9 ± 1.87
GraphSAGE	81.1 ± 0.86	83.3 ± 1.15	50.5 ± 2.26
PGE	81.8 ± 0.59	88.3 ± 0.86	57.8 ± 2.31
TPGL	**82.2** ± 0.53	**88.9** ± 1.02	**59.3** ± 1.37

Table 2: Comparison on Method with PGE, TPGL* and TPGL

Method	MovieLens	PubMed	BlogCatelog
PGE	82.2 ± 0.19	83.3 ± 0.76	53.8 ± 1.23
TPGL*	82.6 ± 0.69	85.7 ± 0.52	55.1 ± 1.61
TPGL	**83.9** ± 0.70	**87.2** ± 0.82	**57.2** ± 1.69

for node classification. Especially for the node classification for BlogCatelog, which is a dataset regarded as very hard on the node classification task, the proposed TPGL had achieved 59.3%, which is significantly better than the benchmark in the work of [5]. Also, the methods using the node property feature such as GCN, GraphSAGE, PGE and TPGL, could achieve a higher F1-score than the methods which could not use the property feature (DeepWalk and Node2vec). In addition, when comparing TPGL with PGE, we could see that after adding the global topological information from TDA, TPGL could improve the node classification results substantially. The results in Table 2 are shown the comparison among TPGL, TPGL* (without TDA) and PGE. To avoid using the same experiment data, we randomly sampled 60% of the data from the original 3 datasets. From the results, we could see that with the proposed TPGL, the node classification F1-score is substantially higher than the other two methods. The experiment indicates that using TDA in the TPGL could substantially improve the embedding performance from the learning.

5.2 Effects of Cluster Number and TDA

As we discussed in Sect. 3, clustering plays an important role in TPGL. In this section, we will further analyse the parameter of the cluster number. Particularly, we are evaluating the performance of TPGL for node classification with different cluster numbers. We ran TPGL 50 times for node classification on MovieLens dataset, using different number of clusters in Kmeans. We also fixed the training epoch to 20 in each run of TPGL for avoiding huge workload in computing. The range on the number of clusters is from 5 to 35.

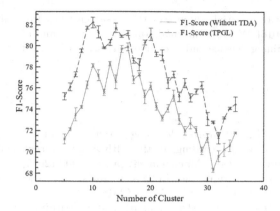

Fig. 1: Effects on cluster number and TDA

Figure 1 reports the results of F1 score on node classification. The results indicate that as the increase of the number of clusters, the F1-score grows first but then drops. As we found the best performance of node classification on `MovieLens` is when cluster number equals to 10. The performance indicated that when the number of clusters is too small, the cluster contains too many noises for distinguishing the similarity between nodes, and the persistence feature vector $P(v_i)$ becomes noisy as well. When the number of clusters is too large, then the fewer node neighbours will render the embedding of z_{v_i} and there are too many dissimilar biases assigned to the property feature of node neighbours. Here, we also tested the TPGL method without including the TDA in the learning steps over the same cluster numbers and bias setting. Figure 1 illustrates the results. In general, the F1-score for TPGL method without including the TDA is significantly lower than TPGL.

6   Conclusions

We proposed the TPGL for topological graph representation learning on property graphs. The TPGL could not only utilize the property feature and graph local topological feature together but also use the topological data analysis to enrich the embedding information. The experiment results from node classification indicated that the proposed TPGL could learn the better node embedding comparing to many other benchmarks. The TPGL is a key step on exploring the graph representation learning by using the applied topology mathematics. With more and more involvement on applied topology mathematics on graph representation learning, the more AI tasks on graph application could be implemented such as large online recommendation and identification.

Acknowledgment. This study was undertaken with the support of the Scheme for Promotion of Academic and Research Collaboration (SPARC) grant SPARC/2018-2019/P616/SL "Intelligent Anomaly Detection System for Encrypted Network Traffic", and we also thank the partial support of this research work from Shannxi Province Key R&D Plan (2018KW-010), National Natural Science Fund of China (Project No. 71871090) and Xinjiang Science and Technology Research fund with Chinese Academy of Sciences.

References

1. Cao, S., Lu, W., Xu, Q.: GraRep: learning graph representations with global structural information. In: Proceedings of the 24th ACM International Conference on Information and Knowledge Management, pp. 891–900 (2015)
2. Goyal, P., Ferrara, E.: Graph embedding techniques, applications, and performance: a survey. Knowl.-Based Syst. **151**, 78–94 (2018)
3. Grover, A., Leskovec, J.: node2vec: scalable feature learning for networks. In: Proceedings of the 22nd ACM SIGKDD International Conference on Knowledge Discovery and Data Mining, pp. 855–864. ACM (2016)
4. Hamilton, W., Ying, Z., Leskovec, J.: Inductive representation learning on large graphs. In: Advances in Neural Information Processing Systems, pp. 1024–1034 (2017)
5. Hou, Y., Chen, H., Li, C., Cheng, J., Yang, M.C.: A representation learning framework for property graphs. In: Proceedings of the 25th ACM SIGKDD International Conference on Knowledge Discovery & Data Mining, pp. 65–73 (2019)
6. Jin, Z., Liu, R., Li, Q., Zeng, D.D., Zhan, Y., Wang, L.: Predicting user's multi-interests with network embedding in health-related topics. In: 2016 International Joint Conference on Neural Networks (IJCNN), pp. 2568–2575. IEEE (2016)
7. Kang, U., Tong, H., Sun, J.: Fast random walk graph kernel. In: Proceedings of the 2012 SIAM International Conference on Data Mining, pp. 828–838. SIAM (2012)
8. Kataoka, T., Shiotsuki, E., Inokuchi, A.: Mapping distance graph kernels using bipartite matching. In: ICPRAM, pp. 61–70 (2017)
9. Kipf, T.N., Welling, M.: Semi-supervised classification with graph convolutional networks. arXiv preprint arXiv:1609.02907 (2016)
10. Lee, J.B., Rossi, R., Kong, X.: Graph classification using structural attention. In: Proceedings of the 24th ACM SIGKDD International Conference on Knowledge Discovery & Data Mining, pp. 1666–1674. ACM (2018)
11. Roweis, S.T., Saul, L.K.: Nonlinear dimensionality reduction by locally linear embedding. Science **290**(5500), 2323–2326 (2000)
12. Shervashidze, N., Vishwanathan, S., Petri, T., Mehlhorn, K., Borgwardt, K.: Efficient graphlet kernels for large graph comparison. In: Artificial Intelligence and Statistics, pp. 488–495 (2009)
13. Vu, H.Q., Li, G., Law, R., Zhang, Y.: Exploring tourist dining preferences based on restaurant reviews. J. Travel Res. **58**(1), 149–167 (2019)
14. Wang, X., Li, G., Jiang, G., Shi, Z.: Semantic trajectory-based event detection and event pattern mining. Knowl. Inf. Syst. **37**(2), 305–329 (2011). https://doi.org/10.1007/s10115-011-0471-8
15. Wang, Z., Li, Q., Li, G., Xu, G.: Polynomial representation for persistence diagram. In: Proceedings of the IEEE Conference on Computer Vision and Pattern Recognition, pp. 6123–6132 (2019)
16. Wasserman, L.: Topological data analysis. Ann. Rev. Stat. Appl. **5**, 501–532 (2018)
17. Zhang, Y., Li, G., Muskat, B., Law, R., Yang, Y.: Group pooling for deep tourism demand forecasting. Ann. Tour. Res. **82**, 102899 (2020)

Measuring Triplet Trustworthiness in Knowledge Graphs via Expanded Relation Detection

Aibo Guo, Zhen Tan, and Xiang Zhao[✉]

Key Laboratory of Science and Technology on Information System Engineering,
National University of Defense Technology, Changsha, China
{aiboguo,tanzhen08a,xiangzhao}@nudt.edu.cn

Abstract. Nowadays, large scale knowledge graphs are usually constructed by (semi-)automatic information extraction methods. Nevertheless, the technology is not perfect, because it cannot avoid introducing erroneous triplets into knowledge graphs. As a result, it is necessary to carry out some screening of the trustworthiness of the triplets in knowledge graphs before putting them into industrial use. In this paper, we propose a novel framework named as KGerd, for measuring triplet trustworthiness via expanded relation detection. Given a triplet (h, r, t), we center our framework on the basis of the classic translation-based mechanism among h, r and t. Besides translation-based relation detection, we introduce two additional types of relation detection approaches, which consider, respectively, to expand the task vertically by leveraging abstract versions of the relation r, as well as laterally by generating connecting paths between the entities h and t. The three detection results are then combined to provide a trustworthiness score for decision making. Comprehensive experiments on real-life datasets demonstrate that our proposed model KGerd is able to offer better performance over its state-of-the-art competitors.

Keywords: Knowledge graph · Triplet trustworthiness · Relation detection · Relation expansion

1 Introduction

Knowledge graph is playing a pivotal role in various applications, such as automatic question answering, information retrieval and so on. Conventional knowledge graph construction relies on manual annotation and expert supervision [5,18], and hence, it tends to be inefficient, and cannot accommodate the rapidly growth of real-world knowledge [5]. Lately, efforts have been dedicated to machine learning based information extraction from unstructured plain texts (including ORE [1,6,11], NELL [4]). By doing this, the scale of knowledge graphs is largely extended, and typical large-scale knowledge graphs include Freebase, DBpedia and WikiData, which respectively comprises billions of knowledge triplets in the form of (h, r, t).

© Springer Nature Switzerland AG 2020
G. Li et al. (Eds.): KSEM 2020, LNAI 12274, pp. 65–76, 2020.
https://doi.org/10.1007/978-3-030-55130-8_6

The current automatic information extraction methods are not perfect in the sense that the resultant triplets usually contain errors or contradictory information [8, 14]. In other words, those information extraction models fail to assess whether the extracted triplets are correct or not, which substantially affects the quality of knowledge graphs if erroneous triplets are injected without trustworthiness screening.

Therefore, to obtain triplets of high-precision, measuring trustworthiness is a critical step, especially in the early stage of knowledge graph construction as the effect of error propagation can be minimized. In this connection, Knowledge Vault [18] introduces a framework for measuring trustworthiness that constructs a probability knowledge base. However, the model is so delicate and complex that it may not be applicable to large-scale knowledge graphs. Later, KGTtm [12] utilizes a crisscrossing neural network to learn latent features, and achieves state-of-the-arts results. A key observation of the model construction is that a knowledge graph is parallel to a social network, which yet necessitates a second thought on the motivation of the model.

As relation is the first-class citizen in knowledge graphs, in this work, we establish a novel framework for measuring trustworthiness of triplets, which is centered on relations in knowledge graphs. We exploit the effect of various forms of the relation in question, and assess the triplet from a couple of orthogonal views. Specifically,

- The translation-based mechanism lays the foundation of our model, which detects the relation between head and tail entities by treating the relation as a translation operation from the head entity to the tail entity;
- The model is vertically extended by following the hierarchy of the relations, which detects the abstracted versions of the relation form an appropriate triplets with the head and tail entities after space projection; and
- The model is also laterally extended by considering possible paths in parallel to the relations, which detects the appropriateness of relating the head entity and the tail entities by using those methods.

Consequently, by putting them together, we construct a new model, namely, KGerd, which takes into consideration of all-round information of relations, and hence, the trustworthiness of the triplet connected by the relation can be well evaluated.

Contribution. In short, the main contribution of this paper can be summarized into three parts:

- We propose a novel framework for measuring triplet trustworthiness by centering on the intrinsic core of the triplets—relation, which considers a triplet from three orthogonal views;
- We propose to vertically extend a triplet by using abstract versions of the relation in the relational hierarchy, in order to detect the appropriateness of the triplet in terms of abstract concepts; and
- We propose to laterally extend a triple by composing paths between the head and tail entities at the same level as the triplet, in order to detect the appropriateness of the triplet through the influence of similar connections.

Comprehensive experiments are conducted to verify the design of the framework, and the results demonstrate that the proposed model outperforms its competitors on real-life datasets.

Organization. Section 2 overviews related work in existing literature, and then, Sect. 3 introduces the trustworthiness detection model. Afterwards, experiments and results analysis are reported in Sect. 4, followed by conclusion in Sect. 5.

2   Related Work

The concept of trustworthiness has been put into use in knowledge graph related tasks. The model proposed in [23] boosts the effect of knowledge representation. However, this approach ignores entities and focuses only on information about relationships. NELL [2] improves the learning effect by constantly iterating to extract templates. Nevertheless, this method is relatively simple and efficient, but its performance is suboptimal, because it ignores semantic information.

Increasing the trustworthiness of knowledge bases has been investigated before. For instance, Knowledge Vault [18] was presented to supplement Freebase by building a probabilistic knowledge base, which is considered to be more accurate, but at the cost of model generalization. CKRL [13] introduced the neural network model and used ConceptNet to give trustworthiness scores for invisible triplets. The disadvantage of this approach is that it does not consider the evidence provided by the knowledge base global information. Notably, KGTtm [12] is an elegant model that directly targets measuring triplets trustworthiness. It is based on neural network to evaluate the trustworthiness of triplets with the confidence score provided by three components. However, it is noted that the model is established on an intuition by comparing knowledge graphs to social networks, which may not be perfectly true, as they are graphs of intrinsically different nature. Thus, better observations on knowledge graphs may drive improvement over the task.

The task of measuring triplet trustworthiness may be cast into a triplet classification—"right" or "wrong". In the early stage, in order to handle this task, researchers used the method of manual detection [4,8,9,12]. The efficiency and accuracy of this method are difficult to guarantee. Now the research starts to turn to automatic detection [5,14,17,19]. In particular, the current popular embedding-based methods for knowledge graphs include TransE [3], TransH [22], TransR [16], TransD [10], PTransE [15], ComplEx [21], etc. This kind of method makes calculation and prediction by mapping entity relation to vector space through relevant embedded matrix.

3   The Expanded Relation Detection Model

In this section, we introduce the proposed model KGerd in four parts, as shown in Fig. 1:

- Translation-based detection (TBD) module evaluates the triplet trustworthiness by itself according to the classic translation-based mechanism;

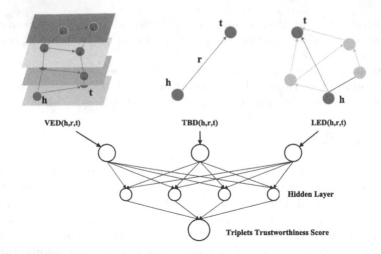

VED(h,r,t)   TBD(h,r,t)   LED(h,r,t)

Hidden Layer

Triplets Trustworthiness Score

Fig. 1. Framework of expanded relation detection

- Vertically-expanded detection (VED) module elevates the relation in question into abstract levels and determines the appropriateness between the abstracted relation and projected entities;
- Laterally-expanded detection (LED) module assess the triplet trustworthiness by considering the other connecting paths between the head and tail entities; and
- A fusion network is used to combine the learned features from the three aforementioned modules to produce a final evaluation of triple trustworthiness.

The basic idea of relation expansion is provided in Fig. 2. In this figure, the nodes and arc in blue represent the triplet in question (h, r, t); by abstracting the relation r into some upper level, r' is some abstract version of r, and h' (resp. t') is the resultant data point after projecting h (resp. t) onto the space where r' is in; and the nodes and arcs in green make paths connecting h and t besides r. Intuitively, the red nodes and arcs extend the relation r from a vertical view, while the green nodes and arcs extend r from a lateral view. Under KGerd, the blue part is handled by the translation-based detection module, the red part is dealt with by the vertically-expanded detection module, and the green part is addressed by the laterally-expanded detection module.

In the sequel, we explain how the translation-based detection, vertically-expanded detection and laterally-expanded detection modules evaluate the appropriateness of the triple, as well as aggregation towards a final trust worthiness score.

3.1   Translation-Based Detection

In order to evaluate the trustworthiness of complete triplets, we use a normalized function based on vector translation to calculate the trustworthiness of

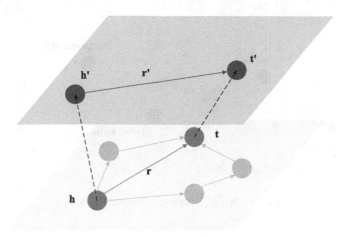

Fig. 2. Sketch of relation expansion

triplets. According to the invariant translation principle, the relation in knowledge graph can be regarded as a translation vector between entities (h, t). As shown in Fig. 3, in a given embedding space, the relation vector represent the translation between entity h and t, which is unchangeable for different entity pairs with the same relation. For example, triplets $(China, capital, Beijing)$ and triplets $(UnitedStates, capital, Washington, D.C.)$. Although the entity vectors $(China, Beijing)$ and $(UnitedStates, Washington, D.C.)$ are different, the relation vector between to entity pairs is the same. If there is a candidate triplet $(China, capital, Washington, D.C.)$, due to the translation between two entities is obviously different with relation $capital$, it is easy to be recognized as a wrong triplet.

Therefore, a positive triplet (h, r, t) should satisfy $\mathbf{h} + \mathbf{r} \approx \mathbf{t}$. Then, the triplet trustworthiness score function is described as $E(h, r, t) = ||\mathbf{h} + \mathbf{r} - \mathbf{t}||$. Overall, the lower score is, the higher the trustworthiness is.

Next, we use the normalization function to convert the triplets score.

$$TBD(h, r, t) = \frac{1}{1 + e^{-\lambda(\theta_r - E(h,r,t))}}, \tag{1}$$

where, θ_r is the parameter related to relationship r (obtained through training). The hyperparameter λ is used for smoothing. This $TBD(\mathbf{h}, \mathbf{r}, \mathbf{t})$is used as the second part of the final neural network input.

3.2 Vertically-Expanded Detection

To obtain more latent features of each relation, we leverage hierarchical information of each relation. For example, There exists the hierarchical structure for each relationship, such as "$/people/profession/people\_with\_this\_profession$". Only use r to represent a relation will omit lots of latent features.

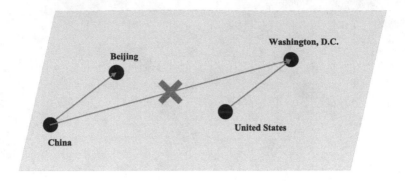

Fig. 3. Sketch of translation-based detection

Considering this relationship $r_1 = A_3/A_2/A_1$, A_1, A_2, and A_3 represent different features with different abstract levels. $r_2 = A_3/A_2$ and $r_3 = A_3$ are the abstract representation of r_1. In this paper, we use r_1, r_2, r_3 to denote the relation with different levels. For the concrete relation r_1, we use TransE to verify the trustworthiness of triplets, the score function can be described as

$$\mathbf{h} + \mathbf{r_1} = \mathbf{t}, \tag{2}$$

For abstract relation r_2 and r_3, we leverage four matrices $\mathbf{M_{h_2}}$, $\mathbf{M_{t_2}}$, $\mathbf{M_{h_3}}$ and $\mathbf{M_{t_3}}$ to extract the corresponding abstract entity features, respectively. the score functions are shown as:

$$\mathbf{M_{h_2}} \cdot \mathbf{h} + \mathbf{r_2} = \mathbf{M_{t_2}} \cdot \mathbf{t}, \tag{3}$$

$$\mathbf{M_{h_3}} \cdot \mathbf{h} + \mathbf{r_3} = \mathbf{M_{t_3}} \cdot \mathbf{t}, \tag{4}$$

where h and t are respectively the head and tail in triplets. It is important to note that each time change the hierarchical relationships, it needs to change the mapping of h and t. The mapping matrices are obtained by training model.

The score for each level of the relationship can be expressed as

$$S_i = ||\mathbf{M_{h_i}} \cdot \mathbf{h} + \mathbf{r_i} - \mathbf{M_{t_i}} \cdot \mathbf{t}||, \tag{5}$$

When $i = 1$, $\mathbf{M_{h_i}}$ and $\mathbf{M_{t_i}}$ are unit matrices. In this paper, we take three levels in the hierarchy. Calculate the score of the hierarchical relationship:

$$VED(h, r, t) = \mathbf{W} \cdot [S_1, S_2, S_3], \tag{6}$$

where $\mathbf{W}$ is a 1×3 weight matrix. $VED(h, r, t)$ is the trustworthiness score obtained by hierarchical information.

3.3   Laterally-Expanded Detection

Considering introducing global information to detect the trustworthiness of triplets. We refer to some existing methods of Embedding. The traditional

TransE method can only calculate the h and r of two direct connections. We hope to make more use of the information of global knowledge graph to prove the trustworthiness of triplets. The PTransE approach fits this need perfectly. A path of multiple relationships $P(h, t) = [p_1, p_2, ..., p_n]$ connecting h and t. Each p represents a path of multiple variables from h to t, $p_i = [r_1, r_2, ..., r_l]$. For each triplets, the score function is

$$G(h, r, t) = E(h, r, t) + E(h, P, t), \qquad (7)$$

$E(h, r, t)$ and TransE are calculated in the same way.

$$E(h, P, t) = \frac{1}{z} \sum_{p \in P(h,t)} R(p \mid h, t) E(h, p, t), \qquad (8)$$

where p represents relation path, $R(p \mid h, t)$ represents the reliability of the relationship path p between entity pairs (h, t) and $Z = \sum_{p \in P(h,t)} R(p \mid h, t)$ is the normalization factor, $E(h, p, t)$ is the energy of the entity pair under the relationship path.

Confidence level of the path to relationships, and put forward the resource allocation algorithm based on path constraint PCRA [15], its basic idea is: assuming that there is a certain amount of resources, entity h outflow from the head, and will flow along a given path p, t the amount of resources used in the end tail entity flow to measure the path p as the reliability of the connection path between h and t. The size of the resource acquired by the tail entity represents the size of the information it can obtain from the original entity. Tail entity resource $R_p(t)$ is used to measure the confidence of path p to entity (h, t), that is $R(p \mid h, t) = R_p(t)$.

And the confidence of the final relationship path $LED(h, r, t) = G(h, r, t)$, which is the triplets trustworthiness fraction at the global level that we need.

3.4   Fusion Network

We build a multi-layer perceptron structure [7] to output the final triplets trustworthiness score. First, we splice the three scores into a vector $\mathbf{f}(\mathbf{h}, \mathbf{r}, \mathbf{t})$:

$$\mathbf{f}(\mathbf{h}, \mathbf{r}, \mathbf{t}) = [VED(h, r, t), TBD(h, r, t), LED(h, r, t)]. \qquad (9)$$

Vector $\mathbf{f}(\mathbf{h}, \mathbf{r}, \mathbf{t})$ is used as the input of fusion estimator, and is transformed by hidden layer. The output layer is set to a binary classifier. When the triplets is correct, $y = 1$. On the contrary, $y = 0$. We use a nonlinear activation function (sigmoid) to calculate $p(y = 1 | \mathbf{f}(\mathbf{h}, \mathbf{r}, \mathbf{t}))$ as,

$$h_i = \sigma(\mathbf{W_{h_i}} \mathbf{f}(\mathbf{h}, \mathbf{r}, \mathbf{t}) + \mathbf{b_{h_i}}), \qquad (10)$$

$$p(y = 1 | \mathbf{f}(\mathbf{h}, \mathbf{r}, \mathbf{t})) = \varphi(\mathbf{W_o} h + \mathbf{b_o}), \qquad (11)$$

where h_i is the i_{th} hidden layer, $\mathbf{W_{h_i}}$ and $\mathbf{b_{h_i}}$ are the parameter matrices to be learned in the i_{th} hidden layer, and $\mathbf{W_o}$ and $\mathbf{b_o}$ are the parameter matrices of the output layer.

Table 1. Statistics of datasets

Dataset	#Rel	#Ent	#Train	#Vaild	#Test
FB15K	1,345	14,951	483,142	50,000	59,071

4 Experiments

In this section, we report the experiment studies and analyze the results in detail.

4.1 Datasets

In this paper, we used a commonly used benchmark from [23] for experiments, which was established on the basis of FB15K (statistics shown in Table 1).

There are three different versions—FB15K-N1, FB15K-N2 and FB15K-N3 in the dataset. FB15K-N1 (resp. FB15K-N2 and FB15K-N3) denotes the dataset in which 10% (resp. 20% and 40%) of the triplets were manually turned into negative triplets, and the detailed number of negative triplets in each dataset is presented in Table 2.

Table 2. Number of negative triplets

Dataset	FB15K-N1	FB15K-N2	FB15K-N3
#Negative	46,408	93,782	187,925

4.2 Overall Results

In order to verify whether our model can effectively distinguish the correct triplets from the wrong ones, we calculate the accuracy and recall of the model output results based on the statistical information of the results.

In Fig. 4, it is obvious that the range of triplets trustworthiness is almost all in [0.8, 1], while the error triplets scores are concentrated below 0.6 or even lower. This is almost consistent with the results we expected in the early stage. The better the score is, the higher the trustworthiness are and vice versa.

In the following calculation of accuracy and recall rate, we incorporate a threshold against trustworthiness score to determine whether a triplet is positive or negative. That is, when the trustworthiness is better than the threshold value, and the triplet is considered to be a positive triplet, and vice versa. Figure 5 shows the results of accuracy and recall.

It can be seen from the figure that with the increasing of the threshold, the precision rate is constantly improving, while the recall rate is decreasing. As the threshold is low, we put a large number of wrong triplets into the right category. This result inevitably leads to an excessively low accuracy. The increase of

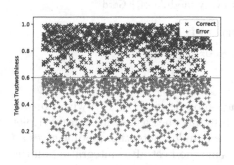 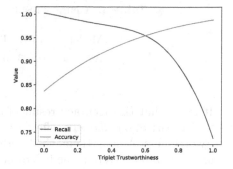

Fig. 4. Scatter plot of the triplets trustworthiness values distribution.

Fig. 5. Various value cures of precision and recall with the triplets trustworthiness values.

threshold value, especially when the threshold value rises above 0.6, can quickly and massively eliminate the wrong triplets. However, the much high threshold can also decrease recall rate.

4.3 Comparing with Other Models

The trustworthiness measuring task could also be regarded as a triplet classification task. In this connection, we would like to compare with other viable methods, which are mainly based on knowledge representation learning. By setting the threshold to 0.6, the comparison results are shown in Table 3.

Table 3. Accuracy results of measuring triplet trustworthiness

Model	MLP	Bilinear	TransE	TransD	TransR	PTransE	KGTtm	KGerd
Accuracy	0.833	0.861	0.868	0.913	0.902	0.941	0.981	**0.985**
F1-score	0.846	0.869	0.876	0.913	0.942	0.942	0.982	**0.984**

As Table 3 shows that our method has better accuracy and more F1-score than other method. Our method is at least 10% more accurate than traditional methods. Compared with the basic TransE, TransH, TransD, TransR and PTransE methods, our model is more accurate because it incorporates more internal semantic information and global information. At the same time, due to our model is scalable, we may get better results by considering other embedding methods.

4.4 Ablation Analysis

To see how each module contributes to the overall result, we analyzed the effects of each detection module of KGerd, and the results are given in Table 4.

Table 4. Accuracy results of every module of KGerd

Model	VED	TBD	LED	KGerd
Accuracy	0.903	0.868	0.917	0.984

It reads that the accuracy result of each module is above 0.8, which proves that each part of our model is effective in fulfilling the task of measuring triplet trustworthiness. Among them, the laterally-expanded detection module gives an accuracy of 0.917. Overall, the KGerd performs the best after integrating all the features that are helpful in deciding whether a triplet is credible.

5 Conclusion

In this paper, we have investigated the problem of measuring triplet trustworthiness, which play a pivotal role in making knowledge graph-based applications trustable. To handle the problem, we propose a novel framework for measuring triplet trustworthiness, namely, KGerd. It comprises three major modules by looking at relation in three different views—itself, vertical expansion and lateral expand, respectively, and the results are aggregated using a fusion network. Experiments on real-life datasets empirically verifies the effectiveness of the proposed model, which is able to outperform its state-of-the-art competitors.

In future, it is of interest to see whether the triplet trustworthiness evaluation could be integrated with the process of triplet extraction (e.g., [20]), then it can become an end-to-end service to obtain credible triplets from plain texts.

Acknowledgement. This work was partially supported by NSFC under grants Nos. 61701454, 61872446, 61902417, and 71971212, and NSF of Hunan Province under grant No. 2019JJ20024.

References

1. Banko, M., Cafarella, M.J., Soderland, S., Broadhead, M., Etzioni, O.: Open information extraction from the web. In: Veloso, M.M., (ed.) IJCAI 2007, Proceedings of the 20th International Joint Conference on Artificial Intelligence, Hyderabad, India, 6–12 January 2007, pp. 2670–2676 (2007)
2. Bollacker, K.D., Cook, R.P., Tufts, P.: Freebase: a shared database of structured general human knowledge. In: Proceedings of the Twenty-Second AAAI Conference on Artificial Intelligence, Vancouver, British Columbia, Canada, 22–26 July 2007, pp. 1962–1963. AAAI Press (2007)
3. Bordes, A., Usunier, N., García-Durán, A., Weston, J., Yakhnenko, O.: Translating embeddings for modeling multi-relational data. In: Burges, C.J.C., Bottou, L., Ghahramani, Z., Weinberger, K.Q., (eds.) Proceedings of the 27th Annual Conference on Neural Information Processing Systems, Lake Tahoe, Nevada, United States, 5–8 December 2013, pp. 2787–2795 (2013)

4. Carlson, A., Betteridge, J., Kisiel, B., Settles, B, Hruschka, E.R., Jr Mitchell, T.M.: Toward an architecture for never-ending language learning. In: Fox, M., Poole, D., (eds.) Proceedings of the Twenty-Fourth AAAI Conference on Artificial Intelligence, AAAI 2010, Atlanta, Georgia, USA, 11–15 July 2010. AAAI Press (2010)
5. Dong, X., Knowledge vault: a web-scale approach to probabilistic knowledge fusion. In: Macskassy, S.A., Perlich, C., Leskovec, J., Wang, W., Ghani, R., (eds.) The 20th ACM SIGKDD International Conference on Knowledge Discovery and Data Mining, KDD 2014, New York, NY, USA, 24–27 August 2014, pp. 601–610. ACM (2014)
6. Fader, A., Soderland, S., Etzioni, O.: Identifying relations for open information extraction. In: Proceedings of the 2011 Conference on Empirical Methods in Natural Language Processing, EMNLP 2011, 27–31 July 2011, Edinburgh, UK, pp. 1535–1545. ACL (2011)
7. Hampshire, J.B., Pearlmutter, B.: Equivalence proofs for multi-layer perceptron classifiers and the Bayesian discriminant function. In: Touretzky, D.S., Elman, J.L., Sejnowski, T.J., Hinton, G.E. (eds.) Connectionist Models, pp. 159–172. Morgan Kaufmann, Burlington (1991)
8. Heindorf, S., Potthast, M., Stein, B., Engels, G.: Vandalism detection in wikidata. In Mukhopadhyay, S., et al. (eds.) Proceedings of the 25th ACM International Conference on Information and Knowledge Management, CIKM 2016, Indianapolis, IN, USA, 24–28 October 2016, pp. 327–336. ACM (2016)
9. Hoffart, J., Suchanek, F.M., Berberich, K., Weikum, G.: YAGO2: a spatially and temporally enhanced knowledge base from wikipedia. Artif. Intell. **194**, 28–61 (2013)
10. Ji, G., He, S., Xu, L., Liu, K., Zhao, J.: Knowledge graph embedding via dynamic mapping matrix. In: Proceedings of the 53rd Annual Meeting of the Association for Computational Linguistics, Volume 1: Long Papers, 26–31 July 2015, Beijing, China, pp. 687–696. The Association for Computer Linguistics (2015)
11. Jia, S., Shijia, E., Li, M., Xiang, Y.: Chinese open relation extraction and knowledge base establishment. ACM Trans. Asian Low Resour. Lang. Inf. Process. **17**(3), 15:1–15:22 (2018)
12. Jia, S., Xiang, Y., Chen, X., Wang, K.: Triple trustworthiness measurement for knowledge graph. In: Liu, L., et al. (ed.) The World Wide Web Conference, WWW 2019, San Francisco, CA, USA, 13–17 May 2019, pp. 2865–2871. ACM (2019)
13. Li, X., Taheri, A., Tu, L., Gimpel, K.: Commonsense knowledge base completion. In: Proceedings of the 54th Annual Meeting of the Association for Computational Linguistics, Volume 1: Long Papers. ACL 2016, 7–12 August 2016, Berlin, Germany. The Association for Computer Linguistics (2016)
14. Liang, J., Xiao, Y., Zhang, Y., Hwang, S.W., Wang, H.: Graph-based wrong IsA relation detection in a large-scale lexical taxonomy. In: Singh, S.P., Markovitch, S., (eds.) Proceedings of the 31th AAAI Conference on Artificial Intelligence, 4–9 February 2017, San Francisco, California, USA, pp. 1178–1184. AAAI Press (2017)
15. Lin, Y., Liu, Z., Luan, H., Sun, M., Rao, S., Liu, S.: Modeling relation paths for representation learning of knowledge bases. In: Màrquez, L., Callison-Burch, C., Su, J., Pighin, D., Marton, Y., (eds.) Proceedings of the 2015 Conference on Empirical Methods in Natural Language Processing, EMNLP 2015, Lisbon, Portugal, 17–21 September 2015, pp. 705–714. The Association for Computational Linguistics (2015)

16. Lin, Y., Liu, Z., Sun, M., Liu, Y., Zhu, X.: Learning entity and relation embeddings for knowledge graph completion. In: Bonet, B., Koenig, S., (eds.) Proceedings of the Twenty-Ninth AAAI Conference on Artificial Intelligence, 25–30 January 2015, Austin, Texas, USA, pp. 2181–2187. AAAI Press (2015)
17. Nickel, M., Murphy, K., Tresp, V., Gabrilovich, E.: A review of relational machine learning for knowledge graphs. Proc. IEEE **104**(1), 11–33 (2016)
18. Qiao, L., Yang, L., Hong, D., Yao, L., Zhiguang, Q.: Knowledge graph construction techniques. J. Comput. Res. Dev. **53**(3), 582–600 (2016)
19. Shi, B., Weninger, T.: Discriminative predicate path mining for fact checking in knowledge graphs. Knowl. Based Syst. **104**, 123–133 (2016)
20. Tan, Z., Zhao, X., Wang, W., Xiao, W.: Jointly extracting multiple triplets with multilayer translation constraints. In: The Thirty-Third AAAI Conference on Artificial Intelligence, AAAI 2019, Honolulu, Hawaii, USA, pp. 7080–7087 (2019)
21. Trouillon, T., Dance, C.R., Gaussier, É., Welbl, J., Riedel, S., Bouchard, G.: Knowledge graph completion via complex tensor factorization. J. Mach. Learn. Res. **18**, 4735–4772 (2017)
22. Wang, Z., Zhang, J., Feng, J., Chen, Z.: Knowledge graph embedding by translating on hyperplanes. In: Brodley, C.E., Stone, P., (eds.) Proceedings of the Twenty-Eighth AAAI Conference on Artificial Intelligence, 27–31 July 2014, Québec City, Québec, Canada, pp. 1112–1119. AAAI Press (2014)
23. Xie, R., Liu, Z., Lin, F., Lin, L.: Does william shakespeare REALLY write hamlet? knowledge representation learning with confidence. In: McIlraith, S.A., Weinberger, K.Q., (eds.) Proceedings of the 32nd AAAI Conference on Artificial Intelligence, New Orleans, Louisiana, USA, 2–7 February 2018, pp. 4954–4961. AAAI Press (2018)

A Contextualized Entity Representation
for Knowledge Graph Completion

Fei Pu, Bailin Yang$^{(\boxtimes)}$, Jianchao Ying, Lizhou You, and Chenou Xu

School of Computer and Information Engineering, Zhejiang Gongshang University,
Hangzhou, China
{pufei,ybl}@zjgsu.edu.cn, Venut_yjc@126.com, thekernelz@163.com,
wstcoow1995@gmail.com

Abstract. Knowledge graphs (KGs) have achieved great success in
many AI-related applications in the past decade. Although KGs contain billions of real facts, they are usually not complete. This problem
arises to the task of missing link prediction whose purpose is to perform
link prediction between entities. Knowledge graph embedding has proved
to be a highly effective technology in many tasks such as knowledge reasoning, filling in the missing links, and semantic search. However, many
existing embedding models focus on learning static embeddings of entities
which pose several problems, most notably that all senses of a polysemous entity have to share the same representation. We, in this paper,
propose a novel embedding method, which is named KG embedding with
a contextualized entity representation ($KGCR$ for short), to learn the
contextual representations of entities for link prediction. $KGCR$ encodes
the contextual representations of an entity by considering the forward
and backward contexts of relations which helps to capture the different senses of an entity when appearing at different positions of a relation or in different relations. Our approach is capable to model three
major relational patterns, i.e., symmetry, antisymmetry, and inversion.
Experimental results demonstrate that $KGCR$ can capture the contextual semantics of entities in knowledge graphs and outperforms existing
state-of-the-art (SOTA) baselines on benchmark datasets for filling in
the missing link task.

Keywords: Relation patterns · Contextual entity representation ·
Link prediction

1 Introduction

Recently, knowledge graphs have attracted great concerns from both academia
and industry in the circumstance that needs to exploit the vast amount of heterogeneous data. KGs are structural representations of data, all facts in KGs are

Supported by the Key R&D Program Project of Zhejiang Province under Grant no.
2019C01004 and Zhejiang Education Department Project under Grant no. Y201839942.

G. Li et al. (Eds.): KSEM 2020, LNAI 12274, pp. 77–85, 2020.
https://doi.org/10.1007/978-3-030-55130-8_7

in form of (head entity, relation, tail entity). KGs enhance many AI-related tasks in a lot of domains such as semantic search and ranking [1], question answering [2], machine reading [3], and natural language processing [4], etc. Since KGs usually are not complete, they still loss many valid triples. Thus, it is important to fill in the missing links of KGs automatically as it is impractical to find all valid triples manually.

Many approaches have been devised to deal with the link prediction problem. Knowledge graph embedding (KGE) approaches are the state-of-art models for link prediction tasks. They aim to represent entities and relations in a low dimensional continuous space while preserving their semantics and inherent structures. The KGE based approaches can be roughly classified into two categories: semantic matching models and translation models using the scoring functions. TransE [5] is a representative translation model, which views relation vectors as translations in the vector space. Given a valid triple (h, r, t), let $\mathbf{h}, \mathbf{r}$ and $\mathbf{t}$ be embedding vectors of h, r and t, respectively, it is expected that $\mathbf{h} + \mathbf{r} \approx \mathbf{t}$. RotatE [6] view each relation as a rotation in the complex vector space from the source entity to the target entity. Further, with a self-adversarial negative sampling technique, which generates negative samples according to the current entity and relation embeddings, RotatE has achieved the best performance so far.

Semantic matching models measure the plausibility by the scoring functions which match latent semantics of entities and relations. DistMult [7] learns embeddings using a bilinear scoring function $\phi(h, r, t) = \mathbf{h}^T M_r \mathbf{t}$, where matrix M_r is a diagonal matrix for reducing the number of parameters. It can successfully infer the composition of relations. ComplEx [8] represents entities and relations by replacing real-valued vectors with complex-valued ones. With the Hermitian product of complex-valued vectors, it can better model asymmetric relations.

So far, most existing embedding models have focused on static representations of entities which pose a problems that all senses of a polysemous entity have to share the same representation. SimplE [9] takes into account the position encodings of the entities. It takes two vectors h_e and t_e as two different embeddings for each entity e. Here, h_e captures e's behaviour as the head of a relation and t_e captures e's behaviour as the tail of a relation. SimplE also considers the inverse of relations and uses them to address the independence of the entity vectors. Given two vectors $h_e, t_e \in \mathbb{R}^d$ as the embedding of each entity e, and two vectors $v_r, v_{r-1} \in \mathbb{R}^d$ for each relation r, the similarity function of SimplE for a triple (e_i, r, e_j) is defined as $\frac{1}{2}(< h_{e_i}, v_r, t_{e_j} > + < h_{e_j}, v_{r-1}, t_{e_i} >)$. It is worth noting that SimplE only encodes the different positions (the head part and the tail part) of each relation r, i.e., two embeddings of h_e and t_e may not be equal to the same entity e at different positions. However, besides the positions of a relation, different relations which an entity is associated with are also the key contexts for the entity. For example, consider two triples $< e_1, r_1, e_2 >$, $< e_1, r_2, e_3 >$, where e_1, r_1 and e_2 represent "apple", "release" and "MacBook", respectively, e_1, r_2 and e_3 represent "apple", "planted on" and "ground", respectively. Since entity e_1 is at the head part of relation $r1$ and $r2$, according to

SimplE, it is encoded with the same embedding vector and even e_1 is associated with two different relations r_1 and r_2. It is obvious that the header entity e_1 of the relation r_1 is distinctly different from the header entity e_1 of the relation r_2. Thus, two senses of a polysemous entity e_1 have to share the same representation in SimplE. So, it creates the need for learning the contextualized representations of entities, which are the most useful information for KG embeddings.

To address the aforementioned problem, when encoding an entity, we need to take into account both the positions of the relation that an entity is in and the relations that an entity is associated with. We introduce KGCR model – an embedding model with contextualized entity representations for the KG completion task. To model the contextual semantics of entities, KGCR is expected to distinguish entities in two categories: (1) entities in different relations; (2) entities at different positions in the same relation. In KGCR, each relation r has two vectors P_r and Q_r, named forward and backward contextual vectors of r, respectively. For each triple $< e_1, r, e_2 >$, the scoring function is defined as $\phi(e_1, r, e_2) = (v_{e_1} \circ P_r) \bullet (v_{e_2} \circ Q_r)$. The contextualized representation of entity e at different positions (head part or tail part) of relation r are different due to $v_e \circ P_r \neq v_e \circ Q_r$ in general. Similarly, for triples $< e_1, r_1, t_1 >$ and $< e_1, r_2, t_2 >$, the contextualized representation of entity e_1 in triple $< e_1, r_1, t_1 >$ is not equal to the one in triple $< e_1, r_2, t_2 >$ due to $v_{e_1} \circ P_{r_1} \neq v_{e_1} \circ P_{r_2}$ in general. For triples $< e_1, r_1, t_1 >$ and $< e_2, r_2, t_1 >$, we also conclude that the contextualized representation of entity t_1 in triple $< e_1, r_1, t_1 >$ does not equal to the one in triple $< e_2, r_2, t_1 >$ because of $v_{t_1} \circ Q_{r_1} \neq v_{t_1} \circ Q_{r_2}$ in general. So, two different cases of relational contexts are considered to embed the entities.

Table 1 summarizes different scoring functions $\phi(h, r, t)$ and parameters in existing SOTA baselines as well as KGCR model. Note that $\| v \|_p$ is the p-norm of the vector v. $< v_h, v_r, v_t > = \Sigma_i v_{h_i} v_{r_i} v_{t_i}$ denotes a tri-linear dot product. "$\circ$" is the Hadmard or element-wise product. "$*$" is a convolution operator. "$\cdot$" denotes a dot product of vectors. g is an activating function. $\hat{v}$ is a 2D reshaping of the vector v. Ω is a set of various filters used in a convolutional neural network. concat is a concatenation operator of vectors. "$\bullet$" denotes the inner product of complex vector (see Sect. 3 for detail).

Our contributions are as follows.

- We present the KGCR model −a novel contextualized entity representation model for knowledge graph completion. Most existing KG embedding based models were static: each entity had a single representation vector, regardless of contextual semantics of entities. Replacing static entity representations with contextualized entity representations achieves improvements on link prediction task.
- KGCR is capable to model three main relation patterns, i.e., symmetry, antisymmetry and inversion, so as to obtain high expressiveness and good generalization.
- KGCR learns the contextualized entity representations without any additional information other than the triples in knowledge graphs.

Table 1. Scoring functions and parameters in existing SOTA baselines as well as KGCR model

Model	Scoring function $\phi(e_1, r, e_2)$	Parameters
TransE	$\parallel v_{e_1} + v_r - v_{e_2} \parallel_p$	$v_{e_1}, v_r, v_{e_2} \in \mathbb{R}^d$
DistMult	$< v_{e_1}, v_r, v_{e_2} >$	$v_{e_1}, v_r, v_{e_2} \in \mathbb{R}^d$
ComplEx	$Re(< v_{e_1}, v_r, \bar{v}_{e_2} >)$	$v_{e_1}, v_r, v_{e_2} \in \mathbb{C}^d$
SimplE	$\frac{1}{2}(< h_{e_1}, v_r, t_{e_2} > + < h_{e_2}, v_{r^{-1}}, t_{e_1} >)$	$h_{e_1}, v_r, t_{e_2}, h_{e_2}, v_{r^{-1}}, t_{e_1} \in \mathbb{R}^d$
RotatE	$- \parallel v_{e_1} \circ v_r - v_{e_2} \parallel$	$v_{e_1}, v_r, v_{e_2} \in \mathbb{C}^d, \quad |r_i| = 1$
ConvE	$g(vec(g(concat(\widehat{v}_{e_1}, \widehat{v}_r) * \Omega))W) \cdot v_{e_2}$	$v_{e_1}, v_r, v_{e_2} \in \mathbb{R}^d$
ConvKB	$concat(g([v_{e_1}, v_r, v_{e_2}] * \Omega)) \cdot \mathbf{w}$	$v_{e_1}, v_r, v_{e_2} \in \mathbb{R}^d$
KGCR	$(v_{e_1} \circ P_r) \bullet (v_{e_2} \circ Q_r)$	$v_{e_1}, v_r, v_{e_2}, P_r, Q_r \in \mathbb{C}^d$

The rest of the paper is organized as follows. First, we outline related work in Sect. 2, then discuss KGCR model in Sect. 3. Finally, we report experimental results in Sect. 4 before concluding remarks in Sect. 5.

2 Related Work

Filling in the missing links of a knowledge graph is important since knowledge graphs are characterized by incompleteness and manually adding new information is costly. KGE based approaches have been widely explored for this task. Among these approaches, the models based on deep neural networks can learn rich semantic embeddings because of their parameter reducing and consideration of sophisticated relations.

NTN [10] utilizes a neural tensor network for relation classification. ConvE [11] firstly use a convolutional neural network model to predict missing links. ConvE contains a single 2D convolution layer, which is capable of extracting more interaction features among different dimensional entries of v_h and v_r focusing on their local relationships. To extract the global relationships among h, r, t rather than h, r, ConvKB [12] uses a convolution operator on same dimensional entries of the entity and relation embeddings. R-GCN [13] is the first framework to apply graph convolutional networks to high dimensional multi-relational data. RSNs [14] proposes recurrent skipping networks to learn representations of long-tail entities. In RSNs, a head entity can directly predict not only its subsequent relation but also its tail entity by skipping its connection similar to residual learning.

Previous work is static: each entity has a single embedding vector. Recently, the newly proposed language models, Transformer and BERT [15,16], have yielded significant improvements on many NLP tasks. BERT has successfully created contextualized word representations, word vectors that are sensitive to the context in which they appear. The success of contextualized word representations suggests that KG embeddings also need to consider the contexts of entities.

3   KGCR Model

Let $\mathscr{E}$ be a set of entities and $\mathscr{R}$ be a set of relations. A knowledge graph $\mathcal{G} = \{(h, r, t)\} \subseteq \mathscr{E} \times \mathscr{R} \times \mathscr{E}$ is defined as a set of factual triples formalized as (the head entity, the relation, and the tail entity) denoted as (h, r, t). The link prediction task is formalized as a ranking problem: given an input triple, the goal is to learn a *scoring* function $\phi : \mathscr{E} \times \mathscr{R} \times \mathscr{E} \mapsto \mathbb{R}$. Its score $\phi(h, r, t)$ is the likelihood indicating that the triple is true.

The procedure to corrupt positive triples [5] is as follows. For a positive triple (h, r, t), we randomly decide to corrupt the head entity or tail entity. If the head entity is selected, we replace h in the triple with an entity h' randomly selected from $\mathscr{E} - \{h\}$ and generate the corrupted triple (h', r, t). If the tail entity is selected, we replace t in the triple with an entity t' randomly selected from $\mathscr{E} - \{t\}$ and generate the corrupted triple (h, r, t').

Notations. We denote the i-th entry of a vector $v_h \in \mathbb{C}^d$ as $[v_h]_i$. Let d denote the embedding dimension.

Let $\circ : \mathbb{C}^d \times \mathbb{C}^d \to \mathbb{C}^d$ denote the Hadamard multiplication (element-wise product) between two vectors, that is

$$[v_h \circ v_t]_i = [v_h]_i \cdot [v_t]_i$$

Let $v_h, v_t \in \mathbb{C}^d$, the inner product between v_h and v_t is

$$< v_h, v_t > = \sum_{j=1}^{d} [v_h]_j \cdot [v_t]_j \tag{1}$$

Here, $[v_h]_j \cdot [v_t]_j$ is the product between two complex numbers.

We define the operator "$\bullet$" on two vectors $u, v \in \mathbb{C}^d$ as follows: let $u = a_u + \mathbf{i}b_u$, $v = a_v + \mathbf{i}b_v$, where $a_u, b_u, a_v, b_v \in \mathbb{R}^d$. Then

$$u \bullet v = < a_u, a_v > + < b_u, b_v >$$

To calculate the product "$\bullet$" between u and v, we first calculate the inner multiplications between their real and imaginary components of u and v accordingly and take the sum of these two inner products.

When encoding an entity, we need to consider both the positions of a relation where the entity is in and the relations that the entity is associated with. We present a KG embedding model with a contextualized entity representation, named KGCR, for the KG completion task. In KGCR, each relation r has two vectors P_r and Q_r, named forward and backward contextual vectors of r, respectively. Then, when e appears at the header of the relation r, the contextualized entity representation of e can be obtained through Hadamard product between v_e and P_r, i.e., $v_e \circ P_r$. Similarly, $v_e \circ Q_r$ is the contextualized entity representation of e when e appears at the tail part of the relation r. Here, relation $r = (P_r, Q_r) \in \mathbb{C}^{2d}$.

Given a triple (h, r, t), the scoring function of KGCR is defined as

$$\phi(h, r, t) = (v_h \circ P_r) \bullet (v_t \circ Q_r)$$

3.1   Optimization

Once we have the scoring function, to optimize the model, we utilize the negative sampling loss functions with self-adversarial training [6]:

$$\mathcal{L} = -log\sigma(\gamma - \phi(h, r, t)) - \Sigma_{i=1}^{n}p(h_i', r, t_i')log\sigma(\phi(h_i', r, t_i') - \gamma)$$

here, σ is the sigmoid function, γ is a margin which is a hyper-parameter and (h_i', r, t_i') is i-th corrupted triple.

Specifically, we sample negative triples from the following distribution:

$$p(h_j', r, t_j'|(h_i, r_i, t_i)) = \frac{exp\; \alpha\phi(h_j', r, t_j')}{\sum_i exp\; \alpha\phi(h_i', r, t_i')}$$

here, α is the temperature of sampling.

We use Adam as the optimizer to minimize loss function $\mathcal{L}$, and during optimizing we constrain the vector modulus to satisfy: $\| P_r \|_2 \leq 1$ and $\| Q_r \|_2 \leq 1$.

3.2   Ability to Model Relational Patterns

KGCR can model three types of relations, namely, symmetry, antisymmetry, and inversion. First, let $\mathcal{G}'$ be a set of corrupted triples in $\mathcal{G}$. Then, the formal definitions of symmetry, antisymmetry and inversion are as follows:

- A relation r is symmetry if for any two entities e_1 and e_2, $(e_1, r, e_2) \in \mathcal{G} \Leftrightarrow (e_2, r, e_1) \in \mathcal{G}$.
- A relation r is antisymmetry if for any two entities e_1 and e_2, $(e_1, r, e_2) \in \mathcal{G} \Rightarrow (e_2, r, e_1) \in \mathcal{G}'$.
- Relation r_2 is the inverse of relation r_1 if for any two entities e_1 and e_2, $(e_1, r_1, e_2) \in \mathcal{G} \Leftrightarrow (e_2, r_2, e_1) \in \mathcal{G}$.

Theorem 1. *For symmetry relation r, it can be encoded by tying the parameters $P_r = Q_r$ in model KGCR.*

Theorem 2. *For antisymmetry relation r, it can be encoded by tying the parameters $P_r \neq Q_r$ in model KGCR.*

Theorem 3. *Suppose relation r_2 is the inverse of relation r_1, this property of r_1 and r_2 can be encoded by making the parameters $P_{r_1} = -Q_{r_2}$ and $Q_{r_1} = -P_{r_2}$ in model KGCR.*

4   Experimental Results

We evaluate KGCR using two widely used benchmarks, i.e., WN18RR and FB15K-237 instead of WN18 and FB15K, since WN18 and FB15K have test leakage via inverse relations [11]. Table 2 summarizes the statistics of the datasets.

We compare KGCR with baselines using the following metrics of correct entities: Mean Rank (MR), the mean of all the predicted ranks. Mean Reciprocal Rank (MRR), the mean of all the inverse of predicted ranks. Hits@n, the proportion of ranks are not larger than n. In the experiments, n usually takes values of 1, 3, 10. The lower MR or the higher MRR or the higher Hits@n, the better performance. We utilize the "Filtered" setting protocol [5], which does not consider any valid triples when sampling negative triples.

Table 2. Summary of datasets

Dataset	#entity	#relation	#training triples	#validation triples	#test triples
FB15K-237	14541	237	272115	17535	20466
WN18RR	40943	11	86835	3034	3134

Experimental Setup. We optimize the KGCR model with Adam optimizer, fine-tune the hyper-parameters over the validation dataset. We set the range of hyperparameters for grid search as shown below: embedding dimension $k \in \{100, 250, 500, 1000\}$, self-adversarial sampling temperature $\alpha \in \{0.5, 1.0, 1.5, 2.0, 2.5\}$, maximal margin $\gamma \in \{3, 6, 9, 12, 18, 24, 30\}$, learning rate $lr \in \{0.00005, 0.0001, 0.0002, 0.0005, 0.001\}$, negative_sample_size $n \in \{64, 128, 256, 512, 1024\}$, and max_steps $max\_steps \in \{60000, 80000, 100000, 150000\}$.

Table 3. Experimental results

	WN18RR				FB15K-237			
		Hits@N				Hits@N		
	MRR	1	3	10	MRR	1	3	10
DistMult	0.444	0.412	0.47	0.504	0.281	0.199	0.301	0.446
ComplEx	0.449	0.409	0.469	0.53	0.278	0.194	0.297	0.45
ConvE	0.456	0.419	0.47	0.531	0.312	0.225	0.341	0.497
TransE	0.243	0.043	0.441	0.532	0.279	0.198	0.376	0.441
ConvKB	0.265	0.058	0.445	0.558	0.289	0.198	0.324	0.471
R-GCN	0.123	0.08	0.137	0.207	0.164	0.10	0.181	0.30
SimplE	0.390	0.378	0.394	0.411	0.169	0.095	0.179	0.327
Rotate	0.476	0.428	0.492	0.571	0.338	0.241	0.375	0.533
KGRC	**0.482**	**0.438**	**0.504**	**0.577**	**0.345**	**0.247**	**0.380**	**0.540**

Table 3 shows that KGCR achieves better performance compared with previous state-of-the-art models on the task of predicting missing links.

5 Conclusion

We present the KGRC model − a novel contextualized entity representation model for knowledge graph completion. KGRC encodes the contextual information of a relation by its forward and backward contextual vectors which is helpful for capturing the different senses of an entity when appearing at different positions of a relation or in different relations. The contextualized representation of the entities has advantages in many downstream tasks, for instance, question answering, semantic search, and neural machine translation. KGCR is capable to model three main relation patterns, namely inversion, symmetry, and antisymmetry, so as to obtain high expressiveness and good generalization.

The results show that for link prediction tasks, KGRC is superior to the existing SOTA baselines on the benchmark dataset. We are gonging to extend KGRC with attention to obtain better link prediction performance in future work.

References

1. Xiong, C.Y., Russell, P., Jamie, C.: Explicit semantic ranking for academic search via knowledge graph embedding. In: Proceedings of the 26th International Conference on World Wide Web, pp. 1271–1279 (2017)
2. Hao, Y.C., Zhang, Y.Z., Liu, K., et al.: An end-to-end model for question answering over knowledge base with cross-attention combining global knowledge. In: Proceedings of the 55th Annual Meeting of the Association for Computational Linguistics, pp. 221–231 (2017)
3. Yang, B.S., Tom, M.: Leveraging knowledge bases in LSTMs for improving machine reading. In: Proceedings of the 55th Annual Meeting of the Association for Computational Linguistics, pp. 1436–1446 (2017)
4. Annervaz, K.M., Chowdhury, S.B.R., Dukkipati, A.: Learning beyond datasets: knowledge graph augmented neural networks for natural language processing. In: Proceedings of the Conference of the North American Chapter of the Association for Computational Linguistics: Human Language Technologies, pp. 313–322 (2018)
5. Antoine, B., Nicolas, U., Alberto, G., et al.: Translating embeddings for modeling multirelational data. In: Proceedings of Advances in Neural Information Processing Systems, pp. 2787–2795 (2013)
6. Sun, Z.Q., Deng, Z.H., Nie, J.Y., Tang, J.: Rotate: knowledge graph embedding by relational rotation in complex space. In: Proceedings of The Seventh International Conference on Learning Representations (2019)
7. Yang, B.S., Yih, W.T., He, X.D., et al.: Embedding entities and relations for learning and inference in knowledge bases. In: Proceedings of the International Conference on Learning Representations, pp. 1–12 (2015)
8. Trouillon, T., Welbl, J., Riedel, S., Gaussier, E. Bouchard G.: Complex embeddings for simple link prediction. In: Proceedings of International Conference on Machine Learning, pp. 2071–2080 (2016)
9. Seyed, M.K., David, P.: Simple embedding for link prediction in knowledge graphs. In: Proceedings of the 32nd Conference on Advances in Neural Information Processing Systems, pp. 4289–4300 (2018)

10. Socher, R., Chen, D.Q., Manning, C.D., Ng, A.: Reasoning with neural tensor networks for knowledge base completion. In: Proceedings of Advances in Neural Information Processing Systems, pp. 926–934 (2013)
11. Dettmers, T., Minervini, P., Stenetorp, P., Riedel, S.: Convolutional 2D knowledge graph embeddings. In: Proceedings of Thirty-Second AAAI Conference on Artificial Intelligence, pp. 1811–1818 (2018)
12. Nguyen, D.Q., Nguyen, T.D., Nguyen, D.Q., et al.: A novel embedding model for knowledge base completion based on convolutional neural network. In: Proceedings of the Conference of the North American Chapter of the Association for Computational Linguistics: Human Language Technologies, pp. 327–333 (2018)
13. Schlichtkrull, M., Kipf, T.N., Bloem, P., van den Berg, R., Titov, I., Welling, M.: Modeling relational data with graph convolutional networks. In: Gangemi, A., et al. (eds.) ESWC 2018. LNCS, vol. 10843, pp. 593–607. Springer, Cham (2018). https://doi.org/10.1007/978-3-319-93417-4_38
14. Guo, L.B., Sun, Z.Q., Hu, W.: Learning to exploit long-term relational dependencies in knowledge graphs. In: Proceedings of the 36th International Conference on Machine Learning, pp. 2505–2514 (2019)
15. Vaswani, A., Shazeer, N., Parmar, N., et al.: Attention is all you need. In: Proceedings of Advances in Neural Information Processing Systems, pp. 5998–6008 (2017)
16. Devlin, J., Chang, M.W., Lee, K., Toutanova, K.: Bert: pre-training of deep bidirectional transformers for language understanding. arXiv preprint arXiv:1810.04805 (2018)

A Dual Fusion Model for Attributed Network Embedding

Kunjie Dong[1], Lihua Zhou[1(✉)], Bing Kong[1], and Junhua Zhou[2]

[1] School of Information, Yunnan University, Kunming 650091, China
`kunjiedong@qq.com`, {`lhzhou,Bingkong`}`@ynu.edu.cn`
[2] School of Public Administration, Yunnan University, Kunming 650504, China
`ynuzjh@163.com`

Abstract. Attributed network embedding (ANE) maps nodes in network into the low-dimensional space while preserving proximities of both node attributes and network topology. Existing methods for ANE integrated node attributes and network topology by three fusion strategies: the early fusion (EF), the synchronous fusion (SF) and the late fusion (LF). In fact, different fusion strategies have their own advantages and disadvantages. In this paper, we develop a dual fusion model named as DFANE. DFANE integrated the EF and the LF into a united framework, where the EF captures the latent complementarity and the LF extracts the distinctive information from node attributes and network topology. Extensive experiments on eight real-world networks have demonstrated the effectiveness and rationality of the DFANE.

Keywords: Network analysis · Attributed network embedding · Fusion strategy · Auto-encoder

1 Introduction

With more and more information becoming available, nodes in real-world networks are often associated with attributed features, for example, papers in academic citation networks generally have a published conference, author, research topic and keywords, which are known as attributed networks [4]. Recently, ANE, aiming to learn the low-dimensional latent representations of nodes which can well preserve the proximities based node attributes and network topology at the same time, has attracted lots of researchers' interests.

It plays an important role in integrating the node attributes and network topology for ANE, because ANE focuses on capturing latent relationships in

Supported by organization of the National Natural Science Foundation of China (61762090, 61262069, 61966036 and 61662086), The Natural Science Foundation of Yunnan Province (2016FA026), the Project of Innovative Research Team of Yunnan Province (2018HC019), Program for Innovation Research Team (in Science and Technology) in University of Yunnan Province (IRTSTYN), and the National Social Science Foundation of China under Grant No. 18XZZ005.

G. Li et al. (Eds.): KSEM 2020, LNAI 12274, pp. 86–94, 2020.
https://doi.org/10.1007/978-3-030-55130-8_8

terms of node attributes and network topology. Existing ANE approaches, based on the different fusion styles, can be divided into three categories: the EF embedding models [2,3], the SF embedding models [1,8,9], and the LF embedding models [6]. The EF and SF allow attributes modeling and topology modeling closely interact each other, but they cannot guarantee the individual distinctive characteristic; the LF trains individual models separately without knowing each other and results are simply combined after training, thus it can guarantee individual characteristic, but it may lose the consistency information of nodes [1].

To fully excavate the relationships and intrinsic essences between node attributes and network topology, we propose a dual fusion model named DFANE for ANE in this paper. DFANE consists of two components, the EF component and the LF component, where the former first concatenates the node attributes and network topology into a united vector on the input layer, and then conducts collaborative training to capture the relationships between node attributes and network topology; and the latter, including the node attributes Auto-Encoder and the network topology Auto-Encoder, first to capture the individual inherent essences from node attributes and network topology without interacting each other, and then achieves information integration via concatenating two types of individual inherent essences obtained by the two independent Auto-Encoders.

To summarize, our main contributions are as follows:

(i) A dual fusion model with the EF and LF components is proposed. The EF component extracts the latent interrelationship, the LF component captures the peculiarity of attributes and topology, and then the unity of two components preserve the consistency and complementarity information.
(ii) We conduct abundant experiments on eight datasets by the tasks of node classification and node clustering. The experimental results demonstrate the effectiveness and rationality of the DFANE.

2   Related Work

In this section, we briefly summarize the development of ANE methods in items of the strategies of fusing node attributes and network topology.

NANE [3] designed a self-feedforward layer to capture weight features in node attributes and network topology, which employed the pairwise constraint based the input data and weights to preserve the local information and the global information respectively. NetVAE [2] adopted a shared encoder to perform co-training and introduced the attributed decoder and networked decoder to reconstruct node attributes and network topology.

DANE [1] captured the latent highly non-linearity in both node attributes and network topology by two symmetrical Auto-Encoders, in which considered the consistency and complementary. ANRL [8] fed the node attributes into the encoder and the decoder reconstructed node's target neighbors instead of node itself under the guidance of network topology. The partial correlation, i.e. nodes with similar attributes may be dissimilar in topology and vice versa, was proposed in PRRE [9], and EM algorithm was utilized to tune two thresholds to define node relations: positive, ambiguous, negative.

LINE [6] considered different roles of each node in network, i.e. node itself and "context" of the other nodes in network, the node's embedding representations were learned by two separate models for preserving the first-order and second-order proximity of network topology respectively. The final representation for each node was obtained by concatenating the distinctive representations learned from two separate models. However, LINE only taken the network topology into account.

3   The Proposed Model

In this section, we first to present the definition of ANE and then develop a dual fusion ANE model.

3.1   Problem Definition

Let $G = (V, E, X)$ be an attributed network with n nodes, where V represents the set of nodes, E represents the set of edges, and $X \in R^{n \times l}$ represents the attributes matrix in which the row-vector $x_i \in R^l$ corresponds to the attributes vector of the node v_i. Besides, the adjacency matrix $M \in R^{n \times n}$ represents the link relationship between nodes, in which the element $m_{ij} > 0$ represents there existing the edge between the nodes v_i and v_j, while $m_{ij} = 0$ represents the edge is absent.

It is necessary to preserve the proximities of both node attributes and network topology in ANE. Let $A \in R^{n \times n}$ be the attributes similarity matrix, and the element $a_{ij} \in A^{n \times n}$ can be measured by the functions of distance similarity of attributes vectors x_i and x_j of nodes v_i and v_j, such as Cosine similarity, Euclidean distance. The Cosine similarity can be calculated by $a_{ij} = \frac{x_i \cdot x_j}{||x_i|| \times ||x_j||}$, where the operator "$\cdot$" represents the dot product, "$\times$" represents the scalar multiplication, "$||x_*||$" represents the L2-norm of the vector.

Based on attributes similarity matrix A, the semantic proximity, reflecting the attributes homogeneity effect amongst nodes, can be computed. For example, the semantic proximity $b_{ij} \in B^{n \times n}$ between the nodes v_i and v_j can be calculated by $b_{ij} = \frac{a_i \cdot a_j}{||a_i|| \times ||a_j||}$, where $a_i = (a_{i1}, \cdots, a_{in}) \in A$ and $a_j = (a_{j1}, \cdots, a_{jn}) \in A$.

The proximities related to the topology of network include the first-order proximity which corresponds to the direct neighbor relationships and the high-order proximity which corresponds to multi-hops neighbor relationships connected by the shared neighbor. The first-order proximity between nodes v_i and v_j can be measured by the value of $m_{ij} \in M$. Specifically, the larger value of m_{ij} indicates the stronger proximity between two nodes. Let $\hat{M}^1$ is the 1-step probability transition matrix amongst nodes, which can be obtained by the row-wise normalization of M; $\hat{M}^t$ be the t-step probability transition matrix, which can be computed by $\hat{M}^1$: $\hat{M}^t = \underbrace{\hat{M}^1 \cdots \hat{M}^1}_{t}$; the neighborhood proximity matrix

$S = \hat{M}^1 + \hat{M}^2 + \cdots + \hat{M}^t$, then the high-order proximity between nodes v_i and v_j can be measured by the similarity of vectors s_i and s_j.

ANE aims to find a map function $f(a_i, s_i) \to h_i$ that maps node vectors a_i and s_i into a unified embedding representation $h_i \in R^d$, such that the node semantic proximities and network topological proximities can be captured and preserved, d is the dimension of the embedding representation and $d \ll l$.

3.2 The Architecture of DFANE

The proposed DFANE model consists of the EF component and the LF component. They fuse node attributes and network topology at different stages to implement ANE. The architecture of DFANE is displayed in Fig. 1.

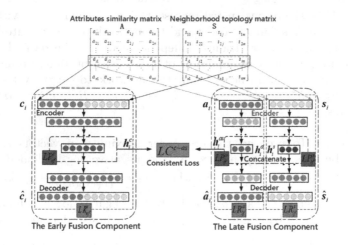

Fig. 1. The architecture of DFANE.

The Early Fusion Component. The EF component is implemented by a deep Auto-Encoder. Let $C = [A, S] \in R^{n \times 2n}$ be a matrix formed by concatenating the attributes similarity matrix $A \in R^{n \times n}$ and the neighborhood topology matrix $S \in R^{n \times n}$, i.e. $c_i = (a_i, s_i) = (a_{i1}, \cdots, a_{in}, s_{i1}, \cdots, s_{in})$, where c_i, a_i and s_i are the i-th row vector of C, A and S respectively. Let the deep Auto-Encoder have $2T - 1$ layers, where the layers $1, 2, \cdots, T$ for the encoder, the layers $T, T+1, \cdots, 2T-1$ for the decoder, the layer T be shared for the encoder and decoder; the vectors $h_{i,t}^c (t = 1, \cdots, T)$ and $y_{i,t}^c (t = 1, \cdots, T)$ be the hidden representations of the node v_i at t-th layer of the encoder and decoder respectively, $h_i^c = h_{i,T}^c \in R^{2d}$ be the desired underlying compact representation of the node v_i, $\hat{c}_i = y_{i,T}^c \in R^{2n}$ be the reconstructed data point from the decoder.

Proximity Loss. To preserve the first-order proximity in the concatenate matrix C, the negative log-likelihood $LP_{ef}^c = - \sum_{m_{ij}>0} \log p_{ij}^c$ should be minimized, where $p_{ij}^c = \frac{1}{(1+\exp(h_i^c \cdot (h_j^c)^T))}$ is the joint probability between the c_i and c_j.

Reconstruction Loss. To preserve the highly non-linear relationship existed in the concatenate matrix C, the reconstruction loss between the input c_i and output $\hat{c}_i$, i.e. $LR_{ef}^c = \sum_{i=0}^n ||\hat{c}_i - c_i||_2^2$, should be minimized to train the Auto-encoder.

The Late Fusion Component. The LF complement consists of two symmetrical Auto-Encoders: the node attributes Auto-Encoder and the network topology Auto-Encoder. Both of them have the same layers architecture with the Auto-Encoder of the EF complement. Let $h_i^a = h_{i,T}^a \in R^d$ and $h_i^s = h_{i,T}^s \in R^d$ be the desired underlying compact representations, $\hat{a}_i = y_{i,T}^a \in R^d$ and $\hat{s}_i = y_{i,T}^s \in R^d$ be the reconstructed representations of the node v_i with respect to the node attributes Auto-Encoder and the network topology Auto-Encoder. The two Auto-Encoders are trained independently, and the desired representation of a node is the concatenation of two representations obtained from the node attributes Auto-Encoder and the network topology Auto-Encoder, i.e. $h_i^{as} = h_i^a \oplus h_i^s = (h_{i1}^a, \cdots, h_{id}^a, h_{i1}^s, \cdots, h_{id}^s) \in R^{2d}$, $\hat{c}_i^{as} = \hat{a}_i \oplus \hat{s}_i = (\hat{a}_{i1}, \cdots, \hat{a}_{in}, \hat{s}_{i1}, \cdots, \hat{s}_{in}) \in R^{2n}$.

Proximity Loss. To preserve the first-order proximity of vectors a_i and a_j in the attributes similarity matrix A, the negative log-likelihood $LP_{lf}^a = - \sum_{m_{ij}>0} \log p_{ij}^a$ should be minimized, where $p_{ij}^a = \frac{1}{(1+\exp(h_i^a \cdot (h_j^a)^T))}$ is the joint probability between the a_i and a_j. Similarly, the negative log-likelihood $LP_{lf}^s = - \sum_{m_{ij}>0} \log p_{ij}^s$ should be minimized to preserve the first-order proximity of vectors s_i and s_j, where $p_{ij}^s = \frac{1}{(1+\exp(h_i^s \cdot (h_j^s)^T))}$ is the joint probability between the s_i and s_j.

Reconstruction Loss. To preserve the semantic proximity associated with node attributes, the reconstruction loss between node attributes encoder's input a_i and decoder's output $\hat{a}_i$, i.e. $LR_{lf}^a = \sum_{i=1}^n ||\hat{a}_i - a_i||_2^2$, should be minimized. Similarly, the reconstruction loss between network topology encoder's input s_i and decoder's output $\hat{s}_i$, i.e. $LR_{lf}^s = \sum_{i=1}^n ||\hat{s}_i - s_i||_2^2$, should be minimized.

In DFANE, we propose a consistency loss function to measure the consistency of the two components, shown as $LC^{c^-as} = \sum_{i=1}^n ||h_i^c - h_i^{as}||_2^2 + \sum_{i=1}^n ||\hat{c}_i - \hat{c}_i^{as}||_2^2$, where the first item indicates the embedding consistency loss and the second item means the reconstruction consistency loss of two components.

3.3   The Total Loss of DFANE

To sum up, the total loss of the DFANE is defined as the $LT = \alpha LP + \beta LR + \gamma LC^{c\bar{\ }as}$, where the proximity loss $LP = LP_{ef}^c + LP_{lf}^a + LP_{lf}^s$ and the reconstruction loss $LR = LR_{ef}^c + LR_{lf}^a + LR_{lf}^s$ are the sum of the proximity loss and the reconstruction loss in the concatenate matrix C, attributes similarity matrix A and neighborhood topology matrix S, respectively. α, β and γ are hyper-parameters used to balance the weights among different losses.

4   Experiment

In this section, we conduct experiments on eight publicly network datasets to evaluate the performance of the DFANE, compare with the several state-of-the-art methods, by the tasks of node classification and node clustering.

4.1   Datasets and Baselines Method

In experiments, three types of real-world networks are used, i.e. WebKB[1] network, Social network and Bibliographic network, where WebKB network consists of Texas, Cornell (corn), Washington (Wash), Wisconsin (Wisc) datasets; Social network contains Hamilton[2] (Hami) and Wiki datasets; Bibliographic network includes Cora and Pubmed[3] (Pubm) datasets.

We select NANE [3], DANE [1], ANRL [8], PRRE [9] as the baselines. NANE adopted the EF strategy to implement ANE; other baselines all employed the SF strategy to implement ANE. Besides, we develop two variants of DFANE, i.e. DFANE-E and DFANE-L, that only including the single complement.

4.2   Parameter Setting

The number of neurons in each layer is arranged in Table 1, which is a template for all baselines in experiments. The first layer of input of encoder and the last layer of output of decoder for node attributes and network topology correspond to the dimension of node attributes and the number of nodes in network.

DFANE contains three hyper-parameters α, β and γ for balancing the proximity loss, reconstruction loss and consistency loss. The value of hyper-parameters tuned through the algorithm of grid search and listed in Table 1. In experiments, both node classification and node clustering use the same parameters. The parameters of the baselines were set the same as the original papers.

---

[1] https://linqs-data.soe.ucsc.edu/public/lbc/.
[2] https://escience.rpi.edu/data/DA/fb100/.
[3] https://linqs.soe.ucsc.edu/data.

Table 1. The architecture of neural networks and the hyper-parameters of DFANE.

Datasets	Methods number of neurons in each layer		Hyper-parameters		
	Node attributes	Network topology	α	β	γ
Texas	1703-200-100-200-1703	187-200-100-200-187	0.1	1	0.001
Corn	1703-200-100-200-1703	195-200-100-200-195	0.001	100	0.01
Wash	1703-200-100-200-1703	230-200-100-200-230	1	1000	0.001
Wisc	1703-200-100-200-1703	265-200-100-200-230	10	1000	200
Hami	144-200-100-200-144	2314-200-100-200-2314	100	100	200
Wiki	4973-256-128-256-4973	2405-256-128-256-2405	1	1000	10
Cora	1433-256-128-256-1433	2708-256-128-256-2708	10	100	500
Pubm	500-256-128-256-500	19717-256-128-256-19717	0.01	50	0.01

4.3   Results and Analysis

Node classification is carried out on the learned node representations, and L2-regularized Logistic Regression [5] is used as the classifier. Then {10%, 20%, 30%, 40%, 50%} labeled nodes are randomly selected as the training set and the remained nodes as the testing set. For node clustering, we adopt K-means algorithm as the clustering method. In experiments, these processes are repeated ten times, and the average performances with respect to the Micro-F1, Macro-F1, AC and NMI [7] are reported for each dataset. Here due to the space constrains we only list the node classification results selected 40% labeled nodes as training set and node clustering results for eight datasets in Table 2, where bold numbers represent the best results; the trend of node classification results with respect to the training rates of {10%, 20%, 30%, 50%} are similar to the 40%.

Table 2. Node classification performance of different methods on eight datasets with training rate is 40% in Micro-F1 and Macro-F1, and node clustering performance of different methods on eight datasets in AC and NMI.

Metrics	Methods	Texas	Corn	Wash	Hami	Wisc	Wiki	Cora	Pubm
Micro-F1	NANE	53.98	44.44	46.38	86.68	48.43	49.41	50.46	57.51
	ANRL	65.49	41.03	65.94	79.99	57.23	72.14	40.00	81.31
	PRRE	72.66	64.27	77.03	93.94	79.56	73.20	82.61	84.3
	DANE	79.38	56.41	73.98	94.09	76.04	78.62	83.93	87.86
	DFANE-L	79.12	57.18	73.98	93.78	77.23	79.07	83.39	87.6
	DFANE-E	82.3	61.54	79.71	**94.26**	76.54	79.51	81.75	85.44
	DFANE	**83.54**	**70.34**	**81.6**	93.97	**81.76**	**79.85**	**84.18**	**88.44**

(continued)

Table 2. (*continued*)

Metrics	Methods	Texas	Corn	Wash	Hami	Wisc	Wiki	Cora	Pubm
Macro-F1	NANE	14.02	12.31	12.67	13.05	25.83	32.03	46.71	52.38
	ANRL	28.73	17.56	30.24	26.2	14.81	55.03	16.8	81.88
	PRRE	52.09	43.22	47.57	58.11	**36.92**	56.30	81.56	83.99
	DANE	53.73	41.35	50.48	54.75	30.58	70.23	82.46	87.6
	DFANE-L	53.46	44.26	48.93	50.57	30.44	70.02	81.84	87.42
	DFANE-E	57.34	46.59	58.56	52.41	30.7	70.15	80.14	85.11
	DFANE	**59.53**	**53.71**	**59.1**	**59.08**	30.52	**70.86**	**82.72**	**88.17**
AC	NANE	5.92	3.24	3.06	10.27	2.54	9.43	6.59	2.68
	ANRL	52.14	40.2	56.09	31.49	39.47	44.56	30.45	63.58
	PRRE	**53.53**	38.41	**60.32**	34.43	50.49	46.06	68.86	64.42
	DANE	38.5	39.08	39.09	**36.85**	32.3	46.32	64.8	64.14
	DFANE-L	35.56	31.62	39.35	36.83	32.19	44.97	66.11	66.20
	DFANE-E	50.91	47.03	48.48	35.69	49.24	34.47	54.22	43.23
	DFANE	50.37	**51.28**	54.22	36.24	**60.86**	**48.07**	**72.66**	**70.8**
NMI	NANE	5.2	4.32	3.26	16.12	4.25	16.13	10.23	3.7
	ANRL	15.19	19.32	20.97	12.08	12.16	44.14	16.71	25.39
	PRRE	24.78	19.87	**34.32**	34.7	27.93	43.52	48.99	28.02
	DANE	8.56	8.94	10.48	37.02	5.98	47.48	50.49	29.09
	DFANE-L	7.58	4.59	11.06	37.39	5.27	46.36	50.55	28.32
	DFANE-E	22.4	13.39	20.15	33.71	23.9	32.39	38.86	13.32
	DFANE	**25.1**	**21.78**	30.33	**38.41**	**38.88**	47.86	**55.93**	**34.67**

From Table 2, we have the observations that DFANE achieves the best performance for most case than baselines. Specifically, DFANE makes the best classification results on 7 of the 8 datasets in Micro-F1 and Macro-F1, and also performs the best clustering results on 5 of the 8 networks in AC, and 7 of 8 networks in NMI, which further verifies the efficiency of the dual fusion strategies outperforms that of the EF strategy (NANE, DAFNE-E), SF strategy (ANRL, PRRE, DANE), and LF strategy (DANE-L).

5   Conclusion

Integrating heterogeneous information of node attributes and network topology is essential for ANE. In this study, we propose a dual fusion model DFANE for ANE. DFANE integrated the EF and the LF into a united framework, and the unity of the EF and the LF captures the consensus of heterogeneous information. This is the first attempt to adopt dual fusion strategies in a united framework. Furthermore, experiment results on the eight real-world networks, with the tasks of node classification and node clustering, have demonstrated the effectiveness of the DFANE.

References

1. Gao, H., Huang, H.: Deep attributed network embedding. In: Twenty-Seventh International Joint Conference on Artificial Intelligence IJCAI 2018, pp. 3364–3370 (2018)
2. Jin, D., Li, B., Jiao, P., He, D., Zhang, W.: Network-specific variational auto-encoder for embedding in attribute networks. In: Twenty-Eighth International Joint Conference on Artificial Intelligence IJCAI 2019, pp. 2663–2669 (2019)
3. Mo, J., Gao, N., Zhou, Y., Pei, Y., Wang, J.: NANE: attributed network embedding with local and global information. In: Hacid, H., Cellary, W., Wang, H., Paik, H.-Y., Zhou, R. (eds.) WISE 2018. LNCS, vol. 11233, pp. 247–261. Springer, Cham (2018). https://doi.org/10.1007/978-3-030-02922-7_17
4. Pfeiffer III, J.J., Moreno, S., La Fond, T., Neville, J., Gallagher, B.: Attributed graph models: modeling network structure with correlated attributes. In: Proceedings of the 23rd international conference on World wide web, pp. 831–842 (2014)
5. Ribeiro, L.F., Saverese, P.H., Figueiredo, D.R.: struc2vec: learning node representations from structural identity. In: Proceedings of the 23rd ACM SIGKDD International Conference on Knowledge Discovery and Data Mining, pp. 385–394 (2017)
6. Tang, J., Qu, M., Wang, M., Zhang, M., Yan, J., Mei, Q.: Line: large-scale information network embedding. In: Proceedings of the 24th international conference on world wide web, pp. 1067–1077 (2015)
7. Yang, Y., Chen, H., Shao, J.: Triplet enhanced autoencoder: model-free discriminative network embedding. In: Twenty-Eighth International Joint Conference on Artificial Intelligence IJCAI 2019, pp. 5363–5369 (2019)
8. Zhang, Z., Yang, H., Bu, J., Zhou, S., Wang, C.: Anrl: attributed network representation learning via deep neural networks. In: Twenty-Seventh International Joint Conference on Artificial Intelligence IJCAI 2018, pp. 3155–3161 (2018)
9. Zhou, S., Yang, H., Wang, X., Bu, J., Wang, C.: Prre: personalized relation ranking embedding for attributed networks. In: the 27th ACM International Conference, pp. 823–832 (2018)

Attention-Based Knowledge Tracing with Heterogeneous Information Network Embedding

Nan Zhang[1], Ye Du[1], Ke Deng[2], Li Li[1(✉)], Jun Shen[3], and Geng Sun[3]

[1] School of Computer and Information Science, Southwest University,
Chongqing, China
{kathy525,duye99}@email.swu.edu.cn,
lily@swu.edu.cn
[2] RMIT University, Melbourne, Australia
ke.deng@rmit.edu.au
[3] University of Wollongong, Wollongong, Australia
{jshen,gsun}@uow.edu.au

Abstract. Knowledge tracing is a key area of research contributing to personalized education. In recent times, deep knowledge tracing has achieved great success. However, the sparsity of students' practice data still limits the performance and application of knowledge tracing. An additional complication is that the contribution of the answer record to the current knowledge state is different at each time step. To solve these problems, we propose Attention-based Knowledge Tracing with Heterogeneous Information Network Embedding (AKTHE). First, we describe questions and their attributes with a heterogeneous information network and generate meaningful node embeddings. Second, we capture the relevance of historical data to the current state by using attention mechanism. Experimental results on four benchmark datasets verify the superiority of our method for knowledge tracing.

Keywords: Knowledge tracing · Network embedding · Attention mechanism · Student assessment

1 Introduction

Knowledge tracing (KT) is a modeling based on student behavior sequence. With the presence of KT, we can exactly grasp the students' mastery and the understanding of knowledge concepts. Accurate knowledge tracing enables us to grasp the current needs of students and to recommend questions accurately. This task means that students can be provided with resources according to their personal needs [1]. Knowledge tracing is the core and key to build an adaptive education system.

There exist two types of knowledge tracing models now: traditional knowledge tracing models and deep knowledge tracing models. Bayesian Knowledge

© Springer Nature Switzerland AG 2020
G. Li et al. (Eds.): KSEM 2020, LNAI 12274, pp. 95–103, 2020.
https://doi.org/10.1007/978-3-030-55130-8_9

Tracing (BKT) [2] is a very famous model among traditional knowledge tracing models. It is a Markov model with hidden variables. Recently, Deep Knowledge Tracing (DKT) [3] based on students' recent learning performance has shown high performance. It can model the relationship between complex knowledge concepts.

Although knowledge tracing based on deep learning has made great progress, there are still some open issues. For example, the sparsity of students' behavior. Specifically, students usually use fragmented time to study, which only spend little time on the platform [4]. As a result, each student may only practice a small part of the test questions in system so that the mastery of a large part of knowledge is still unknown, which affects the knowledge tracing and limits the further application.

Our view of solving sparse problem depends on attribute information of questions. Specifically, the correct rate of the same student answering questions with similar difficulty or discrimination at the same knowledge concept is similar. Some existing work that study attributes of questions is primarily about mining them [5]. Other work try to use attributes of questions in knowledge tracing [6], but they do not specifically apply question attributes to deep knowledge tracing. By analyzing the various types of nodes and the various link relationships in the network, heterogeneous information network (HIN) [7] can accurately distinguish the different meanings in the network and mine more meaningful knowledge. Questions and their attributes can be expressed as a heterogeneous information network. Features of each question are thus captured by network embedding, which can guide us to better understand knowledge states.

In the sequence of student history questions, the contribution of each question to whether the next question can be answered correctly is different. For example, the next question mainly examines the mastery of integer multiplication, then the questions related to this concept in the past have a greater impact on the current state. Inspired by the attention mechanism used in machine translation where the mechanism can be used to capture the words corresponding to the original sentence and make the translation more accurate [8], we apply the attention mechanism to knowledge tracing problem to achieve better results.

In this paper, we focus on modeling the attribute information of questions with heterogeneous information network and generate meaningful question embeddings to model and predict students' knowledge states from a large number of learning logs. Our contributions are as follows:

- We use a heterogeneous information network to model questions and their attributes to learn the effective representations of questions.
- We propose *Attention-based Knowledge Tracing with Heterogeneous Information Network Embedding* to model students' current knowledge states.
- We evaluate our model and results show that our AKTHE model has better performance in knowledge tracing.

The rest of this paper is organized as follows. Section 2 gives a clear description on the problem definition. Section 3 is the detailed descriptions of proposed model. Experimental results on four datasets are shown in Sect. 4. Finally, we conclude the paper in Sect. 5.

2 Problem Definition

Definition 1 (Question Difficulty). Given a question, difficulty refers to the proportion or percentage of correct answers given by students in history, which is also known as the *ease of question*. The difficulty is defined as:

$$D_i = k_i/N_i, \tag{1}$$

where k_i is the number of students who answered question i correctly, and N_i is the total number of students who answered question i.

The value of question difficulty is between 0 and 1. Instead of using difficulty value directly, we categorize every question into one of 10 difficulty levels:

$$Category(i) = \begin{cases} \lfloor D_i \cdot c \rfloor, & \text{if } |N_i| \geq 4; \\ 5, & \text{else,} \end{cases} \tag{2}$$

where $c = 10$, the number of levels.

Definition 2 (Question Discrimination). The discrimination refers to the ability of questions to distinguish academic performance between students. For simplicity, the following definition is used:

$$D_i' = D_i^1 - D_i^2, \tag{3}$$

where D_i' is the discrimination of question i. We rank the total score of all students from high to low, the top 30% students is the high-level group and the last 30% students is the low-level group. D_i^1 and D_i^2 are the difficulty of question i for high-level group and low-level group respectively using (1). Table 1 is a criterion for evaluating the discrimination of question.

Table 1. The division of question discrimination

Discrimination	Bad	General	Good	Nice
Range	$(-1, 0.19)$	$(0.2, 0.29)$	$(0.3, 0.39)$	$(0.4, 1)$

Definition 3 (Heterogeneous Information Network (HIN) [7]). HIN is a directed graph $G = (V, E)$. Meanwhile, each node v and each link e has a type. Let A and R represent the sets of node type and link type respectively where $|A| + |R| > 2$. For each node v, we have $\phi(v) : V \to A$; for each link e, we have $\psi(e) : E \to R$.

Definition 4 (Meta-Path [9]). A meta-path is defined as a path. The specific form is $A_1 \xrightarrow{R_1} A_2 \xrightarrow{R_2} \cdots A_{l-1} \xrightarrow{R_{l-1}} A_l$. It represents a compound relationship between node types A_1 and A_l.

We express questions and their attributes using a HIN in Fig. 1(a) where the node types are *question, difficulty* and *discrimination*. There exist two types of links between nodes: (i) the *question-difficulty* link, and (ii) the *question-discrimination* link. Nodes in the network can be connected into different kinds of meta-paths, e.g., *"question-difficulty-question"* (QDQ) and *"question-discrimination-question"* ($QD'Q$).

3   Methodology

We present *Attention-based Knowledge Tracing with Heterogeneous Information Network Embedding*, called *AKTHE*. Figure 1 represents the overall schematic illustration of AKTHE.

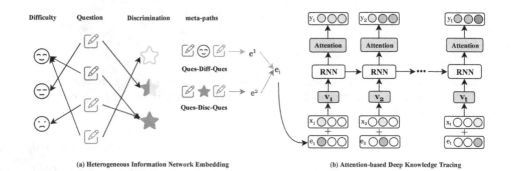

(a) Heterogeneous Information Network Embedding   (b) Attention-based Deep Knowledge Tracing

Fig. 1. The architecture of AKTHE.

3.1   Heterogeneous Information Network Embedding

Heterogeneous information network embedding (HINE) aims to generate meaningful node sequence. To this end, both DeepWalk [10] and node2vec [11] can be applied where a random walker traverses the heterogeneous information network from node to node along the edge between them.

Meta-path Based Random Walk. In the literature of HIN, meta-path [9] can depict the semantic concept accurately. Inspired by this, we use meta-path based random walk method. At step t, the walker moves from node v_t to v_{t+1}, one of its neighbors, at the normalized probability $P(v_{t+1}|v_t)$. Note v_t and v_{t+1} can be of any type. Similarly, Given a HIN $G = (V, E)$ and a meta-path ρ :

$A_1 \xrightarrow{R_1} A_2 \xrightarrow{R_2} \cdots A_t \xrightarrow{R_t} A_{t+1} \cdots \xrightarrow{R_{l-1}} A_l$, the probability that the walker will take one step specified in the path is:

$$P\left(v_{t+1} | v_t, \rho\right) = \begin{cases} \frac{1}{|N_{t+1}(v_t)|}, & (v_t, v_{t+1}) \in E, \phi\left(v_{t+1}\right) = A_{t+1}; \\ 0, & \text{otherwise}, \end{cases} \tag{4}$$

where v_t is the tth node in random walk, as well as $N_{t+1}(v_t)$ is set of first-order neighbor nodes of v_t.

Heterogeneous Skip-Gram. For a given meta-path, we use a fixed-length window to obtain the neighborhood $N(v)$ for node v. Following node2vec [11], we use heterogeneous skip-gram to learn node embeddings for a HIN $G = (V, E)$ by maximizing

$$\max_g \sum_{v \in V} \log P\left(N(v) | g(v)\right), \tag{5}$$

where g is a function to get node embedding.

Embedding Fusion. For a node v, different embeddings on O meta-paths can be represented as a set $\left\{e_q^{(l)}\right\}_{l=1}^{|O|}$, where $e_q^{(l)}$ is the embedding of question q about the lth meta-path. In our model, we only focus on the embeddings of questions, so we only select nodes with the type *question*.

For different representations of a question, we assume that the contribution of each meta-path to this question is equal. Then we use linear function as our fusion function:

$$e_q = \frac{1}{|O|} \sum_{l=1}^{|O|} \left(\mathbf{M}^{(l)} e_q^{(l)} + \mathbf{b}^{(l)}\right), \tag{6}$$

where e_q is the final representation of question q.

3.2 Attention-Based Knowledge Tracing with HINE

We use DKT model as the basic model because it is superiocompared to BKT. In addition, we also added the attention mechanism to our model to learn better distribution of each question which may lead to more accurate prediction.

Input. Formally, an interaction $\{q_t, a_t\}$ can be transformed into a vector x_t. We use $\delta(q_t)$ to represent question q_t, which is a one-hot encoding. x_t is a Cartesian product of q_t and a_t. Before passing into LSTM, the embedding vector x_t and e_t in (6) are integrated by concatenation:

$$\boldsymbol{v}_t = x_t \oplus e_t. \tag{7}$$

Attention Mechanism. We employ the attention mechanism that concatenate the hidden layer and the historical questions $\{q_1, q_2, ..., q_t\}$, which scores the importance of each question in the history for new question q_{t+1}:

$$u_t^i = \mathbf{v}^T \tanh(\mathbf{W}_1 h_t + \mathbf{W}_2 \boldsymbol{\delta}(q_i)), \tag{8}$$

where $\boldsymbol{\delta}(q_i)$ is the one-hot encoding of question q_i in historical questions and h_t is hidden state.

Then attention state can be represented as the weighted sum of all questions in this step by using attention scores as weights. That is,

$$s_t = \sum_{i=1}^{t} a_t^i u_t^i, \ a_t^i = \frac{\exp\left(\mathbf{W}_a u_t^i\right)}{\sum_{j=1}^{t} \exp\left(\mathbf{W}_a u_t^j\right)}. \tag{9}$$

Output. Finally, based on the updated hidden state h_t and attention state s_t, we can get the probability of answering each question correctly:

$$\mathbf{y}_t = \sigma\left(\mathbf{W}(s_t \oplus h_t) + \mathbf{b}\right). \tag{10}$$

Optimization. The purpose of AKTHK is to predict next interaction performance. In this paper, we use the cross-entropy loss function to learn the parameters:

$$\mathcal{L} = -\sum_t \left(a_{t+1} \log\left(\mathbf{y}_t^T \boldsymbol{\delta}\left(q_{t+1}\right)\right) + (1 - a_{t+1}) \log\left(1 - \mathbf{y}_t^T \boldsymbol{\delta}\left(q_{t+1}\right)\right)\right), \tag{11}$$

where $\boldsymbol{\delta}(q_{t+1})$ and a_{t+1} are the one-hot encoding and the true label of next question q_{t+1}.

4   Experiments

In this section, we demonstrate the effectiveness of our proposed AKTHE model by comparing it with baselines on four datasets.

4.1   Experimental Setup

Datasets. For our experiments, we test on four public datasets: ASSISTments2009, ASSISTments2015, Statics2011 and Synthetic-5. The statistical information of datasets is reported in Table 2.

Table 2. Dataset statistics.

Datasets	Students	Exercises	Records
ASSISTments2009	4,151	110	325,637
ASSISTments2015	19,840	100	683,801
Synthetic-5	4,000	50	200,000
Statics2011	333	1,223	189,297

Baselines. To demonstrate performance of our AKTHE model, we compare with multiple baselines in each dataset:

Bayesian Knowledge Tracing (BKT): In BKT [2] model, a hidden variable about knowledge states of students is proposed. BKT is to track changes in state of students' knowledge.

Deep Knowledge Tracing (DKT): DKT [3] is the most widely used knowledge tracing model. Its performance on a variety of open datasets is basically better than traditional models.

Dynamic Key-Value Memory Networks (DKVMN): DKVMN [12] is a state-of-the-art KT model. This model uses a static matrix key to store all knowledge concepts and a dynamic matrix value to store and update learners' mastery of concepts.

4.2 Performance Analysis

The performance of knowledge tracing is customarily evaluated by area under the curve (AUC). The higher the AUC, the better the model performance. The AUC Results of models on all datasets are shown in Table 3.

Table 3. The AUC results on all datasets.

Model	ASSISTments2009	ASSISTments2015	Synthetic-5	Statics2011
BKT	0.623	0.631	0.650	0.730
DKT	0.805	0.725	0.803	0.802
DKVMN	0.816	0.727	0.827	0.828
DKT+Attention	0.827	0.739	0.835	0.834
DKT+HINE	0.818	0.735	0.829	0.832
AKTHE	**0.831**	**0.741**	**0.837**	**0.838**

From Table 3, we can get two points of view. Firstly, compared with the benchmark models, our model has higher AUC, which means a better prediction performance. Concretely speaking, the average AUC obtained by AKTHE on two ASSISTments datasets is 1.8% higher than the most advanced model DKVMN. Compared with DKVMN, the average improvement rates are 1.1% in synthetic-5 and statics2011 datasets. Secondly, models based on neural network perform better than traditional model on four datasets, which means that the neural network can capture more useful information.

The results of ablation experiments show that the combination of HINE or attention with DKT can all improve the prediction effect. Adding attention to DKT contributes to the better results, which means that the attention mechanism provides more useful information.

5   Conclusion

In this paper, we mainly focus on exploring the attribute information of questions in knowledge tracing. We describe questions and their attributes with a heterogeneous information network, which is desirable for this application. With the help of heterogeneous network embedding, the learned node embeddings can be integrated into deep knowledge tracing model. We also add the attention mechanism to deep knowledge tracing model. On four datasets for knowledge tracing, our model has better performance than BKT, DKT and DKVMN.

In the future, we intend to introduce the relationship between concepts contained in the questions by using the method of graph embedding.

Acknowledgements. This research was supported by NSFC (Grants No. 61877051), and Natural Science Foundation Project of CQ, China (Grants No. cstc2018jscx-msyb1042, and cstc2018jscx-msybX0273).

References

1. Daomin, X., Mingchui, D.: Appropriate learning resource recommendation in intelligent web-based educational system. In: ISDEA, pp. 169–173. IEEE (2013)
2. Corbett, A.T., Anderson, J.R.: Knowledge tracing: modeling the acquisition of procedural knowledge. UMUAI **4**(4), 253–278 (1994). https://doi.org/10.1007/BF01099821
3. Piech, C., et al.: Deep knowledge tracing. In: NIPS, pp. 505–513 (2015)
4. Lin, J., et al.: From ideal to reality: segmentation, annotation, and recommendation, the vital trajectory of intelligent micro learning. World Wide Web, **23**, 1747–1767 (2019). https://doi.org/10.1007/s11280-019-00730-9
5. Narayanan, S., Kommuri, V.S., Subramanian, N.S., Bijlani, K., Nair, N.C.: Unsupervised learning of question difficulty levels using assessment responses. In: Gervasi, O., et al. (eds.) ICCSA 2017. LNCS, vol. 10404, pp. 543–552. Springer, Cham (2017). https://doi.org/10.1007/978-3-319-62392-4_39
6. González-Brenes, J., Huang, Y., Brusilovsky, P.: General features in knowledge tracing to model multiple subskills, temporal item response theory, and expert knowledge. In: EDM, pp. 84–91. University of Pittsburgh (2014)
7. Shi, C., Li, Y., Zhang, J., Sun, Y., Philip, S.Y.: A survey of heterogeneous information network analysis. TKDE **29**(1), 17–37 (2016)
8. Bahdanau, D., Cho, K., Bengio, Y.: Neural machine translation by jointly learning to align and translate. arXiv preprint arXiv:1409.0473 (2014)
9. Sun, Y., Han, J., Yan, X., Yu, P.S., Wu, T.: Pathsim: meta path-based top-k similarity search in heterogeneous information networks. PVLDB **4**(11), 992–1003 (2011)

10. Perozzi, B., Al-Rfou, R., Skiena, S.: Deepwalk: online learning of social representations. In: KDD, pp. 701–710. ACM (2014)
11. Grover, A., Leskovec, J.: node2vec: scalable feature learning for networks. In: KDD, pp. 855–864. ACM (2016)
12. Zhang, J., Shi, X., King, I., Yeung, D.Y.: Dynamic key-value memory networks for knowledge tracing. In: WWW, pp. 765–774. International World Wide Web Conferences Steering Committee (2017)

Knowledge Representation

Detecting Statistically Significant Events in Large Heterogeneous Attribute Graphs via Densest Subgraphs

Yuan Li[1], Xiaolin Fan[1], Jing Sun[1(✉)], Yuhai Zhao[2], and Guoren Wang[3]

[1] North China University of Technology, Beijing, China
sunjing8248@163.com
[2] Northeastern University, Shenyang, China
[3] Beijing Institute of Technology, Beijing, China

Abstract. With the widespread of social platforms, event detection is becoming an important problem in social media. Yet, the large amount of content accumulated on social platforms brings great challenges. Moreover, the content usually is informal, lacks of semantics and rapidly spreads in dynamic networks, which makes the situation even worse. Existing approaches, including content-based detection and network structure-based detection, only use limited and single information of social platforms that limits the accuracy and integrity of event detection. In this paper, (1) we propose to model the entire social platform as a heterogeneous attribute graph (HAG), including types, entities, relations and their attributes; (2) we exploit non-parametric scan statistics to measure the statistical significance of subgraphs in HAG by considering historical information; (3) we transform the event detection in HAG into a densest subgraph discovery problem in statistical weighted network. Due to its NP-hardness, we propose an efficient approximate method to find the densest subgraphs based on (k, Ψ)-core, and simultaneously the statistical significance is guaranteed. In experiments, we conduct comprehensive empirical evaluations on Weibo data to demonstrate the effectiveness and efficiency of our proposed approaches.

Keywords: Event detection · Densest subgraphs · Heterogeneous attribute graphs · Statistical significance

1 Introduction

The rapid development of Mobile Internet and big data makes social media an indispensable part of people's lives, work and entertainment. In particular, social platforms such as Weibo and Twitter are experiencing explosive growth, with billions of users creating, accepting and propagating daily observations and ideas. Event detection in social media is a very important and meaningful problem, which can discover the hidden value in social platforms, and further help predict events and their development tendency. For example, the event that

G. Li et al. (Eds.): KSEM 2020, LNAI 12274, pp. 107–120, 2020.
https://doi.org/10.1007/978-3-030-55130-8_10

Wuhan appeared unexplained pneumonia on December 31, 2019 spreads quickly on Weibo. Detecting this event accurately and predicting its development can provide a powerful support for the decision-making of the government.

However, the scale and complex content of social media generates lots of challenges. First, a large amount of data can be collected on social platforms at low cost, but massive data also brings great difficulties for event detection. Second, the language style of data is mostly informal and lacks semantics, so traditional methods cannot handle this limitation well. Finally, social platforms are quite dynamic that the information spreads quickly, and thus the large scale data needs to be processed in real time.

In this paper, we focus on the problem of domain-specific event (e.g. disease outbreaks) detection and forecasting. Most existing work on event detection can be roughly divided into two categories, including content-based detection [2,9] and network structure-based detection [15,19]. However, these methods solely exploit partial of the useful features in social platforms, such as user, content and network structure, and do not consider from the overall perspective of event propagation.

To overcome the aforementioned limitations, we model the social platform as a *heterogeneous attribute graph* (HAG), which integrates content and network structure. HAG describes the relation between all things in the world in the form of a network. It includes entity types and relation types, where entity types include user, blog, topic, link, etc; while relation types include post, repost, comment, etc. Moreover, each entity and relation has its own attributes, for example, a blog may have attributes such as likes, reposts, comments and key words, while post may have attributes such as date. The proposed HAG model urges us to design a novel event detection method.

In this paper, we propose a generalized event detection approach in HAG, which can be applied to any type of social platforms [12]. To detect and predict events more accurately, our approach transforms the event detection into the densest subgraph discovery problem in HAG, which can simultaneously guarantee the statistical significance of the events. Specifically, we first model the entire social platform as a HAG, which includes (1) entity and relation types, such as user, blog, topic, post, repost and comment; (2) each vertex is a specific instance of entity types; (3) each relation is the relation between specific entity instances; (4) entity and relation types have their own attributes, such as followers, fans and regions of user, etc. Then, we exploit non-parametric scan statistics to measure the statistical significance of subgraphs in HAG by comparing data with the past and between peers. Finally, the event detection in HAG is represented as the problem of identifying the densest subgraphs in the statistical weighted network. Due to its NP-hardness, we propose an efficient approximate method to find the densest subgraphs with statistical significance by (k, Ψ)-core based local expansion, where each vertex is contained in at least k h-clique (Ψ) instances.

Our contributions and the structure of the paper are summarized as follows:

- *Model of Heterogeneous Attribute Graph (HAG) (Sect. 3)*. We propose a novel model, called heterogeneous attribute graph to represent any social platforms.

- *Statistical Values of Vertices and Edges in HAG (Sect. 4)*. We compute statistical values of vertices and edges by comparing data with the past and between peers.
- *Non-Parametric Scan Statistics (Sect. 5.1)*. Without any prior distribution, we measure the statistical significance of subgraph by non-parametric scan statistics.
- *Approximate Densest Subgraph Algorithm (Sect. 5.2)*. Since the problem is NP-hard, we propose an efficient approximate algorithm to discover the densest subgraph with statistical significance in the statistical weighted network.

The experimental part is presented in *Sect.* 6, and *Sect.* 7 is the conclusion.

2   Related Works

First, our work is related to event detection, which can be roughly categorized as follows:

(1) *Content-Based Detection*. Some methods [2,8,9,13,16] utilize machine learning techniques for content classification. First, they extract feature vectors in the event training set, and then select a suitable classification algorithm to train an event classifier [5,22]. Others utilize content comparison. They first select the object with known reliability as the pivot, and then discriminate other objects by comparing with it [7,20].

(2) *Network Structure-Based Detection*. Some methods [6,15,19,21] rely on the propagated subgraph snapshot. They first obtain all or part of nodes once or multiple times and test whether these nodes have received event information from the state subgraph. Then, they calculate the node with the maximum probability becoming the event source. Other methods [11,17,18] exploit deployment nodes. They first deploy a small number of observation points in the network to record the time and direction when they initially receive messages from their neighbor nodes, and then infer event source of the current network. Recently, some methods [3,14] combine the content and network structure to detect anomaly events. However, they only identify the connected subgraphs as events and do not consider the cohesiveness of subgraphs.

Our work is also related to heterogenous information networks (HIN) [12] and densest subgraphs [4]. However, the differences are that HIN only considers the types of vertices and edges while our HAG also takes the attributes of vertices and edges into account. In addition, densest subgraph are mostly studied in simple graphs while our work aims to find densest subgraphs in HAG.

3   HAG Model and Problem Definition

In this section, we introduce the model of heterogeneous attribute graph (HAG) and the event detection problem in HAG.

We use $C = \{C_1, ..., C_m\}$ to denote a set of entity types, while $V = V_1 \cup ... \cup V_m$ is a set of entities, such that C_i represents the type of the set of entities V_i.

In the same way, $D \subseteq C \times C = \{D_1, ..., D_n\}$ is a set of relation types. $E \subseteq V \times V = E_1 \cup ... \cup E_n$ represents the sets of relations, where E_i is the set of relations of type D_i. Then, the HAG is defined as follows:

Definition 1 *(HAG). A HAG is defined as a directed graph $G = \{Q, V, E, f\}$ composed of types, entities, relations and attribute information, where $Q = C \cup D = \{Q_1, ..., Q_{m+n}\}$ refers to sets of entity and relation types. The attribute information f of entities and relations are a set of mapping functions. Let $f = \{f_1, ..., f_{m+n}\}$, where $f_i : Q_i \rightarrow R^{q_i}$ defines a q_i-dimensional feature vector $f_i(x)$ for each element x of type Q_i.*

Then, we detail the structure of HAG. Figure 1 illustrates the structure of HAG modeled by Weibo. The entity types selected by our model are user{name id, registration age, region, sun credit, followers, fans and blogs}, blog{key words, emotion, linguistic style, region, likes, reposts and comments}, topic{key words} and link{mentions}; the relation types between entities are user-user{follow types}, user-blog (post, repost, comment, mention){date}, user-topic{date}, user-link{date}, blog-link{mentions}, blog-topic{mentions} and link-topic{mentions}. Each entity and relation has their own attributes and attribute values.

In addition, we divide attributes into two categories, including dynamic attributes and static attributes. Dynamic attributes represent the difference between past and current states of an entity or relation, and static attributes express significant entity or relation within the same entity or relation type. In particular, the content in braces after each entity and relation are its corresponding attributes. Figure 2 is a specific example of HAG about the unexplained pneumonia appeared in Wuhan.

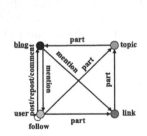

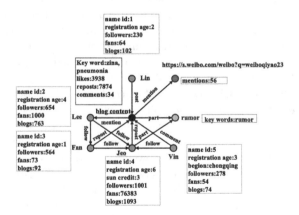

Fig. 1. Structure of HAG

Fig. 2. An example of HAG. The content of blog is the unexplained pneumonia appears in Wuhan on December 31, 2019. Since its content is too long, we abbreviate it as blog content.

Next, we detail the related definitions of clique and clique-based densest subgraphs.

Definition 2 *(Clique). An h-clique (h≥2) instance is a subgraph $S(V_s, E_s)$ in the undirected graph $G(V, E)$, where V_s has h vertices, $V_s \in V$, and $\forall u, v \in V_s$, $(u, v) \in E$.*

Definition 3 *(Clique-degree). The clique-degree of v in a graph $G(V, E)$ w.r.t. h-clique Ψ, i.e., $deg_G(v, \Psi)$ is the number of h-clique Ψ instances containing v.*

Definition 4 *(h-clique-density). The h-clique-density of a graph $G(V, E)$ w.r.t. h-clique $\Psi(V_\Psi, E_\Psi)$ is $\rho(G, \Psi) = \frac{\eta(G, \Psi)}{|V|}$, where $\eta(G, \Psi)$ is the number of instances of Ψ in G.*

Based on the above definitions, we give the problem definition of statistically significant event detection in HAG.

Problem 1. Given a HAG $G = \{Q, V, E, f\}$ and an h-clique $\Psi(V_\Psi, E_\Psi)$ $(h \geq 2)$, we first obtain the statistical weighted network G_c and then identify the subgraphs S from G_c with the highest h-clique-density $\rho(G_c, \Psi)$ and statistical significance as events.

In the following sections, we first detail the calculation of statistical values of vertices and edges and then the discovery of events.

4   Statistical Values of Vertices and Edges in HAG

For the attributes for different entity and relation types, we handle their heterogeneity by calculating the statistical value of each vertex and edge. First, we establish baseline distributions for attributes to indicate their corresponding behavior when no events occur. Then, a statistical value that represents the significance of each vertex or edge (the smaller the value, the more abnormal) is estimated based on the baseline distribution.

In order to estimate the baseline distributions of attributes, it is necessary to collect training sample data for a good distribution estimation. We first define a time granularity (e.g., hour, day, week), and then collect historical observation data. We divide the collected historical observation data into two categories: sufficient observation data and insufficient observation data. Calculations of statistical values for above two data are detailed below.

4.1   Calculation of Statistical Value for Sufficient Observation Data

Sufficient observation data is denoted by $\{x_1, ..., x_T\}$ (x_i is data of x at the i-th time granularity). We first calculate the statistical value of each attribute by comparing historical observations with current observations, and then calculate the statistical value for this vertex or edge through a secondary check. Under the null hypothesis that no events occur, the statistical value explains the proportion that a randomly selected sample from historical observations is greater than or equal to the current observations.

Statistical Value of the Vertex or Edge Attribute. Statistical value of attribute $j \in [1, q_i]$ for the vertex or edge $x \in Q_i$ type is defined as:

$$p(f_{i,j}(x_T)) = \frac{1}{T-1} \sum_{t=1}^{T-1} I(f_{i,j}(x_t) \geq f_{i,j}(x_T)) \tag{1}$$

where $f_{i,j}(x_T)$ refers to the j-th dimension attribute of the vertex or edge $x \in Q_i$ type at the current time T. $p(f_{i,j}(x_T))$ represents the proportion that historical observations $f_{i,j}(x_t)$ are greater than or equal to the current observations $f_{i,j}(x_T)$.

Statistical Value of the Vertex or Edge. Statistical value of the vertex or edge $x \in Q_i$ type is defined as:

$$p(x_T) = \frac{1}{T-1} \sum_{t=1}^{T-1} I(p_{min}(x_t) \leq p_{min}(x_T)) \tag{2}$$

where $p_{min}(x_t) = min_{j=1,...,q_i} p(f_{i,j}(x_t))$ refers to the minimum value of statistical value for all attributes of the vertex or edge x at current moment. $p(x_T)$ is explained as the proportion that the minimum statistical value of historical attribute, $p_{min}(x_t)$, are not larger than the minimum statistical value of current attribute, $p_{min}(x_T)$.

The statistical value of attribute, $p(f_{i,j}(x_T))$, and the minimum statistical value of the attribute, $p_{min}(x_t)$, are not used as the final statistical value, because the aforementioned two statistical values are biased towards vertices or edges with more attributes when performing non-parametric scan statistics [3]. Moreover, the final statistical value follows a uniform distribution on [0, 1] under the null hypothesis.

4.2   Calculation of Statistical Value for Insufficient Observation Data

For insufficient observation data, we compare differences in the same entity or relation type. Observation of the same entity or relation type is denoted by $\{x^1, ..., x^{|Q_i|}\}$. Under the null hypothesis that no events occur, the statistical value reflects the proportion of different currently considered vertex or edge observation randomly selected in the same type that is greater than or equal to the currently considered vertex or edge observation.

Statistical Value of the Vertex or Edge Attribute. Statistical value of attribute $j \in [1, q_i]$ for the vertex or edge $x \in Q_i$ type is defined as:

$$p(f_{i,j}(x)) = \frac{1}{|Q_i|} \sum_{q=1}^{|Q_i|} I(f_{i,j}(x^q) \geq f_{i,j}(x)) \tag{3}$$

where $f_{i,j}(x)$ refers to the j-th dimension attribute of the vertex or edge $x \in Q_i$ type. $p(f_{i,j}(x))$ represents the proportion of other vertex or edge attributes in the same type that is greater than or equal to the currently considered vertex or edge attributes.

Statistical Value of the Vertex or Edge. Statistical value of the vertex or edge $x \in Q_i$ type is defined as:

$$p(x) = \frac{1}{|Q_i|} \sum_{q=1}^{|Q_i|} I(p_{min}(x^q) \leq p_{min}(x)) \tag{4}$$

where $p_{min}(x^q) = min_{j=1,\ldots,q_i} p(f_{i,j}(x^q))$ refers to the minimum statistical value for all attributes of the same entity or relation type. $p(x)$ represents the proportion that other $p_{min}(x^q)$ values are less than or equal to the current $p_{min}(x)$ value. Similarly, the final statistical value calculated in this way follows the uniform distribution on $[0, 1]$.

5   Methodology

In this section, we propose the method to find the statistically significant densest subgraphs in the statistical weighted network. Next, we first detail how to measure the statistical significance of a subgraph.

5.1   Non-parametric Scan Statistics

To measure statistical significance of the subgraph, we apply non-parametric scan statistics [10] in the statistical weighted network. Non-parametric scan statistics can be defined as follows:

$$A(S) = \max_{0 < \alpha \leq \alpha_{max}} \phi(\alpha, V_\alpha(S), V(S)) \tag{5}$$

where S represents a connected subgraph, and significant score $A(S)$ refers to statistical significant score of S. $\alpha_{max}(\alpha_{max} < 1)$ is the maximum statistical significance level, which means the S has at least $1 - \alpha_{max}$ statistical significance. $V(S)$ is the all number of vertices in S, and $V_\alpha(S) = \sum_{v \in V(S)} I(p(v) \leq \alpha)$ is the number of vertices in S with statistical values that are not greater than the confidence level α ($\alpha > 0$). $\phi(\alpha, V_\alpha(S), V(S))$ refers to non-parametric scan statistic, i.e., a method that compares the observed number of α-significant statistical values $V_\alpha(S)$ to the expected number of α-significant statistical values $E[V_\alpha(S)]$. And because when no events occur, statistical values follow a uniform distribution on $[0, 1]$, that is, $E[V_\alpha(S)] = \alpha V(S)$.

Therefore, the BJ statistic [1] can directly compare $V_\alpha(S)$ and $V(S)$, and the mathematical form of BJ statistic is:

$$\phi(\alpha, V_\alpha(S), V(S)) = V(S) \times KL(\frac{V_\alpha(S)}{V(S)}, \alpha) \tag{6}$$

where KL is the Kullback-Liebler divergence, which measures differences between the observed and expected proportions of statistical values less than α:

$$KL(a, b) = a \log \frac{a}{b} + (1 - a) \log \frac{1-a}{1-b} \tag{7}$$

5.2   Approximate Densest Subgraph Discovery Algorithm

Because events in social media usually express as cohesive subgraphs, we want to represent the event as the densest subgraph in the network. In addition, there may exist false positive in the events, and thus we propose to identify the statistically significant densest subgraph by taking both the content and the network topology into consideration.

In specific, we first obtain a statistical weighted network $G_c = C, V, E, p\}$, where the weight of vertex represents its statistical value, as detailed in Sect. 4. (Due to the space limit, we only consider the statistical value of vertex in HAG.) Then, we transform the event detection in HAG to find the densest subgraph with statistical significance measured by non-parametric scan statistics in the statistical weighted network.

Since finding the statistically significant connected subgraph in the statistical weighted network is NP-hard [14], our problem is also NP-hard. Moreover, it is impractical to find the exact clique-based densest subgraph in the large network. And we know using (k_{max}, Ψ)-core as the densest subgraph has an approximate guarantee $\frac{1}{|V_\Psi|}$ [4]. The definition of (k, Ψ)-core is as follows:

Definition 5 *((k, Ψ)-core). A (k, Ψ)-core or H_k is the largest subgraph of a graph $G(V, E)$ such that $\forall v \in H_k$, $deg_{H_k}(v, \Psi) \geq k$, where k $(k \geq 0)$ is an integer, and Ψ is an h-clique.*

H_k has order k, and the clique-core-number of a vertex $v \in V$ or $core_G(v, \Psi)$ is the highest order of a (k, Ψ)-core containing v. k_{max} is the *maximum clique-core-number*.

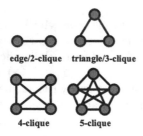

Fig. 3. Examples of clique

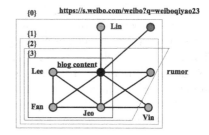

Fig. 4. An simple example of HAG

Example 1. Figure 3 shows 2-clique, 3-clique, 4-clique and 5-clique. And Fig. 4 is the simplified version of Fig. 2. Let Ψ be triangle. Figure 4 shows that the clique-degree of vertex *Lee* is 3; and the 3-clique-density of Fig. 4 is 3/8. In addition, The number k in each brace shows the (k, Ψ)-core, which contained in that region. Since subgraph of {*Lee, Fan, Jeo, blog content*} is 4-clique, and each vertex of it participates in 3 of them, then it is $(3, \Psi)$-core.

Based on (k, Ψ)-core, we propose an efficient approximate algorithm to find the top-r disjoint statistically significant densest subgraphs. The pseudo code is shown in Algorithm 1. It takes G_c as input; the number of seed vertices in each vertex type is K; and the maximum expansion number of seed vertices is Z. The output is the top-r statistically significant densest subgraphs, denoted as $\mathcal{S}$.

The main idea of Algorithm 1 is to make a local expansion from the seed vertex to their neighbours and iteratively add the vertices meeting the statistical value and clique-core-number constraints into the current subgraph. Specifically, we first compute $core_{G_c}(v, \Psi)$ of each vertex v in G_c by (k, Ψ)-core decomposition [4] and sort the vertices by non-increasing order according to $core_{G_c}(v, \Psi)$ (lines 1–2). Then, we select a seed vertex from each vertex type in the residue subgraph R_D and add it into the current subgraph R (line 6). Further, we iteratively find the subgraph S with the highest significant score $A(S)$ in the subgraph induced by R and its neighbor vertices [14], and then shrink S to the

---

Algorithm 1: Approximate densest subgraph discovery (ADSD) algorithm.

Input: A statistical weighted network $G_c=\{C,V,E,p\}$;
      The number of seed vertices in each vertex type, K (default=5);
      The maximum expansion number of seed vertices, Z (default=log($|V|$));
Output: Top-r statistically significant densest subgraphs, $\mathcal{S}$;

1 Compute clique-core-number of each vertex v, i.e., $core_{G_c}(v, \Psi)$;
2 Sort the vertices in G_c by non-increasing order according to their clique-core-numbers;
3 Set $\alpha_{max}=0.05$, $\mathcal{S}=\emptyset$, $R_D=V$, $R=\emptyset$;
4 **for** $c\in[1,...,m]$ **do**
5   **for** $i\in[1,...,K]$ **do**
6         $R=R\cup\{v_i\}$, where v_i is a significant vertex with the largest $core_{G_c}(v_i, \Psi)$ in type c of R_D;
7   **for** $z\in[1,...Z]$ **do**
8   sort the vertices of R in increasing order of their clique-core-numbers;
9            $mincore$ is the minimum $core_{G_c}(v, \Psi)$ in the current subgraph R;
10           $V_n=\{v_n\in V\backslash R:\exists v\in R,\{(v_n,v)\in E\}\}$;
11           $\langle S,A(S)\rangle$=HighestScoreSubgraph(V_n,R,α_{max}) [14]; //$A(S)$ is the highest significant score of S, whose vertices have the smallest statistical value;
12   S_d=DS($S,R,mincore$); //detailed in Algorithm 2;
13   **if** $S_d\backslash R\neq\emptyset$ **then**
14 | $R=S_d$;
15   **else**
16 | break;
17         $R_D=R_D\backslash R$, $\mathcal{S}=\mathcal{S}\cup\{R\}$, $R=\emptyset$;

18 Return Top-r of $\mathcal{S}$;

Algorithm 2: Densest Subgraph (DS) algorithm.

Input: The most significant subgraph S, the current subgraph R;
    The minimum $core_{G_c}(v, \Psi)$ in R, $mincore$;
Output: The densest subgraph S_d;
1 $maxcore$ is the maximum $core_{G_c}(v, \Psi)$ of all vertices in $S \backslash R$;
2 $S_d = R$;
3 **if** $maxcore \geq mincore$ **then**
4   **for** $v \in S \backslash R$ **do**
5   **if** $core_{G_c}(v, \Psi) = maxcore$ **then**
6           $S_d = S_d \cup \{v\}$

7 Return S_d;

---

densest subgraph S_d whose maximum $core_{G_c}(v, \Psi)$ is not less than the minimum $core_{G_c}(v, \Psi)$ in R by Algorithm 2 (lines 8–12). In each iteration, R is updated by S_d. When R cannot be further expanded, the iteration terminates early (lines 13–16). By this algorithm, the obtained subgraph is guaranteed to have the maximum significant score $A(S)$ and $core_{G_c}(v, \Psi)$.

Complexity Analysis. In Algorithm 1, compute $core_{G_c}(v, \Psi)$ in graph G_c needs $O(|V| \cdot \binom{d-1}{h-1})$ time, where d is the maximum degree of G_c and h is the number of vertices for clique. In the worst case, the function of HighestScore-Subgraph takes $O((|V_n| + |R|)^2)$ and Algorithm 2 takes $O(|S|)$. Therefore, the overall running time of Algorithm 1 is $O(|V| \cdot \binom{d-1}{h-1} + m \cdot K \cdot Z \cdot ((|V_n| + |R|)^2 + |S|))$.

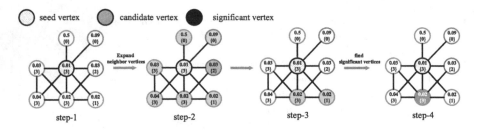

Fig. 5. Illustration of Algorithm 1, the values in the circles are statistical value and the clique-core-number, respectively.

Example 2. Figure 5 shows how to expand from the seed vertex in one iteration. First, the vertex with solid lines is the seed vertex. Then, we determinate its neighbor vertices and then select the vertices (orange translucent) with the minimum statistical value. Finally, we identify the vertices (orange solid) with the minimum statistical value and maximum clique-core-number.

6   Experiments

We conduct extensive experiments to evaluate the effectiveness and efficiency of the proposed methods by using the real Weibo data. All algorithms are implemented in Python. All the experiments are conducted on a Windows Server with Intel(R) i5-3337U Core(TM) 1.8 GHz CPU and 8 GB main memory.

6.1   Datasets and Metrics

Datasets. We select rumor event detection on Weibo for evaluation. Since Weibo provides the official rumor-busting service, we collected all the Weibo data about rumors from December 1, 2019 to March 30, 2020. The collected data contains $213k$ blogs, $145k$ users, 1601 topics, 16384 links and 204 events ($k=10^3$).

Metrics. We choose the following four metrics for evaluation:

- The *lead time* to detection, which is the number of days between the predicted event when the event has not yet occurred and the first occurrence of the event.
- The *lag time* to detection, which is the number of days between the first occurrence of the event and its detection.
- The *coefficient* can be seen as a combination of precision and recall applied to graph setting. Let $coeffcient = |E \cap E^*|/|E \cup E^*|$, where E is the set of event-related entities, and E^* is the set of entities labelled by event detection technique.
- The *runtime* of the proposed and compared algorithms.

Next, we compare our ADSD algorithm with NPHGS proposed in [3] for effectiveness and efficiency evaluation under different measurements.

6.2   Evaluation on the Weibo Rumor Dataset

For the above two algorithms, we have three parameters. Specifically, K is the number of seed vertices, α_{max} is maximum statistical significance level, and Ψ is h-clique.

Varying Seed Vertices (K). We set parameter $\alpha_{max} = 0.05$. Figure 6 shows the comparison for both algorithms on four metrics when varying the number of seed vertices. First, we can see that our ADSD algorithm outperforms NPHGS in the average *coeffcient*, *lead time*, and *lag time*. This is because events in social media usually express as cohesive subgraphs. In addition, the *runtime* also becomes faster, because we ignore the vertices that do not appear in the dense region of the graph at each iteration.

Figure 6 also shows the comparison on four metrics with the change of parameter K. we can see that when $K = 15$, *coeffcient*, *lead time*, *lag time* are the best. This is because along with K becoming larger, the results first contains more significant and dense subgraphs, but when the number of selected seed vertices

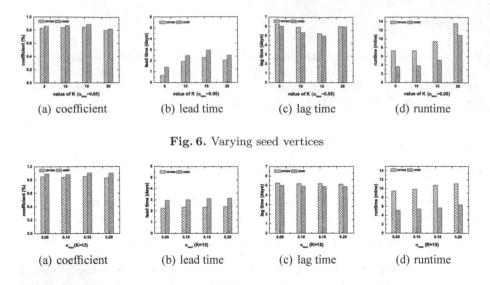

Fig. 6. Varying seed vertices

Fig. 7. Varying α_{max}

are too many, the quality of the results goes down. In addition, the *runtime* keeps increasing. This is because the larger the K is, the larger the detection range is, which leads to the increase of *runtime*.

Varying α_{max}. We set parameter $K = 15$ for both algorithms. Figure 7 shows the comparison on four metrics for our ADSD and NPHGS. We can see that for different parameter α_{max}, our ADSD always outperforms NPHGS on four metrics. This is because we consider the cohesive subgraph.

Next, Fig. 7 also shows that the increasing of α_{max}, does not affect the performance of *coefficient*, *lead time* and *lag time* too much, which is just slightly improved. This is because α_{max} is artificially set, and Eq. (5) can automatically optimize α. In addition, the reason for the increasing of *runtime* is that the larger α_{max}, more vertices need to be considered.

Varying Data Size. We vary data size in the dataset by randomly removing some rumors. In Fig. 8, we evaluate our ADSD and NHPGS under different data size. The number on x-axis represents the percentage of the current data size to the original. We can see that our ADSD outperforms NPHGS in all four metrics.

In addition, with the increasing of data size, the performance on *coefficient*, *lead time* and *lag time* becomes better, but takes more runtime. This is because the number of detected vertices about rumors increases.

Varying h-clique (Ψ). To conduct the ADSD algorithm, we need to select h-clique. Thus, we evaluate the performance metrics for our ADSD at various h-clique. In Fig. 9, along with the increasing of h, the reason for the increased *coeffcient* is that events exists as densest subgrahs, but the *lead time*, *lag time*

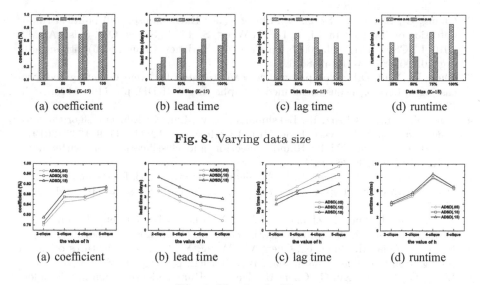

Fig. 8. Varying data size

(a) coefficient   (b) lead time   (c) lag time   (d) runtime

Fig. 9. Varying h-clique

delay and *runtime* increases first and then decreases. This is because when h becoming larger, we have to detect denser cliques, which takes longer time, while when h exceeds a certain range, there may not be such dense cliques in HAG, so the detection will be terminated early.

7 Conclusion

Event detection is an important problem in social media. Existing methods combine the content and network structure to detect connected subgraphs as events, however they overlook the cohesiveness of subgraphs. In this paper, we propose to model events as statistically significant densest subgraphs in HAG, and design the ADSD algorithm. Extensive experimental results on the Weibo dataset demonstrate the effectiveness and efficiency of our proposed approaches.

Acknowledgment. This research is partially supported by the National NSFC (61902004, 61672041, 61772124, 61732003, 61977001), National Key Research and Development Program of China (2018YFB1004402), Project of Beijing Municipal Education Commission (KM202010009009) and the Start-up Funds of North China University of Technology.

References

1. Berk, R.H., Jones, D.H.: Goodness-of-fit statistics that dominate the kolmogorov statistics. Probab. Theory Relat. Fields **47**(1), 47–59 (1979). https://doi.org/10.1007/BF00533250

2. Chang, C., Zhang, Y., Szabo, C., Sheng, Q.Z.: Extreme user and political rumor detection on Twitter. In: Li, J., Li, X., Wang, S., Li, J., Sheng, Q.Z. (eds.) ADMA 2016. LNCS (LNAI), vol. 10086, pp. 751–763. Springer, Cham (2016). https://doi.org/10.1007/978-3-319-49586-6_54
3. Chen, F., Neill, D.B.: Non-parametric scan statistics for event detection and forecasting in heterogeneous social media graphs. In: SIGKDD, pp. 1166–1175. ACM (2014)
4. Fang, Y., Yu, K., Cheng, R., Lakshmanan, L.V., Lin, X.: Efficient algorithms for densest subgraph discovery. Proc. VLDB Endowment 12(11), 1719–1732 (2019)
5. Hamidian, S., Diab, M.T.: Rumor detection and classification for twitter data. CoRR abs/1912.08926 (2019)
6. Ji, F., Tay, W.P., Varshney, L.R.: An algorithmic framework for estimating rumor sources with different start times. IEEE Trans. Signal Process. 65(10), 2517–2530 (2017)
7. Jin, Z., Cao, J., Guo, H., Zhang, Y., Wang, Y., Luo, J.: Rumor detection on twitter pertaining to the 2016 U.S. presidential election. CoRR abs/1701.06250 (2017)
8. Kwon, S., Cha, M., Jung, K., Chen, W., Wang, Y.: Prominent features of rumor propagation in online social media. In: DMCEEE, pp. 1103–1108 (2013)
9. Liang, G., He, W., Xu, C., Chen, L., Zeng, J.: Rumor identification in microblogging systems based on users' behavior. TCSS 2(3), 99–108 (2015)
10. Neill, D.B., Lingwall, J.: A nonparametric scan statistic for multivariate disease surveillance. Adv. Dis. Surveill. 30(6), 106–110 (2007)
11. Pinto, P.C., Thiran, P., Verrerli, M.: Locating the source of diffusion in large-scale networks. CoRR abs/1208.2534 (2012)
12. Shi, C., Li, Y., Zhang, J., Sun, Y., Yu, P.S.: A survey of heterogeneous information network analysis. IEEE Trans. Knowl. Data Eng. 29(1), 17–37 (2017)
13. Sun, S., Liu, H., He, J., Du, X.: Detecting event rumors on sina weibo automatically. In: Ishikawa, Y., Li, J., Wang, W., Zhang, R., Zhang, W. (eds.) APWeb 2013. LNCS, vol. 7808, pp. 120–131. Springer, Heidelberg (2013). https://doi.org/10.1007/978-3-642-37401-2_14
14. Tam, N.T., Weidlich, M., Zheng, B., Yin, H., Hung, N.Q.V.: From anomaly detection to rumour detection using data streams of social platforms. VLDB 12(9), 1016–1029 (2019)
15. Wang, C.: Research on identifying information source in networks. Ph.D. thesis, University of Science and Technology of China, Hefei, China (2016)
16. Wu, K., Yang, S., Zhu, K.Q.: False rumors detection on sina weibo by propagation structures. In: ICDE, pp. 651–662 (2015)
17. Xu, W., Chen, H.: Scalable rumor source detection under independent cascade model in online social networks. In: MASN, pp. 236–242. IEEE Computer Society (2015)
18. Zhang, Y., Zhang, X., Zhang, B.: Observer deployment method for locating the information source in social network. J. Softw. 25, 2837–2851 (2014)
19. Zhang, Z., Xu, W., Wu, W., Du, D.-Z.: A novel approach for detecting multiple rumor sources in networks with partial observations. J. Comb. Optim. 33(1), 132–146 (2015). https://doi.org/10.1007/s10878-015-9939-x
20. Zhao, Z., Resnick, P., Mei, Q.: Enquiring minds: early detection of rumors in social media from enquiry posts. In: WWW, pp. 1395–1405. ACM (2015)
21. Zheng, L., Tan, C.W.: A probabilistic characterization of the rumor graph boundary in rumor source detection. In: ICDSP, pp. 765–769. IEEE (2015)
22. Zubiaga, A., Liakata, M., Procter, R.: Learning reporting dynamics during breaking news for rumour detection in social media. CoRR abs/1610.07363 (2016)

Edge Features Enhanced Graph Attention Network for Relation Extraction

Xuefeng Bai[1] (ID), Chong Feng[1(✉)], Huanhuan Zhang[2], and Xiaomei Wang[3]

[1] School of Computer Science and Technology, Beijing Institute of Technology,
Beijing, China
aaronbai1995@gmail.com, fengchong@bit.edu.cn
[2] China Academy of Electronics and IT of CETC, Beijing, China
huanhuanz_bit@126.com
[3] Institutes of Science and Development, Chinese Academy of Sciences,
Beijing, China
wangxm@casid.cn

Abstract. Dependency trees of sentences contain much structural information that is useful for capturing long-range relations between words in the text. In order to distill the useless information, the pruning strategy is introduced into the dependency tree for preprocessing. However, most hard-pruning strategies for selecting relevant partial dependency structures are too rough and have poor generalization performance. In this work, we propose an extension of the graph attention network for relation extraction task, which makes use of the whole dependency tree and its edge features. The graph attention layer in our model can implicitly prune the neighbor nodes of each node by assigning different weights according to the content. The edge feature information makes the pruning strategy trainable and non-discrete. Our model can be viewed as a soft-pruning approach strategy that automatically learns the relationship between different nodes in the full dependency tree. The results on various datasets show that our model utilizes the structural information of the dependency tree better and gets the state-of-the-art results.

Keywords: Relation extraction · Edge enhanced graph attention network · Dependency trees

1 Introduction

Relation extraction is proposed to find semantic relation exists between entities in sentences. It's a primary task for many downstream applications. Existing models for relation extraction can be categorized into two main classes: sequence-based models [18,22] and dependency-based models [2,9].

Most existing graph neural models for dependency-based relation extraction task rely on hard-pruning strategies, which help to distill useless dependency information. One popular way is to reduce the parse tree to the shortest dependency path between the entities [19]. Another general approach is to perform

© Springer Nature Switzerland AG 2020
G. Li et al. (Eds.): KSEM 2020, LNAI 12274, pp. 121–133, 2020.
https://doi.org/10.1007/978-3-030-55130-8_11

bottom-up or top-down computation along with the subtree below the lowest common ancestor (LCA) of the entities [8]. However, this kind of rule-based hard-pruning strategies may be too rough to retain some important information in the full tree. Taking the best path-centric strategy [23] currently as an example, Fig. 1 shows a complete tree. The shortest dependency path between entities "bark" and "salicin" is highlighted in bold (edges and tokens). The root node of the LCA subtree of entities is "contains". All the solid edges indicate tokens $K = 1$ away from the subtree and the final result of the path-centric approach. Obviously, the further description of "salicin" has been cut off. This makes the relationship between entities "bark" and "salicin" misjudged as "Content-Container", which is actually "stuff-object".

In order to utilize the full dependency tree and enable the neural network to automatically select useful information to retain based on the content of the text node, we replace the GCNs with the graph attention networks, which [15] proposed to perform node classification of graph-structured data. The attention mechanism has the advantage in the dynamic weight generation of the variable-length sequence, which the convolution network and the fully connected network can't match. The GAT model computes the hidden representations of each node in the graph by attending over its neighbours following the masked self-attention strategy. So the model can implicitly prune the neighbor nodes of each node by assigning different weights according to the content.

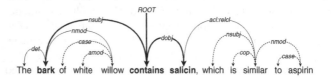

Fig. 1. Dependency tree example

Figure 1 also shows that the results of the dependency analysis include not only the dependency adjacencies of tokens in the sentence but also the types of dependencies. For example, "nmod" stands for "noun compound modifier", "nsubj" stands for "nominal subject" and so on. This part of information can also be used to adjust the importance of the edge dynamically. To utilize the dependency better, we merge the type information as the edge feature into attention coefficients as multi-channel embeddings. This method makes the pruning strategy trainable and non-discrete.

Our contributions can be summarized as follows: We promote a variant of graph attention network, Edge feature Enhanced GAT (EEGAT) to encode the dependency tree and extract the entity-centric feature representations for relation classification. Our model takes sentence-level attention information into account when aggregating node information from neighbours nodes and assigning different weights to different words. Compared with more commonly used

GCNs, the GAT-based model retains more original information of the dependency tree and reduces the number of layers because of more adjacencies. Our EEGAT model incorporates edge features on the dependency tree into the adjacency matrix can be viewed as a soft-pruning strategy. Compared to the ordinary GAT which applies neighbour mask mechanism only on binary edge indicators, our model improves the representation of adjacencies by trainable and non-discrete values. Our model achieves the state-of-the-art results on the NYT, TACRED and SemEval 2010 Task8 dataset. Also, our model can be applied over dependency trees efficiently in parallel, unlike recurrent networks and tree-structured networks.

2 Relate Work

2.1 Relation Extraction

Traditional works are mostly statistical methods. Kernel-based approaches, including tree-based kernels [21] and dependency path-based kernels [2]. Recent works mostly leverage different neural networks to extract relations. Sequence-based approaches use convolutional neural networks [18,22], recurrent neural networks [24,25], the combination of both [17] and transformer [16].

Dependency-based approaches try to incorporate structural information into the neural models. They are shown to improve relation extraction performance by capturing long-distance relations. [19] encoded the shortest dependency path by neural models. [7] applied a recursive network over subtrees and a CNN over the path. [8] applied the LSTM model over the LCA subtree of two entities. [9] extend the tree LSTM model [13] over two DAGs split from the dependency graph. Some graph network models also perform well as dependency-based models. [12] use graph recurrent networks to directly encode the whole dependency graph without breaking it. [23] apply graph convolutional network over the dependency tree pruned by a path-centric strategy.

2.2 Graph Attention Network

Self-attention mechanism is referred to the attention mechanism used to compute the representation of a single sequence. It has been proved useful for tasks such as machine reading [3], machine translation [14], and so on.

[15] apply the attention mechanism on graph structure and present graph attention networks, leveraging masked self-attention layers to address the shortcomings of prior methods based on graph convolutions or their approximations. [5] propose a Relation-aware Graph Attention Network to model multi-type inter-object relations via a graph attention mechanism, to learn question-adaptive relation representations for visual question answering. [1] propose Masked Graph Attention Network, allowing nodes directionally attend over other nodes' features under the guidance of label information in the form of mask matrix, to make use of rich global mutual information.

3   Model

We construct graph attention networks with edge features on the whole dependency tree. Then we introduce the whole model architecture for relation
extraction.

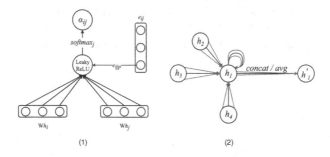

Fig. 2. Graph self-attention mechanism

3.1   Edge Feature Enhanced GAT (EEGAT) Layer

We use graph attention layer as the sole layer to build all the GAT architectures
used in our experiments. The input of each graph attention layer is set of node
features:

$$\mathbf{h} = \{\vec{h}_1, \vec{h}_2, ..., \vec{h}_N\}, \vec{h}_i \in \mathbb{R}^F \qquad (1)$$

where N is the number of nodes, and F is the dimension of each node. The
corresponding output is a new set of node features:

$$\mathbf{h'} = \{\vec{h}'_1, \vec{h}'_2, ..., \vec{h}'_N\}, \vec{h}'_i \in \mathbb{R}^{F'} \qquad (2)$$

where F' is the dimension of each node in the output set. In order to transform
the input features into higher-level features to obtain sufficient expressive power,
we add one learnable linear transformation, which is parameterized by a weight
matrix $W \in \mathbb{R}^{F' \times F}$. Then the self-attention $a : \mathbb{R}^{F' \times F}$ is performed on the nodes,
to computes attention coefficients by nonlinear transformation:

$$f_{ij} = f\left(\mathrm{W}\vec{h}_i, \mathrm{W}\vec{h}_j\right) = LeakyReLU\left(\vec{U}^\top \left[\mathrm{W}\vec{h}_i || \mathrm{W}\vec{h}_j\right]\right) \qquad (3)$$

where $||$ represents concatenation and $\vec{U}^\top \in \mathbb{R}^{2F'}$ is a weight vector. Attention
coefficients f_{ij} represent the importance of node j features to node i. In most
general version of the attention mechanism, the model calculates coefficients
between every node and every other node, ignoring all structural information.
That is a great loss for graph processing problems.

   In the vanilla GAT, they incorporate graph-structured information into the
self-attention model by applying masked attention: only compute f_{ij} between

node i and every node j in its neighborhood set $(j \in \mathcal{N}_i)$. In self-attention mechanism, f_{ij} is normalized across all choices of neighbor nodes by $softmax$ function to make coefficients easily comparable across nodes:

$$a_{ij} = softmax_j(f_{ij}) = \frac{exp(f_{ij})}{\sum_{k \in \mathcal{N}_i} exp(f_{ik})} \qquad (4)$$

The dependency trees contain not only the connection structure information between tokens but also the dependency relationship category information between tokens. This part is ignored in the previous graph models, which are based on the dependency tree. The results of *Stanford CoreNLP Dependency-ParseAnnotator*[1] contain not only the head and tail token but also the types of dependency relations. We embed them into P dimensional vector $e_{ij} \in \mathbb{R}^P$ representing the dependency edge from token i to token j. We denote the p channel of e_{ij} as e_{ij}^p and merge it into the calculation of attention coefficients. So Eq. 4 is rewritten as (illustrated by Fig. 2 (left)):

$$a(x_i, x_j, e_{ij}^p) = softmax_j(f_{ij}) = \frac{exp(f_{ij}) \times e_{ij}^p}{\sum_{k \in \mathcal{N}_i} exp(f_{ik}) \times e_{ik}^p} \qquad (5)$$

The normalized attention coefficients are used to compute a linear combination of the features corresponding to them. After potentially applying a nonlinearity σ, the final output features for every node is expressed as:

$$\vec{h}_i^{\prime p} = \sigma \left(\sum_{j \in \mathcal{N}_i} a(x_i, x_j, e_{ij}^p) W^p \vec{h}_j \right) \qquad (6)$$

Like many attention mechanisms, we extend the edge enhanced graph attention to multi-head attention so that the learning process of self-attention is better stabilized. We consider the multiple dimensional edge feature as multi-channel signals. We construct a separate attention operation for each channel and concatenate the as the multi-head attention mechanism. Specifically, the model contains P independent attention mechanisms executing the transformation of Eq. 6, and then concatenate their features, resulting in the following output feature representation:

$$\vec{h}_i^{\prime} = \overset{K}{\underset{k=1}{\big\|}} \sigma \left(\sum_{j \in \mathcal{N}_i} a(x_i, x_j, e_{ij}^p) W^p \vec{h}_j \right) \qquad (7)$$

where $\|$ represents concatenation, a_{ij}^p is the attention coefficient computed and normalized by the p-th attention mechanism, and W^P is the corresponding input linear transformation's weight matrix. As the result of concatenation, the final output of graph attention layer, h', will consist of PF' feature (embedding dimension) for each node. The aggregation process of a multi-head graph attentional layer is illustrated by Fig. 2 (right).

---

[1] https://stanfordnlp.github.io/CoreNLP/.

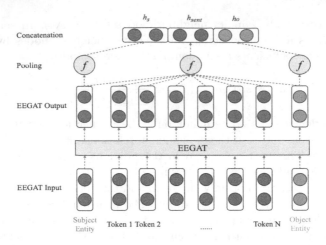

Fig. 3. EEGAT model for relation extraction

3.2   EEGAT Model for Relation Extraction

We define the input of extraction model as sequence $W = [w1, w2, ..., wn]$, where w_i indicates the i_{th} token. In the same way, the subject entity and object entity are defined as $W_{sub} = [W_{s_1}, ..., W_{s_n}]$ and $W_{obj} = [W_{o_1}, ..., W_{o_n}]$, where s_n and o_n are the lengths of entities respectively. The output of model is the softmax vector representing the probabilities corresponding to all the relation $r_i \in \mathcal{R}$ which is the predefined relation set. Then the most likely relation is chosen. The whole structure is illustrated by Fig. 3.

The L-layers EEGAT network accept word vector as input and return each token's hidden representations as output, which exchange information with its neighbors no more than L edges apart on the dependency tree. We use the full dependency tree without any pruning as the input of EEGAT layer to construct adjacency matrix and indicate the neighboor words of each word. We apply a max pooling function f on these representations to get a sentence representation for relation extraction:

$$h_{sent} = f(\mathbf{h}^{(L)}) = f(\text{EEGAT}(\mathbf{h}^{(0)})) \tag{8}$$

where $\mathbf{h}^{(L)}$ denotes the output of the Lth EEGAT layer, and $f : \mathbb{R}^{d \times n} \to \mathbb{R}^d$. The representations of subject and object entities is crucial for relation classification, so we apply function f as well:

$$h_s = f(\mathbf{h}^{(L)}_{s_1:s_n}), \quad h_o = f(\mathbf{h}^{(L)}_{s_1:s_n}) \tag{9}$$

The concatenation of these three vectors is fed through a feed-forward neural network.

$$h_{final} = FFNN([h_{sent}; h_s; h_o]) \tag{10}$$

The output h_{final} is fed through a linear layer and a softmax function to get the probability distribution over relations set.

3.3   Contextualized Layer

Graph-based models improve the ability of leveraging dependency relation but weaken the use of sentence sequence information. The input word vectors of EEGAT or GCN layer do not contain contextual information about word order or disambiguation. Adding a bi-directional long short-term memory (BiLSTM) layer before the graph network is proven useful [23] for improving the performance. So we use a BiLSTM network as a pre-encoder and use its output representations as the input $\mathbf{h}^{(0)}$ in the original model (Table 1).

Table 1. Hyperparameter setup.

Hyperparameter	TACRED	NYT	SemEval
BiLSTM layer/hidden size	1/200	1/200	1/200
EEGAT layer/hidden size	2/200	2/200	1/200
Word embedding/Dimension	GloVe vectors [10]/300		
POS & NER tags embedding/Dimension	Random initialized vector/30		
Edge features embedding	Random initialized vector		
Edge features embedding dimension	3	3	2
Dropout rate of BiLSTM/EEGAT	0.5/0.3		
Learning rate/Anneal factor	1.0/0.9	1.0/0.9	0.5/0.95
Train epoch	100	100	150

4   Experiments

4.1   Baseline Models

We compare our models with several dependency-based models and neural sequence models.

(1) *Dependency-based Models*:

The logistic regression classifier (LR) [24], which adds other lexical features into dependency features.

Shortest Path LSTM (SDP-LSTM) [20], which combines a neural model with the shortest path between the subject and object entities in the dependency tree.

Tree-structured neural model (TreeLSTM) [13], which applies the LSTM model on the arbitrary tree structures (the dependency trees of sentences).

Contextualized Graph Convolution Network (C-GCN) [23], which applies graph convolutional layers over optional RNN layers, using the path-centric pruned dependency tree as the adjacency matrix.

Contextualized Attention Guided GCN (C-AGGCN) [4], which replaces pruned dependency trees with attention matrix in C-GCN.

Graph Attention Network (GAT) and Contextualized GAT (C-GAT), which applies the vanilla graph attention layer above the pathcentric pruned dependency tree.

(2) *Neural sequence model*: **Position-aware Attention LSTM**[24], which employs a position-aware attention mechanism over LSTM outputs (PA-LSTM).

4.2   Datasets

We perform experiments on three relation extraction datasets of the sentence-level relation extraction task: (1) **NYT:** NYT dataset [11] was generated by aligning Freebase relations with the New York Times corpus (distant supervision method). There are 53 possible relationships, including a special relation NA.

(2) **TACRED:** Introduced in [24], TACRED dataset contains over 106K instances and introduces 41 relation types and no relation type to describe the relations between the mention pairs in instances. Subject mentions in TACRED are categorized into person and organization, while object mentions are categorized into 16 fine-grained types(e.g., date and location).

(3) **SemEval2010 Task 8:** The SemEval dataset is public and is widely used in recent relation extraction work. It is much smaller with 8,000 examples for training and 2,717 for testing. It contains 9 relations with two directions (subject and object mention) and a special *Other* class. We simplify the entity type of each example to the type mentioned in relation property. Each sample of SemEval dataset consists of one sentence, the relation between two entities and alternative comment which we discard while preprocessing the dataset.

We use *Stanford CoreNLP*[2] to complete the tokenize and dependency parse tasks on the NYT and SemEval2010 datasets, in order to build dependency tree.

Table 2. Results on TACRED, NYT, SemEval dataset.

Model	TACRED (*F1*)	NYT (*F1*)	SemEval (*F1*)
LR [24]	59.4	–	73.9
SDP-LSTM [20]	58.7	–	80.9
Tree-LSTM [13]	62.4	–	83.1
PA-LSTM [24]	65.1	–	82.9
C-GCN [23]	66.4	71.07 ± 0.06	84.1
C-AGGCN [4]	69.0	71.62 ± 0.09	85.3
GAT	65.93 ± 0.07	71.32 ± 0.12	84.29 ± 0.21
C-GAT	69.02 ± 0.19	72.44 ± 0.06	86.31 ± 0.16
EEGAT	66.27 ± 0.17	71.63 ± 0.09	85.03 ± 0.15
C-EEGAT	$\mathbf{69.64 \pm 0.03}$	$\mathbf{73.58 \pm 0.05}$	$\mathbf{87.14 \pm 0.09}$

*Bold marks highest number among all model.

*Some improvements are smaller, we added standard error.

---

[2] https://stanfordnlp.github.io/CoreNLP/.

4.3 Experimental Setup

We tune the hyperparameters based on the results of testing on the development set. For NYT and TACRED dataset, we use the same data split and the same development set used in [6] and [23], respectively. For SemEval dataset, since there was no canonical split development set on the SemEval dataset, we randomly split out 500 examples from the training set to form the development set.

For vanilla GAT and C-GAT, we set the multihead number to 3 based on experiments. When training, we use Stochastic Gradient Descent (SGD).

Table 3. Results of edge feature dimension.

Model	TACRED ($F1$)	NYT ($F1$)	SemEval ($F1$)
C-EEGAT ($P = 1$)	68.5	71.8	85.7
C-EEGAT ($P = 2$)	69.1	72.9	**87.1**
C-EEGAT ($P = 3$)	**69.6**	**73.6**	86.5
C-EEGAT ($P = 4$)	68.3	72.5	85.6
C-EEGAT ($P = 5$)	68.0	71.3	85.4

*Bold marks highest number among all model.

5 Results

5.1 Results with Different Models

We report our results on the TACRED test set in Table 2. We observe that the F1 score of our EEGAT model performs better than all other dependency-based models. Since C-GCN and C-AGGCN already show their superiority over other dependency-based models and PA-LSTM, we mainly compare our EEGAT model with them. After adding a bidirectional LSTM network to capture the contextual representations, Our C-EEGAT model achieves an F1 score of 69.64, 73.58 and 87.14 respectively, which outperforms C-GCN by 3.24, 2.51, 3.04 F1 points and C-AGGCN by 0.64, 1.96, 1.84 F1 points. The vanilla C-GAT still outperforms C-GCN by 2.61, 1.37, 2.21 points, which indicates that the GAT model is better at distinguishing relevant from irrelevant information for learning a better graph representation adjacency matrix. The performance gap between C-GAT and C-EEGAT shows that our edge feature enhancing strategy is effective for soft-pruning the dependency tree.

The results indicate that our model can not only make full use of the structural information in dependency trees but also deal with the noise caused by distant supervision effectively. Notably, by properly extracting structural dependency information, our model outperforms the previous dependency path-based model (SDP-LSTM, GCN, AGGCN).

Overall, compared with the two main models of C-GCN and C-AGGCN, the improvement of our C-EEGAT model on the two datasets of NYT and SemEval is more obvious. In order to further explore the reasons for the model improvement, we also added edge features to C-GCN and C-AGGCN. The F1 points of Edge Enhanced C-GCN on three datasets are 66.35 ± 0.27, 71.14 ± 0.16 and 84.7 ± 0.28. The F1 points of Edge Enhanced C-AGGCN on three datasets are 68.77 ± 0.19, 71.92 ± 0.11 and 85.52 ± 0.03. These results shows that the main reason for the improvement of our C-EEGAT model is that the model learns different weights for different dependent edges instead of introducing more priori information.

5.2   Performance with Different Edge Feature Dimension

Table 3 shows the effect of different edge feature dimensions P. We experimented with $P \in \{1, 2, 3, 4, 5\}$. The result indicates that the multi-channel edge feature embedding mechanism does improve the performance of our C-EEGAT model on relation extraction task. The best number of multi-channel setting is 3, 3 and 2 on TACRED, NYT and SemEval datasets respectively.

Table 4. Results of EEGAT and GAT using path-centric pruning

Model	TACRED ($F1$)	NYT ($F1$)	SemEval ($F1$)
C-EEGAT (Full tree)	**69.6**	**73.6**	**87.1**
C-GAT ($K = 0$)	68.5	71.9	85.8
C-GAT ($K = 1$)	69.0	72.4	86.3
C-GAT ($K = 2$)	68.3	71.6	85.6

*Bold marks highest number among all model.

Table 5. Ablation study for C-EEGAT model.

Model	TACRED ($F1$)	NYT ($F1$)	SemEval ($F1$)
Best C-EEGAT	69.6	73.6	87.1
- h_s, h_o, Feedforward (FF)	68.9	72.7	85.7
- LSTM Layer	68.3	72.1	84.5
- Dependency Tree	67.4	71.5	83.4
- Edge Feature	66.8	70.8	82.1
- FF, LSTM, Tree, Edge	59.6	62.8	77.2

*Bold marks highest number among all model.

6   Discussion

6.1   Performance with Pruned Trees

Our model takes the full dependency tree as the adjacency matrix. We would like to compare our soft-prune strategy (edge features) with the hard-prune strategy. Table 4 shows the performance of path-centric pruning [23] on GAT and the full dependency tree. The K means that the pruned trees include tokens that are up to distance K away from the dependency path in the LCA subtree. Firstly, we can observe that all the GAT based models (C-EEGAT and C-GAT) outperform the C-GCN model, which relies heavily on the path-centric pruning. That may indicate that our GAT based model can learn better representations of the graph than GCN based models even just on the hard pruned dependency tree. Then we can see that the performance of C-EEGAT without pruned trees (full tree) outperforms all other C-GAT with pruned trees. That may further explain that the edge feature-based "soft-pruning" approach can make use of full tree better than "hard-pruning" approach.

6.2 Ablation Study

We do some experiments using the best-performing C-EEGAT model on the TACRED dataset to examine the contributions of each component in C-EEGAT model (hs, ho and feed-forward (FF), LSTM layer, dependency tree). Table 5 shows the results. (1) The entity hidden representations and the feed-forward layer raise the F1 score a little on three datasets. (2) Without the LSTM layer, the F1 score result drop by 1.3, 1.5 and 2.6 respectively. (3) When we remove the dependency tree (retaining the adjacent matrix as I), the score drops by 2.2, 2.1 and 3.7 F1 score respectively. (4) When we remove the edge feature, the score drops by 2.8, 2.8 and 5.0 F1 score respectively. (5) F1 score drops by 10.0, 10.8 and 9.9 respectively when we remove all the components mentioned above.

7   Conclusion

We introduce edge feature enhanced graph attention mechanisms into dependency-based relation extraction. Experimental results prove that the masked self-attention mechanism performs better than graph convolutional network on extracting information from sentences when considering the structure of dependency trees. We further improve the performance by incorporating edge features into the calculation of attention coefficients, which makes full use of the connection relationships and categories in the dependency analysis results. Overall, our work improves the effectiveness of graph neural networks in relation extraction tasks.

There are multiple exploring directions for future work. One basic problem we would like to solve is how to extraction adjacency information from the dependency tree automatically without manual designed pruning strategies.

Acknowledgements. This work was supported by the National Key R&D Program of China (No. 2017YFB1002101), and the Joint Advanced Research Foundation of China Electronics Technology Group Corporation (CETC) (No. 6141B08010102), and the National Natural Science Foundation of China (No. U19B2026). This work was also funded by the strategic research project of the Development Planning Bureau of the Chinese Academy of Sciences, the Development of Technical Structure Graph and

Advanced Technology (GHJ-ZLZX-2019-42). The corresponding author of this article is the second author Chong Feng.

References

1. Bao, L., Ma, B., Chang, H., Chen, X.: Masked graph attention network for person re-identification. In: CVPR Workshops (2019)
2. Bunescu, R.C., Mooney, R.J.: A shortest path dependency kernel for relation extraction. In: HLT/EMNLP (2005)
3. Cheng, J., Dong, L., Lapata, M.: Long short-term memory-networks for machine reading. In: EMNLP (2016)
4. Guo, Z., Zhang, Y., Lu, W.: Attention guided graph convolutional networks for relation extraction. In: ACL (2019)
5. Li, L., Gan, Z., Cheng, Y., Liu, J.: Relation-aware graph attention network for visual question answering. In: 2019 IEEE/CVF International Conference on Computer Vision (ICCV), pp. 10312–10321 (2019)
6. Liu, T., Wang, K., Chang, B., Sui, Z.: A soft-label method for noise-tolerant distantly supervised relation extraction. In: EMNLP (2017)
7. Liu, Y., Wei, F., Li, S., Ji, H., Zhou, M., Wang, H.: A dependency-based neural network for relation classification. ArXiv abs/1507.04646 (2015)
8. Miwa, M., Bansal, M.: End-to-end relation extraction using LSTMs on sequences and tree structures. ArXiv abs/1601.00770 (2016)
9. Peng, N., Poon, H., Quirk, C., Toutanova, K., tau Yih, W.: Cross-sentence n-ary relation extraction with graph LSTMs. Trans. Assoc. Comput. Linguist. **5**, 101–115 (2017)
10. Pennington, J., Socher, R., Manning, C.D.: GloVe: global vectors for word representation. In: EMNLP (2014)
11. Riedel, S., Yao, L., McCallum, A.: Modeling relations and their mentions without labeled text. In: Balcázar, J.L., Bonchi, F., Gionis, A., Sebag, M. (eds.) ECML PKDD 2010. LNCS (LNAI), vol. 6323, pp. 148–163. Springer, Heidelberg (2010). https://doi.org/10.1007/978-3-642-15939-8_10
12. Song, L., Zhang, Y., Wang, Z., Gildea, D.: N-ary relation extraction using graph state LSTM. In: EMNLP (2018)
13. Tai, K.S., Socher, R., Manning, C.D.: Improved semantic representations from tree-structured long short-term memory networks. In: ACL (2015)
14. Vaswani, A., et al.: Attention is all you need. In: NIPS (2017)
15. Velickovic, P., Cucurull, G., Casanova, A., Romero, A., Liò, P., Bengio, Y.: Graph attention networks. ArXiv abs/1710.10903 (2017)
16. Verga, P., Strubell, E., McCallum, A.: Simultaneously self-attending to all mentions for full-abstract biological relation extraction. In: NAACL-HLT (2018)
17. Vu, N.T., Adel, H., Gupta, P., Schütze, H.: Combining recurrent and convolutional neural networks for relation classification. ArXiv abs/1605.07333 (2016)
18. Wang, L., Cao, Z., de Melo, G., Liu, Z.: Relation classification via multi-level attention CNNs. In: ACL (2016)
19. Xu, K., Feng, Y., Huang, S., Zhao, D.: Semantic relation classification via convolutional neural networks with simple negative sampling. ArXiv abs/1506.07650 (2015)
20. Xu, Y., Mou, L., Li, G., Chen, Y., Peng, H., Jin, Z.: Classifying relations via long short term memory networks along shortest dependency paths. ArXiv abs/1508.03720 (2015)

21. Zelenko, D., Aone, C., Richardella, A.: Kernel methods for relation extraction (2003)
22. Zeng, D., Liu, K., Lai, S., Zhou, G., Zhao, J.: Relation classification via convolutional deep neural network. In: COLING (2014)
23. Zhang, Y., Qi, P., Manning, C.D.: Graph convolution over pruned dependency trees improves relation extraction. In: EMNLP (2018)
24. Zhang, Y., Zhong, V., Chen, D., Angeli, G., Manning, C.D.: Position-aware attention and supervised data improve slot filling. In: EMNLP (2017)
25. Zhou, P., et al.: Attention-based bidirectional long short-term memory networks for relation classification. In: ACL (2016)

MMEA: Entity Alignment for Multi-modal Knowledge Graph

Liyi Chen[1], Zhi Li[1], Yijun Wang[2], Tong Xu[1(✉)], Zhefeng Wang[2], and Enhong Chen[1]

[1] School of Data Science, University of Science and Technology of China, Hefei, China
liyichencly@gmail.com, zhili03@mail.ustc.edu.cn,
{tongxu,cheneh}@ustc.edu.cn
[2] Huawei Technologies, Shenzhen, China
{wangyijun13,wangzhefeng}@huawei.com

Abstract. Entity alignment plays an essential role in the knowledge graph (KG) integration. Though large efforts have been made on exploring the association of relational embeddings between different knowledge graphs, they may fail to effectively describe and integrate the multi-modal knowledge in the real application scenario. To that end, in this paper, we propose a novel solution called Multi-Modal Entity Alignment (MMEA) to address the problem of entity alignment in a multi-modal view. Specifically, we first design a novel multi-modal knowledge embedding method to generate the entity representations of relational, visual and numerical knowledge, respectively. Along this line, multiple representations of different types of knowledge will be integrated via a multi-modal knowledge fusion module. Extensive experiments on two public datasets clearly demonstrate the effectiveness of the MMEA model with a significant margin compared with the state-of-the-art methods.

Keywords: Multi-modal knowledge · Entity alignment · Knowledge graph

1 Introduction

Knowledge graph (KG), which is composed of relational facts with entities connected by various relations, benefits lots of AI-related systems, such as recommender systems, question answering, and information retrieval. However, most KGs are constructed for specific purposes and monolingual settings, which results in the separate KGs with gaps of different descriptions for even the same concepts. Therefore, entity alignment techniques are urgently required to integrate the distinct KGs by linking entities referring to the same real-world identity.

Along this line, many efforts have been made in exploring the associations of distinct KGs and querying knowledge completely by entity alignment. In general, prior arts could be roughly grouped into two categories, i.e., similarity-based methods [9,12] and embedding-based methods [3,20]. Early studies mostly focus

© Springer Nature Switzerland AG 2020
G. Li et al. (Eds.): KSEM 2020, LNAI 12274, pp. 134–147, 2020.
https://doi.org/10.1007/978-3-030-55130-8_12

on the attribute similarity, such as string similarity [12] and numeric similarity [16]. However, these methods often suffer from the attribute heterogeneity, which makes the entity alignment error-prone [15]. Recently, in view of the rapid development of knowledge graph embedding, many researchers have attempted to utilize embedding-based models for the entity alignment problem [10,15]. In spite of the importance of prior arts, existing researches mainly focus on the semantics or concept knowledge graphs alignment but largely ignore the multi-modal knowledge from the real scenarios.

Indeed, in real-world application scenarios, knowledge is usually summarized in various forms, such as relational triples, numerical attributes and images. These distinct knowledge forms not only can play an important role as extra pieces of evidence for the KG completion, but also highly support the entity alignment task. For instance, Fig. 1 illustrates a toy example of entity alignment for multi-modal knowledge graphs, in which the image of "Fuji" can clearly demonstrate that the entity type is the mountain. Moreover, the similar images and numerical attributes (such as "Height" and "Latitude") can be helpful for aligning the same entity between two KGs. Unfortunately, it is not trivial to leverage multi-modal knowledge to the entity alignment problem. On the one hand, the alignment task is challenging in terms of computational complexity, data quality, and acquisition of prior alignment data in large-scale knowledge graphs. On the other hand, the inevitable heterogeneity among different modalities makes it difficult to learn and fuse the knowledge representations from distinct modalities. Therefore, traditional techniques may fail to deal with this task.

To conquer these challenges, in this paper, we propose a novel solution called Multi-Modal Entity Alignment (MMEA) for modeling the entity associations of multi-modal KGs and finding entities referring to the same real-world identity. To be specific, we first propose a multi-modal knowledge embedding method to discriminatively generate knowledge representations of three different types of knowledge, i.e., relational triples, visual contents (images) and numerical attributes. Then, to leverage multi-modal knowledge for the entity alignment task, a multi-modal fusion module is designed to integrate knowledge representations from multiple modalities. Extensive experiments on two large-scale real-world datasets demonstrate that MMEA not only provides insights to take advantages of multi-modal knowledge in the entity alignment task, but also outperforms the state-of-the-art baseline methods.

2   Related Work

Generally, the related work can be classified into two perspectives, i.e., entity alignment and multi-modal knowledge graph.

2.1   Entity Alignment

Actually, the entity alignment problem has been one of the major studies in the knowledge graph area for a long time. Early researchers mainly focus on

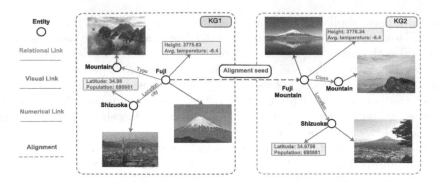

Fig. 1. A toy example of entity alignment between multi-modal knowledge graphs

exploring the content similarity to align the entities between different KGs. LD-Mapper [12] utilizes entity nearest neighbor similarity and string similarity. RuleMiner [9] refines a set of matching-rules with an Expectation-Maximization algorithm. SILK [16] measures entity similarity with string equality and similarity, numeric similarity and so on.

Recently, it is notable that entity alignment based on knowledge graph embedding representation becomes popular in the area. The current methods often embed entity to a low-dimensional space and measure the similarity between entity embeddings. Embedding-based methods concentrate on the semantics or concept so that they have a better analysis of knowledge. IPTransE [20] is an iterative method through joint knowledge embedding. BootEA [14] iteratively labels possible entity alignments as the training data, and employs an alignment editing method to reduce the error accumulation during the iterations. SEA [10] utilizes an awareness of the degree difference in adversarial training and incorporates the unaligned entities to enhance the performance. KDCoE [2] adds entity descriptions for entity alignment with a semi-supervised learning method for joint training. Furthermore, there are several methods utilizing attributes to strengthen the performance of entity alignment model. AttrE [15] uses a large number of attribute triples to generate character embeddings, and employs the relationship transitivity rule. IMUSE [6] achieves entity alignment and attribute alignment with an unsupervised method, and employs bivariate regression to merge alignment results. Additionally, GCN [17] uses relations to build the structures of graph convolutional networks and combines relations and attributes. However, these methods ignore the multi-modal knowledge from the real scenarios.

2.2 Multi-modal Knowledge Graph

In diverse domains, researchers study multi-modal learning in order to extract semantic information from various modalities. Multi-modal information such as structural and visual features is significant for entity alignment. PoE [7] is proposed to find entity alignment in multi-modal knowledge graphs through extracting relational, latent, numerical and visual features. In addition, the most relevant task to our multi-modal entity alignment is multi-modal knowledge representation. Considering visual features from entity images for knowledge representation learning, IKRL [19] integrates image representations into an aggregated image-based representation via an attention-based method. MKBE [11] models knowledge bases that contain a variety of multi-modal features such as links, images, numerical and categorical values. It applies neural encoders and decoders which embed multi-modal evidence types and generate multi-modal attributes, respectively. [8] proposes a multi-modal translation-based method, which defines the energy of a knowledge graph triple as the sum of sub-energy functions that leverages structural, visual and linguistic knowledge representations. On the whole, multi-modal knowledge graph is still a novel problem, and the entity alignment has not been fully discussed.

3   Methodology

In this section, we formally introduce the entity alignment task for multi-modal knowledge graphs (KGs) and give an overview of our proposed model, i.e., Multi-Modal Entity Alignment (MMEA). Then, we describe the details of MMEA.

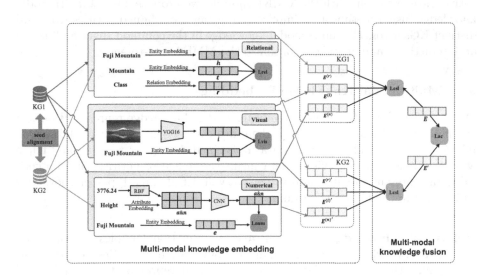

Fig. 2. The framework overview of MMEA.

3.1   Preliminaries and Technical Framework

Notation and Problem Definition. A multi-modal knowledge graph can be noted as $G = (\widehat{E}, R, I, N, X, Y, Z)$, where $\widehat{E}, R, I, N$ denote the sets of entities, relations, images and numerics, and X, Y, Z denote the sets of relational triples, entity-image pairs and numerical triples, respectively. With multi-modal knowledge embedding, we denote $\mathbf{E}^{(r)}$, $\mathbf{E}^{(i)}$, $\mathbf{E}^{(n)}$ as entity embeddings for relational, visual and numerical information, respectively.

The task of entity alignment refers to matching entities describing the same thing in the real world from different knowledge graphs, which is beneficial for people to acquire knowledge completely, and it is not necessary to find related information of the same entity from multiple knowledge graphs anymore. Let $G_1 = (\widehat{E}_1, R_1, I_1, N_1, X_1, Y_1, Z_1)$ and $G_2 = (\widehat{E}_2, R_2, I_2, N_2, X_2, Y_2, Z_2)$ be two different KGs. $H = \left\{ (e_1, e_2) | e_1 \in \widehat{E}_1, e_2 \in \widehat{E}_2 \right\}$ denotes the set of aligned entities across knowledge graphs.

Framework Overview. In this paper, we propose a multi-modal model for entity alignment, namely Multi-Modal Entity Alignment (MMEA) model, which can automatically and accurately align the entities in two distinct multi-modal knowledge graphs. As illustrated in Fig. 2, our proposed MMEA consists of two major components, i.e., *Multi-Modal Knowledge Embedding* (MMKE) and *Multi-Modal Knowledge Fusion* (MMKF). In the MMKE module, we extract the relational, visual and numerical information to complement the absence of useful entity features. Then, with the MMKF module, we propose a novel multi-modal knowledge fusion method to minimize the distance of aligned entities from two distinct KGs across the multi-modal knowledge in the common space and design an interactive training stage to optimize the MMEA end-to-end.

3.2 Multi-modal Knowledge Embedding

Multi-modal knowledge plays a significant part in knowledge representations. In our multi-modal knowledge graph, there are three types of data modality, i.e., relational, visual and numerical data. Relational data refer to relational triple with entity associations, visual data mean the image of entities, and numerical data represent the attribute value. We will detail three types of knowledge embedding in the following section.

Relational Knowledge Representations. Relational triples are the main part of KGs, which are essential to judge the association of entities from different KGs. Under the relational data, we adopt the most representative translational distance model: TransE [1]. Given a fact $(h, r, t) \in X$, h and t can be associated by r in a low-dimensional continuous vector space. The process named translation adjusts the distance between $\mathbf{h} + \mathbf{r}$ and $\mathbf{t}$ in the space constantly, in order that $\mathbf{h}+\mathbf{r}$ is equal to $\mathbf{t}$ as much as possible when (h, r, t) holds. In multi-relational data, there are certain structural similarities. Such as ("Fuji", "Location city", "Shizuoka") and ("Eiffel", "Location city", "Paris") in the embedding space, we have "Shizuoka" − "Fuji" ≈ "Paris" − "Eiffel". Through the relationship "Location city", we can acquire "Eiffel" + "Location city"

$\approx$ "Paris" from "Fuji" + "Location city" $\approx$ "Shizuoka" automatically. The scoring function which we take to be L_2-norm is defined as follows:

$$f_{rel}(h, r, t) = -||\mathbf{h} + \mathbf{r} - \mathbf{t}||_2^2. \tag{1}$$

To learn the entity embeddings from relational data, we apply the margin-based [18] loss function with $\gamma > 0$ over the training set:

$$L_{rel} = \sum_{\tau^+ \in D^+} \sum_{\tau^- \in D^-} max(0, \gamma - f_{rel}(\tau^+) + f_{rel}(\tau^-)). \tag{2}$$

Here, D^+ and D^- are positive and negative examples sets, respectively. Given a positive example $\tau^+ = (h, r, t)$, we supplement the set of positive examples through the exchange strategy. The exchange strategy means that if h has been aligned by $\bar{h}$ in the other knowledge graph, $(\bar{h}, r, t)$ will be expanded into the set D^+. For t, the exchange strategy generates $(h, r, \bar{t})$ in D^+ identically. The supplementary relational triples benefit linking two diverse knowledge graphs in the unified low-dimensional continuous vector space. The definition of D^- is described as follows:

$$D^- = \left\{ (h', r, t) \, | h' \in \widehat{E} \wedge h' \neq h \wedge (h, r, t) \in D^+ \wedge (h', r, t) \notin D^+ \right\}$$
$$\cup \left\{ (h, r, t') \, | t' \in \widehat{E} \wedge t' \neq t \wedge (h, r, t) \in D^+ \wedge (h, r, t') \notin D^+ \right\}.$$

Negative examples sampled by replacing the head or tail entities of real relational triples at random are arranged to approximate the partition function.

Visual Knowledge Representations. Sometimes the relational structure information of knowledge graphs can cause ambiguity. When finding the entity aligned with "Fuji" in the other knowledge graph, "Fuji Mountain" and "Fujifilm" exist. The visual features characterize the appearance of the entity more intuitively and vividly than relational knowledge, and we can distinguish "Fuji Mountain" from "Fujifilm" because the one is a mountain, and the other one is a company logo. Therefore, visual data serve as a vital part of multi-modal knowledge graphs and visual features disambiguate the relational information to some extent.

In order to extract visual features, we achieve the vectorization of images and each of entity images is embedded into a vector. However, image vectors can not be directly applied in this scene, hence we project them to associate with entity embedding vectors. We learn embeddings for images according to the VGG16 [13] model. The model pre-trained on the ILSVRC 2012 dataset derived from ImageNet [5] is applied in our model. The filters in a stack of convolutional layers have the receptive fields of 3×3. We develop 13 convolutional layers which have different depths in various architectures. They are followed by 3 fully-connected layers, but we remove the last fully-connected layer and the softmax layer, then obtain the 4096-dimensional embeddings for all entity images. Given a pair $(e^{(i)}, i) \in Y$ in the visual knowledge, we use the following score function to utilize visual features:

$$f_{vis}(e^{(i)}, i) = -||\mathbf{e}^{(i)} - tanh(vec(\mathbf{i}))||_2^2, \tag{3}$$

where $vec(\cdot)$ denotes the projection, and $tanh(\cdot)$ is a kind of activation function. Based on the above score function, we minimize the following loss function to optimize the visual knowledge representations:

$$L_{vis} = \sum_{(e^{(i)}, i) \in Y} \log \left(1 + \exp \left(-f_{vis}(e^{(i)}, i)\right)\right). \tag{4}$$

Numerical Knowledge Representations. The numerical triple is denoted as $(e^{(n)}, a, n) \in Z$ in the numerical data, where a denotes the attribute key, and n denotes the numerical value. Attribute keys and corresponding numerical values form the *key-value pairs* to describe entities. Formally, relational structures only model the translation between head entities and tail entities while numerical features supplement the information between some entities which can not be constituted of a relational fact in the knowledge graphs. For instance, the "height" of "Fuji" is 3775.63 and the "height of "Fuji Mountain" is 3776.24, hence we deduce that they are likely to refer the same thing for entity alignment.

First of all, we deal with numeric since continuous value needs special treatment. Sparse numerical data demands to be fitted to a simple parameter distribution, and the radial basis function (RBF) [4] meets our requirement exactly. The RBF network is able to approximate any non-linear function and handle the issues of analyzing data regularity. It has good generalization ability and has a fast speed of convergence.

We convert numerical information to embeddings in high-dimensional spaces with applying a radial basis function as follows:

$$\phi \left(n_{(e^{(n)}, a_i)}\right) = \exp \left(\frac{-\left(n_{(e^{(n)}, a_i)} - c_i\right)^2}{\sigma_i^2}\right), \tag{5}$$

where c_i denotes the radial kernel center, σ_i denotes the variance and they are both vectors. Firstly, all corresponding numerical values for each attribute key will be normalized. After normalization, c_i and σ_i can be computed in the RBF neural network through the supervised method.

In addition, we intend to extract features from attribute keys and corresponding numerical values of entities, which indeed form the *key-value pairs*. We concatenate the embedding of an attribute key and its numerical vector got from the RBF layer. This process generates a new $2 \times d$ matrix denoted by $\mathbf{M} = \langle \mathbf{a}, \phi(n_{(e^{(n)}, a)}) \rangle$. Then we define the score function to measure the plausibility of the embeddings:

$$f_{num}(e^{(n)}, a, v) = -||\mathbf{e}^{(n)} - tanh(vec(\mathrm{CNN}(tanh(\mathbf{M})))\mathbf{W})||_2^2, \tag{6}$$

where CNN denotes l convolutional layers, and $\mathbf{W}$ means a fully-connected layer. We reshape the feature map to a vector, then project it to the embedding space. The loss function is given as follows:

$$L_{num} = \sum_{(e^{(n)},a,n)\in Z} \log\left(1 + \exp\left(-f_{num}(e^{(n)}, a, v)\right)\right), \tag{7}$$

where Z denotes the set of numerical triples in the numerical data. Exchanging aligned entities in the involved numerical triples, because they refer to the same real-world object across different knowledge graphs and they own the same numerical features. If a numerical triple (e, a, n) exists and $(e, \bar{e})$ appears in the seed entity alignment, $(\bar{e}, a, n)$ is added to Z.

3.3   Multi-modal Knowledge Fusion

Information from different independent sources under different modalities complements each other. Commonly, multi-modal features tend to correlate, which provide additional redundancy for better robustness. The features in the three types of modality could not be directly extracted to one space, therefore we propose a *Multi-Modal Knowledge Fusion* (MMKF) module to integrate knowledge representations from multiple modalities. MMKF migrates multi-modal knowledge embeddings from separate spaces to a common space. Common space learning enables multi-modal features to benefit from each other. It enhances the complementarity of multiple modalities which improves the accuracy of the task of entity alignment. The loss function is designed as follows:

$$L_{csl}(\mathbf{E}, \mathbf{E}^{(r)}, \mathbf{E}^{(i)}, \mathbf{E}^{(n)}) = \alpha_1||\mathbf{E} - \mathbf{E}^{(r)}||_2^2 + \alpha_2||\mathbf{E} - \mathbf{E}^{(i)}||_2^2 + \alpha_3||\mathbf{E} - \mathbf{E}^{(n)}||_2^2, \tag{8}$$

where $\mathbf{E}$ denotes the entity embeddings in the common space, and $\mathbf{E}^{(r)}$, $\mathbf{E}^{(i)}$ and $\mathbf{E}^{(n)}$ are the entity embeddings in the spaces of relational, visual and numerical knowledge, respectively. Besides, α_1, α_2 and α_3 are ratio hyper-parameters for each type of knowledge.

Since aligned entities have identical meaning in different knowledge graphs, it is intuitive for us to make those aligned entities closer in the common space. The distance between aligned entities is calculated as $||\mathbf{e}_1 - \mathbf{e}_2||$, where $\mathbf{e}_1$, $\mathbf{e}_2 \in \mathbf{E}$. Taking the distance into account, we adapt the alignment constraint approach in the common space to minimize the mapping loss:

$$L_{ac}(\mathbf{E}_1, \mathbf{E}_2) = ||\mathbf{E}_1 - \mathbf{E}_2||_2^2, \tag{9}$$

where $\mathbf{E}_1$ and $\mathbf{E}_2$ denote embeddings of entities in the sets of $\widehat{E}_1$ and $\widehat{E}_2$, which are defined as follows:

$$\widehat{E}_1 = \left\{e_1 | e_1 \in KG_1 \wedge e_1 \in \widehat{E} \wedge (e_1, e_2) \in H\right\}$$
$$\widehat{E}_2 = \left\{e_2 | e_2 \in KG_2 \wedge e_2 \in \widehat{E} \wedge (e_1, e_2) \in H\right\},$$

where H denotes the set of aligned entities across different knowledge graphs.

For the purpose of making up for imbalance among different types of knowledge, we design an interactive training stage which learns embeddings of three multi-modal (relational, visual and numerical) knowledge and optimizes the common space learning during an epoch, repeatedly. We constrain all entity embeddings with L_2 normalization to regularize embedding vectors. Firstly, we train image embeddings from VGG16 and obtain the 4096-dimensional embeddings for all the entities. Then, at each step, the parameters are updated by L_{rel}, L_{vis}, L_{num}, L_{csl} and L_{ac}.

4   Experiments

In this section, we evaluate MMEA on two real-world datasets, and demonstrate that MMEA provides insights to take advantages of multi-modal knowledge in the entity alignment task and outperforms the baselines which were shown to achieve state-of-the-art performance for entity alignment.

4.1   Experimental Settings

Datasets. In our experiments, we use two multi-modal datasets which were built in [7], namely FB15K-DB15K and FB15K-YAGO15K. FB15K is a representative subset extracted from the Freebase knowledge base. Aiming to maintain an approximate entity number of FB15K, DB15K from DBpedia and YAGO15K from YAGO are mainly selected based on the entities aligned with FB15K. Table 1 depicts the statistics of multi-modal datasets. Each dataset provides 20%, 50%, and 80% reference entity alignment as training sets, respectively.

Evaluation Metrics. We utilize cosine similarity to calculate the similarity between two entities and employ Hits@n, MRR, and MR as metrics to evaluate all the models. Hits@n means the rate correct entities rank in the top n according to similarity computing. MR denotes the mean rank of correct entities and MRR denotes the mean reciprocal rank of correct entities. The higher values of Hits@n and MRR explain the better performance of the method, while the lower value of MR proves it.

Implementation Details. All the experiments are tuned for both datasets. For MMEA we initialize the embeddings of KGs in each type of knowledge with Xavier initializer and restrain their lengths to 1. The dimensions of all the embeddings are set as 100. We adopt the mini-batch method with the batch size of 5000. We start to valid every 10 epochs after 300 epochs and stop the training when the metric MRR is declining continually in the valid set. We set all the learning rates to 0.01 except that the learning rate of common space learning is 0.004. In addition, the max epochs are set as 600. More specifically, γ in the relational knowledge representation is 1.5. In the numerical knowledge representation, l and the number of filters are both set as 2. The kernel size is 2×4. α_1, α_2, and α_3 in the common space learning are selected as $\{1, 0.01, 1\}$ on FB15K-DB15K dataset and $\{1, 1, 0.01\}$ on FB15K-YAGO15K dataset, respectively. We optimize all the above loss functions using stochastic gradient descent (SGD).

Table 1. Statistics of multi-modal datasets.

Datasets	Entities	Relations	Attributes	Relational triples	Numerical triples	Images	Links
FB15K	14951	1345	116	592213	29395	13444	–
DB15K	12842	279	225	89197	48080	12837	12846
YAGO15K	15404	32	7	122886	23532	11194	11199

4.2   Compared Methods

To demonstrate that MMEA framework outperforms the state-of-the-art entity alignment models, we compare it with the following methods:

- **TransE** [1] is a typical translational method for knowledge graph embedding. We perform this method in the entity alignment task by sharing the parameters between aligned entities.
- **MTransE** [3] learns the translation matrix to map the aligned entities from different knowledge graphs in the unified space. It acquires a great deal of seed alignment, otherwise the translation matrix will be inaccurate.
- **IPTransE** [20] obtains entity embeddings through employing an iterative and parameter sharing method. Additionally, soft alignment and multi-step relation paths are utilized to align entities from different KGs.
- **SEA** [10] served as a semi-supervised method realizes the adversarial training with an awareness of the degree difference and leverages both labeled entities and the abundant unlabeled entity information for the alignment.
- **GCN** [17] adopts GCNs to encode the structural information of entities, and combine relation and attribute embeddings for the entity alignment task.
- **IMUSE** [6] generates lots of high-quality aligned entities with an unsupervised method. Besides, a bivariate regression is utilized to merge the alignment results of relations and attributes better.
- **PoE** [7] combines the multi-modal features and measures the credibility of facts by matching the underlying semantics of the entities and mining the relations contained in the embedding space. Regarding computing the scores of facts under each modality, it learns the entity embeddings for entity alignment.

4.3   Results and Analyses

We partition the datasets to compare the results of all models. For each dataset, we use the 20%, 50%, 80% data as training sets, and the remains are treated as testing sets, respectively.

Table 2. 20% alignment results on two datasets. (R.: Relational knowledge, N.: Numerical knowledge, V.: Visual knowledge)

Models		FB15K-DB15K					FB15K-YAGO15K				
		Hits@1	Hits@5	Hits@10	MR	MRR	Hits@1	Hits@5	Hits@10	MR	MRR
R.	MTransE	0.359	1.414	2.492	1239.465	0.0136	0.308	0.988	1.783	1183.251	0.011
	IPTransE	3.985	11.226	17.277	387.512	0.0863	3.079	9.505	14.443	522.235	0.07
	TransE	7.813	17.95	24.012	442.466	0.134	6.362	15.11	20.254	522.545	0.112
	PoE-1	7.9	–	20.3	–	0.122	6.4	–	16.9	–	0.101
	SEA	16.974	33.464	42.512	191.903	0.255	14.084	28.694	37.147	207.236	0.218
R. + N.	GCN	4.311	10.956	15.548	810.648	0.0818	2.27	7.209	10.736	1109.845	0.053
	IMUSE	17.602	34.677	43.523	182.843	0.264	8.094	19.241	25.654	397.571	0.142
R. + N. + V.	PoE-lni	12.0	–	25.6	–	0.167	10.9	–	24.1	–	0.154
	MMEA	**26.482**	**45.133**	**54.107**	**124.807**	**0.357**	**23.391**	**39.764**	**47.999**	**147.441**	**0.317**

Performance Comparison. Table 2 lists the results of all the models with 20% alignment data on FB15K-DB15K and FB15K-YAGO15K datasets. The results for PoE are taken from [7]. From the overview, our proposed MMEA achieves the state-of-the-art performance for entity alignment. Specifically, there are several observations. First, MMEA performs better than all the other methods. Compared with these methods, Hits@1, Hits@5, Hits@10, MRR are at least improved by 8.88%, 10.456%, 10.584%, 0.093 and 9.307%, 11.07%, 10.852%, 0.099, and MR is at least decreased by 58.036 and 59.795 on two datasets. The results indicate that MMEA is more suitable for multi-modal knowledge graphs from the real scenarios. Second, solutions with multi-modal knowledge generate better results than solutions with a single modality in most cases. Both MMEA and PoE-lni outperform MTransE, IPTransE, TransE and PoE-l, which indicates that as increasing numerical and visual knowledge leads to improvements, the effects of multi-modal knowledge have been proven. Third, MMEA outperforms PoE-lni absolutely, suggesting our modeling for multi-modal knowledge is more effective, and multi-modal fusion method with common space is better.

Figure 3 shows the experimental results with different test splits on FB15K-DB15K and FB15K-YAGO15K datasets. In most cases, especially when only 20% alignment data is split to the training set, MMEA with a significant margin compared with the state-of-the-art methods could make full use of limited data. Moreover, it demonstrates the robustness and effectiveness of MMEA once again.

Ablation Study. To further validate the effectiveness of multi-modal knowledge in the task of entity alignment, we design two variants for ablation study, namely

MMEA-R and MMEA-RN. MMEA-R is a variant of MMEA with only relational knowledge, and MMEA-RN is a variant of MMEA with relational and numerical knowledge. According to the experimental results, MMEA outperforms both MMEA-R and MMEA-RN, which reveals that multi-modal knowledge complements the absence of useful entity features, and MMEA provides insights to take advantages of multi-modal knowledge. Moreover, it is obvious that our multi-modal knowledge fusion method could leverage multi-modal knowledge for entity alignment.

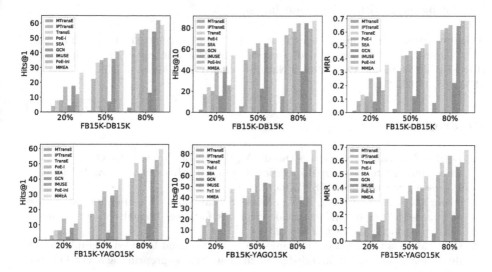

Fig. 3. Experimental results with different test splits on two datasets.

In summary, all above evidences demonstrate that MMEA framework has a good ability to find entities referring to the same real-world identity from different KGs by taking full advantages of multi-modal knowledge and achieves state-of-the-art performance for entity alignment (Table 3).

Table 3. Ablation study.

Models		FB15K-DB15K					FB15K-YAGO15K				
		Hits@1	Hits@5	Hits@10	MR	MRR	Hits@1	Hits@5	Hits@10	MR	MRR
20%	MMEA-R	24.957	43.084	51.581	143.171	0.340	22.199	38.563	46.493	160.576	0.305
	MMEA-RN	26.209	44.982	53.759	125.874	0.355	23.091	39.589	47.689	154.908	0.314
	MMEA	**26.482**	**45.133**	**54.107**	**124.807**	**0.357**	**23.391**	**39.764**	**47.999**	**147.441**	**0.317**
50%	MMEA-R	40.95	61.362	69.721	58.093	0.505	39.161	56.696	63.956	65.848	0.477
	MMEA-RN	41.436	61.691	70.089	54.53	0.51	39.509	56.959	63.979	65.255	0.48
	MMEA	**41.653**	**62.1**	**70.345**	**54.257**	**0.512**	**40.263**	**57.231**	**64.51**	**62.969**	**0.486**
80%	MMEA-R	58.256	80.192	86.466	14.557	0.679	58.803	77.078	83.132	15.308	0.672
	MMEA-RN	58.411	80.355	86.76	14.493	0.681	59.377	78.22	83.34	14.745	0.68
	MMEA	**59.034**	**80.405**	**86.869**	**14.129**	**0.685**	**59.763**	**78.485**	**83.892**	**14.512**	**0.682**

5   Conclusion

In this paper, we proposed a novel solution for the entity alignment task in multi-modal knowledge graphs, which integrated multiple representations of different types of knowledge based on knowledge embedding. Moreover, a multi-modal fusion method was designed through common space learning to migrate features under different knowledge spaces. Extensive experiments on two real-world datasets demonstrated the robustness and effectiveness of our solution for multi-modal entity alignment, which outperformed several state-of-the-art baseline methods with a significant margin.

Acknowledgments. This research was partially supported by grants from the National Key Research and Development Program of China (Grant No. 2018YFB1402600), and the National Natural Science Foundation of China (Grant No. 61703386, U1605251).

References

1. Bordes, A., Usunier, N., Garcia-Duran, A., Weston, J., Yakhnenko, O.: Translating embeddings for modeling multi-relational data. In: Advances in Neural Information Processing Systems, pp. 2787–2795 (2013)
2. Chen, M., Tian, Y., Chang, K.W., Skiena, S., Zaniolo, C.: Co-training embeddings of knowledge graphs and entity descriptions for cross-lingual entity alignment. arXiv preprint arXiv:1806.06478 (2018)
3. Chen, M., Tian, Y., Yang, M., Zaniolo, C.: Multilingual knowledge graph embeddings for cross-lingual knowledge alignment. arXiv preprint arXiv:1611.03954 (2016)
4. Chen, S., Cowan, C.F., Grant, P.M.: Orthogonal least squares learning algorithm for radial basis function networks. IEEE Trans. Neural Netw. **2**(2), 302–309 (1991)
5. Deng, J., Dong, W., Socher, R., Li, L.J., Li, K., Fei-Fei, L.: ImageNet: a large-scale hierarchical image database. In: 2009 IEEE Conference on Computer Vision and Pattern Recognition, pp. 248–255. IEEE (2009)
6. He, F., et al.: Unsupervised entity alignment using attribute triples and relation triples. In: Li, G., Yang, J., Gama, J., Natwichai, J., Tong, Y. (eds.) DASFAA 2019. LNCS, vol. 11446, pp. 367–382. Springer, Cham (2019). https://doi.org/10.1007/978-3-030-18576-3_22
7. Liu, Y., Li, H., Garcia-Duran, A., Niepert, M., Onoro-Rubio, D., Rosenblum, D.S.: MMKG: multi-modal knowledge graphs. In: Hitzler, P., et al. (eds.) ESWC 2019. LNCS, vol. 11503, pp. 459–474. Springer, Cham (2019). https://doi.org/10.1007/978-3-030-21348-0_30
8. Moussailly-Sergieh, H., Botschen, T., Gurevych, I., Roth, S.: A multimodal translation-based approach for knowledge graph representation learning. In: Proceedings of the Seventh Joint Conference on Lexical and Computational Semantics, pp. 225–234 (2018)
9. Niu, X., Rong, S., Wang, H., Yu, Y.: An effective rule miner for instance matching in a web of data. In: Proceedings of the 21st ACM International Conference on Information and Knowledge Management, pp. 1085–1094 (2012)

10. Pei, S., Yu, L., Hoehndorf, R., Zhang, X.: Semi-supervised entity alignment via knowledge graph embedding with awareness of degree difference. In: The World Wide Web Conference, pp. 3130–3136 (2019)
11. Pezeshkpour, P., Chen, L., Singh, S.: Embedding multimodal relational data for knowledge base completion. arXiv preprint arXiv:1809.01341 (2018)
12. Raimond, Y., Sutton, C., Sandler, M.B.: Automatic interlinking of music datasets on the semantic web. LDOW **369** (2008)
13. Simonyan, K., Zisserman, A.: Very deep convolutional networks for large-scale image recognition. arXiv preprint arXiv:1409.1556 (2014)
14. Sun, Z., Hu, W., Zhang, Q., Qu, Y.: Bootstrapping entity alignment with knowledge graph embedding. In: IJCAI, pp. 4396–4402 (2018)
15. Trisedya, B.D., Qi, J., Zhang, R.: Entity alignment between knowledge graphs using attribute embeddings. Proceedings of the AAAI Conference on Artificial Intelligence, vol. 33, pp. 297–304 (2019)
16. Volz, J., Bizer, C., Gaedke, M., Kobilarov, G.: Discovering and maintaining links on the web of data. In: Bernstein, A., et al. (eds.) ISWC 2009. LNCS, vol. 5823, pp. 650–665. Springer, Heidelberg (2009). https://doi.org/10.1007/978-3-642-04930-9_41
17. Wang, Z., Lv, Q., Lan, X., Zhang, Y.: Cross-lingual knowledge graph alignment via graph convolutional networks. In: Proceedings of the 2018 Conference on Empirical Methods in Natural Language Processing, pp. 349–357 (2018)
18. Wu, C.Y., Manmatha, R., Smola, A.J., Krahenbuhl, P.: Sampling matters in deep embedding learning. In: Proceedings of the IEEE International Conference on Computer Vision, pp. 2840–2848 (2017)
19. Xie, R., Liu, Z., Luan, H., Sun, M.: Image-embodied knowledge representation learning. arXiv preprint arXiv:1609.07028 (2016)
20. Zhu, H., Xie, R., Liu, Z., Sun, M.: Iterative entity alignment via joint knowledge embeddings. In: IJCAI, pp. 4258–4264 (2017)

A Hybrid Model with Pre-trained Entity-Aware Transformer for Relation Extraction

Jinxin Yao[1], Min Zhang[1,2(✉)], Biyang Wang[1], and Xianda Xu[1]

[1] East China Normal University, Shanghai, China
mzhang@sei.ecnu.edu.cn
[2] Shanghai Key Laboratory of Trustworthy Computing, Shanghai, China

Abstract. Distantly supervised relation extraction is an efficient method to extract novel relational facts from unstructed text. Most previous neural methods adopt Convolutional Neural Network (CNN) or Recurrent Neural Network (RNN) to encode sentences. However, CNN is difficult to learn long-range dependencies and the parallelization of training RNN is precluded by its sequential nature. In this paper, we propose a novel hybrid model that combines Piece-wise Convolutional Neural Network (PCNN) and Entity-Aware Transformer to extract local features and learn the dependencies between distant positions jointly. The entity-aware Transformer is able to take semantic and syntax information under consideration and acquire entity-specific representations. The inner-sentence attention mechanism is then used over Transformer to alleviate the noise caused by irrelevant words. We concatenate outputs of PCNN and Transformer with word embeddings of entity mentions and then send them to the classifier, which can boost the performance of our model further. A transfer learning based strategy is applied, where the entity-aware Transformer is initialized with a priori knowledge learned from the related task of entity typing to improve the robustness of our model. The experimental results on a large-scale benchmark dataset show that our hybrid model with the pre-training strategy gets AUC score of 0.432 and outperforms the state-of-the-art baselines.

Keywords: Relation extraction · Transformer · Transfer learning

1 Introduction

Relation extraction (RE) is a fundamental task in information extraction and benefits many downstream applications in Natural Language Processing (NLP) field such as knowledge graph construction. RE aims to identify the semantic relations between entity pair in raw text. Traditional supervised RE methods need a large amount of training data but the process of labelling data is human-intensive. Mintz et al. [16] proposed Distant Supervision to generate training data automatically by aligning text with existing Knowledge Bases (KBs). Distant supervision assumes that if an entity pair has a relation in KBs, any sentence

© Springer Nature Switzerland AG 2020
G. Li et al. (Eds.): KSEM 2020, LNAI 12274, pp. 148–160, 2020.
https://doi.org/10.1007/978-3-030-55130-8_13

which contains both entities might express that relation. As not all sentences that mention two entities can exactly express such relation, this heuristic approach inevitably leads to wrong label problem, e.g., distant supervision will label the sentence "*Steve_Ballmer replaced Bill_Gates as CEO of Microsoft.*" and entity pair (*Bill_Gates, Microsoft*) with the relation /company/founders.

A lot of methods [10,17,19] have been proposed for relation extraction, but they utilize hand-crafted features derived from NLP tools which can be erroneous. Recently, neural models have shown promising power on RE. Zeng et al. [25] adopt CNN to learn sentence representations and achieve better performance than feature based methods. Zhou et al. [26] employ bidirectional Long Short-Term Memory Networks to encode sentences effectively. He et al. [8] utilize dependency parses with tree-GRU networks to take advantage of long-range dependencies. However, CNN is restricted to learn dependencies between distant tokens by its structure and RNN computes the current token's representation requiring the previous token's representation as input, which limits RNNs to execute in parallel. Also, dependency parse trees of sentences which are obtained with NLP tools may be noisy and then misguide the relation extractor.

Sentences annotated automatically by distant supervision often contain irrelevant words that mean noise for RE so robustness is crucial for relation extractors. Most previous neural methods train the extractor with single RE task and initialize parameters randomly where models are not robust enough against noise. Initializing neural networks of target task with the parameters learned from relevant task by transfer learning can improve the robustness of the model. Entity type is capable to guide relation extraction since it imposes soft constraints of relations so Entity Typing (ET) that aims to classify entities into a set of types can be adopted as relevant task for relation extraction. For example, the relation between *Microsoft* (a company entity) and *Bill_Gates* (a person entity) may be the relation /company/founders instead of the relation /location/contains. However, most previous works handle two tasks separately and ignore the strong relatedness between them.

In this paper, we propose a novel hybrid model for distantly supervised relation extraction by combining Piece-wise Convolutional Neural Network (PCNN) with Entity-Aware Transformer which is able to learn long-range dependencies and concentrate on entities. Entity-aware Transformer compute each token's representation using global information and can be executed in parallel. Inner-sentence attention mechanism over Transformer is used to subdue noisy words. Word embeddings of entities are employed directly and can boost the model performance. We design a transfer learning based strategy to initialize Transformer with parameters learned from entity typing task.

Our contributions of this paper can be summarized as follows:

- We develop a novel hybrid neural model which combines PCNN and entity-aware Transformer to extract local and global features simultaneously.
- Entity-aware Transformer network is initialized with the parameters learned from entity typing task to enhance the robustness of our model.

– Experiments conducted on a public dataset show the effectiveness of our method and our hybrid model outperforms the state-of-the-art baselines.

2   Related Work

Traditional supervised RE methods [6,23] mainly utilize human-designed features and require a lot of annotated data. To reduce the need of human labor, Mintz et al. [16] propose distant supervision which labels sentence with KBs. As distant supervision assumption cannot hold for all sentences, Riedel et al. [17] adopt multi-instance learning to handle the wrong label problem. To deal with entity pairs which have multiple relations, Hoffmann et al. [10] and Surdeanu et al. [19] employ multi-instance multi-label learning. These methods rely on the quality of features derived from NLP tools and suffer from error propagation.

With the significant improvement of deep learning, many neural models have been proposed. Zeng et al. [24] design piece-wise convolutional neural network to extract semantic features from sentences. To mitigate the wrong label problem, Lin et al. [13] apply attention mechanism over sentences which can de-emphasize noisy sentences. Zhou et al. [26] utilize bidirectional Long Short-Term Memory Networks to model sentences and keep useful information with word-level attention. Vashishth et al. [20] proposed RESIDE that utilizes Graph Convolutional Neural Network (GCN) over dependency parse trees to encode syntax information and employs side information from KBs as additional supervision.

Transformer proposed by Vaswani et al. [21] has shown the effectiveness of extracting semantic and syntax features in recent research [4]. Verga et al. [22] propose Bi-affine Relation Attention Networks that include a modification of Transformer network to extract relations from biological text. Alt et al. [1] utilize a pre-trained language model, the OpenAI Generative Pre-trained Transformer for long-tail relations. In our model a single entity-aware Transformer block is designed and it's architecture is shown in Fig. 1.

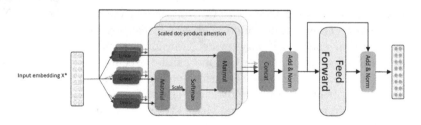

Fig. 1. The architecture of proposed entity-aware transformer.

Transfer learning can improve the performance of objective task with the transferring knowledge learned from related tasks. Liu et al. [15] initialize the model with parameters that are trained in the related task to improve the stability of the model. In this work we adopt a transfer learning strategy to initialize the parameters in entity-aware Transformer with entity typing as the related task to improve the robustness of our model reasonably. Unlike Liu et al. [15] where the entire model is pre-trained, we only pre-train the Transformer block that is suitable for entity typing task, which can save time in pte-training.

3   Methodology

In this section, we are going to introduce the framework of our methodology. The notations and problem definition of distantly supervised relation extraction will be given firstly. Afterwards, the neural relation extractor that includes PCNN and entity-aware Transformer will be described in details. Parameter transfer learning for Transformer with entity typing as related task is in the last part. Figure 2 shows the overall architecture of our model. We simplify the Transformer block for brevity and the architecture of Transformer is in Fig. 1.

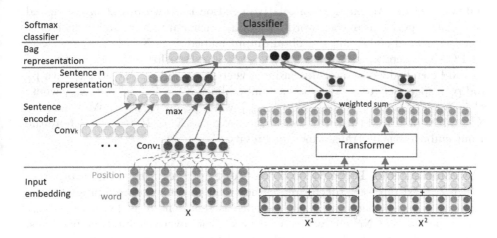

Fig. 2. The framework of our methodology.

3.1   Notations

In multi-instance learning paradigm, we split training data into multiple entity-pair bags $\pi = \{B_1, B_2, B_3, \cdots\}$. For a bag $B = \{s_1, s_2, \cdots, s_n\}$ where each sentence s_i contains the same entity pair (e_1, e_2), each sentence is denoted as a word sequence $s = \{w_1, w_2, \cdots, w_m\}$. The task of distantly supervised RE is to predict the relation between (e_1, e_2) for each bag.

3.2   Neural Relation Extractor

Input Layer of neural relation extractor aims to transform raw text into input embeddings for following layers, which make use of both semantic and positional information of words.

Word embedding [9] is used to represent each word with a low-dimensional vector. For a sentence with m words $s = \{w_1, w_2, \cdots, w_m\}$, we transform each word w_i into a real-valued vector $\mathbf{w}_i \in \mathbf{R}^{d_w}$.

Steve_Ballmer replaced Bill_Gates as CEO of Microsoft .

Fig. 3. An example for relative position.

Position embedding [25] can encode the relative distances between each word and two entities. An example for relative position is shown in Fig. 3. We embed two relative positions of each word into d_p-dimensional vectors and concatenate them with word embeddings to obtain input embeddings $\mathbf{X} = \{\mathbf{x}_1, \mathbf{x}_2, \cdots, \mathbf{x}_m\}$ for PCNN, where $\mathbf{x}_i \in \mathbf{R}^d$, $d = d_w + d_p \times 2$. For Transformer, two positions of w_i will be embedded into d_w-dimensional vectors $\mathbf{p}_i^1$ and $\mathbf{p}_i^2$. We add $\mathbf{x}_i$ with $\mathbf{p}_i^1$ and $\mathbf{p}_i^2$ respectively to get two distinct embeddings $\mathbf{X}^1 = \{\mathbf{x}_1^1, \mathbf{x}_2^1, \cdots, \mathbf{x}_m^1\}$ where $\mathbf{x}_i^1 = \mathbf{w}_i + \mathbf{p}_i^1$ and $\mathbf{X}^2 = \{\mathbf{x}_1^2, \mathbf{x}_2^2, \cdots, \mathbf{x}_m^2\}$ where $\mathbf{x}_i^2 = \mathbf{w}_i + \mathbf{p}_i^2$. We transpose $\mathbf{X}^1$ and $\mathbf{X}^2$ and feed them to Transformer. We use $\mathbf{X}^* \in \mathbf{R}^{m \times d_w}$ to denote the input embeddings of Transformer for brevity.

Convolution and Piece-Wise Max-Pooling are utilized to incorporate nearby context into each token's representation and get sentence representations with the pooling operation. Each sentence is represented as $\{\mathbf{x}_1, \mathbf{x}_2, \cdots, \mathbf{x}_m\}$ after input layer. CNN slides a convolution kernel over the input embeddings, and we use $\mathbf{CNN}(\cdot)$ to denote a convolutional layer with window size l:

$$\mathbf{c}_i = \mathbf{CNN}(\mathbf{x}_{i-\frac{l-1}{2}}, \cdots, \mathbf{x}_{i+\frac{l-1}{2}})$$

Where $\mathbf{c}_i \in \mathbf{R}^k$ is the hidden representation of word w_i and k is the number of convolution kernels. If the convolution kernel goes beyond the sentence boundaries, $\mathbf{x}_j$ will be taken to zero. Each sentence can be divided into three segments by two entities and then piece-wise max pooling operation over hidden representations will be conducted:

$$[\mathbf{s}^{(1)}]_j = \max\{[\mathbf{c}_i]_j\}, 1 \leq i \leq i_1$$
$$[\mathbf{s}^{(2)}]_j = \max\{[\mathbf{c}_i]_j\}, i_1 < i \leq i_2$$
$$[\mathbf{s}^{(3)}]_j = \max\{[\mathbf{c}_i]_j\}, i_2 < i \leq m$$

Where i_1 and i_2 are the positions of two entities and $[\cdot]_j$ is the j-th value of the vector. We stack $\mathbf{s}^{(1)}$, $\mathbf{s}^{(2)}$ and $\mathbf{s}^{(3)}$ to get sentence representation $\mathbf{s} \in \mathbf{R}^{3 \times k}$:

$$\mathbf{s} = [\mathbf{s}^{(1)}; \mathbf{s}^{(2)}; \mathbf{s}^{(3)}]$$

Entity-Aware Transformer and Inner-Sentence Attention. Entity-aware Transformer that includes Multi-head Self-Attention layer and Feed-Forward layer is capable to capture long dependency between two words without regard to their distance. The relative position embeddings can offer sequential information of the sentence that indicates the order of words and can make Transformer be aware of two entities at the same time.

Attention mechanism can be regarded as a mapping of a query and key-value pairs to an output. Attention mechanism weighs the importance of each value using the scaled dot-product between query and key and then perform a weighted sum over values. The softmax function is used to normalize the weights. For self-attention, the query $\mathbf{Q}$, key $\mathbf{K}$ and value $\mathbf{V}$ are the same input embeddings $\mathbf{X}^*$. Multi-head self-attention use separate linear projections to transform query, key and value to d_w/h dimension respectively and the projections will be performed self-attention in parallel. This process will be executed h times and h is the number of heads. The Multi-head self-attention layer is defined as follows:

$$\text{Attention}(\mathbf{Q}, \mathbf{K}, \mathbf{V}) = \text{softmax}(\frac{\mathbf{Q}\mathbf{K}^T}{\sqrt{d_w}})\mathbf{V}$$

$$\mathbf{H}_i = \text{Attention}(\mathbf{Q}\mathbf{W}_i^Q, \mathbf{K}\mathbf{W}_i^K, \mathbf{V}\mathbf{W}_i^V)$$

Where $\mathbf{W}_i^Q, \mathbf{W}_i^K, \mathbf{W}_i^V \in \mathbf{R}^{d_w \times (d_w/h)}$ are parameter matrices for i-th linear projection. The outputs of each attention head $\mathbf{H}_i$ are then concatenated to get hidden representations $\mathbf{H} \in \mathbf{R}^{m \times d_w}$. We use residual connections [7] that add the output of multi-head self-attention layer and its input. Layer normalization [2] is also employed, denoted as $\text{LN}(\cdot)$:

$$\mathbf{H}' = \text{LN}(\mathbf{H} + \mathbf{X}^*)$$

Where $\mathbf{H}' \in \mathbf{R}^{m \times d_w}$. The following component in Transformer block is the feed-forward layer which is implemented with two successive width-1 convolution layer. Meanwhile, residual connection and layer normalization are also adopted:

$$[\mathbf{T}^{(1)}]_i = \text{ReLU}(\mathbf{CNN}([\mathbf{H}']_i))$$

$$[\mathbf{T}^{(2)}]_i = \text{ReLU}(\mathbf{CNN}([\mathbf{T}^{(1)}]_i))$$

$$\mathbf{T} = \text{LN}(\mathbf{T}^{(2)} + \mathbf{H}')$$

Where $[\mathbf{T}^{(1)}]_i \in \mathbf{R}^{d_w \times 4}$, $[\mathbf{T}^{(2)}]_i \in \mathbf{R}^{d_w}$ and $\mathbf{T} \in \mathbf{R}^{m \times d_w}$. Since there are irrelevant words in long sentences we apply the following inner-sentence attention over hidden representations to keep relational features and reduce the weights of noisy words which are not related to the target relation:

$$\mathbf{a} = \mathbf{v}^{wa}\tanh(\mathbf{W}^{wa}\mathbf{T}^T)$$

$$\mathbf{s}^* = \text{softmax}(\mathbf{a})\mathbf{T}$$

Where $\mathbf{W}^{wa} \in \mathbf{R}^{d_{wa} \times d_w}$ and $\mathbf{v}^{wa} \in \mathbf{R}^{d_{wa}}$ are attention parameters.

Bag Representation and Classifier. We will achieve sentence representations $\{s_1, s_2, ..., s_n\}$ after PCNN layer and apply inter-sentence attention over s_i to make use of all informative sentences in a bag. The valid sentences which exactly express the target relation will obtain high scores while noisy sentences which are labelled wrongly will get little attention. Here is the inter-sentence attention:

$$g = \sum_{i=1} \alpha_i s_i$$

$$\alpha_i = \frac{\exp(a_i)}{\sum_k \exp(a_k)}, a_i = v^{sa} \tanh(\mathbf{W}^{sa} s_i)$$

where $\mathbf{W}^{sa} \in \mathbf{R}^{d'_{sa} \times 3k}$ and $v_{sa} \in \mathbf{R}^{d'_{sa}}$ are attention parameters. We use the same inter-sentence attention with different parameters over $\{s_1^*, s_2^*, ..., s_n^*\}$ to generate the entity-specific representations g^1 and g^2. We omit the formulas for brevity. Word embedding of two entity mentions denoted as e_1 and e_2 are also employed. We get the final bag representation g' that is defined as follows and send it to softmax classifier:

$$g' = [g; g^1; e_1; g^2; e_2]$$
$$p(r|B) = \text{softmax}(\mathbf{W}^R g' + b^R)$$

Where $\mathbf{W}^R$ is weight matrix and b^R is bias vector, r is the relation of bag B which is labelled by KB, $p \in \mathbf{R}^o$, o is the number of relation classes. Dropout [18] is utilized over g, g^1 and g^2 respevtively to prevent overfitting. The following cross entropy loss function will be minimized during the training process:

$$J(\theta) = -\frac{1}{|\pi|} \sum_{i=1}^{|\pi|} \log p(r_i|B_i)$$

3.3 Parameter Transfer Learning

As entity type is capable to lead relation extraction, entity typing is selected as the related task by transfer learning. Similar to distantly supervised RE, we predict types of two entities with a bag of sentences. Furthermore, an entity may have multiple types so we treat entity typing as a multi-label classification problem. Entity-aware Transformer and inner and inter-sentence attention are utilized for entity typing and the output is fed to the sigmoid classifier layer:

$$p^i = \sigma(\mathbf{W}^i[g^i; e_i] + b^i), i \in \{1, 2\}$$

Where σ denotes sigmoid function, $\mathbf{W}^i$ is weight matrix and b^i is bias vector, $p^i \in \mathbf{R}^t$ is the score vector of each class and t is the number of entity classes. The loss function for entity typing task is defined as follows:

$$J_e(\theta_e) = -\frac{1}{|\pi|} \sum_i^{|\pi|} (\sum_i (-\frac{1}{t} \sum_{k=1}^t y_k^i \log p_k^i)), i \in \{1, 2\}$$

Where θ_e represents parameters used in entity typing task. We train the entity-aware Transformer until convergence and keep only the parameters in multi-head self-attention layer and feed-forward layer.

4   Experiments

4.1   Dataset and Evaluation Metrics

The New York Times (NYT) dataset is proposed by Riedel et al. [17], and is widely used in the previous research [5,12,13,24]. The dataset was built by aligning Freebase [3] relations with the NYT corpus. Sentences in NYT of the years 2005–2006 are used as training set while sentences of 2007 compose testing set. This dataset has 53 relations including NA relation that means Freebase doesn't cover the relations between the entity pair. We use 38 types from the first hierarchy of FIGER [14] as the labels for entity typing. The training set has 570,088 sentences, 233,064 entity pairs and 18,252 relational facts and testing set has 172,448 sentences, 96,678 entity pairs and 1,950 relational facts. The entity mention in the sentence is treated as one token.

Following the recent works, we evaluate our model using held-out evaluation. We draw the Precision/Recall (P/R) curves for all methods and report the top-N precision (P@N) metric in our experiments.

4.2   Hyper-parameter Settings

All of the hyper-parameters in our experiments are listed in Table 1. Word embeddings used in experiments were relased by Lin et al. [13][1]. Before we train our hybrid model with pre-trained entity-aware Transformer, we also pre-train the sentence encoder of PCNN with relation extraction task.

4.3   Overall Performance

In this section, we compare the performance of our model with that of baselines to show the effectiveness of our method. Baselines we compared are listed below:

- **Mintz** [16] designs lexical and syntactic features for distant supervision RE.
- **MultiR** [10] is a probabilistic and graphic model under multi-instance learning paradigm.
- **MIML** [19] adopts multi-instance multi-label learning.
- **PCNN** [24] proposes piece-wise convolution neural network and select the most valid instance in one bag for multi-instance learning.
- **PCNN+ATT** [13] utilizes PCNN as sentence encoder and sentence-level attention in each bag to alleviate wrong label problem.
- **BGWA** [11] develops a Bi-GRU based model with word and sentence level attention.

---

[1] https://github.com/thunlp/NRE.

Table 1. Hyperparameter settings.

Parameters	Value
Word dimension d_w	50
Relative position dimension d_p	5
PCNN filter number k	230
PCNN window size l	3
Inter-sentence attention dimension d'_{sa}	300
Inner-sentence attention dimension d_{wa}	50
Inter-sentence attention dimension d_{sa}	50
Optimization strategy	SGD
Learning rate	0.2
Batch size	160
Dropout rate	0.5

- **RESIDE** [20] uses Bi-GRU to model sentences and GCN to encode syntax information, incorporating additional knowledge from KBs.

P/R curves generated by our hybrid model and various previous RE methods are reported in Fig. 4(a) to show the performance of our method. Our model is denoted as **Hybrid+TL** that means the entire hybrid model and the entity-aware Transformer is pre-trained by transfer learning.

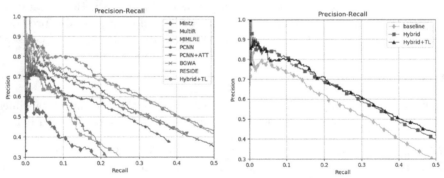

(a) Comparison of our Hybrid+TL model and various methods.

(b) Comparison between our models and a re-implemented baseline.

Fig. 4. Precision/Recall curve of our model and baselines on NYT dataset.

From the results we can see that Hybrid+TL obtains higher precision over most of recall range. Feature-based methods can achieve high precision when recall is pretty small, but with the increase of recall precision of feature-based

Table 2. AUC values and P@N(%) of our models and previous methods.

Method	AUC	P@100	P@200	P@300	Mean
Mintz	0.107	51.8	50.0	44.8	48.9
PCNN+ATT	0.341	78.2	75.1	72.8	75.4
BGWA	0.340	75.2	74.1	71.4	73.6
RESIDE	0.415	81.8	75.4	74.3	77.2
Hybrid-WE	0.412	79.0	77.5	76.3	77.6
Hybrid	0.417	**85.1**	**82.1**	77.1	**81.4**
Hybrid+TL	**0.432**	**85.1**	79.6	**79.4**	**81.4**

methods declines sharply, which indicates that error propagation caused by NLP tools will mislead the relation classifier significantly and human-designed features are limited for RE. The performance of attention-based models is better than PCNN, showing that attention mechanism can distinguish valid instances from noisy data. Furthermore, additional knowledge is helpful for relation extraction. Our model utilizes priori knowledge learned from entity typing task and achieves state-of-the-art performance.

To show the superiority of our model more specifically, we select PCNN+ATT as the baseline and re-implement it with our inter-sentence attention. The results are reported in Fig. 4(b) where **Hybrid** is the proposed model without parameter initialization by transfer learning. From the comparison we can see that the proposed methods outperform the baseline in a large margin showing the hybrid model has the ability to learn more precious sentence representations. Our model can extract not only local but also global features effectively from raw text which is more challenging for previous methods. The proposed transfer learning strategy that takes advantage of the priori knowledge can improve the performance further.

We list the Area Under Curve (AUC) values and P@N scores in Table 2. Consistent with the P/R curve, our methods achieve the best performance on AUC and P@N scores. Hybrid is more powerful than **Hybrid-WE** which excludes word embeddings of entity mentions in bag representation, showing word embeddings of entities that contain useful information can facilitate relation extraction. Hybrid can make more accurate predictions without any external knowledge than RESIDE which make use of side information such as relation alias and entity types. External knowledge can be utilized in a reasonable way with our transfer learning strategy to improve the robustness of our model and boost the performance also.

4.4   Case Study

Two real examples are presented in Table 3. The baseline (PCNN+ATT) classifies the first sentence that expresses the relational fact (*Selena_Fox*, place_lived, *Wisconsin*) and the second sentence that expresses (*Germany*, contains, *Jena*)

Table 3. Two samples from NYT test set.

Sentence	Baseline	Hybrid+TL
At least 11 families will be immediately affected by the V.A.'s decision, said the Rev. $[Selena\_Fox]_{e_1}$, senior minister of Circle Sanctuary, a Wiccan church in $[Wisconsin]_{e_2}$	NA (0.640)	place_lived (0.682)
In a recent ranking of the most competitive and economically promising regions of $[Germany]_{e_1}$, 3 of the top 20 cities are in the east: Dresden, Potsdam and $[Jena]_{e_2}$	NA (0.984)	contains (0.726)

to NA relation with the score of 0.640 and 0.984 respectively. The proposed model is able to predict the correct relations with high scores although two entities are far away in both samples. This case study shows that our hybrid model can learn long-range dependencies more effectively and consider syntax information located in sentences.

5   Conclusion

In this paper, we propose a neural hybrid model by combining piece-wise convolutional neural network and entity-aware Transformer for distantly supervised relation extraction. Our methods take advantage of CNN and Transformer to extract local and global semantic features simultaneously and can execute in parallel. For Transformer, we utilize relative positions to make model focus on entities and provide the information of words order. Inner-sentence attention are designed to alleviate the impact caused by noisy words. Word embeddings of entities containing important features of entities can also benefit relation extraction. A transfer learning strategy is proposed to leverage priori knowledge learned from the related task and improve the robustness of our model. We select entity typing as the related task due to the observation that entity types offer soft constraints of relations which can guide relation extraction reasonably. The experimental results on a well-studied benchmark dataset validate the effectiveness of our methods as our model significantly outperforms the state-of-the-art methods. For future, we plan to modify the proposed approach with multi-task learning and explore the joint extraction of entities and relations.

Acknowledgement. This work is partially supported by the NSFC Project (No. 61672012).

References

1. Alt, C., Hübner, M., Hennig, L.: Fine-tuning pre-trained transformer language models to distantly supervised relation extraction. arXiv preprint arXiv:1906.08646 (2019)
2. Ba, J.L., Kiros, J.R., Hinton, G.E.: Layer normalization. arXiv preprint arXiv:1607.06450 (2016)
3. Bollacker, K., Evans, C., Paritosh, P., Sturge, T., Taylor, J.: Freebase: a collaboratively created graph database for structuring human knowledge. In: Proceedings of the 2008 ACM SIGMOD International Conference on Management of Data, pp. 1247–1250 (2008)
4. Devlin, J., Chang, M.W., Lee, K., Toutanova, K.: BERT: pre-training of deep bidirectional transformers for language understanding. arXiv preprint arXiv:1810.04805 (2018)
5. Feng, J., Huang, M., Zhao, L., Yang, Y., Zhu, X.: Reinforcement learning for relation classification from noisy data. In: Thirty-Second AAAI Conference on Artificial Intelligence (2018)
6. GuoDong, Z., Jian, S., Jie, Z., Min, Z.: Exploring various knowledge in relation extraction. In: Proceedings of the 43rd Annual Meeting on association for Computational Linguistics, pp. 427–434. Association for Computational Linguistics (2005)
7. He, K., Zhang, X., Ren, S., Sun, J.: Deep residual learning for image recognition. In: Proceedings of the IEEE Conference on Computer Vision and Pattern Recognition, pp. 770–778 (2016)
8. He, Z., Chen, W., Li, Z., Zhang, M., Zhang, W., Zhang, M.: SEE: syntax-aware entity embedding for neural relation extraction. In: Thirty-Second AAAI Conference on Artificial Intelligence (2018)
9. Hinton, G.E., et al.: Learning distributed representations of concepts. In: Proceedings of the Eighth Annual Conference of the Cognitive Science Society, Amherst, MA, vol. 1, p. 12 (1986)
10. Hoffmann, R., Zhang, C., Ling, X., Zettlemoyer, L., Weld, D.S.: Knowledge-based weak supervision for information extraction of overlapping relations. In: ACL, pp. 541–550 (2011)
11. Jat, S., Khandelwal, S., Talukdar, P.: Improving distantly supervised relation extraction using word and entity based attention. arXiv preprint arXiv:1804.06987 (2018)
12. Ji, G., Liu, K., He, S., Zhao, J.: Distant supervision for relation extraction with sentence-level attention and entity descriptions. In: Thirty-First AAAI Conference on Artificial Intelligence (2017)
13. Lin, Y., Shen, S., Liu, Z., Luan, H., Sun, M.: Neural relation extraction with selective attention over instances. In: ACL, vol. 1, pp. 2124–2133 (2016)
14. Ling, X., Weld, D.S.: Fine-grained entity recognition. In: Twenty-Sixth AAAI Conference on Artificial Intelligence (2012)
15. Liu, T., Zhang, X., Zhou, W., Jia, W.: Neural relation extraction via inner-sentence noise reduction and transfer learning. arXiv preprint arXiv:1808.06738 (2018)
16. Mintz, M., Bills, S., Snow, R., Jurafsky, D.: Distant supervision for relation extraction without labeled data. In: Proceedings of the Joint Conference of the 47th Annual Meeting of the ACL and the 4th International Joint Conference on Natural Language Processing of the AFNLP: Volume 2, vol. 2, pp. 1003–1011 (2009)

17. Riedel, S., Yao, L., McCallum, A.: Modeling relations and their mentions without labeled text. In: Balcázar, J.L., Bonchi, F., Gionis, A., Sebag, M. (eds.) ECML PKDD 2010. LNCS (LNAI), vol. 6323, pp. 148–163. Springer, Heidelberg (2010). https://doi.org/10.1007/978-3-642-15939-8_10

18. Srivastava, N., Hinton, G., Krizhevsky, A., Sutskever, I., Salakhutdinov, R.: Dropout: a simple way to prevent neural networks from overfitting. J. Mach. Learn. Res. 15(1), 1929–1958 (2014)

19. Surdeanu, M., Tibshirani, J., Nallapati, R., Manning, C.D.: Multi-instance multi-label learning for relation extraction. In: Proceedings of the 2012 Joint Conference on Empirical Methods in Natural Language Processing and Computational Natural Language Learning, pp. 455–465. Association for Computational Linguistics (2012)

20. Vashishth, S., Joshi, R., Prayaga, S.S., Bhattacharyya, C., Talukdar, P.: RESIDE: improving distantly-supervised neural relation extraction using side information. arXiv preprint arXiv:1812.04361 (2018)

21. Vaswani, A., et al.: Attention is all you need. In: Advances in Neural Information Processing Systems, pp. 5998–6008 (2017)

22. Verga, P., Strubell, E., McCallum, A.: Simultaneously self-attending to all mentions for full-abstract biological relation extraction. arXiv preprint arXiv:1802.10569 (2018)

23. Zelenko, D., Aone, C., Richardella, A.: Kernel methods for relation extraction. J. Mach. Learn. Res. 3(Feb), 1083–1106 (2003)

24. Zeng, D., Liu, K., Chen, Y., Zhao, J.: Distant supervision for relation extraction via piecewise convolutional neural networks. In: EMNLP, pp. 1753–1762 (2015)

25. Zeng, D., Liu, K., Lai, S., Zhou, G., Zhao, J., et al.: Relation classification via convolutional deep neural network (2014)

26. Zhou, P., et al.: Attention-based bidirectional long short-term memory networks for relation classification. In: Proceedings of the 54th Annual Meeting of the Association for Computational Linguistics (Volume 2: Short Papers), pp. 207–212 (2016)

NovEA: A Novel Model of Entity Alignment Using Attribute Triples and Relation Triples

Tao Sun$^{(\boxtimes)}$, Jiaojiao Zhai$^{(\boxtimes)}$, and Qi Wang

Qilu University of Technology (Shandong Academy of Sciences), Jinan 250353, China
suntao0906@163.com, qlu717@126.com, yining1104@sina.com

Abstract. Entity alignment is the foundation of knowledge fusion, which can find the alignment relationships between entities in heterogeneous knowledge graphs. However, traditional methods rely on external information and need to construct data features manually. Meanwhile, embedding models do not fully utilize the pertinent information of attributes in the knowledge graphs, which limit the role of attribute information in entity alignment. Considering the shortcomings of existing methods, this paper proposes a novel model named NovEA that using attribute triples and entity triples in the knowledge graphs to complete the entity alignment task together. Besides, for attribute triples, we propose a method that can automatically generate the optimal attribute according to the data characteristics to constrain the result of attribute triples alignment and improve the accuracy of entities in alignment. Finally, we use a binary regression method to measure the similarity of the combination results of structure and attribute. Our research on real datasets shows that the NovEA model has a significant improvement in entity alignment compared with the most advanced methods.

Keywords: Knowledge graph · Entity alignment · Relation triple · Attribute triple

1 Introduction

Knowledge Graphs (KG) are used in many fields at present, such as Entertainment, Geography [1], Industry [2]. They play an important role in Information Retrieval, Recommendation System, Machine Understanding and Question Answering System. For knowledge graphs, different knowledge is stored between different knowledge graphs in the same field, and there are numerous repetitions of this knowledge, which can also complement each other. Therefore, we can integrate such a knowledge graph to form a more unified knowledge graph. To integrate these knowledge graphs, a basic problem is to find out the entities that exist in different knowledge graphs but represent the same meaning, which is usually called entity alignment. Due to the diversity of expression and structure of knowledge in different knowledge graphs, it presents considerable challenges.

In many knowledge graphs, the Resource Description Framework (RDF) [16] has been widely regarded as a flexible data model representing a large knowledge base. For

G. Li et al. (Eds.): KSEM 2020, LNAI 12274, pp. 161–173, 2020.
https://doi.org/10.1007/978-3-030-55130-8_14

large-scale RDF [17] graphs, achieving efficient and scalable query processing becomes critical. Knowledge usually exists in the form of the RDF triples. We divide it into relation triples and attribute triples. They are independent in knowledge graphs, but describe the structure information and attribute information of the entities respectively. Traditional research of entity alignment is based on the similarity between entity attributes. These methods rely on the external information of the entities and need to build a large number of features and design matching rules manually. Later, the embedding-models are proposed for the task of entity alignment, which are based on the representation-learning method. These models do not depend on the content information of the knowledge graph and they are not responsive to the dataset with sparse relational triples. Recently, some researches use attribute information of entities to align entities. JAPE [8] jointly embeds the structure of two KGs into a unified vector space, and further refines it by using the attribute correlation between KGs. However, it does not make full use of the attribute value information, but simplifies the attribute value to datatype, such as (*Zhang Ziyi, Birthdate, 1979-02-09*) to (*Zhang Ziyi, Birthdate, Datetime*). AttrE [9] proposes to combine the structural information and attribute information of entities based on embedding model, and uses a large number of attribute triples in the knowledge graphs to generate attribute feature embedding, and finally realizes entity alignment. Although the model considers attribute values, it does not find that unique entities need different attributes to compare, and the same attribute has a different importance to different entities. For example, time attribute is an important attribute for the film knowledge graph, but it may not be so essential for crop knowledge graph. Therefore, we need to select the optimal attribute for entity alignment according to the attribute types and the importance of attribute values contained in different datasets. In many triples, entity relations are not always of high quality (such as small quantity and incomplete relations), which may damage the accuracy of entity alignment results.

Because of the shortcomings of traditional methods and the embedding models for entity alignment, this paper proposes a novel model called NovEA to align entities from the perspective of structure and attribute respectively by using the relationship triples and attribute triples in the knowledge graphs. According to the characteristics of the knowledge graph dataset, the NovEA model sorts the attributes according to the priority to get the optimal attributes and then constrains the results of attribute alignment. Finally, we use the binary regression to measure the similarity results from the structure and the attribute. It can dynamically adjust the weight according to the number of relation triples and attribute triples in the knowledge graphs and improve the effect of entity alignment better. Specifically, we first align and rename the triples in the two knowledge graphs through predicates, so that the entities and relationships can be in the same vector space, and then embed the relationship and attribute of the relationship triples and attribute triples respectively. When the attribute triples are aligned, we first align the entities with the values of the common attributes, and then proceed according to the matched entities Row attributes are aligned, and attribute types are prioritized according to the domain characteristics of the knowledge graphs. Finally, the alignment results are weighted according to the priority of attributes. We conducted experiments on real-world datasets and proved to be significantly improved over other methods. To sum up, our contributions in this paper are as follows:

- Given insufficient utilization of attribute triples, the NovEA model is proposed by using attribute triples and relation triples at the same time. When merging alignment results, we use binary regression to dynamically fit the similarity weight between relationships and attributes.
- Attributes alignment and entities alignment are iteratively carried out when attribute triples are aligned, and the selected attribute alignment results are weighted according to the domain characteristics of the knowledge graphs.
- Experimented on real-world datasets and proved a significant improvement over other models.

The rest of this paper is organized as follows. Section 2 summarizes the related works. Section 3 presents the details of the NovEA model. Section 4 presents the experimental process and results. In Sect. 5, we conclude the paper.

2 Related Work

Entity alignment methods are divided into two parts: the traditional entity alignment methods and the embedding models for entity alignment.

2.1 The Traditional Entity Alignment Method

The traditional entity alignment method is mainly used for supervised machine learning models, attribute similarity matching ways to align the entity. Scharffe et al. [10] proposed an entity alignment model based on sequence alignment fuzzy string matching, word relationship, and classification similarity. Volz et al. [11] proposed to allow users to define rules using standardized grammar, including string similarity, numeric similarity, date similarity to achieve entity alignment. Niu et al. [12] used the expectation-maximization algorithm to optimize artificial defined entity matching rules. Although the traditional alignment methods using the entity attribute information, they need to design different attributes to different categories of entities similarity calculation function, it would cost a lot of manpower and increase the workload. The expressions of attributes are discrete, ignoring the properties of semantic similarity and limiting the effect of entity aligned.

2.2 The Embedding Models for Entity Alignment

In recent years, knowledge graph embedding models has gradually become the mainstream. They are mainly based on the method of presentation learning. TransE [3] represents each triple (h, r, t) as a vector from head entity h to tail entity t, that is $h + r = t$. By learning the vector representation of entities and relationships in knowledge graphs, the semantic similarity between entities can be obtained according to the relationship triples, but the disadvantage of this method is that it can only model one-to-one simple relationships and can not deal with multiple complex relationships. Then there are a series of improved models such as TransR [4], TransH [5], TransA [6], etc. The SEEA [7] treats entity alignment as a special cross-network relationship, achieving entity alignment through self-learning. Sun et al. [15] proposed an iterative entity alignment method based on bootstrapping to transform the entity alignment problem into a classification problem. Since the learned embedding has the highest likelihood of entity alignment, the limit-based objective function is used to make the trained embedding more discriminative.

However, most of these methods only focus on how to encode relational triples in a better way and ignore attribute triples. Especially for the lack of relationships between entities, if only align relation triples entity, the effect is not enough good. To make use of attribute triples, Sun et al. [8] proposes a cross-language entity alignment JAPE embedded joint properties to maintain alignment model, it uses two models of structure embedding and attributes embedding to learn the embedding of KGs. Structure embedding is to model the relationship structure of KGs from the perspective of relationship triples. Attribute embedding attempts to cluster the attributes that are commonly used to describe entities. Finally, we embed all entities into two KGs for a uniform vector space. However, JAPE [8] ignores the attribute values of attribute triples and does not make full use of attribute information. AttrE [9] model proposes to combine the structural information and attribute information of entities based on the embedding model, and uses a large number of attribute triples in knowledge graph to generate attribute feature embedding and finally realize entity alignment. It uses the frequency ratio of relationships and attributes as the weight of entity alignment, but lacks the ability to capture the fact that entities cannot sort multiple attribute types based on different datasets during alignment. IEAJKE [14] algorithm does not consider the important role of entity semantic integration and attribute weight in entity alignment, so the experimental results need to be improved.

3   The NovEA Model

3.1   The NovEA Model Overview

As shown in Fig. 1, this paper proposes a novel model called the NovEA, which uses both the relationship triples and attribute triples in the knowledge graphs for entity alignment. First, the predicates are aligned and renamed uniformly so that the relationships and entities can be embedded in the same vector space. Then use the relationship triples and attribute triples to embed the structure and attribute respectively, when aligning attribute triples, we first use the value of the common attribute to align the entity, and then align the attribute according to the matching entity. At this time, the attribute types are prioritized according to the domain characteristics of the knowledge graph, and the similarity result of the attribute triples alignment is weighted according to the priority of the attribute. Finally, a binary regression algorithm is used to combine the relation triples and attribute triples alignment results to dynamically learn their respective weights, which improves the accuracy of the entity alignment process. Next, we will introduce our method in detail.

3.2   Predicate Alignment

In order to embed entities and relationships into the same vector space during the subsequent structural embedding process, this section first use the unified naming method to

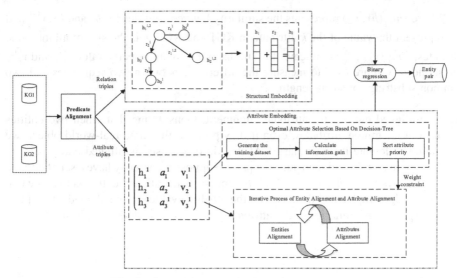

Fig. 1. Overall framework of the NovEA model.

merge the two knowledge graphs through predicates. Here we use the method proposed by AttrE [8] to rename the predicates.

3.3 Attribute Embedding

This section mainly consists of the following two parts including the Iterative Process of Entity Alignment and Attribute Alignment and Optimal Attribute Selection Based On Decision-Tree. The latter can impose weight constraints on the former iteration process according to the characteristics of the dataset, which can improve the speed and quality of the attribute alignment process.

1. **The Iterative Process of Entity Alignment and Attribute Alignment:** We determine the percentage of common attributes and attribute values between attribute triples of two entities. According to this percentage, we can measure the similarity between entities.
(1) When using attribute values to align entities, we believe that all common attributes of two entities have the same weight. We use the Eq. 1 and Eq. 2 to calculate the similarity between two entities:

$$sim_A(h_1, h_2) = \frac{1}{n} \sum_{k}^{n} sim_V(v_k^{G_1}, v_k^{G_2}) \tag{1}$$

$$sim_V(v_k^{G_1}, v_k^{G_2}) = \frac{lcesim(v_k^{G_1}, v_k^{G_2})}{leven(v_k^{G_1}, v_k^{G_2}) + lcesim(v_k^{G_1}, v_k^{G_2})} \tag{2}$$

Where $sim_A(h_1, h_2)$ represents the similarity between entities for h_1 and h_2, v_k^{G1} and v_k^{G2} represent the value of the k attribute in KG1 and KG2, n is the size of all the same attributes, $sim_V(v_k^{G1}, v_k^{G2})$ indicates the similarity between property value v_k^{G1} and v_k^{G2}, *Leven* and *lcssim* are used to measure the difference between two sequences and two common substrings of string length.

(2) Use aligned entities for attribute alignment. Considering that when two entities have no common attributes, even if they point to the same real-world object, we cannot calculate their similarity. At this time, we use the aligned entity pairs to find more possible aligned attribute pairs. Suppose we already have a set of aligned entity pairs $T_h = (h_1, h_2 \ldots \ldots h_n)$, h_i represents the aligned entity pairs in the two knowledge graphs, and then uses all the attribute values of the aligned entity pairs to express the similarity of the attribute pairs, and calculates the similarity using formula 3.

$$sim_A(a_i^{G1}, a_j^{G2}) = \frac{1}{c} \sum_{m=1}^{c} sim_v(v_i^m, v_j^m) \qquad (3)$$

Where sim_v represents the similarity of attribute values, c represents the size of aligned entities. We illustrate the above process with Example 1.

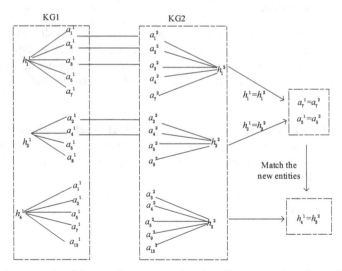

Fig. 2. An example of the iterative process of entity alignment and attribute alignment.

Example 1: Suppose we know that $a_m^{G1} = a_m^{G2}$ (it is expressed as $a_m^1 = a_m^2$), m is a natural number and $m \in [1, 5]$. We do not know the rest of unaligned attribute information. KG1 and KG2 are known to have entities h_1^1, h_2^1, h_4^1 and h_1^2, h_2^2, h_3^2 respectively from

Fig. 2. Except for the entity pair (h_4^1, h_3^2), the remaining entity pairs all have at least one pair of common attributes. By calculating the similarity of these common attributes, we can obtain two aligned entity pairs (h_1^1, h_1^2) and (h_2^1, h_2^2). But we don't know (h_4^1, h_3^2) if it's aligned. The alignment attributes (a_7^1, a_7^2) and (a_5^1, a_5^2) are obtained from the alignment entity pair (h_1^1, h_1^2) and (h_2^1, h_2^2) respectively. At this point, we use the result of the above attribute alignment information based on the aligned entity. Then carry out the iterative process of entity alignment, we use the alignment attribute to align the new entity and get the new alignment pair (h_4^1, h_3^2).

2. **Optimal Attribute Selection Based on Decision-Tree:** We choose the method based on decision-tree to select the optimal attribute, so as to constrain the attribute alignment process in the iterative process of attribute alignment and attribute alignment of attribute triples. First, we construct a positive triple from the aligned entity set M in the training set $T_p = (h, R, v)$. Then randomly replace the head and tail entities of the positive triples to get the negative triple $T_n = (h', R, v')$, we labeled the positive and negative triples, the positive triple is labeled 1, the negative triple is labeled 0, and the collection $T_R = T_p \cup T_n$ is constructed. We choose attribute similarity as the candidate values to select the optimal attribute. The attribute value of the entity in the triple is often not unique. If it is used directly, the attribute information will be scattered too much, which reduces the accuracy of the optimal attribute selection. Therefore, for single-valued attributes, we directly use the absolute value of the difference between the attribute values of the T_R head and tail entities as the feature value, and for attributes that contain multiple attribute values, the Jaccard Coefficients of the corresponding attribute values of the T_R head and tail entities are used as the feature values, as shown in formula 4:

$$M = (M_{i,j}), \qquad 1 \le i \le |T_R|$$

$$M_{i,j} = \frac{|V(h_i, a_j) \cap V(t_i, a_j)|}{|V(h_i, a_j) \cup V(t_i, a_j)|} \quad a_j \text{ are multi-valued attributes} \qquad (4)$$

$$|v(h_i, a_j) \cap v(t_i, a_j)| \quad a_j \text{ is single value attribute}$$

Where $M_{i,j}$ represents the value in column i-th and row j-th of M, $v(h_i, a_j)$ and $v(t_i, a_j)$ represents the j-th attribute value set of the head and tail entity of the i-th triple in T_R, $v(h_i, a_j)$ and $v(t_i, a_j)$ represent the j-th attribute value of the head-tail entity of the i-th triple in T_R.

Then we use the Information-Gain to calculate the purity of the information. The greater the information gain, the higher the purity of the dataset divided by this attribute and the higher the priority of this attribute. We calculate the information gain of each attribute of each candidate entity and output the information gain in ascending order. Equation 5 is as follows:

$$Gain(M, a_i) = \max_{t \in T_{a_i}} Gain(D, a_i, t) = \max_{t \in T_{a_i}} \left(Ent(M) - \sum \frac{|M_t|}{|M|} Ent(M_t) \right) \qquad (5)$$

Where $Gain(M, a_i)$ represents the information gain value of the attribute type a_i in the aligned dataset M, the value is larger, attribute classification effect of a_i is better, $Ent(M)$ represents information entropy, p_k represents the proportion of type K samples in M, Eq. 6 is expressed as below:

$$Ent(M) = -\sum_{k=1}^{K} p_k \log_2^{p_k} \tag{6}$$

The optimal attribute selection based on Decision-Tree is to select attributes with strong classification ability from attribute triples. We use the first n attributes with the largest information gain selected by the best attributes as the constraint attributes of the entity alignment task, which are used in the subsequent embedding process.

3.4 Structural Embedding

We use triples aligned with predicates, using relationship triples and training sets for structural embedding, and learn vector representations of entities and relationships, given relation triple $tr = (h, r, t)$, h is the head entity and t is the tail entity, we expect $h + r = t$. To measure the rationality of tr, structural embedding model optimizes margin-based ranking loss [13], making the positive triples score lower than the negative triples.

$$O_{SE} = \sum_{tr \in Tr} \sum_{tr' \in Tr'} (f(tr) - \alpha(tr')) \tag{7}$$

Where $f(tr) = \|h + r - t\|_2^2$ is the score function, Tr represents all positive triples, Tr' represents the set of related negative triples generated by replacing its head or tail with random entities (but not with both entities), therefore we can learn the approximate vector representation of entities on KGs, and the entity similarity measure after structure embedding is shown in Eq. 8.

$$Sim_{SE}(h_i^{G_1}, h_j^{G_2}) = \cos(h_{i,}^{G_1} h_j^{G_2}) \tag{8}$$

3.5 Entity Alignment

Through the Decision-tree based optimal attribute selection process, the first n attributes with the largest information gain are selected as the optimal constraint attributes of attribute triple alignment, the final attribute embedding similarity is obtained w_i represents the constraint weight of attribute a_i, the value range is $(0, 1)$. The priority attribute weight is higher, the corresponding W value is higher.

$$sim_{AE} = \sum_{i=1}^{n} w_i sim_A \tag{9}$$

Finally, we comprehensively measure the final similarity of the two entities in terms of relationship and attributes. Linearly weight the similarity calculated by the relationship triples and attribute triples. We use Eq. 10 to express the final similarity of the entity pair $h_i \in G_1$ and $h_j \in G_2$:

$$sim(h_i^{G_1}, h_j^{G_2}) = \beta Sim_{SE}(h_i^{G_1}, h_j^{G_2}) + (1 - \beta)sim_{AE}(h_i^{G_1}, h_j^{G_2}) \tag{10}$$

Where sim_{SE} is the similarity calculated using relational triples, sim_{AE} represents the entity similarity calculated by the attribute triple selected by the optimal attribute, β is dynamically selected by the number of relationships and attributes in the datasets to maintain balance. For example, some datasets have a large proportion of relationship triplets and more emphasis on the relationship of entities, so attributes are given higher weight. Some datasets have a large proportion of attribute triples, indicating that more emphasis is placed on the attributes of the entity, so the relationship is given a higher weight.

4 Experiment

This section describes our dataset, evaluation indicators and comparison methods, parameter setting and experimental results.

4.1 Datasets

We evaluated our method on two real knowledge graph datasets, namely IMDB-YAGO and DBP-YAGO. They are the physical alignment of IMDB and YAGO and DBP and YAGO respectively, and the relevant datasets information can be found at http://web dam.inria.fr/paris/. The overall statistics of dataset are shown in Table 1.

Table 1. Statistics for Datasets

Dataset		Entities	Attribute triples	Relations triples
IMDB-YAGO	IMDB	3661	11254	23415
	YAGO	3670	11624	28451
DBP-YAGO	DBP	33627	184672	36906
	YAGO	30628	173309	38451

4.2 Evaluation Indicators and Comparison Methods

Indicators: In order to evaluate the performance of the method, we use Hits @ K and Mean Rank (MR) to evaluate the performance of the method. Hits @ K reflects the proportion of correctly aligned entities in the top K. Higher Hits @ K and lower Mean Rank indicate better performance. In addition, we also use precision (P), recall rate (R), and F1 value to measure our optimal attribute selection process based on Decision-Tree. F1 is equal to Eq. 11.

$$F_1 = 2 \cdot \frac{P \cdot R}{P + R} \tag{11}$$

Where P represents the precision, R represents the recall rate.

Comparison Methods: Compare our method with TransE [3], JAPE [8], and AttrE [9]. The JAPE [8] model jointly embeds the structure of two KGs into a unified vector space, and further refines it using the correlation of attribute in KGs. The AttrE [9] model uses a large number of attribute triples existing in the knowledge graph to generate attribute character embedding. Attribute character embedding transfers the entity embedding from two knowledge graphs to the same space by calculating the attribute-based similarity between entities. We do not use transitivity rules here.

4.3 Parameter Setting

In the experiment, we set the NovEA model parameters as follows: The value range of the margin value γ is among {0.5, 1.0, 1.5}, the value range of the dimension d of the entity or relationship vector is {50 100 150 200}, the range of the learning rate λ is {0.001, 0.005, 0.01}, and the range of training times n is {500 1000 2000 3000}. Through a grid search, the experimental results show that the optimal parameters are: γ = 1.0, λ = 0.001, n = 3000.

4.4 Experimental Results

First, we give the experimental process of optimal attribute selection based on Decision-tree, the optimal attribute types of IMDB-YAGO and DBP-YAGO are (*title, year, author*) and (*name, time, location*) respectively, and the corresponding weight parameters are (1, 0.75, 0.5) and (1, 0.61, 0.35).

Table 2. Experimental results of optimal attribute selection based on Decision-Tree

Method	IMDB-YAGO			DBP-YAGO		
	P	R	F	P	R	F
TransE	87.06	63.79	73.63	92.97	88.58	93.49
JAPE	90.01	75.58	82.17	95.21	90.23	92.65
AttrE	85.03	**82.62**	84.67	95.98	90.55	93.18
NovEA	**94.39**	80.26	**86.79**	**97.25**	**94.45**	**95.82**

Precision(P), Recall rate(R), F1. As it is known from Table 2, P and R of our model are improved by 4.3% and 4% respectively compared to JAPE [8]. This is because this model makes the attribute embedding of the entity more accurate by selecting the optimal attribute type. Compared with the TransE [3] model, it is improved by 7.3% and 5%. because the TransE [3] model only aligns the structural information of the triples, ignoring the effect of attributes on entity alignment. The experimental results show that our research method has improved the accuracy rate P and F1 value, indicating that our model can improve the effect of entity alignment. In addition, the experimental results on two different types of datasets also show that our method can automatically select the corresponding optimal attributes according to the characteristics of the datasets.

Top-K and Mean Results. As shown in Table 3, we use Hits @ K and Mean Rank to measure the performance of our model. It can be seen from the experimental results that TransE [3] does not perform well on the two datasets, because it embeds knowledge graphs in different domains into different vector spaces, and cannot capture the entity similarity between knowledge graphs. JAPE [8] and AttrE [9] are better than TransE [3], because they not only rely on entities and relationships for alignment, but also take into account the role of attributes. However, JAPE simply reduces the value of the attribute triples to the attribute type, and does not really use the attribute value. Our model not only embeds relationships and entities in the same space, but also consider the role of attribute values in attribute triples. The experimental results show that our model can improve the effect of entity alignment.

In addition, the results of the model on Hits@1 are slightly different on the two datasets. This is because the number of relationship triplets and attribute triplets are different between the two datasets. IMDB-YAGO averages 1 relationship triplet for 2 attribute triplets, which is almost the same, while DBP-YAGO averages 6 relationships. The triples and one attribute triple, so it is better to only use the relationship to its entity performance than to use the attribute and relationship at the same time.

Table 3. Experimental results were compared by Hits@ K and Mean Rank

Method	DBP-YAGO			IMDB-YAGO		
	Hits@1	Hits@10	MR	Hits@1	Hits@10	MR
TransE	3.21	8.07	19331	2.36	6.29	21004
JAPE	54.63	52.98	8065	51.04	53.14	9234
AttrE	80.23	90.21	725	82.62	90.65	663
NovEA	**87.36**	**93.65**	**104**	**88.65**	**94.35**	**94**

Dynamic Combination. We use binary regression to dynamically learn the weights of similarity measures from both aspects of relationship and attributes. At the same time, we give two combined results with the same similarity weights (static combination). As shown in Table 4, the experimental results show that dynamic combination does perform better than static combination, because the learning weights contain information about the different importance of entity relationships and attributes when performing entity alignment.

Table 4. Comparison of static and dynamic combinations

	Static combination	Dynamic combination
DBP-YAGO	85.67	**87.36**
IMDB-YAGO	87.11	**88.65**

5 Conclusion

The NovEA model is using the relation triples and attribute triples in the knowledge graph to align the entities from the perspective of structure and attribute respectively. According to the characteristics of the knowledge graph dataset, we get the optimal attributes based on the method of Decision-Tree and then constrain the result of attribute embedding. Finally, we use the binary regression model to dynamically get the optimal attributes from the perspective of structure and attribute to measure the similarity results. We have conducted experiments on real-world datasets and proved that it is significantly improved than other methods. But the NovEA model depends on the aligned entity data. In actual datasets, it is usually difficult to find enough aligned entities. The next work will consider unsupervised algorithms for entity alignment.

Acknowledgments. This work was supported in part by Shandong Provincial Natural Science Foundation, China (No. ZR2017LF019).

References

1. Kai, S., Yun, Z., Jia, S.: Progress and challenges on entity alignment of geographic knowledge bases. ISPRS Int. J. Geo-Inf. **8**, 77 (2019)
2. Sun, T., Wang, Q.: Multi-source fault detection and diagnosis based on multi-level knowledge graph and Bayesian theory reasoning. In: 31th International Conference on Software Engineering and Knowledge Engineering, pp. 177–180. KSI, Pittsburgh (2019)
3. Bordes, A., Usunier, N., Garcia-Duran, A., Weston, J., Yakhnenko, O.: Translating embeddings for modeling multi-relational data. In: Advances in Neural Information Processing Systems, pp. 2787–2795 (2013)
4. Wang, Z., Zhang, J., Feng, J., Chen, Z.: Knowledge graph embedding by translating on hyperplanes. In: 28th AAAI Conference on Artificial Intelligence. AAAI Press (2014)
5. Lin, Y., Liu, Z., Sun, M., Liu, Y., Zhu, X.: Learning entity and relation embeddings for knowledge graph completion. In: AAAI, vol. 15, pp. 2181–2187 (2015)

6. Xiao, H., Huang, M., Hao, Y., Zhu, X.: TransA: an adaptive approach for knowledge graph embedding. Comput. Sci. (2015)
7. Guan, S., et al.: Self-learning and embedding based entity alignment. Knowl. Inf. Syst. **59**(2), 361–386 (2018). https://doi.org/10.1007/s10115-018-1191-0
8. Sun, Z., Hu, W., Li, C.: Cross-lingual entity alignment via joint attribute-preserving embedding. In: d'Amato, C., et al. (eds.) ISWC 2017. LNCS, vol. 10587, pp. 628–644. Springer, Cham (2017). https://doi.org/10.1007/978-3-319-68288-4_37
9. Trsedya, B.D., Qi, J., Zhang, R.: Entity alignment between knowledge graphs using attribute embeddings (2019)
10. Scharffe, F., Liu, Y., Zhou, C.: RDF-AI: an architecture for RDF datasets matching, fusion and interlink (2009)
11. Volz, J., Bizer, C., Gaedke, M., Kobilarov, G.: Discovering and maintaining links on the web of data. In: Bernstein, A., Karger, D.R., Heath, T., Feigenbaum, L., Maynard, D., Motta, E., Thirunarayan, K. (eds.) ISWC 2009. LNCS, vol. 5823, pp. 650–665. Springer, Heidelberg (2009). https://doi.org/10.1007/978-3-642-04930-9_41
12. Raimond, Y., Sutton, C., Sandler, M.: Automatic interlinking of music datasets on the semantic web. LDOW (2008)
13. He, F., et al.: Unsupervised entity alignment using attribute triples and relation triples. In: Li, G., Yang, J., Gama, J., Natwichai, J., Tong, Y. (eds.) DASFAA 2019. LNCS, vol. 11446, pp. 367–382. Springer, Cham (2019). https://doi.org/10.1007/978-3-030-18576-3_22
14. Zhu, H., Xie, R., Liu, Z., Sun, M.: Iterative entity alignment via joint knowledge embeddings. In: 31th AAAI Conference on Artificial Intelligence. AAAI (2017)
15. Sun, Z., Hu, W., Qu, Y.: Bootstrapping entity alignment with knowledge graph embedding. In: 27th International Joint Conference on Artificial Intelligence, IJCAI, Stockholm, Sweden, pp. 4396–440 (2018)
16. Wang, X., Wang, S., Xin, Y., Yang, Y., Li, J., Wang, X.: Distributed pregel-based provenance-aware regular path query processing on RDF knowledge graphs. World Wide Web **23**, 1–32 (2019). https://doi.org/10.1007/s11280-019-00739-0
17. Xu, Q., Wang, X., Li, J., Zhang, Q., Chai, L.: Distributed subgraph matching on big knowledge graphs using pregel. IEEE Access **7**, 116453–116464 (2019)

A Robust Representation with Pre-trained Start and End Characters Vectors for Noisy Word Recognition

Chao Liu[1,2], Xiangmei Ma[1,2], Min Yu[1,2(✉)], Xinghua Wu[1,2], Mingqi Liu[1], Jianguo Jiang[1], and Weiqing Huang[1]

[1] Institute of Information Engineering, Chinese Academy of Sciences, Beijing, China
`yumin@iie.ac.cn`
[2] School of Cyber Security, University of Chinese Academy of Sciences, Beijing, China

Abstract. Powered by the advanced neural network, many tasks in the field of natural language processing could be completed by the network models. As the noise in the input text can affect negatively on the performance of these model, researchers are gradually paying more attention to the word recognition, which is placed before the downstream task to accomplish those tasks better. Text noise, in terms of words, usually includes random insertion, deletion, swapping, or keyboard errors. They belongs to the category of *Out-of-vocabulary* (OOV) in the word-level language model. Using a vector to represent all OOV words is the most common technique. However, such a representation may cause information loss as it overlooks the meaning of the word. In this paper, we propose a reasonable and effective representation method for the noisy words identifying, based on the semantic correlation and dependency of words. When modeling character level dependency, we imitate the process of human recognizing noisy words, paying more attention to the start and end characters while ignoring the order of internal characters. To get a better embedding representation of noisy words, we train a neural network to predict the start and end characters, and then combine the predicted start and end character vectors with other character vectors into the whole word representation. Empirical results on the publicly available `Penn Treebank` datasets have shown that our proposed noisy word representation improves the accuracy of word recognition.

Keywords: Text processing · Noisy word representation · Word recognition · Character-level dependence · Word-level dependence

1 Introduction

Many tasks can be completed by the network models in the field of natural language processing [4,12,13]. Many decision makers will make the next decision

Supported by National Natural Science Foundation of China (No. 71871090).

based on the data processing and analysis results [23,24]. So people have higher expectations on the robustness of machine learning models. However, when dealing with text, word noise is a common phenomenon that includes character random insertion, deletion, swapping, or keyboard errors. Many pre-trained word vectors, such as `word2vec` [17], `Glove` [18], refer to the noisy words as OOV words and use a uniform vector to represent them. As a result, noisy words with different meanings are classified as an unknown category. This directly results in the loss of text information. Most existing natural language processing systems are vulnerable to these noises. Some are even fooled by them. Emails with precise misspellings can bypass the spam detection system [6]. Some specially designed sentences can stimulate the language model to give offensive language [5,14,26]. To reduce the influence of noisy words on NLP models, the existing solutions could be categorized into two main groups.

The first one is to improve the robustness of the model to the noise by adversarial training [2,21]. Belinkov et al. [2] choose to improve the robustness of their translation system by adding natural and synthetic noise in the training time. But it increases the difficulty of the model training and requires reasonable generation methods for the additional noise data meanwhile. The other way is to place a word recognition in front of NLP models [19] to reduce the interference of noise data on the model. Special word recognition in front of the model targets to recover the noisy text. And then the clean text is sent to the downstream task model so that the training of the downstream model is offloaded with handling noisy data. This method requires a robust word recognition. Character-level word recognition [20,25] is receiving more and more attention because of the OOV problem in the word-level models. Study [3] has shown that the start and end characters of a word are very important for recognizing a word. But exiting word recognitions don't make full use of them reasonably.

In order to solve the problem mentioned above, we propose a method PSEC that uses **P**re-trained **S**tart and **E**nd **C**haracters vectors to generate a reasonable and effective representation for the noisy words which would be used for the robust word recognition task. To do this, we treat the start and end characters differently to the others in a word. We firstly build a classifier to predict the start and end characters of words, and take the hidden state as the start and end character embedding vectors. Next, we get the new noisy word representation by combining these pre-trained vectors with the whole word character embeddings.

During the experiments, aiming to verify the validity of the start and end character embedding vectors, besides PSEC, we use PSC and PEC to get noisy word representation, PSC donates getting noisy word representation only using pre-trained start character vector and PEC donates only using pre-trained end character vector. We mainly experiment on 4 different types of noise with PSC, PEC, and PSEC: permutation(W_PER, rearranging the character order of the word), deletion (W_DEL, deleting a character in the word), insertion (W_INS, inserting a character to the word), and substitution (W_SUB, exchanging two characters in the word). And our experiments results have shown that the robust noisy word representation with pre-trained character vectors is effective for word recognition. In conclusion, our contributions are shown as follows:

- We propose a method, PSEC, that uses pre-trained start and end characters vectors to generate a robust representation for noisy word.
- We carried out an experiment to evaluate and demonstrate the performance of the proposed method. And the experiments indicate that our method PSEC could help improve word recognition accuracy.

The rest of the paper is conducted as follows: The related work is introduced briefly Sect. 2. Our model and the proposed method are described in detail in Sect. 3. The analysis of the experiments are shown in Sect. 4. Finally, Sect. 5 offers our conclusion and future work.

2   Related Work

We will review the current work on word recognitions and then introduce the related mechanism we are about to use in this section.

2.1   Word Recognition

The task of word recognition is to correct the wrong words in the input text. There are two methods for word recognition: the traditional method and neural network based method.

Once a traditional word recognition detects a noisy word, it would generate a set of candidate words that are similar to the misspelled word. Then it ranks the candidate words according to the probability that they are intended word for the misspelled word. At last, the word recognition determines only the most likely candidate word as the correction result. The most important step for traditional word recognition is candidate generation [7,9]. Candidate generation is a complex process because it has to compute the similarity between the noisy word and each word in the correct words dictionary.

There are many approaches for word recognition using neural nets as well. These word recognition usually work with characters. scRNN [20] proposes a semi-character recurrent neural network structure for word recognition, whose principle is based on the work [3]. As long as the start and end characters of a word are unchanged, humans can easily recognize the word and ignore the random characters inside. scRNN uses the one-hot representation for the start and end characters while using the Bag-of-words model [8] for the middle characters, and this improves the performance of the word recognition by improving the tolerance to the word. In fact, prior knowledge essential for humans to recognize these noisy words can be broken down into two types of dependencies: the character-level and the word-level dependencies. In terms of character-level dependencies, the common character combinations would be taken into account and rearranged. This step can narrow the selection of correct words, while cannot determine the exact word. Word-level dependencies work at this point, it will contact the context of the target word based on bi-directional long short-term memory (BiLSTM) to recognize the exact word further. Wang et al. [25] propose MUDE model to learn these two dependencies and complete the task of word recognition. However, MUDE does not distinguish the start and end characters from the others in a word while learning the dependencies between characters. A study [3] has shown that the start and end characters of a word are very important for recognizing a word. It means that MUDE may not entirely utilize the information of the start and end characters of a word.

2.2 Attention Mechanism

Word recognition is expected to correct the word despite the wrong ordering of the characters, and this requires the model to be able to learn the character level dependency. Attention mechanism could help in this process.

Human visual mechanism is the inspiration of attention mechanism. Generally speaking, when humans observe objects, they usually don't look at the whole scene from beginning to end, but at a specific part of the scene according to the requirements. As for the neural network model, some parts may be much more important for better output, so more attention should be paid to them. In the field of NLP, attention mechanism is often used as the decoder in an encoder-to-decoder structure. It changes the strategy in the traditional decoder that each input is given the same weight. Attention mechanism was originally used for machine translation [1], and it has become an important concept in the field of neural networks. There are two common ways proposed by Luong [15] to calculate the attention mechanism, global attention and local attention. Vaswani [22] proposed an attention-only structure to deal with problems related to sequence models, such as machine translation. Transformer is commonly used in NLP [11,13], and the pre-trained language model Bert [4] is completed on the structure of a multi-layer bidirectional Transformer encoder.

3   Model

During the test time, the architecture of our model is offered in Fig. 1. The whole architecture could be divided into four parts. The first part and the last part correspond to input and output respectively. The input is a word which needs to be recognized, and output is the result after word recognition which is expected to be the correct word. In the second part, we use PSEC method to get the robust noisy word representation. In the third part, the noisy word representation we get in the third part would be sent to the BiLSTM for predicting the correct word.

3.1   Predict the Start and the End Characters

As mentioned before, the start and the end characters are important to recognize a word while the ordering of the internal characters could be ignored.

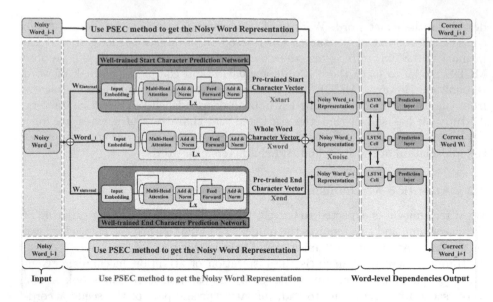

Fig. 1. Model architecture

Existing methods fix the start and end characters and simulate the disorder of the characters [20], but they did not consider the cases of the wrong start and end characters, and a fixed start and end characters will cause information loss. Based on this, we design a sub-network to predict the start and end characters. This prediction task can be converted to classification work. The internal order of many noisy words may be destroyed. We hope that we can still accurately predict the start and end characters in this case. So we need to ignore the inner order of words to make predictions, the sequential model such as RNN network is no longer suitable for our task. While attention mechanism regards the distance between the input information at each moment as one step and directly interactively calculates the input information at any two moments without considering their ordering. This structure exactly meets our network requirements. So we raise a classification model which is based on attention mechanism for predicting the start and end characters.

Let $S = (w_1, w_2, w_3, \cdots, w_n)$ denote a noisy sentence which has n words, and let w_i denote the i_{th} word in S. For each word, we process it with a fixed-length m. If its length is longer than m, we will replace it with its first m characters. If its length is shorter than m, we will do padding operation. So we represent w_i as $w_i = (c_{i_1}, \cdots, c_{i_m})$, where c_{i_1} is the start character and c_{i_m} is the end character. Our target is to predict the correct start and end characters for a given word. We assume $w_{Sinternal}$ as the word w without the end character and $w_{Einternal}$ as the word without the start character. We train the start and the end nets separately but with the same model. We will take the start character prediction network as an example to describe the model. The classification model's architecture is illustrated in Fig. 2.

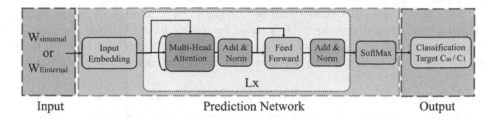

Fig. 2. Prediction model of the start or end character

For the start character prediction network, the input should be $w_{Einternal} = (c_2, c_3, \cdots, c_m)$, where m is the word length, c_i is the i_{th} character, and the classification task target is c_1. $w_{Einternal}$ could not be input directly to the network. We firstly map each character c_i to an one-hot representation o_i, and then get a d_k-dimensional character embedding $token_i$ as follows:

$$token_i = Eo_i \tag{1}$$

where $E \in R^{N \times d_k}$ is the randomly initialized embedding matrix, and the total class of the target characters is N. Therefore after "Input Embedding" layer as shown in Fig. 2, the input $_{Einternal} = (token_2, token_3, \cdots, token_m)$ replaces $w_{Einternal}$ as the prediction network input. In order to learn the character dependence in the word and capture the internal structure of the word, for each character token in the word, attention score would be calculated between this token and others. Generating three vectors is the first step to calculate the attention score from the input vectors. That is, for each character token, we would create a key vector, a value vector, and a query vector. And these three vectors are created by character embedding multiplied by three weight matrices. The keys, values, and queries of character tokens would be packed together into matrices K, V and Q and compute the matrix of outputs as:

$$\text{Attention}\,(Q, K, V) = \text{softmax}\left(\frac{QK^T}{\sqrt{d}}\right) V \tag{2}$$

The formula for multi-head attention as follows. Multi-head attention allows the model to attend to information from different representation subspaces in different locations.

$$\text{MultiHead}\,(Q, K, V) = \text{Concat}(\text{head}_1, \ldots, \text{head}_h)\, W^O$$
$$\text{where head}_i = \text{Attention}\left(QW_i^Q, KW_i^K, VW_i^V\right) \tag{3}$$

Where $\mathrm{W}^O, \mathrm{W}^Q, \mathrm{W}^K, \mathrm{W}^V$ are parameter matrices. According to the practice of Vaswani et al. [15], we choose to add a feed forward layer to x too, by assuming that x is the output after multi-head attention as follows:

$$FFN(x) = \max(0, xW_1 + b_1)W_2 + b_2 \tag{4}$$

Note that a deep structure could be composed by a number of layers mentioned above stacked together. After those layers, we have obtained the representation vector $Emd_{Einternal}$ of the input $w_{Einternal}$. Our model is built for the multi-classification task, so the softmax layer is connected next for prediction.

$$p_{start} = \text{softmax}(W \cdot Emd_{Einternal}) \tag{5}$$

where p_{start} is the probability matrix for the prediction target, $W \in \mathbb{R}^{d_k \times N}$ is a trainable parameter, d_k is the dimension of $Emd_{Einternal}$, and N is the number of classes. Cross-entropy is chose for the cost function to train the model.

3.2 Noisy Word Representation and Word Recognition

A high-quality word embedding is important for natural language process [10] and it helps subsequent word recognition tasks as well. In the section previously, we have trained a classification model. It is a supervised learning process and could learn the characteristic distribution of the start and end characters. So we can get the hidden state from the pre-trained model as the embedding vectors for the start and end character. In practice, we use the hidden state before softmax function in the well-trained classification models as the start and the end character embedding vector x_{start} and x_{end}. In this way, x_{start} and x_{end} are with the most likely start or end character prediction for a given noisy word.

These two character embedding vectors are not enough to represent the whole word information because the target of the pre-trained networks is predicting the start and end characters. It mainly contains the information of the character distribution of the start and end characters instead of the whole information of the word. So we encode the whole word characters with the same structure with the classifier but without the softmax layer. We get x_{word} after the whole word encoding. We generate our noisy word representation x_{noise} in two ways. In each way, x_{noise} has different calculation methods according to the different methods to get the noisy word representation. The first method is to add the start or end character vector and word vectors directly. The second method is to control the influence of the start or end character vector on the noisy word representation by tanH function:

$$Add: x_{noise} = \begin{cases} x_{start} + x_{word}, & PSC \\ x_{word} + x_{end}, & PEC \\ x_{start} + x_{word} + x_{end}, & PSEC \end{cases} \tag{6}$$

$$TanH: x_{noise} = \begin{cases} tanH(W_1 x_{start}) + x_{word}, & PSC \\ x_{word} + tanH(W_2 x_{end}), & PEC \\ tanH(W_1 x_{start}) + x_{word} + tanH(W_2 x_{end}), & PSEC \end{cases} \tag{7}$$

Where W_1 and W_2 are trainable parameters. Here x_{noise} is a noisy word representation with not only the character dependency but also the information in the most likely start or end characters.

In the training phase of our model, after we get the noisy word representation, we send it to two networks. The one is the Gated Recurrent Unit (GRU) for character decoding. The other one is BiLSTM for predicting the correct word. Because of the limitation of the page, and this is not the core part of our method, we will not discuss their details here. Each network computes a loss with cross-entropy loss function. We calculate the total loss of the modal as follows:

$$total\_loss = loss_{BiLSTM} + \beta loss_{GRU} \tag{8}$$

where $loss_{BiLSTM}$ is the loss of word prediction, $loss_{GRU}$ is the loss of the decoding process, β which is a hyper parameter and the contribution of the decoding model could be controlled by it. Note that the decoding model would be removed in test time.

4 Experiments

4.1 Experiments Settings

Data. For the word recognition work, We choose part of **Penn Treebank** (PTB)'s [16] data as our recognition dataset. PTB is a widely used dataset in NLP. 2,499 stories are picked from 98,732 stories in Wall Street Journal for PTB to do syntactic annotation. The releases of PTB include the raw text for each story. We choose part of them to make for our dataset. Our training, validation and testing dataset respectively contains 6528, 931 and 940 sentences. We extract 14,826 correct words from Bert's dictionary and recognition dataset as the dataset of pre-trained prediction models. Each of them has a length longer than 3. We firstly experiment on start nosie and end noise types with methods PSC and PEC. The start noise has three types: Start_DEL(deleting the start character randomly), Start_INS(inserting a character before the start character) and Start_SUB(replacing the start character with a random character). In the same way, end noise has three types noise of End_DEL, End_INS, and End_SUB. Then we experiment on 4 different types of noise with PSC, PEC, and PSEC: W_PER, W_DEL, W_INS, W_SUB. The noise may operate on the whole characters of words. Take the word "exchange" for example, noise type W_PER could be "geahxcen", W_DEL type could be "excange", W_INS type could be "yexchange" and W_SUB type could be "exghange". We make these noise data from our recognition dataset by the same method to scRNN and MUDE noise data.

Baselines. By comparing it with two strong and commonly used baselines, scRNN [20] and MUDE [25]. We can see the effectiveness of PSEC. scRNN is based on recurrent neural network. MUDE is a word recognition model which has achieved the best results so far on spell correction tasks. Our experiment did not select more baselines because they have been totally surpassed by MUDE.

Implementation Details. Our pre-trained model and word recognition model with PSC, PEC, and PSEC are all implemented with Pytorch. Pre-trained model based on the structure of Transformer encoder. We set 20 for the fixed length for each word. The layer of Transformer encoders is set to 2 and the number of attention heads is set to 8. The hidden size that represents the character representation's dimension, is 512. We use RMSprop optimizer with the learning rate of 0.0001 and the training batch size of 64. For the whole word recognition model, 650 is the number of hidden units of word representations.

4.2   Results and Analysis

As mentioned before, we use 14826 correct words to train the start and end character prediction network and we get the accuracy **93.3%** and **91.5%** separately. We verify the validity of the predicted start and end character embedding vector firstly on the Start Noise data and End Noise data. The results are shown in Table 1 and Table 2. According to the results in the tables, the accuracy of word recognition can be improved to a certain extent by adding the corresponding pre-trained character vector information when the start or end character is wrong.

Table 1. Word recognition accuracy (%) with different error types of Start_Noise. Bold numbers represent the best performance.

Method		Start_SUB	Start_INS	Start_DEL
scRNN		68.11	81.10	79.42
MUDE		88.45	91.93	**91.01**
PSC (our)	Add	**91.81**	**93.59**	90.87
	TanH	91.62	92.73	90.12

Table 2. Word recognition accuracy (%) with different error types of End_Noise. Bold numbers represent the best performance.

Method		End_SUB	End_INS	End_DEL
scRNN		69.46	80.45	78.98
MUDE		87.30	91.44	90.50
PEC(our)	Add	**90.78**	**92.87**	91.00
	TanH	90.21	92.45	**91.06**

This shows we have gotten a better noisy word representation in word recognition task because we add the start or end character information to it. It is a remarkable fact that the performance of the two baselines on the noise of SUB type is the worst, but PSC and PEC have significantly improved the recognition accuracy of this noise type. As for the method to generate noisy word representation, it seems that adding directly would have better results for Start or End Noise. Besides this, there is another phenomenon worth noting. The accuracy curve of PSC for the start noise has an upward inflection point after some epochs as shown in Fig. 3, while PEC's accuracy curve for the end noise does not have this phenomenon. To some extent, this may explain that the effect of the start and end character information on the word is not the same.

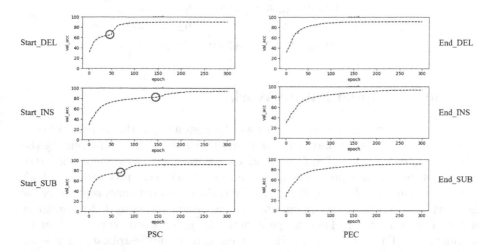

Fig. 3. Accuracy Curves of PSC for Start_noise and PEC for End_noise

We have shown above that adding the start character or the end character information alone is helpful for the word recognition task. Based on this, we conduct experiments with adding both start and end character information on the W_Noise data, and the results compared with baselines scRNN and MUDE are shown in Table 3. We have carried out experiments with three models PSC, PEC, and PSEC respectively to prove the effectiveness of the pre-trained start and end character embedding vectors using in the noisy word representation. And for each model, we used two methods to generate the noisy word embedding. From the results, it is obvious that no matter which method is used to generate noisy word embedding, the word recognition accuracy of PSEC model is higher than that of MUDE. From PSC and PEC results, it is shown again that the noisy words representation generated by the method of adding directly is more conducive to the recognition of words, which is consistent with the results of the experiments of the start and end Noise. And it is also shown that adding the information of the start or end character alone can help improve the accuracy of the model while adding them meanwhile can improve the recognition results of the model to a greater extent. And for PSEC, tanH generating method is better than adding directly.

Table 3. Word recognition accuracy (%) with different error types of W_Noise. Bold numbers represent the best performance.

Method		W_PER	W_SUB	W_INS	W_DEL
scRNN		87.76	57.31	81.21	71.02
MUDE		95.59	82.42	91.66	86.57
PSC (our)	Add	**96.07**	84.66	91.65	88.74
	TanH	95.42	84.56	90.85	88.67
PEC (our)	Add	95.52	85.05	92.43	88.63
	TanH	93.06	83.51	91.50	86.78
PSEC (our)	Add	95.71	87.00	92.88	88.67
	TanH	95.63	**87.69**	**93.56**	**88.95**

5   Conclusion and Future Work

We focus on improving the accuracy of word recognition in this paper. We consider the impacts of start and end characters information when representing the noisy word, and propose PSEC method to generate a robust noisy word representation based on a pre-trained start and end character prediction network for word recognition. Specifically, we firstly trained two networks separately for the start and end character prediction and obtained the state before softmax as the character vector. Then we generated the noisy word representation by two methods. The extensive experiments indicate that the robust noisy word representation by our method PSEC can improve word recognition accuracy. In future work, we are going to explore the different effects of the start and end character information on the words.

References

1. Bahdanau, D., Cho, K., Bengio, Y.: Neural machine translation by jointly learning to align and translate. arXiv preprint arXiv:1409.0473 (2014)
2. Belinkov, Y., Bisk, Y.: Synthetic and natural noise both break neural machine translation. arXiv preprint arXiv:1711.02173 (2017)
3. Davis, M.: Aoccdrnig to a rscheearch at cmabrigde uinervtisy (2003)
4. Devlin, J., Chang, M.W., Lee, K., Toutanova, K.: BERT: pre-training of deep bidirectional transformers for language understanding (2018)
5. Dinan, E., Humeau, S., Chintagunta, B., Weston, J.: Build it break it fix it for dialogue safety: robustness from adversarial human attack (2019)
6. Fumera, G., Pillai, I., Roli, F.: Spam filtering based on the analysis of text information embedded into images. J. Mach. Learn. Res. **7**(Dec), 2699–2720 (2006)
7. Harish Kumar, R.: Spelling correction to improve classification of technical error reports (2019)
8. Hertel, M.: Distributional structure. Papers in Structural and Transformational Linguistics, pp. 775–794 (1970)
9. Hertel, M.: Neural language models for spelling correction (2019)

10. Jiang, J., et al.: Sentiment embedded semantic space for more accurate sentiment analysis. In: Liu, W., Giunchiglia, F., Yang, B. (eds.) KSEM 2018. LNCS (LNAI), vol. 11062, pp. 221–231. Springer, Cham (2018). https://doi.org/10.1007/978-3-319-99247-1_19

11. Jiang, J., et al.: CIDetector: semi-supervised method for multi-topic confidential information detection. In: European Conference on Artificial Intelligence (2020)

12. Law, R., Li, G., Fong, D.K.C., Han, X.: Tourism demand forecasting: a deep learning approach. Ann. Tour. Res. **75**, 410–423 (2019)

13. Liu, C., et al.: A two-stage model based on BERT for short fake news detection. In: Douligeris, C., Karagiannis, D., Apostolou, D. (eds.) KSEM 2019. LNCS (LNAI), vol. 11776, pp. 172–183. Springer, Cham (2019). https://doi.org/10.1007/978-3-030-29563-9_17

14. Liu, H., Derr, T., Liu, Z., Tang, J.: Say what i want: towards the dark side of neural dialogue models. arXiv preprint arXiv:1909.06044 (2019)

15. Luong, M.T., Pham, H., Manning, C.D.: Effective approaches to attention-based neural machine translation. arXiv preprint arXiv:1508.04025 (2015)

16. Marcus, M., Santorini, B., Marcinkiewicz, M.A.: Building a large annotated corpus of English: the penn treebank (1993)

17. Mikolov, T., Chen, K., Corrado, G., Dean, J.: Efficient estimation of word representations in vector space. arXiv preprint arXiv:1301.3781 (2013)

18. Pennington, J., Socher, R., Manning, C.D.: GloVe: global vectors for word representation. In: Proceedings of the 2014 Conference on Empirical Methods in Natural Language Processing (EMNLP), pp. 1532–1543 (2014)

19. Pruthi, D., Dhingra, B., Lipton, Z.C.: Combating adversarial misspellings with robust word recognition. arXiv preprint arXiv:1905.11268 (2019)

20. Sakaguchi, K., Duh, K., Post, M., Van Durme, B.: Robsut wrod reocginiton via semi-character recurrent neural network. In: Thirty-First AAAI Conference on Artificial Intelligence (2017)

21. Vaibhav, V., Singh, S., Stewart, C., Neubig, G.: Improving robustness of machine translation with synthetic noise. In: Proceedings of the 2019 Conference of the North American Chapter of the Association for Computational Linguistics: Human Language Technologies, Volume 1 (Long and Short Papers), pp. 1916–1920 (2019)

22. Vaswani, A., et al.: Attention is all you need. In: Advances in Neural Information Processing Systems, pp. 5998–6008 (2017)

23. Vu, H.Q., Li, G., Law, R., Zhang, Y.: Travel diaries analysis by sequential rule mining. J. Travel Res. **57**(3), 399–413 (2018)

24. Vu, H.Q., Li, G., Law, R., Zhang, Y.: Exploring tourist dining preferences based on restaurant reviews. J. Travel Res. **58**(1), 149–167 (2019)

25. Wang, Z., Liu, H., Tang, J., Yang, S., Huang, G.Y., Liu, Z.: Learning multi-level dependencies for robust word recognition. arXiv preprint arXiv:1911.09789 (2019)

26. Wolf, M.J., Miller, K., Grodzinsky, F.S.: Why we should have seen that coming: comments on microsoft's tay "experiment", and wider implications. ACM SIGCAS Comput. Soc. **47**(3), 54–64 (2017)

Intention Multiple-Representation Model for Logistics Intelligent Customer Service

Jingxiang Hu[1] , Junjie Peng[1,2,3] , Wenqiang Zhang[4,5], Lizhe Qi[4],
Miao Hu[1], and Huanxiang Zhang[1]

[1] School of Computer Engineering and Science, Shanghai University, Shanghai, China
jjie.peng@shu.edu.cn
[2] Shanghai Institute for Advanced Communication and Data Science,
Shanghai University, Shanghai, China
[3] Lab of Intelligent Technology and Systems, Shanghai Computer Society,
Shanghai, China
[4] Academy for Engineering and Technology, Fudan University, Shanghai, China
wqzhang@fudan.edu.cn
[5] School of Computer Science and Technology, Fudan University, Shanghai, China

Abstract. With the development of the Internet, more and more people express their ideas on the internet in the form of short text. And a question with the same intention can be expressed in different ways. So it is necessary and important to understand the meaning of short text when we want to provide intelligent service to customer. Many studies have focused on the short texts based on public data sets. However, little studies have been carried out or can be effectively used in some specific fields. Taking logistics Intelligent Customer Service (ICS) as an example, the short texts has the above characteristics. To solve this issue about intention multiple-representation in logistics, a self-attention-based model, that is, One question to Many question (O2M) is proposed. On the basis of classification task, the model can learn the mapping relation from customer questions to standard questions. And it consists of three parts: standard questions domain, customer questions domain and selector. For the two domains, they learn semantic patterns of their own questions. And the selector becomes the bridge between them. Extensive experiments were carried out on logistics corpus. And the results show that the model is effective and the accuracy of the model is higher than that of traditional neural network models.

Keywords: Logistics · Intelligent customer service · Domain knowledge · Intention multiple-representation · Logistics corpus

1 Introduction

The development of the Internet has spawned and developed the logistics company. In this field, manual customer service is required to answer users' questions. And customer attendants are faced with great work pressure. Fortunately,

© Springer Nature Switzerland AG 2020
G. Li et al. (Eds.): KSEM 2020, LNAI 12274, pp. 186–193, 2020.
https://doi.org/10.1007/978-3-030-55130-8_16

the ICS alleviates the above phenomenon. However, the quality of ICS fails to meet people's requirements. Therefore, it is urgent to improve the understanding ability of ICS.

Short text matching is the basic work in the actual ICS. And it represents the semantic relation between two short texts. In other words, the semantic relation between two short texts is either the same intention or not. In recent years, more and more people prefer to use deep learning to solve the short text matching. However, it is not efficient when applying this method to real ICS. It means that the system have to compare the customer question with all standard questions to obtain the real answer that the customer wants.

To solve the above problem, we take different domain and classification task as consideration. And a model, named O2M is proposed. It is a classification model, and it means that the customer question belongs to the same category as the corresponding standard question. Results of O2M are carried out on logistics corpus, and the results show that O2M is better than that of other neural network models.

2   Related Work

It is difficult and important to full semantic mining for short text. And more and more people have focused on the short text problems while little studies have focused on logistics short text. However, these methods are instructive for dealing with problems in logistics.

Nowadays, more and more people prefer to use deep learning method to solve problems. And on the basis of text vectorization, the similarity calculation has been widely studied. Chen et al. [1] enhanced the ratiocinative capability of LSTM by local inference, and they added syntactic parsing information to improve the performance of their model. And Liu et al. [11] designed a model on the basis of Chen et al. [1], and they added siamese network to ESIM, and they claimed that the model increase the utilization of inner information in sentences. Wang et al. [16] proposed BiMPM model, and they provided a variety of matching method to improve the capability of neural network.

Besides, in Natural Language Processing (NLP), text classification has been widely used. In this subject, Joulin et al. [5] designed an efficient baseline for text classification. Zhou et al. [18] designed Att-BLSTM based on BiLSTM, and they claimed the model can mine the most important semantic information in a sentence. Lai et al. [9] applied a recurrent structure to learn the contextual information, and said it can reduce the noise when compare to traditional window-based neural network. Liu et al. [10] proposed three different mechanisms to improve the performance of their model, and they tried to solve the problem of insufficient data. Xie et al. [17] designed a feature-enhanced fusion model on the basis of LSTM and CNN, and they added the certain semantic context information into the embedding to achieve their better performance.

Finally, as many scholars have set a lot research on NLP, a new feature extractor, named Transformer [15], appears in NLP world. And scholars said that it is considered to be the suitable solution for text. Sergio et al. [13] proposed Stacked DeBERT, and they designed an encoding scheme in BERT [2] to solve the problem with incomplete data. As for logistics, there exist business-related word in questions, and this semantic information can be capture by Transformer because of its self-attention mechanism.

3 Method

3.1 O2M Model

As for each customer question, it belongs to the same category as the corresponding standard question. Besides, Jieba [14] is used to transform a question into words list and the word vector is produced by Word2vec [12]. And the model structure is as Fig. 1 shows.

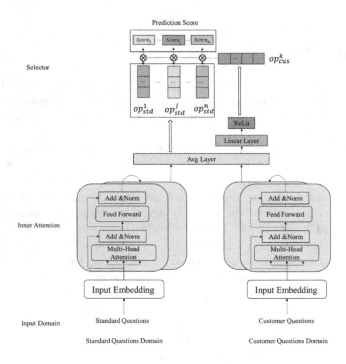

Fig. 1. Model structure

Standard and Customer Questions Domain. In this paper, the definition of the standard question is: a normal question that contains logistics business-related words and its known synonyms. And a stacked Transformer Block is used to obtain the synonyms attention. The Transformer Block has the multi-head mode. And for each head of a Transformer Block, query vector $Q \in \mathbb{R}^{l \times d_{model}}$, key vector $K \in \mathbb{R}^{l \times d_{model}}$ and value vector $V \in \mathbb{R}^{l \times d_{model}}$ of questions embedding $W \in \mathbb{R}^{l \times d_{model}}$, as in Eq. (1), where d_{model} is the embedding dimension and l is the length of question. And the output $A \in \mathbb{R}^{l \times d_{model}}$ is deduced as Eq. (2) shows, and it is the score of all words in a question.

$$Q = W \times W_Q, K = W \times W_K, V = W \times W_V \tag{1}$$

$$A = softmax((Q \times K^T)/\sqrt{d_{model}}) \times V \tag{2}$$

Where W_Q, W_K, W_V are the linear matrices. And the result of each head is concatenated together and then fed into a fully connected feed-forward network, as Eq. (3) shows. And dropout [3] and add-norm layer are adapted to enhance the performance.

$$MulH = [A_1; A_2; ...; A_h] \times W_O, FFN(x) = relu(x \times W_1 + b_1) \times W_2 + b_2 \tag{3}$$

MulH is the result of Multi-head. W_O, W_1, W_2 are the linear matrices, A_i is the i_{th} head score of a block. h is the number of head of a block, and b_1, b_2 are bias value. *relu* is the activation function.

The two domains do not share parameters. The output op_{std}, op_{cus} of two domains is defined as follows. x_{std} and x_{cus} is the standard and customer questions embedding. *Avg* is the average layer.

$$op_{std} = Avg(FFN(x_{std})) \tag{4}$$

$$op_{cus} = Avg(FFN(x_{cus})) \tag{5}$$

Selector. After the two domains extract their own feature attention, this section produces prediction score *score* as Eq. (6) shows.

$$score = f_{selector}(op_{cus}, op_{std}; W), f_{selector} = op_{cus}^k \circ W_k \circ op_{std}^T \tag{6}$$

$f_{selector}$ is a select function. $W \in \mathbb{R}^{d_{model} \times d_{model}}$ is the linear matrix. W_k is the k_{th} row of W, and op_{cus}^k the k_{th} question in customer question domain.

The cross entropy loss function is widely used in classification, which is usually defined as follows:

$$CrossEntropy(r, p) = -\sum_i r(i) \times log(p(i)) \tag{7}$$

Where p is the prediction score, r is the real score and i is the index of the value of r, p.

4   Experiment

4.1   Baseline of Experiment

The logistics standard questions data set is manually annotated, and it consists of nine categories, such as Check, Send, Receive the package and so on. For each category, it contains some real customer questions which is provided by logistics company. And the data processing consists of four parts: word segmentation, wrong words modification, deduplication and synonyms of standard questions.

After the data processing, the data set contains nine categories and 2k customer questions. And for each category of questions, 70% for training and 30% for testing.

4.2   Experimental Environment

For each domain, it uses two Transformer blocks [15] stacked together. Each block has 8 attention heads. The maximum length of the sentence is 32. The dimension of the word vector is set to 256, and the size of each batch is $n \times ns$, n is an integer with initial value 3. ns is the number of categories. In training, stochastic optimization [7] is used as the optimizer with $\beta_1 = 0.9$, $\beta_2 = 0.98$ and $\epsilon = 10^{-9}$. And the learning rate is governed by Eq. (8) with initial value 0 [8]:

$$lr = \frac{1}{\sqrt{d_{model}}} \times min(\frac{1}{\sqrt{step}}, step \times \frac{1}{\sqrt{warmup^3}}) \tag{8}$$

Where $step$ is the training step and $warmup$ is the warmup step with initial value 4000. The differences between O2M and other algorithms are explained from $Accuracy$ and $F_1 - score$.

4.3   Experimental Results and Analysis

Overall Performance. Extensive experiments are carried out on logistics corpus. And the word embedding matrix of CNN-based or RNN-based model is the same as O2M. The second column $Para$ is the size of models, and the same below. And the left column is the $F_1 - Score$ of the first category, that is, Check the package.

From Table 1, the proposed O2M model achieves better performance than other methods. The NB shows the superficial combinations of words, and its performance is not good. Most of models (2)–(7) are CNN-based or RNN-based. And the accuracy of them shows an upward trend than NB. For about CNN-based models, they utilize convolving filters to process local features. Especially, DPCNN uses shortcut connections to train deep networks. RNN-based models often mean LSTM or recurrent architecture. TextRNN models text with different strategies of sharing information while TextRCNN combines the structure characteristics of CNN and RNN. The next group of models (8)–(9) are Transformer-based, and they performance better than other models because of

Table 1. Performance for classification

Model	Para	Accuracy	$F_1 - score$
(1) Native Bayes (NB)	–	0.64	–
(2) DPCNN [4]	0.42M	0.69	0.78
(3) FastText [5]	1.55M	0.71	0.81
(4) TextCNN [6]	0.44M	0.72	0.85
(5) TextRCNN [9]	0.48M	0.65	0.81
(6) TextRNN [10]	0.46M	0.65	0.81
(7) TextRNN-Att [18]	0.46M	0.69	0.83
(8) Transformer [15]	0.52M	0.73	0.71
(9) O2M	1.05M	**0.75**	**0.86**

their self-attention mechanism. In a word, these models only focus on customer questions and do not consider other relevant questions while O2M considers two domains. And O2M builds a bridge between the two domains. So O2M performance better than others.

Ablation Studies. In order to verify the major components that are important to help O2M achieves this performance, we remove some parts of O2M. And the results between O2M with different component are listed in Table 2.

Table 2. Performance of O2M

Model	Para	Accuracy
O2M-Customer (Transformer)	0.52M	0.73
O2M-NoSelector	1.05M	0.70
O2M-full	1.05M	0.75

O2M-Customer means the model is tested with customer questions domain. And O2M-NoSelector means the model is tested without selector. If standard questions domain and selector are removed, the accuracy drops to 73% when compares with O2M-full. It means that they can help O2M to learn additional semantic patterns and the mapping relations. If selector is removed, the performance of O2M degrades. It shows that this two parts learn their own semantic patterns. However, there is no mapping relations between them. So the model can not fully capture the semantic relation between the two domains.

5   Conclusion

Aiming at the problem of insufficient understanding ability of logistics ICS, a self-attention-based model is put forward. With the classification task, the model achieves good results for the above problems. Meanwhile, in order to capture the implicit relationship of standard questions and customer questions, the model builds a selector to do that. Results show that the model is more efficient than the traditional neural network models, and it has application prospect in interactive robot, intelligent customer service and so on.

In the future, the domain knowledge graph will be explored to try to maximize the ratiocinative capability of the model and the optimal strategy of selector.

Acknowledgements. The authors would like to thank the funding from National Natural Science Foundation of China (Grant no. 61572305) and the resources and technical support from the High performance computing Center of Shanghai University, and Shanghai Engineering Research Center of Intelligent Computing System (No. 19DZ2252600). And it is especially grateful for the data and industry knowledge support from YTO express company.

References

1. Chen, Q., Zhu, X., Ling, Z., Wei, S., Jiang, H., Inkpen, D.: Enhanced LSTM for natural language inference. In: Proceedings of the 55th Annual Meeting of the Association for Computational Linguistics, ACL 2017, Vancouver, Canada, 30 July–4 August 2017, Volume 1: Long Papers, pp. 1657–1668 (2017). https://doi. org/10.18653/v1/P17-1152
2. Devlin, J., Chang, M., Lee, K., Toutanova, K.: BERT: pre-training of deep bidirectional transformers for language understanding. CoRR abs/1810.04805 (2018). http://arxiv.org/abs/1810.04805
3. Hinton, G.E., Srivastava, N., Krizhevsky, A., Sutskever, I., Salakhutdinov, R.: Improving neural networks by preventing co-adaptation of feature detectors. CoRR abs/1207.0580 (2012). http://arxiv.org/abs/1207.0580
4. Johnson, R., Zhang, T.: Deep pyramid convolutional neural networks for text categorization. In: Barzilay, R., Kan, M. (eds.) Proceedings of the 55th Annual Meeting of the Association for Computational Linguistics, ACL 2017, Vancouver, Canada, 30 July–4 August 2017, Volume 1: Long Papers, pp. 562–570. Association for Computational Linguistics (2017). https://doi.org/10.18653/v1/P17-1052
5. Joulin, A., Grave, E., Bojanowski, P., Mikolov, T.: Bag of tricks for efficient text classification. In: Lapata, M., Blunsom, P., Koller, A. (eds.) Proceedings of the 15th Conference of the European Chapter of the Association for Computational Linguistics, EACL 2017, Valencia, Spain, 3–7 April 2017, Volume 2: Short Papers, pp. 427–431. Association for Computational Linguistics (2017). https://doi.org/ 10.18653/v1/e17-2068
6. Kim, Y.: Convolutional neural networks for sentence classification. In: Moschitti, A., Pang, B., Daelemans, W. (eds.) Proceedings of the 2014 Conference on Empirical Methods in Natural Language Processing, EMNLP 2014, 25–29 October 2014, Doha, Qatar, A meeting of SIGDAT, a Special Interest Group of the ACL, pp. 1746–1751. ACL (2014). https://doi.org/10.3115/v1/d14-1181
7. Kingma, D.P., Ba, J.: Adam: A method for stochastic optimization. In: Bengio, Y., LeCun, Y. (eds.) 3rd International Conference on Learning Representations, ICLR 2015, San Diego, CA, USA, 7–9 May 2015, Conference Track Proceedings (2015). http://arxiv.org/abs/1412.6980

8. Klein, G., Kim, Y., Deng, Y., Senellart, J., Rush, A.M.: OpenNMT: open-source toolkit for neural machine translation. In: Proceedings of ACL (2017). https://doi.org/10.18653/v1/P17-4012

9. Lai, S., Xu, L., Liu, K., Zhao, J.: Recurrent convolutional neural networks for text classification. In: Bonet, B., Koenig, S. (eds.) Proceedings of the Twenty-Ninth AAAI Conference on Artificial Intelligence, 25–30 January 2015, Austin, Texas, USA, pp. 2267–2273. AAAI Press (2015). http://www.aaai.org/ocs/index.php/AAAI/AAAI15/paper/view/9745

10. Liu, P., Qiu, X., Huang, X.: Recurrent neural network for text classification with multi-task learning. In: Kambhampati, S. (ed.) Proceedings of the Twenty-Fifth International Joint Conference on Artificial Intelligence, IJCAI 2016, New York, NY, USA, 9–15 July 2016, pp. 2873–2879. IJCAI/AAAI Press (2016). http://www.ijcai.org/Abstract/16/408

11. Liu, Y., et al.: An enhanced ESIM model for sentence pair matching with self-attention. In: Proceedings of the Evaluation Tasks at the China Conference on Knowledge Graph and Semantic Computing (CCKS 2018), Tianjin, China, 14–17 August 2018, pp. 52–62 (2018). http://ceur-ws.org/Vol-224020..../paper09.pdf

12. Mikolov, T., Chen, K., Corrado, G., Dean, J.: Efficient estimation of word representations in vector space. In: 1st International Conference on Learning Representations, ICLR 2013, Scottsdale, Arizona, USA, 2–4 May 2013, Workshop Track Proceedings (2013). http://arxiv.org/abs/1301.3781

13. Sergio, G.C., Lee, M.: Stacked DeBERT: all attention in incomplete data for text classification. CoRR abs/2001.00137 (2020). http://arxiv.org/abs/2001.00137

14. Sun, J., et al.: "jieba" chinese text segmentation (2012). https://github.com/fxsjy/jieba

15. Vaswani, A., et al.: Attention is all you need. In: Advances in Neural Information Processing Systems 30: Annual Conference on Neural Information Processing Systems 2017, 4–9 December 2017, Long Beach, CA, USA, pp. 5998–6008 (2017). http://papers.nips.cc/paper/7181-attention-is-all-you-need

16. Wang, Z., Hamza, W., Florian, R.: Bilateral multi-perspective matching for natural language sentences. CoRR abs/1702.03814 (2017). http://arxiv.org/abs/1702.03814

17. Xie, J., et al.: Chinese text classification based on attention mechanism and feature-enhanced fusion neural network. Computing **102**(3), 683–700 (2020). https://doi.org/10.1007/s00607-019-00766-9

18. Zhou, P., et al.: Attention-based bidirectional long short-term memory networks for relation classification. In: Proceedings of the 54th Annual Meeting of the Association for Computational Linguistics, ACL 2016, 7–12 August 2016, Berlin, Germany, Volume 2: Short Papers. The Association for Computer Linguistics (2016). https://doi.org/10.18653/v1/p16-2034

Identifying Loners from Their Project Collaboration Records - A Graph-Based Approach

Qing Zhou$^{(\boxtimes)}$, Jiang Li , Yinchun Tang, and Liang Ge

College of Computer Science, Chongqing University, Chongqing 400044, China
{tzhou,lijfrank,yinchun,geliang}@cqu.edu.cn

Abstract. Identification of lonely students is important because loneliness may lead to sickness, depression, and even suicide for college students. Loneliness scales are the general instruments used to identify loners, but it usually fails when loners try to conceal their real conditions in the questionnaires. In this paper, we propose a framework for the identification of loners based on their project collaboration records, a relatively more objective data source than student's self-reports. Considering that collaborative relationships among students are highly informative for the identification of loners, we employ Graph Neural Networks to model the complex patterns of student interactions. Furthermore, we propose a Graph-based Over-sampling Technique (GOT) to address the class-imbalanced problem for graph-structured data. Experiments on a real-world dataset show that our proposed method can identify loners with high accuracy.

Keywords: Loneliness · Collaboration · Graph Neural Networks · Class imbalance

1 Introduction

Loneliness, a complex emotional condition associated with the lack of social connection with other people, is emerging as a public health problem. Loneliness can impact seriously on people's mental health, increasing their levels of depression and stress [13]. College students are vulnerable to loneliness, as most of them have to leave their family and friends to form a new social network. Recent studies indicate that there is a negative connection between loneliness and academic motivation and academic performance [14]. And it is found that loneliness is an important predictor of depression, suicide ideation, and suicide behavior [3]. If lonely college students are timely identified, proper intervention or assistance can be provided to them, thus reducing the risk of many serious accidents.

To identify loners, psychologists have developed some questionnaire-based loneliness scales, e.g., UCLA scale [12], for the measurement of loneliness. However, loners may avoid reporting their real conditions, which makes the information collected by these questionnaires inaccurate. In this paper, we propose

© Springer Nature Switzerland AG 2020
G. Li et al. (Eds.): KSEM 2020, LNAI 12274, pp. 194–201, 2020.
https://doi.org/10.1007/978-3-030-55130-8_17

an approach for unobtrusive diagnosis of loneliness based on student collaboration records. We found that the collaboration patterns of loner students are significantly different from the others. Therefore, we may employ some machine learning techniques to build a model of these patterns and use the model for the identification of lonely students.

There have been some studies on modeling collaboration networks [9,16]. Nevertheless, these methods generally treat each sample as an independent object during model training and therefore neglect complex relationships between samples. Another problem in the identification of loners is the serious imbalance between two types of students, which may lead to a bias classifier preferring regular students to loner. Recently, the success of Graph Neural Networks (GNNs) in many fields has demonstrated its effectiveness on modeling behaviors in networks [17], thanks to its capability in processing non-Euclidean data. In this paper, we proposed a framework for the identification of loner from student collaboration records based on GNNs. We also proposed a method to tackle the class-imbalanced problem for training Graph Neural Networks. Our contributions can be summarized as follows:

- We propose a framework for the identification of loner students from their project collaboration records. The framework is based on GNNs, which capture the interaction patterns among students.
- We propose a method to alleviate the class-imbalanced problem for loner identification. The method is specially designed for graph-structured data.
- We conduct extensive experiments on a real-world dataset, and show that our proposed method can consistently boost the performance in identifying lonely students based on GNNs approaches.

2   Related Work

The interactions among students can be roughly described with many network metrics, such as betweenness centrality and closeness centrality of each node. These network metrics, together with other features of students, can be used to train classifiers for students or student groups by employing some machine learning techniques. Crespo and Antunes [5] proposed a method of predicting team performance based on a network of students. For each student, a social score is calculated by using the PageRank algorithm [2] or its variants. Becheru and Popescu [1] investigated course collaboration patterns among students in a project-based learning scenario. Three social networks were constructed upon student interactions in both blogs and microblogs. Nevertheless, these methods can not directly deal with non-Euclidean data, it has to be mapped to a simpler representation, which leads to important information may be lost.

Although network metrics can sketch the outline of a social network, massive detailed information in the network is lost due to the statistical processing. Recently, there has been a surge of applications of Graph Neural Networks

(GNNs) in various fields, such as intelligent transportation, recommender system, and bioinformatics. Unlike general deep learning approaches, GNN is a particular type of neural network which can directly operate on graph-structured data and model the complex interaction and interdependency between nodes.

The studies of GNNs generally fall into two categories, spectral-based approaches and spatial-based approaches. More recently, Kipf and Welling [10] proposed a semi-supervised GNN classifier named Graph Convolution Network (GCN). For a graph $\mathcal{G} = (N, E)$, where N and E denote node set and edge set, GCN updates node features according to following rule:

$$X^{(l+1)} = \rho \left(\tilde{D}^{-\frac{1}{2}} \tilde{A} \tilde{D}^{-\frac{1}{2}} X^{(l)} W^{(l)} \right) \tag{1}$$

where $X^{(l)}$ denotes node features in l-th GNN layer, and $X^{(0)}$ is initiated with the origin node features, $\tilde{A}$ is the adjacency matrix of $\mathcal{G}$ with additional self-connection edges for each node, $\tilde{D}$ is a diagonal matrix with $\tilde{D}_{ii} = \sum_j \tilde{A}_{ij}$, $W^{(l)}$ is the trainable weight matrix in l-th layer, and $\rho(\cdot)$ denotes an activation function.

Unlike spectral GNN, spatial-based approaches define no explicit filter on graph signals. It operates directly on the graph nodes and their neighbors. Hamilton et al. [7] introduced GraphSAGE, which aggregates the information from fixed-size neighborhood to update the representation of the center node. The aggregating function includes mean, LSTM [8] or pooling aggregator over the graph.

Another successful spatial GNN is Graph Attention Network (GAT) [15], which integrates the attention mechanism into the message propagation step. Specifically, it learns a transformation matrix to determine the impact of different neighbors on center nodes.

3   Methodology

3.1   Framework

We propose a framework for the identification of lonely students (See Fig. 1). The framework consists of several phases: feature and relationship extraction, graph construction, class-imbalanced processing, model training, and evaluation. In the extraction phase, we build a social network based on student collaboration records and extract network-related features. We also extract some achievement-related features from student academic records. Then a collaboration graph is constructed, which is composed of node set (i.e., students), edge set (i.e., collaborative relationship among students), and node features (i.e., student features). Then we apply a graph-based over-sampling operation on the graph to address the class-imbalanced problem. Finally, we train and evaluate graph neural networks on a dataset, and select the model with the best prediction performance for the identification of loner.

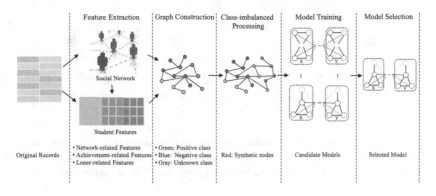

Fig. 1. Overall framework for identification of loners.

3.2 Graph Construction

In this phase, a graph $\mathcal{G} = (N, A, X, Y)$ is constructed, where N is a node set, denoting a collection of students in course collaboration records; A is an adjacent matrix, with A_{ij} denotes whether student i and j have collaborated or not (1 for 'yes' and 0 for 'no'); $X \in \mathbb{R}^{|N| \times F}$ is a student feature set, including all features extracted from student collaboration records and academic records. $Y \in \mathbb{R}^{|N|}$ is a student label set, indicating whether the student is a loner or not. The graph is then used as the foundation for training Graph Neural Networks. Table 1 summarizes the statistics of the graph constructed from a real-world dataset.

Table 1. Statistics of graph data.

Nodes	Edges	Features per node	Classes	Number of positive nodes (loner)	Number of negative nodes
312	1827	25	2	35	277

3.3 Graph Based Over-Sampling Technique

The identification of the lonely students is a highly imbalanced classification problem, that is, the ratio of loner to ordinary students is very low. To address the class-imbalanced problem, we may apply some over-sampling operations, such as SMOTE [4], to synthesize a sufficient number of minority samples before training a model. Nevertheless, these operations do not suit for GNN, because all the synthetic samples are isolated while GNN demands that these synthetic samples be connected to the original graph. To solve this problem, we propose a Graph-based Over-sampling Technique (GOT). The process of GOT can be outlined as follows. First, we synthesize minority nodes by applying some over-sampling technique; second, we build the intra-connections among the synthetic

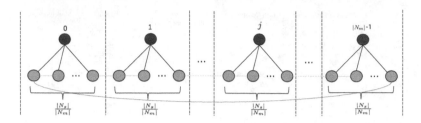

Fig. 2. The illustration of inter-connection and intra-connection. The blue and green nodes represent the original and synthetic minority nodes. The black and the brown edges represent inter-connections and intra-connections. (Color figure online)

nodes; third, we build the inter-connection among the synthetic nodes and the original nodes; and finally, all synthetic nodes and new connections are combined with the original graph to form a new graph, which is used as the material for training GNNs.

We first synthesize minority nodes with some off-shelf over-sampling operation, e.g., SMOTE algorithm. SMOTE synthesizes new instances by interpolating existing samples. Here, we simply adopted linear interpolation operations as follows:

$$\mathbf{x}_n = \mathbf{x}_i + (\mathbf{x}_j - \mathbf{x}_i) \times \eta, \mathbf{x}_j \in \mathcal{N}(\mathbf{x}_i), \tag{2}$$

where $\mathbf{x}_i \in \mathbb{R}^F$ denotes i-th sample of the minority class, $\mathbf{x}_j$ is a neighboring sample of $\mathbf{x}_i$, $j = 1, 2, ..., k$, and η is a random number belonging to $[0, 1]$, and $\mathbf{x}_n$ is the synthesized sample.

The intra-connection is an operation on the synthesis nodes. Suppose all the synthetic nodes are sorted in some order, then for each synthetic node, we create an edge to its following node. The last synthetic node is also connected to the first synthetic node. The process of inter-connection, where synthetic nodes are connected to original nodes, is described as follows. First, $|N_s|$ synthesized nodes are partitioned into $|N_m|$ groups, where $|N_m|$ is the number of the original minority nodes. Then, each group of synthesized nodes are connected to one original node. Specifically, we create an edge between the k-th node of the j-th group and the j-th original minority node, where $j = 0, 1, ..., |N_m| - 1$, $k = 0, 1, ..., \frac{|N_s|}{|N_m|} - 1$. The process of building intra-connection and inter-connection is illustrated in Fig. 2.

We note that SMOTE is just one of off-shelf techniques for over-sampling. And even SMOTE has multiple variants. For instance, K-means SMOTE is one of its variants based on the clustering technique [6]. It clusters all nodes into k groups using the k-means algorithm, and then over-sampling clusters with a high proportion of minority class by using SMOTE. We will compare the impact of both SMOTE and K-means SMOTE on the identification of loner.

4   Data and Experiment

4.1   Data Processing

The dataset was collected from project reports of 67 courses taught from 2014 to 2016 in a university in Chongqing, China. All the students majored in computer science, and student identifiers were encrypted before data analyses for privacy protection. Statistics about course collaborations records are summarized in Table 2.

Table 2. Summary of course collaboration records.

Number of collaboration	Number of teams	Number of students	Number of courses	Number of teams per student	Number of course per student
3533	1260	312	67	≈11	≈33

We do not use the UCLA scale [12] for loners labeling because lonely students may conceal their real feeling and conditions due to the social stigmas associated with loneliness, which probably lead to unreliable labels. Instead, we invited 32 postgraduates to help us label loners. These postgraduates are among the 312 students in course collaboration records. They have completed their bachelor degree in 2017, and are currently pursuing a master's degree in Chongqing University.

We extract three types of student features, i.e., network-related features, achievement-related features, and loner-related features. The network-related features are constructed based on the collaboration network, where each vertex represents a student and each edge indicates a collaborative relationship between two students. The achievement-related features include the average and the standard deviation of course scores of a student, and the number of courses that the student failed to pass. The loner-related features are those reflecting the characteristics of loners, e.g., the number of partners who has cooperated with the student for only one or two times, or the average academic score of the teammates of the student, or the number of teams the student has joined.

4.2   Classifiers and Parameter Settings

We choose both non-graph classifiers and GNNs for model training and comparison in the experiments. The non-graph classifiers include Decision Tree (DT), Logistic Regression (LR), Gaussian Naive Bayes (GNB). These classifiers take only student features as inputs. Instead, a GNN takes collaboration relationships among students as well as student features as inputs. We choose GCN, GAT and GraphSAGE. For each GNN model, we train it for 200 epochs by using Adam optimizer with a learning rate of 0.005. Considering the highly imbalanced characteristics of the problem, we adopt Macro-F1 score as the metrics to evaluate the classifier performance.

4.3   Classifiers Performance and Model Selection

Table 3 shows the performance of various classifiers with different over-sampling operations. The introduction of SMOTE operation improves the performance of all classifiers except GCN. This demonstrates the effectiveness of over-sampling operation. Noticeably, the Macro-F1 of GraphSAGE-LSTM has been increased from 0.473 to 0.814 after SMOTE operation, achieving the highest performance of all classifiers. The GraphSAGE variants, by making full use of the graph-structured information about student interactions, consistently outperforms non-graph classifiers.

Table 3. Impact of over-sampling operations on classifier performance. Best performance of a classifier is in bold type.

Classifiers	DT	LR	GNB	GCN	GAT	GraphSAGE -mean	GraphSAGE -LSTM	GraphSAGE -meanpool
No over-sampling	0.451	0.613	**0.741**	0.454	0.462	0.469	0.473	0.470
SMOTE	**0.550**	**0.775**	**0.741**	0.438	**0.471**	**0.767**	0.814	0.743
K-means SMOTE	0.458	0.588	0.464	**0.470**	0.470	0.500	0.444	0.492
GOT (SMOTE)	N/A	N/A	N/A	0.458	0.451	**0.767**	**0.844**	**0.844**
GOT (K-means SMOTE)	N/A	N/A	N/A	**0.470**	0.470	0.642	0.814	**0.844**

Surprisingly, another two GNNs, i.e., GCN and GAT, achieved poorer performance than non-graph classifiers. The poor performance is mainly caused by the over-smoothing effect of GNNs, i.e., neighboring nodes generally have closed representation, and a node is inclined to be labeled as the same class to its neighboring nodes [11]. Due to the over-smoothing effect, GCN and GAT tend to classify a loner student as the same class to his/her neighboring nodes, i.e., their teammates, who are probably regular students. In GraphSAGE, however, alleviating the influence of over-smoothing effect due to sampling technique.

The impact of our proposed over-sampling operations, GOT, is also shown in Table 3. After introducing GOT, the performance of GraphSAGE-LSTM and GraphSAGE-meanpool are both increased from around 0.470 to 0.844, reaching the highest prediction performance among all classifiers. These two GraphSAGE variants perform well for the identification of loners, thanks to the effectiveness of GOT in processing class-imbalanced data on graphs.

5   Conclusion

Identification of loners is prone to failure due to biased information collected by questionnaires. We propose a framework for identifying lonely students based on a more objective data source, i.e., project collaboration records. We also present a graph-based over-sampling technique named GOT to address the class-imbalanced problem on graphs. Experiments on real-world dataset show that our proposed method can identify loners with an average accuracy of above 84%. We expect to put this method into practice after testing its effectiveness in a broader scope.

References

1. Becheru, A., Popescu, E.: Using social network analysis to investigate students' collaboration patterns in eMUSE platform (2017). https://doi.org/10.1109/ICSTCC.2017.8107045
2. Brin, S., Page, L.: The anatomy of a large-scale hypertextual web search engine. Comput. Netw. ISDN Syst. **30**(1), 107–117 (1998). https://doi.org/10.1016/S0169-7552(98)00110-X. Proceedings of the Seventh International World Wide Web Conference
3. Chang, E., et al.: Loneliness under assault: Understanding the impact of sexual assault on the relation between loneliness and suicidal risk in college students. Pers. Individ. Differ. **72**, 155–159 (2015). https://doi.org/10.1016/j.paid.2014.09.001
4. Chawla, N., Bowyer, K., Hall, L., Kegelmeyer, W.: SMOTE: synthetic minority over-sampling technique. J. Artif. Intell. Res. (JAIR) **16**, 321–357 (2002). https://doi.org/10.1613/jair.953
5. Crespo, P., Antunes, C.: Predicting teamwork results from social network analysis. Expert Syst. **32**, 312–325 (2013). https://doi.org/10.1111/exsy.12038
6. Douzas, G., Bação, F., Last, F.: Improving imbalanced learning through a heuristic oversampling method based on k-means and SMOTE. Inf. Sci. **465**, 1–20 (2018). https://doi.org/10.1016/j.ins.2018.06.056
7. Hamilton, W.L., Ying, R., Leskovec, J.: Inductive representation learning on large graphs. arXiv e-prints arXiv:1706.02216 (2017)
8. Hochreiter, S., Schmidhuber, J.: Long short-term memory. Neural Comput. **9**, 1735–1780 (1997). https://doi.org/10.1162/neco.1997.9.8.1735
9. Khalid, N.H., Ibrahim, R., Selamat, A., Abdul Kadir, M.R.: Collaboration patterns of researchers using social network analysis approach. pp. 001632–001637 (2016). https://doi.org/10.1109/SMC.2016.7844473
10. Kipf, T.N., Welling, M.: Semi-supervised classification with graph convolutional networks. CoRR abs/1609.02907 (2016). http://arxiv.org/abs/1609.02907
11. Li, Q., Han, Z., Wu, X.: Deeper insights into graph convolutional networks for semi-supervised learning. In: AAAI-18 AAAI Conference on Artificial Intelligence, pp. 3538–3545 (2018)
12. Russell, D., Peplau, L., Ferguson, M.: Developing a measure of loneliness. J. Pers. Assess. **42**, 290–294 (1978). https://doi.org/10.1207/s15327752jpa4203_11
13. Sanchez, W., Martínez-Rebollar, A., Campos, W., Estrada Esquivel, H., Pelechano, V.: Inferring loneliness levels in older adults from smartphones. J. Amb. Intell. Smart Environ. **7**, 85–98 (2015). https://doi.org/10.3233/AIS-140297
14. Skues, J., Williams, B., Oldmeadow, J., Wise, L.: The effects of boredom, loneliness, and distress tolerance on problem internet use among university students. Int. J. Ment. Health Addict. **14**(2), 167–180 (2015). https://doi.org/10.1007/s11469-015-9568-8
15. Veličković, P., Cucurull, G., Casanova, A., Romero, A., Lió, P., Bengio, Y.: Graph attention networks (2017)
16. Wood, G.: The structure and vulnerability of a drug trafficking collaboration network. Soc. Netw. **48**, 1–9 (2017). https://doi.org/10.1016/j.socnet.2016.07.001
17. Zhou, J., et al.: Graph neural networks: a review of methods and applications. arXiv Learning (2018). http://arxiv.org/abs/1812.08434

Node Embedding over Attributed Bipartite Graphs

Hasnat Ahmed[1], Yangyang Zhang[1,3], Muhammad Shoaib Zafar[1],
Nasrullah Sheikh[2], and Zhenying Tai[3(✉)]

[1] School of Computer Science and Engineering, Beihang University (BUAA),
Beijing 100191, China
hasnatsatti@buaa.edu.cn, zhangyy@act.buaa.edu.cn, mszafar.dcs@outlook.com
[2] University of Trento, Trento, Italy
nasrullah.sheikh@unitn.it
[3] Beijing Advanced Innovation Center for Big Data and Brain Computing,
Beihang University, Beijing, China
taizy@act.buaa.edu.cn

Abstract. This work investigates the modeling of attributes along with
network structure for representation learning of the bipartite networks.
Most of the attributed network representation learning (NRL) works
consider the homogeneous type network only; However, these methods,
when apply to bipartite type networks, may not be beneficial to learn
an informative representation of nodes for predictive analysis. Hence, we
propose a BIGAT2VEC framework that examines the internode relationships in the form of direct and indirect relations between two different as
well as the same node type of bipartite network to preserve both structure and attribute context. In BIGAT2VEC, learning is enforced on two
levels: (1) direct inter-node relationship between nodes of different type
(either through the edge or attribute similarities perspective) by minimizing the probabilities through KL divergence; (2) indirect inter-node
relationship within same node type (either through 2nd order neighborhood proximity and attributes similarities perspective) by employing shallow neural network model through maximizing the probabilities.
These two levels are separately optimized, and we leverage its learned
embeddings through late fusion to further execute the network mining
tasks such as link prediction, node classification (multi-class and mul-
tilabel), and visualization. We perform extensive experiments on various datasets and evaluate our method with several baselines. The results
show the BIGAT2VEC efficacy as compare to other (non)attributed representation learning methods.

Keywords: Attributed bipartite graphs · Network embedding · Link
prediction · Classification

1 Introduction

The bipartite network is a distinct class of network to model the relationship
between two entity types, such as there should be no direct relationship between

G. Li et al. (Eds.): KSEM 2020, LNAI 12274, pp. 202–210, 2020.
https://doi.org/10.1007/978-3-030-55130-8_18

the nodes of the same entity type. Recently, to perform predictive analysis on a bipartite network, authors [1] have developed a dedicated approach to learn the representations of nodes, but limited to preserving the pure network aspect only. Additionally, most of the existing works of attributed network embedding focus on homogeneous networks (where nodes are of the same type) [2, 4–6]; when apply to attributed bipartite networks, their performance can be suboptimal due to network class difference. The main contributions of our work are to learn more effective node representation to achieve two goals: (1) efficacy of incorporation of node attributes, mainly evaluate with BiNE [1]; (2) performance comparison with other (non) attributed NRL methods to highlight the difference of class of a network.

In this work, we consider the attributed bipartite networks and propose a BIGAT2VEC framework that exploits the direct and indirect relationships through the structure as well as attributes property. We execute extensive experiments on various datasets for link prediction, classification, and visualization task to illustrate the effectiveness of our proposed work.

2   Problem Definition

Given an attributed bipartite graph $G = (X, Y, E, X_A, Y_A)$, where X and Y represent the two disjoint set of nodes, and E, set of edges, serves the relationship purpose. X_A and Y_A are set of attributes. The aim is to map each node $(\Phi : X \cup Y \rightarrow R^d)$ to a low-dimensional embedding vector. That learned representation preserves the structural as well as attributes context. The d shows the dimension of the learned representation. We represent the embedding vector for nodes in a graph as x_i and y_j respectively.

3   BIGAT2VEC Framework

We narrate the details of our proposed framework BIGAT2VEC in this section. In both direct and indirect relationship modeling, we incorporate the attributes aspect by adopting a $TF\text{-}IDF$ model, which further follows the pairwise cosine similarity function, such as $\cos(x_i, y_j, h)$ with similarity threshold h check.

3.1   Direct Inter-node Relationship Modeling

We recognize those edges as direct edges which are extracted through pure structure and attributes similarities function among nodes of different types, and named it direct relationship. We obtain two induced edge list $E_s^G = (X, Y, W)$ and $E_a^G = (X, Y, W)$ from G with its strength w_{ij}. For E_s^G, w_{ij} value is identified from the network and if the network is unweighted its value is 1. For E_a^G, w_{ij} is the similarity value. To model the local pairwise proximity from these two induced edgelist E_s^G and E_a^G, we employ 1st order proximity as previous works [1,8] and define the joint probability between nodes x_i and y_j as

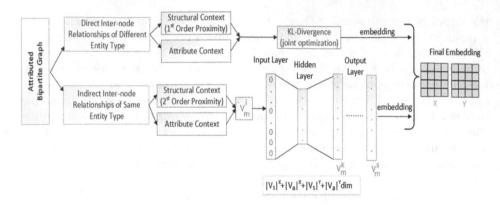

Fig. 1. Architecture of BIGAT2VEC

$P_s(i,j) = \frac{w_{ij}}{\sum_{e_{ij} \in E_s^G} w_{ij}}, P_a(i,j) = \frac{w_{ij}}{\sum_{e_{ij} \in E_a^G} w_{ij}}$, where w_{ij} depicts the strength of relation and identified by e_{ij} from E_s^G or E_a^G. The intuition is same that nodes having a stronger connection in structure as well as attributes holds higher probability to be co-occurred.

To jointly estimate the direct inter-node relations for both aspects in the embedding space, we follow the same word2vec inner product settings [1,8] and utilize sigmoid function σ to transform the interaction value to probability space for $\hat{P}_s(i,j)$ and $\hat{P}_a(i,j)$ as $\frac{1}{1+exp(-x_i^T y_j)}$. We adopt the KL-divergence method, which minimizes their differences from their probability distributions and defines as:

$$- (D_{KL}(P_s||\hat{P}_s) + D_{KL}(P_a||\hat{P}_a)), \tag{1}$$

$$\begin{aligned} D_{KL}(P_s||\hat{P}_s) &= \sum_{e_{ij} \in E_s^G} P_s(i,j) \log \frac{P_s(i,j)}{\hat{P}_s(i,j)}, \\ D_{KL}(P_a||\hat{P}_a) &= \sum_{e_{ij} \in E_a^G} P_a(i,j) \log \frac{P_a(i,j)}{\hat{P}_a(i,j)}. \end{aligned} \tag{2}$$

We optimize the defined direct inter-node relations by utilizing the stochastic objective function, as mentioned in BiNE explicit objective function [1].

3.2 Indirect Inter-node Relationship Modeling

We establish three stages to exploit the indirect relations with respect to structure and attributes of same entity type nodes.

Network Generation: Given an attributed bipartite graph G, we generate four homogeneous graphs $G_s^X = (V_s^X, E_s)$, $G_a^X = (V_a^X, E_a)$ and $G_s^Y = (V_s^Y, E_s)$ and $G_a^Y = (V_a^Y, E_a)$. G_s^X and G_s^Y are formulated by following the idea of [1].

i.e. G_s^X and G_s^Y hold with structural aspect. The idea is to get those nodes which shares the neighbors in same entity type to preserve the 2nd-order neighborhood proximity. To model the node's associated additional informa-tion i.e. attributes, similarity matrix $[S]^{(|X| \times |X|)} = Sim(x_i, x_{others}, h)$, and $[S]^{(|Y| \times |Y|)} = Sim(y_j, y_{others}, h)$ are established to transform them into other two homogeneous graphs G_a^X, G_a^Y.

Random Walk Sequence Generation: We perform short random walk ran-dom walks on these four generated homogeneous graphs $G_s^X, G_a^X, G_s^Y, G_a^Y$. For each node of a graph G_s^X, G_s^Y and G_a^X, G_a^Y, the same number of walks along with its walk length are performed to target the 2nd-order structural neighborhood and the attribute context. We obtain four corpora Q, R, S, T through this step.

Representation Learning: Following the various embedding approaches [2,5,8], we obtain corpora of node sequences and employ Skipgram [3], as illus-trated in Fig. 1. We aim to target these indirect inter-node relations and pre-serve these indirect proximities in the latent representations. For each context as per entity type i.e. X and Y, a node $V_m \in V_s^X | V_a^X | V_s^Y | V_a^Y$ is selected and feed to Skipgram [3]; where input node V_m is the target node, considered as one-hot encoded vector $\{0,1\}^{|V_s^X U V_a^X U V_s^Y U V_a^Y|}$. The output layer gives the $2c$ multinomial distributions of associated context nodes to the given input node. The context size is represented as c, which holds the number of predicted nodes before or after the target node.

Given a target node as per entity type, the objective is to observe its indirect relations through maximizing the probabilities. Following the previous studies [5,6], it is known that the probabilities of context nodes with respect to given input target node are independent of each other. We describe the objective function, which counters structural and attribute aspect as per entity type and if $V_a^X | V_a^Y = 0$, it will become Deepwalk [5].

$$
L_{ir} = \sum_{q \in Q} \sum_{i=1}^{|q|} \log p(q_{-c} : q_c | q_i) + \sum_{r \in R} \sum_{i=1}^{|r|} \log p(r_{-c} : r_c | r_i)
$$
$$
+ \sum_{s \in S} \sum_{i=1}^{|s|} \log p(s_{-c} : s_c | s_i) + \sum_{t \in T} \sum_{i=1}^{|t|} \log p(t_{-c} : t_c | t_i)
$$

(3)

In Eq. 3, the first two term are related to structural and attribute context for entity X type and other terms are for entity Y type with similar settings; whereas $q_{-c} : q_c$, $r_{-c} : r_c$, $s_{-c} : s_c$ and $t_{-c} : t_c$ indicates the sequence of nodes inside a contextual window of length $2c$ in random walks contained in corpus Q, R, S and T, respectively. To compute the probabilities, we employ the hierarchical soft-max and use the Huffman algorithm [3].

3.3   Late Fusion of Learned Latent Representations

We learn the latent representation of nodes separately, which are preserving the defined direct and indirect relations of an attributed bipartite graph. To make it

Table 1. Data statistics

Dataset	$|X|$	$|Y|$	$|E|$	$\#X^A -$ with features	$\#Y^A -$ with features	$\#Lbls$
UNC-Friendship	9695	7128	235207	6-2641	6-2641	–
MovieLens (1M)	6040	3952	1,000,209	3-48	1-18	18
Yelp	1518169	188593	5,996,993	9-45	3-2413	7

more informative and effective representations, we resort to the idea of joining these learned embeddings through the '*python numpy*' concatenation function.

4   Experimental Setup

In this section, we present the details of the experimental setup with the used real-world datasets (statistics mentioned in Table 1).

4.1   Datasets and Preprocessing

UNC-Friendship[1]**:** This pre-processed dataset depicts the student's social friendship network of the American University of North Caroline at Chapel-Hill (UNC). We build a bipartite network with the help of gender attribute, i.e., male and female. In the dataset, we have removed those nodes which do not contain gender information or whose linkage information is not present. For each user, considered six attributes are status, major, second major, dorm, high school, and class year.

MovieLens-1M[2]**:** This dataset contains the user-movie bipartite network where edge weight holds the rating given by the user on a movie. For user type entity, the attributes are age, gender, and occupation. For movie type entity, genres attribute is taken which hold tags like action, drama, adventure, etc.

Yelp[3]**:** It is a social networking service where users rate businesses. To deal with this large network, we calculate node pairwise attribute similarities through locality sensitive hashing (LSH). User attributes are review_count, average_stars, useful, funny, cool, fans, compliment_hot, compliment_cool, compliment_funny, whereas attributes related to businesses are a city, state, and categories. Besides, we assign seven labels to businesses through particular keywords that exist in the categories attribute. Labels are displayed in Fig. 2, which contains the visualization results.

---

[1] https://github.com/lizi-git/ASNE/tree/master/data.
[2] https://grouplens.org/datasets/movielens/1m/.
[3] https://www.yelp.com/dataset.

Table 2. Link prediction performance

Algorithm	UNC-Friendship		MovieLens-1M		Yelp	
	AUC-ROC	AUC-PR	AUC-ROC	AUC-PR	AUC-ROC	AUC-PR
DeepWalk	65.51%	58.91%	79.12%	77.91%	75.57%	54.30%
Node2vec	65.93%	58.64%	83.14%	82.21%	77.76%	58.66%
BiNE	75.16%	73.89%	85.07%	83.56%	86.18%	80.07%
TADW	50.01%	73.15%	60.20%	75.55%	–	–
Gat2Vec	65.90%	58.63%	82.66%	80.85%	80.06%	64.58%
TriDNR	66.63%	59.03%	82.07%	79.75%	81.32%	65.46%
BIGAT2VEC	**75.73%**	**74.15%**	**85.25%**	**83.71%**	**86.41%**	**80.14%**

4.2   Baselines

We compare BIGAT2VEC against the pure network as well as attributed based representation learning methods.

- **DeepWalk** [5]: It learns node embeddings by performing uniform random walks in a homogeneous network and then applies the word2vec model.
- **Node2vec** [2]: This method introduces biased random walks strategy in corpus generation of node sequences for homogeneous networks.
- **BiNE** [1]: It deals with the plain structure of the bipartite network in the context of joint modeling of both implicit and explicit relations of the node.
- **Text Associated DeepWalk (TADW)** [9]: It factorizes a network and associated text in the matrix to learn d-dimensional representations of nodes.
- **Gat2Vec** [6]: It jointly models the structural and additional associated information of the homogeneous network.
- **TriDNR** [4]: This method targets the nodes correlations from the structure and node-word/label-word perspectives of a homogeneous network.

4.3   Parameter Settings

Our method involves various parameters, and we set them as follows: direct and indirect relations embedding size $d_i = 64$; final embedding size $d = 128$; number of walks $= 10$, walk length $= 80$, and window size $c = 5$ are used in $G_s^X, G_a^X, G_s^Y, G_a^Y$; similarity threshold check $h = [0.5 - 0.8]$. For fairness of comparison, we use the same value for common parameters and best values for the rest of the parameters, as reported in their respective methods.

4.4   Performance Comparison

Link Prediction. We follow the same procedure [1,2] for the link prediction task. For each node, observed links are considered as positive instances, and equally number of unconnected node pairs are randomly generated as negative instances. Afterward, to avoid overfitting, we randomly sample 10 folds

Table 3. Performance of multi-label classification on MovieLens-1M

Metric	T_R	DeepWalk	Node2vec	BiNE	TADW	Gat2Vec	TriDNR	**BIGAT2VEC**
MICRO-F1	10%	76.12%	77.16%	37.30%	41.39%	77.25%	77.32%	**92.39%**
	30%	81.58%	82.32%	37.38%	44.98%	82.02%	81.96%	**95.31%**
	60%	83.76%	84.51%	38.91%	47.70%	84.18%	84.06%	**96.14%**
MACRO-F1	10%	61.52%	62.84%	9.36%	30.77%	63.06%	62.16%	**88.28%**
	30%	74.27%	75.47%	9.73%	43.23%	75.74%	75.98%	**92.71%**
	60%	79.12%	79.76%	10.11%	47.61%	79.75%	79.15%	**95.65%**

Table 4. Performance of multi-class classification on Yelp

Metric	T_R	DeepWalk	Node2vec	BiNE	Gat2Vec	TriDNR	**BIGAT2VEC**
MICRO-F1	10%	29.40%	32.48%	23.88%	93.43%	91.87%	**95.26%**
	30%	30.28%	34.10%	23.98%	93.89%	92.31%	**95.81%**
	60%	30.64%	34.76%	24.32%	94.23%	93.81%	**95.88%**
MACRO-F1	10%	21.43%	25.64%	15.48%	92.12%	90.16%	**94.15%**
	30%	21.57%	26.10%	15.57%	93.35%	91.89%	**95.55%**
	60%	21.90%	26.38%	15.76%	94.07%	92.33%	**95.73%**

of the train, test set with a ratio of 60:40. We assess the result on each test set and report the average score. We employ two metrics: Area under the Precision−Recall curve (AUC−PR) and ROC curve (AUC−ROC) in this task. We treat the learned embedding vectors of x_i and y_j as feature vectors with their observed link information, i.e., 1 or 0 and feed into logistic regression classifier. The link prediction task performance of our method, along with collective baselines, is composed in Table 2. Overall, the proposed BIGAT2VEC model performs better than other algorithms in terms of both metrics. It helps us to draw two key observations: (1) considering attributes aspect in learning the node representation certainly has a positive impact when compared to BiNE [1]. (2) network type difference does matter when dealing with other attributed network embedding methods. We are unable to run TADW [9] on the Yelp dataset due to its computationally expensive singular value decomposition operation.

Node Classification. In a classification task, we consider the learned representation of nodes related to entity type Y only, i.e., movies and businesses. Afterwards, we randomly select a training sample of size $T_R \in \{10\%, 30\%, 60\%\}$. The rest is taken for testing purpose. Furthermore, for each training ratio, we randomly generate 10 folds of train-test set and report the averaged results. We perform multi-label classification on movielens, and multi-class classification on yelp by applying the one-vs-rest logistic regression classifier *liblinear*. Micro-F1 and Macro-F1 evaluation metrics are adopted same as previous works [4,6]. Table 3, and Table 4 show the results where BIGAT2VEC beats all other methods; which highlights the modeling of rich associated attributes effect as compare to BiNE, and a class of network property as measure to other attributed NRL methods like Gat2Vec [6], and TriDNR [4]. The performance of DeepWalk [5]

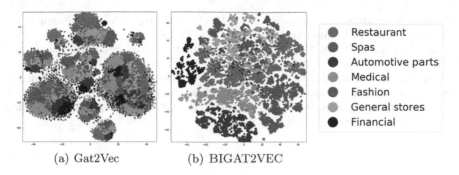

(a) Gat2Vec   (b) BIGAT2VEC

Fig. 2. Visualization of businesses in Yelp. Color of a node indicates the labels of businesses. (Color figure online)

and Node2vec [2] algorithms point out the network sparsity property whereas BiNE [1] additionally points out its poor impact of modeling structural relations for classification purpose.

Visualization. We project the learned representations of business entity type nodes Y in the visualization task by employing t-SNE [7]. We demonstrate the nodes co-relationship by projecting their embeddings, i.e., 128 dimension space to 2-D dimension space. Figure 2 shows the visualization results. The visualization result of Gat2Vec [6] is not very meaningful; in fact, all businesses are not properly clustered. Overall, BIGAT2VEC performs quite well and generate a meaningful layout, and some clearly defined clusters as per business labels are displayed.

5   Conclusion

In this paper, we propose the BIGAT2VEC framework where attributes proximity is preserved along with structural proximity for the bipartite network. The experimental results ensure the efficacy of our work in essential mining tasks such as link prediction, classification (multi-class and multi-label), and visualization.

References

1. Gao, M., Chen, L., He, X., Zhou, A.: BiNE: bipartite network embedding. In: SIGIR, pp. 715–724 (2018)
2. Grover, A., Leskovec, J.: node2vec: scalable feature learning for networks. In: SIGKDD, pp. 855–864 (2016)
3. Mikolov, T., Chen, K., Corrado, G., Dean, J.: Efficient estimation of word representations in vector space. arXiv preprint arXiv:1301.3781 (2013)
4. Pan, S., Wu, J., Zhu, X., Zhang, C., Wang, Y.: Tri-party deep network representation. In: IJCAI, vol. 101, pp. 1895–1901 (2016)
5. Perozzi, B., Al-Rfou, R., Skiena, S.: DeepWalk: online learning of social representations. In: SIGKDD, pp. 701–710 (2014)

6. Sheikh, N., Kefato, Z., Montresor, A.: GAT2VEC: representation learning for attributed graphs. Computing **101**(3), 187–209 (2019)
7. Tang, J., Liu, J., Zhang, M., Mei, Q.: Visualizing large-scale and high-dimensional data. In: WWW, pp. 287–297 (2016)
8. Tang, J., Qu, M., Wang, M., Zhang, M., Yan, J., Mei, Q.: LINE: large-scale information network embedding. In: WWW, pp. 1067–1077 (2015)
9. Yang, C., Liu, Z., Zhao, D., Sun, M., Chang, E.: Network representation learning with rich text information. In: IJCAI (2015)

FastLogSim: A Quick Log Pattern Parser Scheme Based on Text Similarity

Weiyou Liu[1,2], Xu Liu[1,2], Xiaoqiang Di[1,2,3]($\boxtimes$), and Binbin Cai[3]

[1] School of Computer Science and Technology, Changchun University of Science and Technology, Changchun, China
dixiaoqiang@cust.edu.cn
[2] Jilin Province Key Laboratory of Network and Information Security, Changchun, China
[3] Information Center, Changchun University of Science and Technology, Changchun, China

Abstract. Logs completely record all system events which can be used to reveal network security issue and analyse user behaviour. Since logs are stored in the form of unstructured data and there is no universal log retention standard, they can hardly be analysed directly. Most of the existing log parsers focus on the parsing accuracy and ignore the time performance while parsing the large-amount of logs. Therefore, this paper proposes FastLogSim, a fast log parsing scheme based on text similarity. To simplify the parsing workload, we perform deduplication on the logs after removing the key variable values to obtain the template. Subsequently, the similarity is computed to merge the similar templates and then obtain the log pattern. FastLogSim not only reduces the number of templates that need to be parsed from tens of millions to dozens, but also greatly improves the speed of pattern extraction. We evaluated FastLogSim on four real public log datasets. The experimental results show that when the FastLogSim process tens thousands of logs, it performs almost the same time as the mainstream log parser. However, when the number of logs exceeds ten million, FastLogSim is three times faster than previous state-of-the-art parsers. Hence, FastLogSim is appropriative for large-scale log pattern mining.

Keywords: Log parser · Text similarity · Log pattern

1 Introduction

The system log records every detail of the system's operation and plays a vital role in log mining which can be used to detect network anomaly [2,6,9], fault

Supported by the Department of Science and Technology of Jilin Province grant NO. 20190302070GX, the Education Department of Jilin Province grant NO. JJKH20190598KJ, Jilin Education Science Planning Project (GH180148) and Jilin Province College and University "Golden Course" Plan Project (Network protocol and network virus virtual simulation experiment).

© Springer Nature Switzerland AG 2020
G. Li et al. (Eds.): KSEM 2020, LNAI 12274, pp. 211–219, 2020.
https://doi.org/10.1007/978-3-030-55130-8_19

diagnosis [16], user behavior analysis [12], and attack source tracing. However, with the rapid development of IT, log data generated by networks and systems has shown near-explosive growth, which overwhelms the security maintenance personnel and makes forensics time-consuming. Therefore, this paper concentrates on how to quickly parse large-scale logs.

The main challenge of log parsing comes from the high diversity, high velocity and high volume. Since logs are mostly stored as unstructured or semi-structured format, the first step of parsing is to convert the raw logs into structured data and process variable. Raw logs generally consist of time, id, event level and event content. The second step of parsing is to extract log pattern from event content. It has attracted many scholars' attention, and the existing parsing schemes can be generally divided into three categories: log-source-code-based analysis [17], log-content-based analysis [1,2,7,8] and temporal-relationship-based analysis [13]. All of them have obtained some outstanding achievements, but the log source codes are often difficult to get, and when dealing with large-scale log, the processing efficiency still needs to be improved, especially when dealing with real-rime analysis.

In order to accelerate the log parsing, a fast log parsing scheme, FastLogSim, is proposed. We use regular expressions to recognise the possible tokens of log and mark them with specific symbols (such as <ip>, <path>, and so on) instead of the same symbol <*> that is commonly used in the current works [1,4,5]. It can quickly and accurately recover the original log information according to the extracted log pattern and variable values, which is convenient for the log review. And then we perform deduplication [11] on the templates that contain token symbols, which reduces the number of logs that need to be parsed from tens of millions to dozens. Afterwards, to merge the templates that are structure-subordinate or just different at few words, the similarity is computed based on templates coupled with the specific token symbols.

As the log patterns do not have uniform format so far, it is difficult to evaluate the parsing performance. Hence, researchers need to manually extract patterns as ground-truth [5] to calculate the accuracy rate. Manually marking the original log is a time-consuming and repetitive task, but in our proposed method, only need to mark dozens of logs after deduplication and then corresponds to parsing result. We evaluate the pattern effectiveness and time efficiency on four public log datasets and compare with four state-of-the-art parsers [1,7,14,15]. The experimental results show that FastLogSim can flexibly control the number of log patterns to a large extent and its processing speed has obvious advantages. When handling with thousands of logs, the parsing time of different parsers is close, but as the number of logs increases, the advantages of the proposed scheme are gradually obvious. Especially, when dealing with ten millions of logs, parsing speed is three times faster than mainstream parsers. Furthermore, FastLogSim outperforms them in terms of both efficiency and effectiveness. The main contributions are summarized as follows.

– FastLogSim's processing speed is fast, due to the number reduction of object logs for extracting patterns. The parsing time of each log is only microseconds, which is possible for real-time log processing.

– FastLogSim considers semantics and uses the text similarity to judge log similarity, which ensures semantics after the log parsing.

– FastLogSim is generally applicable and can be applied to parsing system log generated by different devices and systems.

The remainder of this paper is organized as follows: We introduce the background and related work in Section 1. Section 2 describes our scheme. Section 3 presents our experimental design and results. We make discussion and conclusion in Section 4.

2 Methodology

In this section, we briefly introduce FastLogSim, the entire structure of Fast-LogSim is described in Fig. 1.

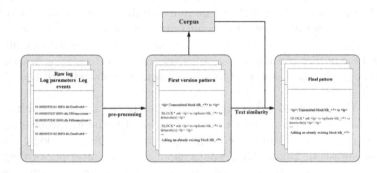

Fig. 1. FastLogSim scheme

2.1 Step1: Log Preprocessing

In the log preprocessing phase, we distinguish between log parameters with log events in raw log messages, and then process variables in log events. Although the lengths of log are various, the length of log parameters in homologous log is fixed. Then Log events can be obtained by specifying the location index of each log event in the entire log. The log event contains constants and variables. In the preprocessing stage, most variables can be represented by identifiers through domain knowledge and regularization expression.

In the previous study [1–3,9], coarse-grained variable substitution was performed during the preprocessing phase, replacing the token containing numbers with a uniform identifier (<*>). We believe that fine-grained preprocessing of raw log can reduce the workload of subsequent parsing, and its time complexity is still O(n) (n is the number of logs). Although there is no standard format for log events, some variables exist in common format, such as website, ip address, file path, software, network protocol, etc. These variables can be matched by regular expressions.

Table 1. Variable handling

Variable	Regularization expression	Constant
Ip address	$((25[0\text{-}5]\|2[0\text{-}4]\backslash d\|((1\backslash d2)\|([1\text{-}9]?\backslash d)))\backslash.)\{3\}$ $(25[0\text{-}5]\|2[0\text{-}4]\backslash d\|((1\backslash d2)\|([1\text{-}9]?\backslash d)))$	<ip>
Website	$([a\text{-}zA\text{-}Z0\text{-}9\backslash.\backslash\text{-}]^*)(com\|net\|hk\|cn\|wpad)(\backslash:[0\text{-}9]^*)?$	<url>
File path	$(\backslash/)(.^*)(\backslash/)(.^*)$	<path>
Software	$([a\text{-}zA\text{-}Z0\text{-}9\backslash.\backslash\text{-}]^*)(exe)$	<app>

In FastLogSim we take precedence over fixed-format variables and then handle variables containing numbers. As shown in the Table1, FastLogSim replaces some variables with a fixed format with the specified constants to avoid errors in the pattern extraction caused by different variables being represented by the same identifier. It is worth noting that file path and the website include the same punctuation marks '/ ' and '-', we use the site-specific top-level domain name to determine whether the token is a website, and then determine whether it is a file path. Finally, the token containing numbers is replaced with the identifier (<*>).

We save the processed log event and its index in the raw log as a dataframe format. It can be seen from logs that have finished variables processing which exists a large amount of duplicate data. By removing duplicate logs, dozens of logs can be obtained. Taking HDFS data as an example, after performing above operations, 54 logs can be obtained. At this time, the scale of objects to be extracted pattern is reduced from the original 10 million to 54, which reduces the time for pattern extraction. Dozens of logs are called as the pattern of first version.

2.2   Step2: Calculate Text Similarity

In this step, FastLogSim uses a TF-IDF-based text similarity to filter similar patterns and selects the first version of the pattern obtained as a corpus. We will classify each log in the first version of the pattern as a sparse vector, then train TF-IDF model. The formula of TF-IDF is shown in Eq. (1). #word is the number of target words in raw log event, #total is the number of target words in all log events. #L is the number of patterns of the first version, and $\#L_{word}$ is the number of patterns containing this target word.

$$TF - IDF = TF * IDF = \frac{\#word}{\#total} * \log(\frac{\#L}{1 + \#L_{word}}) \tag{1}$$

After training TF-IDF model, the text similarity between each pattern of the first version and that of the corpus is calculated. The obtained similarity is a decimal between 0 and 1, similarity between two same words will be 1. Therefore, we combine two patterns that satisfy the text similarity ranging from the defined threshold to 1. When there are multiple patterns that can be merged, put all the operable pattern indexes into a list. The left part of Table 2 shows some examples of the indexes that can be combined by setting threshold to 0.75.

Table 2. Pattern index

Merged index	Ergodic index
[20,28,45]	[20,28],[20,45],[28,45]
[23,33,52,53]	[23,33],[23,52],[23,53],[33,52],[33,53],[52,53]

2.3   Step3: Merge Similar Patterns

According to the merged pattern index obtained in step two, the pattern is automatically combined by recognizing new variable and longest common subsequence (LCS). For each group of merged patterns, the indexes to be merged are divided into two-tuples by means of permutation and combination. We divide the three cases of the merge pattern and traverse all two-tuples.

Case 1: The length of patterns are same. Two patterns with the same length are divided into a list of multiple tokens by split() function, and different tokens at the same position are treated as new variables. Add these new variables to the list named new_variable where all variables will be stored. After traversing all patterns and collecting all variables, merge the patterns.

Case 2 and Case 3 merge patterns with different lengths. We first use the longest common subsequence algorithm to determine whether a pattern is a subsequence of another pattern.

Case 2: If outcome is true. We merge these two patterns directly and use subsequence as the final pattern.

Case 3: If outcome is false. Since the longest common subsequence algorithm uses each letter as an element instead of a token, this does not suit. We choose to split two patterns into two lists of tokens and select the shared token as final pattern.

After the traversal is completed, the variables in the pattern in case one are processed by the complete variable list (new_variable), replaced by identifier($<*>$). The pattern that does not satisfy the merge condition is added to the final pattern directly.

3   Experiment

Since each parser algorithm is designed differently, the number of patterns obtained will be different. We reproduce four mainstream log-based parsers:

Spell, LogSig, SLCT, and IPLOM. Spell [1] uses the longest common subsequence to parse the log as a stream. Through a 3-Step hierarchical partitioning process IPLoM [7] partitions log data into its respective clusters. By searching the most representative messages signatures, LogSig [14] can handle various types of log data. SLCT [15] extracts log patterns through a simple clustering algorithm. We set the threshold for text similarity in FastLogSim to 0.75 to parse four log datasets.

3.1 Log Dataset

We evaluated FastLogSim on four public real log datasets, which are summarized in Table 3. HDFS and Zookeeper are log datasets collected distributed system; proxifier collects logs of independent software; Openstack is generated by an open source cloud infrastructure, we select some logs that are marked as normal in Openstack. All the experiments are conducted on an Ubuntu 16.04 LTS machine with Intel Xeon (R)W-2123, 3.6 GHz CPU, GeForce GTX TITAN Xp COLLECTORS EDITION GPU and 12 GB VRAM.

Table 3. Log dataset

Dataset	Log messages	Length	Size
HDFS	1175629	6–29	1540999 KB
Openstack	137074	6–32	39557 KB
Zookeeper	74380	8–27	10183 KB
Proxifier	11108	10–27	2486 KB

3.2 Efficiency Analysis

FastLogSim is designed to parse large-scale log data sets. We have five parsers processing HDFS data sets in the same environment. Figure 2(a) shows the running time of five parsers. The experimental results prove the obvious advantages of FastLogSim in processing large-scale logs, and the running speed is three times faster than the others.

We also used the other three datasets to validate the generality of FastLogSim. Figure 2(b) shows the parsing time of the five parsers on these datasets. In addition to the fact that FastLogSim is slightly slower than IPLoM on the Proxifier dataset, FastLogSim is able to parse different types of logs and has an advantage in resolution speed. Due to the improvement of FastLogSim in dealing with variables, it can get an appropriate number of patterns when dealing with different datasets.

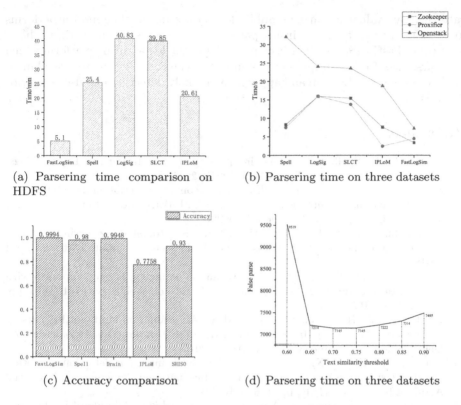

(a) Parsering time comparison on HDFS

(b) Parsering time on three datasets

(c) Accuracy comparison

(d) Parsering time on three datasets

Fig. 2. Experiment analysis

3.3 Effectiveness Analysis

We calculate the accuracy rate of FastLogSim on HDFS to verify the effectiveness, the comparisons are shown in Fig. 2(c). Our FastLogSim method works remarkably better than IPLoM and SHISO [10], we also observe that Spell, Drain [4] and FastLogSim can achieve high parsing accuracy (more than 99%).

Therefore, setting an appropriate threshold can ensure the effectiveness of FastLogSim. We compare parsing results of seven thresholds from 0.6 to 0.9. Considering that the log amount of HDFS dataset is more than 10 million, we calculate the false parse for intuitive performance reflection. The result is shown in Fig. 2(d), when the threshold of text similarity sets between 0.7 and 0.75, the parser performs best. The false alarm rate in these seven groups of experiment is less than 0.1%, which can prove FastLogSim can complete log parser effectively.

4 Conclusion

This paper proposes a new log parsing framework, FastLogSim. It de-duplicates the similar logs based on text similarity in preprocessing stage, and then extracts patterns. By evaluating on four public log data sets, FastLogSim outperforms some state-of-the-art parsers in log parsing speed. In the future work, we hope to improve FastLogSim so that it can be used in the log auditing platform which can parse logs from multiple devices in data center. Furthermore, we plan to further develop FastLogSim and apply it to anomaly detection and user behavior analysis.

References

1. Du, M., Li, F.: Spell: online streaming parsing of large unstructured system logs. IEEE Trans. Knowl. Data Eng. **31**(11), 2213–2227 (2019)
2. Fu, Q., Lou, J., Wang, Y., Li, J.: Execution anomaly detection in distributed systems through unstructured log analysis. In: Ninth IEEE International Conference on Data Mining, pp. 149–158, December 2009
3. He, P., Zhu, J., He, S., Li, J., Lyu, M.R.: Towards automated log parsing for large-scale log data analysis. IEEE Trans. Dependable Secure Comput. **15**(6), 931–944 (2018)
4. He, P., Zhu, J., Zheng, Z., Lyu, M.R.: Drain: an online log parsing approach with fixed depth tree. In: IEEE International Conference on Web Services (ICWS), pp. 33–40, June 2017
5. He, P., Zhu, J., He, S., Jian, L., Lyu, M.R.: An evaluation study on log parsing and its use in log mining. In: 46th Annual IEEE/IFIP International Conference on Dependable Systems and Networks (DSN) (2016)
6. Liu, W., Liu, X., Di, X., Qi, H.: A novel network intrusion detection algorithm based on fast Fourier transformation. In: 1st International Conference on Industrial Artificial Intelligence (IAI), pp. 1–6, July 2019
7. Makanju, A., Zincir-Heywood, A.N., Milios, E.E.: A lightweight algorithm for message type extraction in system application logs. IEEE Trans. Knowl. Data Eng. **24**(11), 1921–1936 (2012)
8. Makanju, A.A., Zincir-Heywood, A.N., Milios, E.E.: Clustering event logs using iterative partitioning. In: Proceedings of the 15th ACM SIGKDD International Conference on Knowledge Discovery and Data Mining, KDD 2009, pp. 1255–1264. ACM, New York (2009)
9. Min, D., Li, F., Zheng, G., Srikumar, V.: Deeplog: anomaly detection and diagnosis from system logs through deep learning. In: The 2017 ACM SIGSAC Conference (2017)
10. Mizutani, M.: Incremental mining of system log format. In: IEEE International Conference on Services Computing, pp. 595–602 (2013)
11. Ortona, S.: An analysis of duplicate on web extracted objects. In: Companion Publication of International Conference on World Wide Web Companion (2014)
12. Poggi, N., Muthusamy, V., Carrera, D., Khalaf, R.: Business process mining from e-commerce web logs. In: Daniel, F., Wang, J., Weber, B. (eds.) BPM 2013. LNCS, vol. 8094, pp. 65–80. Springer, Heidelberg (2013). https://doi.org/10.1007/978-3-642-40176-3_7
13. Saad, K., Simon, P.: Eliciting and utilising knowledge for security event log analysis: an association rule mining and automated planning approach. Expert Syst. Appl. **113**(116–127), S0957417418304226 (2018)
14. Tang, L., Li, T., Perng, C.S.: Logsig: generating system events from raw textual logs. In: Proceedings of the 20th ACM International Conference on Information and Knowledge Management, CIKM '11, pp. 785–794. ACM, New York (2011)

15. Vaarandi, R.: A data clustering algorithm for mining patterns from event logs (2003)
16. Wong, W.E., Debroy, V., Golden, R., Xu, X., Thuraisingham, B.: Effective software fault localization using an RBF neural network. IEEE Trans. Reliab. **61**(1), 149–169 (2012)
17. Xu, W., Huang, L., Fox, A., Patterson, D., Jordan, M.I.: Detecting large-scale system problems by mining console logs. In: Proceedings of the ACM SIGOPS 22Nd Symposium on Operating Systems Principles, SOSP 2009, pp. 117–132. ACM, New York (2009)

Knowledge Management for Education

Robotic Pushing and Grasping Knowledge Learning via Attention Deep Q-learning Network

Zipeng Yang[1,2,3] and Huiliang Shang[1,2,3(✉)]

[1] Academy for Engineering and Technology, Fudan University, Shanghai, China
{yangzp19,shanghl}@fudan.edu.cn
[2] Engineering Research Center of AI and Robotics, Ministry of Education, Shanghai, China
[3] Shanghai Engineering Research Center of AI and Robotics, Shanghai, China

Abstract. Robotic grasping is a fundamental manipulation in multiple robotic tasks, which has great research significance. It is challenging to perform robotic grasping in cluttered environments due to the occlusion and stacking of objects. We propose an attention deep Q-learning network for robotic grasping with the assistance of pushing actions with non-sparse rewards. The attention network improves the performance of deep Q-learning network by weighting feature channels. The robot use pushing actions to dilute dense objects to create space for grasping. The pushing and grasping knowledge are learned by trial and error in a self-supervised way. To evaluate the robotic grasping performance, we present an overall performance metric, which contains three evaluation factors: task completion rate, grasping success rate and action efficiency. The experimental environment is established on the V-REP simulation software to verify our proposed model. The results show that our pushing strategy can not only improve robotic grasping performance but also avoid unnecessary pushing actions to improve action efficiency. At the same time, ablation studies prove the effectiveness of the attention mechanism. Our proposed method can achieve overall performance of 82.33% for robotic grasping.

Keywords: Robotic grasping · Deep reinforcement learning · Visual attention mechanism

1 Introduction

In recent years, robots have been widely applied in manufacturing, service industry and other fields, which are playing an increasingly important role. Among all robotic tasks, robotic grasping is significant in that it will improve work efficiency and save manpower in many scenarios such as garbage sorting. Robotic grasping is a fundamental manipulation that has been studied for decades. The current research focus is how to make robots to learn grasping knowledge in cluttered environments. This means that the position and information of the target objects are unknown in advance, the manipulator robot autonomously determines the grasping manipulation by learning with the help of the sensing device.

© Springer Nature Switzerland AG 2020
G. Li et al. (Eds.): KSEM 2020, LNAI 12274, pp. 223–234, 2020.
https://doi.org/10.1007/978-3-030-55130-8_20

Robotic grasping methods are mainly divided into analytical methods and empirical methods [1]. The analysis methods are not concerned because of the complicated calculation and limited versatility. The learning-based empirical method has extremely high research significance. Learning grasping knowledge through supervised learning is a common idea, which has good performance for single-object grasping tasks. However, in actual application scenarios, the demand is more for multi-object grasping tasks. The complexity of the scene makes it difficult to make the training datasets. Deep reinforcement learning is a learning method that does not require the training datasets. The manipulator robot learn grasping knowledge by trial and error. Moreover, closely arranged objects can make grasping difficult in cluttered environments. It is a good choice to dilute dense objects by pushing actions, which will create space for grasping manipulation and improve the success rate of grasping. Simultaneous learning pushing and grasping knowledge by deep reinforcement learning will help solve the problem of robotic grasping in cluttered environments.

In this work, we propose a method of deep reinforcement learning to enable the manipulator robot to learn pushing knowledge and grasping knowledge synchronously. To sum up, the main contribution of our work as follows:

- We propose an attention deep Q-learning network for robotic grasping in cluttered environments. The visual attention mechanism is added to the traditional deep Q-learning network to fully extract scene feature information, which can improve the effectiveness of knowledge learning.
- To solve the problem of difficulty in grasping dense objects, we propose a push strategy with non-sparse rewards, which is to make dense objects sparse by pushing to create space for grasping. Deep Q-learning network needs to be rewarded according to actions during the learning process. The rewards for pushing is non-sparse, which based on the degree of improvement in the possibility of grasping success after the pushing action.

To the best of our knowledge, our work is the first to propose an attention deep Q-learning network for robotic grasping with the assistance of pushing actions with non-sparse rewards. The experimental results show that our proposed method has a significant improvement in task completion rate, grasping success rate and action efficiency. The overall performance for robotic grasping is 82.33%, which is better than other comparison methods.

2   Related Work

2.1   Robotic Grasping

In early years, analytical methods were the major interests of robotic grasping research, which directly analyze the geometric structure of the target object and then perform grasping manipulation according to force closure or shape closure. However, the analysis and calculation are time-consuming and it is not realistic to obtain the parameter information of the objects. Considering the shortcomings of analytical methods, robotic grasping research turns to data-driven methods [2]. Song *et al.* [3] first use object detector for 2D object recognition to get bounding boxes of objects, which are used as the mask to filter out the corresponding point clouds from the sence point cloud. The grasping pose is gained by template matching with CAD-model database. Zeng *et al.* [4] segment and label multiple views of a scene and then fit pre-scanned 3D object models to the resulting segmentation to get the 6D object pose for robitic grasping. In actual conditions, it is tedious and inefficient to establish a database for each grasped object.

Recently, deep learning has developed rapidly and achieved significant results in many areas such as object detection. Inspiringly, many researchers have applied deep learning to robotic grasping. Lenz *et al.* [5] propose a cascade system with a two-level deep network. The first-level network is used to get all possible grasping detection boxes and the second-level network is used to score these boxes to get the best one. Considering that the two-level grasping detection method takes a long time, Redmon *et al.* [6] propose an end-to-end grasping detection method, which can directly generate the best grasping detection box. Different from object detection, there can be multiple grasping poses for one object. The optimal grasping pose of the same object may be different in different scenes. Making training datasets to make the manipulator robot to learn grasping knowledge is the major challenge.

More recently, some research work formulate robotic grasping as reinforcement learning problem. Kalashnikov *et al.* [7] introduce a scalable self-supervised vision-based reinforcement learning framework which leverage over 580 k grasp attempts with multi-robot for training. The hardware cost and time cost of training are extreme expensive. Zeng *et al.* [8] present a framework for learning pushing and grasping knowledge which combines deep reinforcement learning with affordance-based manipulation. The training process is efficient that only needs about 3 k attempts. However, the pushing strategy with sparse rewards lacks pertinence to promote grasping success. Our reward function for pushing is non-sparse, which based on the degree of improvement in the possibility of grasping success after the pushing action.

2.2 Visual Attention Mechanism

The attention mechanism has made important breakthroughs in the fields of computer vision and natural language processing in recent years, which can effectively improve the performance of the model. Spatial attention is to locate the attention target in the image and perform some transformations such as affine transformation. Google Deep-Mind [9] propose Spatial Transformer Network (STN) to perform target localization and transformation in one stage. Channel attention is to focus attention on the extracted feature channel dimensions. SENet [10] is a typical channel attention model that models the importance of each feature channel. Convolutional Block Attention Module [11] is one of spatial-channel mixed attention models which merge spatial attention and channel attention. Considering that the overly complicated deep network structure will affect the learning effect of reinforcement learning, we apply the attention model of SENet structure to deep Q-learning network.

3   Method

3.1   Markov Decision Process

The basic theory of reinforcement learning is the Markov decision process: at time t, the agent selects and executes an action a_t in the action space according to the state s_t, then the environment state change to s_{t+1} and the agent is given reward $R(s_t, a_t)$ according to the reward function. The goal of reinforcement learning is to learn an optimal strategy to get the maximum cumulative reward value.

In order to use deep reinforcement learning to solve the robotic grasping problem, the process of grasping and pushing can be formulated as the Markov decision process. Due to the high dimensionality of the state and limited action space, the value-based deep reinforcement learning method Deep Q-Learning Network is suitable to solve this problem. The details of the state representation, action space and reward functions are described in the next sections.

3.2   State Representation and Action Space

The problem solved by our work is robotic grasping in cluttered environments. Information perception of workspace scenes is a particularly important part. We choose RGB-D camera, which can acquire RGB information and depth information, as the sensing device of the system. When the camera is fixed directly above the workspace, the occlusion of the manipulator robot will interfere with the acquisition of scene information. To solve this matter, we refer to the practice of [8]: fix the camera obliquely above the workspace, project the RGB-D data on to a 3D point cloud and then orthographically back-project upwards in the gravity direction to get the heightmaps of RGB-D channels. The RGB heightmap and the D heightmap constitute the representation of state s_t at time t.

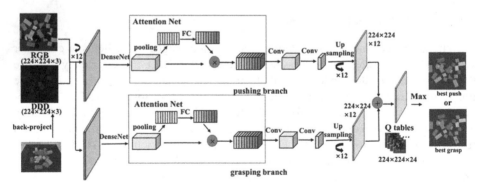

Fig. 1.   Overview of our network structure. There are two network branches with the same structure for pushing and grasping respectively. One branch infers Q-table with the same resolution (224 × 224) as heightmaps. The heightmaps are rotated into 12 orientations to represent different orientations of pushing and grasping and then fed into two branches successively to infer overall Q-Table (2 × 12 × 224 × 224), which corresponds to the action space. The action corresponding to the maximum Q value is the optimal action in the current state.

Usually, the manipulator robot has 6 degrees of freedom, which means that the gripper of the manipulator robot can change in x, y, z position and in roll, pitch, yaw. In our work, we only consider top-down grasps, so the grasping pose is reduced to 4 degrees of freedom (x, y, z, yaw). When x and y are determined, z can be determined according to the heightmap of the D channel. Therefore, the action space for the grasping action is as follows:

$$a_g = \{(x_g, y_g, k_g) | 0 \leq x_g, y_g < 224, 0 \leq k_g < 12\} \tag{1}$$

where x_g, y_g denote the pixel coordinates of the grasping target point in the heightmaps of RGB-D channels and k_g denotes the rotation angle parameter along the z-axis. The grasping position can be obtained by performing coordinate transformation on the pixel coordinates. Combined with the actual situation, we discretize the rotation angle along the z-axis, The grasping orientation along the z-axis can be obtained by $\frac{k_g}{12} \times 2\pi$.

Similarly with the grasping action space, the action space for the pushing actions is as follows:

$$a_p = \{(x_p, y_p, k_p) | 0 \leq x_p, y_p < 224, 0 \leq k_p < 12\} \tag{2}$$

where x_p, y_p denote the pixel coordinates of the starting point of pushing in the heightmaps of RGB-D channels and k_p denotes the pushing orientation parameter along the z axis. During the pushing phase, the closed two-finger gripper move linearly in the direction of $\frac{k_p}{12} \times 2\pi$ from the starting point to push object to be suitable for grasping.

In summary, the total action space can be obtained by merging the grasping action space and the pushing action space:

$$a = \{(\psi, x, y, k) | \psi \in \{grasp, push\}, 0 \leq x, y < 224, 0 \leq k < 12\} \tag{3}$$

3.3   Network Structure and Attention Mechanism

Since the state is the heightmaps of RGB-D channels from the workspace scene, it is difficult to establish Q-table for high-dimensional state. Using deep neural network to establish Q function to fit and obtain the Q value of different actions is a great choice. Our deep neural network has two branches with the same structure for pushing and grasping respectively (see Fig. 1). Each network branch consists of three parts: DenseNet, attention network and fully convolutional network.

Image Preprocess. The RGB-D images collected by the camera need to be preprocessed to generate heightmaps used as the input of the network. The heightmaps have a pixel resolution of 224×224, which has the same resolution as the image in ImageNet. The D heightmap needs to be expanded into three channels (DDD) to standardize the input format.

DenseNet. DenseNet [12] is composed of dense blocks and translation layers. Each node of the dense blocks includes a batch normalization layer, an activation function layer and a 3×3 convolutional layer, while the translation layer includes a 1×1

convolutional layer and a 2×2 pooling layer. We use DenseNet for preliminary feature extraction of heightmap images, using growth rate of 32, dense block configuration parameters of (6, 12, 24, 16). One DenseNet takes as input the RGB channels of the heightmaps to obtain color feature channels, while the other takes as input the DDD channels of the heightmaps to obtain depth feature channels. Color feature channels and depth feature channels are concatenated as the overall feature channels to input the next attention network. Note that we pre-train DenseNet on ImageNet [13] to improve training efficiency.

Attention Network. To improve the performance of the network model, we add the channel attention mechanism to weight the feature channels. The structure of the attention network is squeeze-and-excitation, which refers to the experience of [10]. The squeeze operation uses adaptive average pooling to turn each 2D feature channel into a real number, which has a global receptive field to some extent. The excitation operation uses two fully connected layers and corresponding activation functions to explicitly model interdependencies between channels. The weights obtained by the excitation operation are multiplied by the corresponding original feature channels to output the attention feature channels.

Fully Convolutional Network. Fully convolutional network is used to generate Q-table with different actions in the current state. The attention feature channels pass through two 1×1 convolutional layers interleaved with activation functions and batch normalization and then output a 224×224 Q-table by upsampling. Q-table coordinates correspond to pixel coordinates of the heightmaps. We rotate the heightmaps into 12 orientations to represent different orientations of pushing and grasping. Thus, inputting the heightmaps of the current state can output the $24 \times 224 \times 224$ Q-table, which corresponds to the action space.

3.4 Reward Functions

The self-supervised knowledge learning method based on deep reinforcement learning needs to be rewarded according to the actions of the robot. In our work, we design two different reward rules for the pushing and grasping actions respectively. The reward of the grasping action is based on whether the grasping is successful. The robot automatically determines the grasping success by the force feedback of the gripper. We define the sparse reward function for grasping actions:

$$R_g(s_t, a_t) = \begin{cases} 1 \ if \ grasp \ successful \\ 0 \ else \end{cases} \tag{4}$$

Pushing is an auxiliary action of grasping, which is used to change the environment and promote the success rate of grasping. The pushing reward function proposed by [8] does not consider whether a push enables future grasps, it only encourages the system to make pushes that cause change. We propose a non-sparse pushing reward function, which sets the reward value according to the degree of improvement in the possibility of grasping success after the pushing action. Q value indicates the expected reward of taking action a_t in the state s_t at time t, so the probability of grasping successful is positively correlated with the Q value. The pushing reward value is obtained according to the change in the maximum Q value of the grasping actions in the local area around the push point before and after the pushing action (see Fig. 2). The non-sparse reward function for pushing actions is defined as follows:

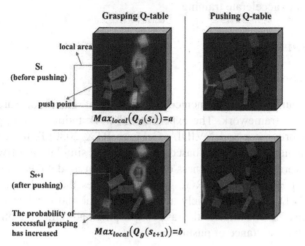

Fig. 2. Non-sparse rewards for pushing actions. In grasping Q-table (visualized with heat maps), find the maximum Q value in the local area of the push point before pushing. Then, do it again after pushing and calculate reward value using $R_p(s_t, a_t) = \frac{1}{1+e^{-(b-a)}}$. The local area is defined as a square space with a side length of 100 pixels centered on the push point and only the part within the workspace is considered.

$$R_p(s_t, a_t) = \begin{cases} \frac{1}{1+e^{-(Max_{local}(Q_g(s_{t+1}))-Max_{local}(Q_g(s_t)))}} & \textit{if push successful} \\ 0 & \textit{else} \end{cases} \qquad (5)$$

3.5 Network Training

The deep neural network, which acts as the Q function, establishes the mapping of the state to the Q values of different actions. The robot chooses the action corresponding to the maximum Q value in the Q-table to execute and has a certain probability to randomly explore, the exploration probability gradually decreases with iteration. The training goal is to minimize the temporal difference error between the target Q value and the actual Q value iteratively. We establish the target Q function according to the Q-learning algorithm:

$$Q_{target} = R(s_t, a_t) + \gamma \max_{a'} Q(s_{t+1}, a') \qquad (6)$$

where a' represents all actions in the action space, γ represents the future reward discount, we set as 0.5 in this work. The loss function for network training is smooth mean absolute error loss:

$$loss = \begin{cases} 0.5 \times \left(Q(s_t, a_t) - Q_{target}\right)^2 & if \left|Q(s_t, a_t) - Q_{target}\right| < 1 \\ \left|Q(s_t, a_t) - Q_{target}\right| - 0.5 & otherwise \end{cases} \tag{7}$$

We use the stochastic gradient descent as the optimizer of the network model, while using momentum to accelerate training.

4   Experiments

4.1   Simulation Setup

The algorithm in our work is implemented with the help of related software development kits in the PyTorch framework. The experimental workstation is equipped with Intel Core i7-9700 K processor and NVIDIA GeForce RTX 2080 Ti for computing. We build the experimental environment based on V-REP [14] simulation software in Ubuntu 16.04 system to verify the effectiveness of our algorithm model (see Fig. 3). The UR5 manipulator robot equipped with RG2 gripper perceives the workspace environment through the RGB-D camera obliquely above. After trial and error training, it makes autonomous decisions to complete the task of grasping and removing all objects in the workspace with the assistance of pushing actions.

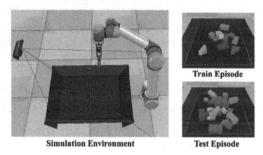

Fig. 3. Simulation setup and task episode. Our experimental environment is built on V-REP simulation software. One training task episode contains 10 randomly generated toy blocks. During the test phase, the number of toy blocks change to 20, so the test task is more challenging due to more occlusion and stacking.

4.2   Task Episode

In the training phase, we randomly place 10 toy blocks in the workspace to construct one training task episode. The color and shape of the toy blocks are chosen random-ly. When all the toy blocks are grasped and removed or 10 consecutive action at-tempts fail to complete the grasp, the training task episode is reconstructed in the same way. The manipulator robot learns pushing and grasping knowledge based on trial and error and reward feedback. The network model training is completed after 3000 action attempts. In the test phase, we randomly generate 5 different test task episodes, each episode is tested 20 times, so the test task contains 100 test task episodes. Unlike the training task episode, the number of toy blocks is 20, which means that the test task is more challenging due to more occlusion and stacking (see Fig. 3).

4.3 Performance Evaluation

To evaluate the performance of our proposed algorithm, we design an overall performance metric, which contains three evaluation factors: task completion rate, grasping success rate and action efficiency.

Task Completion Rate. The sign of task episode completion is that all toy blocks are grasped and the workspace is emptied. The task completion rate represents the ratio of completed test task episodes to the total test task episodes:

$$\Phi_1 = \frac{n}{N} \times 100\% \tag{8}$$

where n represents the number of completed test task episodes, N represents the total number of test task episodes ($N = 100$ in our work).

Grasping Success Rate. The grasping success rate represents the ratio of successful grasps to the grasping attempts. Note that we only consider the grasping success rate in the completed test task episodes. We define the grasping success rate as follows:

$$\Phi_2 = \frac{\sum_{i=0}^{n} \frac{s_i}{S_i}}{n} \times 100\% \tag{9}$$

where s represents the number of successful grasps in one test task episode, S represents the total number of grasping attempts in one test task episode.

Action Efficiency. In practical applications, we hope that the robot can complete the task faster, that is, perform fewer actions to grasp all the toy blocks. When there are m toy blocks in the workspace, at least m actions (grasps) are required to clear the workspace. Similarly, we only consider it in the completed test task episodes. We define the action efficiency function as follows:

$$\Phi_3 = \frac{\sum_{i=0}^{n} \frac{m_i}{M_i}}{n} \times 100\% \tag{10}$$

where m represents the number of toy blocks in the workspace ($m = 20$ in our work), M represents the actual number of actions (containing pushing and grasping actions) in one test task episode.

Overall Performance. The overall evaluation metric is defined as a weighted sum of the above three factors:

$$\Phi = \omega_1 \Phi_1 + \omega_2 \Phi_2 + \omega_3 \Phi_3 \tag{11}$$

In our work, we choose the weight parameters as: $\omega_1 = 0.4$, $\omega_2 = 0.3$, $\omega_3 = 0.3$.

4.4 Results and Discussion

In our work, we propose an attention deep Q-learning network (ADQN) and a non-sparse (NS) reward pushing strategy to help improve the quality of grasping. Hence, we call our model ADQN-NS. Our model training process is efficient and only takes about 5 h each time. We use comparative experiments to verify the performance of the model.

Pushing Strategy Comparison. A single grasping strategy is difficult to solve the problem of object grasping in cluttered environments. To compare the impact of the pushing strategy on the completion quality of the grasping task, we test three methods with different pushing strategies on the same test task episodes. The first method only uses grasping actions in the task without pushing actions (no pushing) and we call it ADQN-NP. The second method uses a sparse reward pushing strategy (ADQN-S), this means that if the pushing action is successful, the reward value is 0.5, otherwise the reward value is 0 [8]. The last method is our proposed ADQN-NS model.

Pushing strategy comparison results are shown in Table 1. The method ADQN-NP can complete most of the test task episodes, however, without the assistance of the pushing actions, the grasping success rate is relatively low, and the task episodes require a large number of grasp attempts to complete. The action efficiency is equivalent to grasping success rate due to no pushing actions. The method ADQN-S increase the grasping success rate with the assistance of pushing actions. However, since the reward strategy for pushing actions is not based on improving the grasping success rate, it makes some pushing actions unnecessary. At the same time, some invalid pushing actions will push the object out of the workspace and affect the task completion rate. Our proposed method ADQN-NS is ahead of the other two methods in terms of task completion rate, grasping success rate and action efficiency. The pushing strategy make the robot push objects to provide space for grasping, while avoiding some unnecessary pushing actions to improve action efficiency. The overall performance is 82.33%, which is 13.35% higher than ADQN-NP and 6.07% higher than ADQN-S.

Table 1. Results on pushing strategy comparison.

Method	Task completion rate (%)	Grasping success rate (%)	Action efficiency (%)	Overall performance (%)
ADQN-NP	91.0	54.3	54.3	68.98
ADQN-S	87.0	71.8	66.4	76.26
ADQN-NS	**97.0**	**73.5**	**71.6**	**82.33**

Ablation Studies for Attention. We use the attention network to weight feature channels to fully extract scene feature information. To verify the effectiveness of the attention network, we conduct ablation studies. The attention network on the ADQN-NS is removed and we call the remaining parts DQN-NS. The feature channels extracted by DenseNet are directly transmitted to the fully convolutional network.

Table 2 shows the results of ablation studies for attention. The task completion rate, grasping success rate and action efficiency all decline without the attention network. The results show that the feature channels weighted by the attention network help to improve the performance of the deep Q-learning network.

Table 2. Results of ablation studies for attention.

Method	Task completion rate (%)	Grasping success rate (%)	Action efficiency (%)	Overall performance (%)
DQN-NS	93.0	71.0	70.0	79.50
ADQN-NS	**97.0**	**73.5**	**71.6**	**82.33**

5 Conclusion

In this work, an attention deep Q-learning network for robotic grasping with the assistance of pushing actions with non-sparse rewards is proposed. The effect of the attention network is to weight the feature channels of the state extracted by DenseNet and then send them to the fully convolutional network to generate the Q-table of the action space for Q-learning algorithm. The proposed pushing strategy creates space for grasping to increase the grasping success rate. Simultaneously, giving non-sparse rewards according to the improvement of the possibility of grasping success can avoid unnecessary pushing actions and improve the action efficiency. Experiments prove that our proposed model can effectively improve the robotic grasping performance. In the future work, we will train and test the proposed model in a real robot environment. Then, we will to use other deep reinforcement learning methods (such as DDPG) and visual attention mechanisms to explore the impact on robotic grasping performance.

References

1. Sahbani, A., Elkhoury, S., Bidaud, P.: An overview of 3D object grasp synthesis algorithms. Robot. Auton. Syst. **60**(3), 326–336 (2012)
2. Bohg, J., Morales, A., Asfour, T., Kragic, D.: Data-driven grasp synthesis - a survey. IEEE Trans. Robot. Autom. **30**(2), 289–309 (2014)
3. Song, K., Chang, Y., Chen, J.: 3D vision for object grasp and obstacle avoidance of a collaborative robot. In: International Conference on Advanced Intelligent Mechatronics, pp. 254–258. IEEE (2019)
4. Zeng, A., Yu, K.T., Song, S., Suo, D., Xiao, J.: Multi-view self-supervised deep learning for 6D pose estimation in the Amazon picking challenge. In: IEEE International Conference on Robotics and Automation. IEEE (2017)
5. Lenz, I., Lee, H., Saxena, A.: Deep learning for detecting robotic grasps. Int. J. Robot. Res. **34**(4–5), 705–724 (2015)

6. Redmon, J., Angelova, A.: Real-time grasp detection using convolutional neural networks. In: IEEE International Conference on Robotics and Automation, pp. 1316–1322. IEEE (2015)
7. Kalashnikov, D., et al.: QT-Opt: scalable deep reinforcement learning for vision-based robotic manipulation. arXiv preprint arXiv: 1806.10293 (2018)
8. Zeng, A., Song, S., Welker, S., Lee, J., Rodriguez, A., Funkhouser, T.: Learning synergies between pushing and grasping with self-supervised deep reinforcement learning. In: IEEE/RSJ International Conference on Intelligent Robots and Systems, pp. 4238–4245. IEEE (2018)
9. Jaderberg, M., Simonyan, K., Zisserman, A., Kavukcuoglu, K.: Spatial transformer networks. In: Annual Conference on Neural Information Processing Systems, pp. 2017–2025. MIT Press (2015)
10. Hu, J., Shen, L., Sun, G., Albanie, S.: Squeeze-and-excitation networks. In: IEEE Trans. Pattern Anal. Mach. Intell. (2017)
11. Woo, S., Park, J., Lee, J.-Y., Kweon, I.S.: CBAM: convolutional block attention module. In: Ferrari, V., Hebert, M., Sminchisescu, C., Weiss, Y. (eds.) ECCV 2018. LNCS, vol. 11211, pp. 3–19. Springer, Cham (2018). https://doi.org/10.1007/978-3-030-01234-2_1
12. Huang, G., Liu, Z., Der Maaten, L.V., Weinberger, K.Q.: Densely connected convolutional networks. In: IEEE Conference on Computer Vision and Pattern Recognition, pp. 2261–2269. IEEE (2017)
13. Deng, J., Dong, W., Socher, R., Li, L., Li, K., Feifei, L.: ImageNet: a large-scale hierarchical image database. In: IEEE Conference on Computer Vision and Pattern Recognition, pp. 248–255. IEEE (2009)
14. Rohmer, E., Singh, S., Freese, M.: V-REP: a versatile and scalable robot simulation framework. In: IEEE/RSJ International Conference on Intelligent Robots and Systems, pp. 1321–1326. IEEE (2013)

A Dynamic Answering Path Based Fusion Model for KGQA

Mingrong Tang[1], Haobo Xiong[1], Liping Wang[1(✉)], and Xuemin Lin[1,2]

[1] Shanghai Key Laboratory of Trustworthy Computing, East China Normal
University, Shanghai, China
{51184501148,51194501192}@stu.ecnu.edu.cn, lipingwang@sei.ecnu.edu.cn
[2] The University of New South Wales, Sydney, Australia
lxue@cse.unsw.edu.au

Abstract. The Knowledge Graph Question Answering (KGQA) task
is useful for information retrieval systems, intelligent customer service
systems, etc., which has attracted the attention of a large number
of researchers. Although the performance of KGQA has been further
improved by introducing the Deep Learning models, there are still some
difficulties to be solved, such as the representation of questions and
answers, the efficient construction way of candidate path set, etc. In
this paper, we propose a complete approach for KGQA task. Firstly,
we devise a novel candidate path generation process, which effectively
improves computation performance by reducing the number of candidate
paths corresponding to a question and at the same time guarantees the
accuracy of results. Secondly, considering the textual expression diversity
of questions and stochastic of candidate paths, we present four models
to learn semantic features of Chinese sequence with different focuses.
Finally, in order to combine the advantages of each presented model,
we propose a dedicated fusion policy which can get the most suitable
path from the path set predicted by our presented models. We conduct
experiments on Chinese Knowledge Base Question Answering (CKBQA)
dataset. Experiment results show that our approach achieves better performance than the best one published in CCKS2019 competition.

Keywords: Knowledge Graph Question Answering · Bert · Path
generation · Model fusion

1 Introduction

With emergence of many high-quality and large-scale knowledge graphs, such as
Freebase, Wikidata, YaGo, etc., the research about KGQA is becoming more and
more popular. KGQA can return one or more nodes in the knowledge graph as
correct answers for a question described by natural language, which can greatly
reduce the difficulty of query and improve the usability of knowledge graph.
Recently, KGQA has been widely applied in information retrieval systems, intelligent customer service systems, etc., which has become the main branch of study
on knowledge graph.

G. Li et al. (Eds.): KSEM 2020, LNAI 12274, pp. 235–246, 2020.
https://doi.org/10.1007/978-3-030-55130-8_21

At present, the mainstream methods for KGQA can be grouped into the following two categories: semantic parsing-based methods and information retrieval-based methods. Semantic parsing-based methods [3,16] aim to convert a natural language question into an intermediate logical form, and then into a structured query statement that can be executed in the knowledge graph. Information retrieval-based methods [20] refer to extracting mentions in the question, linking mentions to corresponding topic entities in the knowledge graph, and then taking the connected entities as candidate answers, lastly ranking candidate answers for choosing the right answer. With the rapid development of Deep Learning, the neural network-based methods make the performance of the KGQA task get further improvement and no longer be limited by rules and templates. The neural network-based methods [1,2] mainly learn the semantic vectors of questions and candidate answers, and then model complex question-answering tasks into relatively simple similarity computation tasks. For those studies, researchers usually make an effort to effectively represent questions and candidate answers in a vector space. As a result, the technology of semantic information extraction and representation has been successively applied to KGQA including word embedding models such as word2vec [13] and Glove [14], and pre-trained language representation models such as GPT [15], Bert [4] and XLnet [21].

Challenges. For Knowledge Graph Question Answering task, there are two main challenges. One is how to efficiently construct a candidate answer set. When knowledge graph is relatively dense, there are lots of candidate paths related to the topic entity causing a lot of redundant calculations. As a result, it is difficult to compute the similarity between a question and all candidate paths. To solve above problem, the answer paths are usually limited to two hops in previous work. The other is how to effectively learn the representation of questions and candidate answers. To obtain more semantic information, most of the previous works represent answers from different aspects [8], but generally we can only represent answers from one aspect of answer path limited by the uncertainty of knowledge graph.

In current research, as the expression of questions and candidate paths is diverse, it's very difficult to learn all the semantic features between questions and paths through only one model. Moreover, for the dynamic candidate path generation process studied in this paper, there is a new challenge in the fusion processing. Specifically, for a question, the candidate path set from different models is different, therefore, we can't fuse results by calculating the average score of every candidate paths directly.

To handle above challenges, the corresponding solutions are proposed. Firstly, inspired by beam search [19] idea, we present a process for dynamically expanding candidate path set. In this way, path set is expanded hop by hop, and the number of candidate paths corresponding to a question is reduced effectively. Secondly, to extract semantic features of questions and candidate paths from different focuses, we design four models incorporating Bert including Basic global model, Path diversity model, Implicit relation model, and Path decomposition model. At last, we propose a suitable model fusion policy, which gets the candidate path with the highest score as the final right answer path by two steps including obtaining the top m paths of each model and calculating the average score of the candidate paths after filling in the missing values.

Contributions. Our contributions are summarized as follows:

1) We present a novel dynamic candidate path generation process named DPG, which iteratively generates candidate path hop by hop, and automatically terminate when postconditions are satisfied.
2) Considering the textual expression diversity of questions and stochastic of candidate paths, we present four models to learn semantic representations of Chinese sequence with different focuses including Basic global model, Path diversity model, Implicit relation model, and Path decomposition model.
3) To combine the advantages of each presented model, we propose a path similarity-based fusion policy which gets the most suitable path from the differentiated path set predicted by proposed models.
4) We construct extensive experiments and evaluate our approach on the CKBQA dataset. The experiment results demonstrate that our approach achieves better performance than the best one published in CCKS2019 competition.

2   Related Work

Semantic Parsing-Based Methods. [12] parses a natural language question into a new semantic representation form called DCS. This method can solve the problem of missing labeled data, but requires more professional linguistic knowledge and is difficult to generalize. [10] parses a question into a query subgraph, and then performs subgraph matching to obtain the correct answer. This method can understand certain implicit relationships but still needs to construct a mapping dictionary and constraint rules. With the popularity of the encoder-decoder models in the field of translation, these models are introduced to the question answering task which converts a question answering task to a translation task. Through the encoding and decoding model, [6] translates questions into logical forms that can be executed in the knowledge graph and effectively deal with the conversational question answering task. With the development of knowledge graph embedding technology, [18] introduces knowledge graph embedding technology to solve the problem of entity and relationship mapping and then construct graph-structured queries. In addition, [7] believes that improving the accuracy of the semantic parsing-based method should not be completely dependent on the design of algorithms and proposes a system named DialSQL, which can identify potential errors in a generated query and return simple multiple-choice Questions over multiple turns to users, and then use the user feedback to modify the query. In general, the semantic parsing-based method can resolve lots of simple and complex questions, but it involves much traditional linguistic knowledge, which is difficult to understand.

Information Retrieval-Based Methods. [22] firstly gets a topic entity in the question, then gets candidate answers related to the topic entity, and ranks them based on their relationship. In order to reason over knowledge graph, [23] proposes an end-to-end variational learning model that can handle noise in problems and learns multi-hop inference simultaneously. With the development of representation learning technology, more and more researchers use the method of vector modeling to deal with the question answering task. This method converts a question answering task into a similarity computation task between the question vector and the candidate answer vector. [5] uses a multi-column convolutional neural network to conduct distributed representation learning of questions and answers from three aspects including answer path, answer context, answer type so that the representation can contain more effective features than the previous. [17] uses tree-structured long short-term memory networks to learn the representation of sequences, which is helpful to analyze the implicit relationship and intent. Considering that most of the works pay more attention to the representation of the candidate answer end and ignore the question representation, [8] introduces an end-to-end neural network model to dynamically represent questions according to the diverse candidate answer aspects via cross-attention.

Based on information retrieval methods, we first locate a topic entity in the knowledge graph by named entity recognition and then dynamically generate and rank candidate paths hop by hop. To rank paths, we design four models which can learn from multi focuses. Finally, we fuse the outputs of our models and get the most suitable answer path.

3   Our Approach

Usually, the knowledge graph (KG) is regarded as a directed graph, in which the nodes are real-world entities and the edges are their relations. In KG, each directed edge, along with its head entity and tail entity, make up a triple, expressed as (h, r, t), which is also named as a fact. The knowledge graph question answering task (KGQA) aims to input a natural language question and then return an entity set A as right answers. The architecture of our approach for KGQA just as shown in Fig. 1. The proposed solution can be divided into three steps: candidate generation, similarity computation and results fusion. The following sections explain the details of each step.

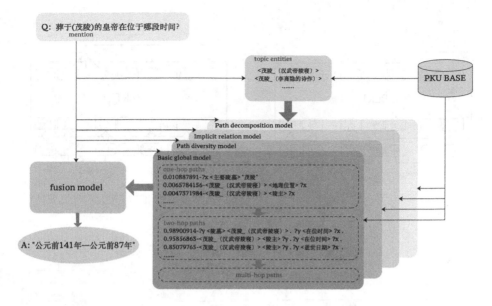

Fig. 1. The overview of the dynamic answering path based fusion model.

3.1   Candidate Generation

Ideally, we use all entities in the knowledge graph as candidate answers, and rank them. However, in practice, this is very time-consuming and unnecessary, so we develop a process for dynamic candidate path generation named DPG.

We first obtain a topic entity set E by named entity recognition and then build the candidate answer path set P according to the topic entities located in the knowledge graph. For the named entity recognition task, we first perform BIO sequence labeling by utilizing Bert, bidirectional LSTM [9] and CRF [11], and then link the identified mentions to topic entities through accurate matching and fuzzy matching. After obtaining the topic entity set, we dynamically generate candidate paths P in an expanded strategy. We first add candidate paths of one hop around the topic entity into set P, then calculate the similarity score. After that, we select top k paths in one-hop set to generate two-hop candidate paths and calculate their similarity score. By the same method, the hop of candidate path can be continuously increased, and the candidate path set P can be expanded by new generated paths. When hop count threshold is reached or scores of all new generated paths are no longer increased, the expansion process is terminated.

3.2   Similarity Computation Model

Considering the diversity of questions and candidate paths, we devise four models to compute similarity scores of questions and paths from different focuses. As a pre-trained language representation model, Bert achieves state-of-the-art results

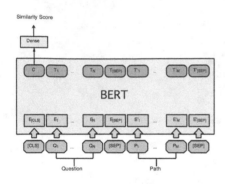
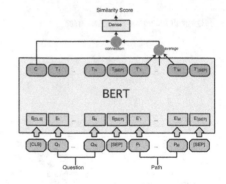

(a) The architecture of Basic global model

(b) The architecture of Path diversity model

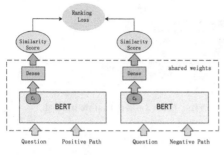
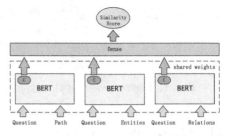

(c) The architecture of Implicit relation model

(d) The architecture of Path decomposition model

Fig. 2. The overview of our proposed models. (a) is Basic global model which takes into account the global representation of questions and paths. (b) is Path diversity model which pays more attention on the expression of paths. (c) is Implicit relation model which is used to extract implicit relation. (d) is Path decomposition model which can learn weights of entity and relation.

on many natural language processing tasks, including the similarity computation task. So we develop our similarity computation model based on Bert to handle the downstream question answering task.

Basic Global Model. This is the simplest similarity computation model based on Bert which takes into account the global representation of questions and paths. In this model, we only get the hidden representation vector from Bert corresponding to first input token ([CLS]) into next layer. The architecture of Basic global model is shown in Fig. 2(a). This model can handle simple questions with one-hop answer path like question "清华大学的校训是什么？", but limited by input information, it usually fails in some complex question with multi-hop answering path like question "维力医疗的高管中谁是复旦大学毕业的？".

Path Diversity Model. For a question with extra semantic content, the expression of paths is usually diverse. In some cases, the expression of answer paths is sequential such as path "< 维力医疗 > < 独立董事 > ?y . ?y < 上任时间 > ?x ."

for question "维力医疗的独立董事的上任时间？", and in a few cases, the expression of answer paths can be reversed such as path "?y < 陵墓 > < 茂陵_(汉武帝陵寝) > . ?y < 在位时间 > ?x ." for question "葬于茂陵的皇帝在位于哪段时间？". Generally, it is easy to deal with the first type of questions, but the second type of questions are difficult to answer. To deal with the complexity of answer path, we propose a new model named Path diversity model, in which we pay more attention to the expression of paths by feeding the hidden representation vector of path token into the next layer. We can see the detail in Fig. 2(b).

Implicit Relation Model. Implicit relation extraction has always been a challenge in KGQA tasks. For the question "《红楼梦》中贾琏妻子的丫鬟是谁？", the right answer path is "?y < 丈夫 > < 贾琏_(小说《红楼梦》人物) > . ?y < 丫鬟 > ?x .". In practice, the simple model is hard to give the relation "< 丈夫 >" a high score. For such complex questions, we propose a model named Implicit relation model by employing the triplet loss function presented below.

$$Loss = max(s(q, n) - s(q, p) + \gamma, 0) \tag{1}$$

where $s(q, p)$ calculates the similarity score between a question and a positive candidate. $s(q, n)$ calculates the similarity score between a question and a negative candidate. The γ is a positive number range from 0 to 1 that ensures a margin between positive candidates and negative candidates. With this loss function, the inputs of Implicit relation model are question, positive path, and negative path. The architecture of model is shown in Fig. 2(c).

Path Decomposition Model. Considering the distinct importance of entity and relation, we design a model named Path decomposition model to extract the similarity features of questions and paths, questions and entities, questions and relationships respectively, and then automatically learn their weights. The architecture of this model is shown in Fig. 2(d).

3.3 Results Fusion

The four models devised above cover the different conditions in KGQA tasks. To achieve the most suitable answer, a path similarity based fusion policy is proposed. Ideally, we expect to get four scores for a candidate path from four models and select the candidate path with the highest average score as the final right answer path. However, the candidate paths generated by our proposed process DPG are related to the requirement of similarity computation model. In other words, the top m paths are different for the same question under different models. To solve the above problem, we propose a novel solution for the best path decision.

Firstly, for each question, top m candidate paths in each model are selected, here m is a parameter. Secondly, for each candidate path absented in model i but recommended by other models, the prediction from model i to the candidate path is missing. We define Eq. 2 to count the missing scores which denotes evaluation score of model i to the candidate path.

$$Score = max(min(M_i, Q_j) * k_1 - k_2, 0) \tag{2}$$

where $min(M_i, Q_j)$ represents the minimum score in top m path set given by model i to question j, and k_1 and k_2 are adjustable parameters. Thirdly, after filling in the missing values, we calculate average score of each candidate path from different models. Finally, we use the candidate path with the highest average score as the final right answer path of the question.

4   Experiments

We use PKU BASE as our knowledge graph. It has 66,191,767 triples with 25,437,419 nodes and 408,261 relations. To evaluate our proposed models, we do experiments on Chinese Knowledge Base Question Answering (CKBQA) dataset which includes 2998 samples for training, 766 samples for validating, and 766 samples for testing. To compare with others, we use average F1 score as the final evaluation matrix.

4.1   Setting

In the candidate path generation step, considering that most of the right answer paths for questions in CKBQA dataset are in two hops, we set the hop threshold to 2. In the similarity computation step, we use Adam as our optimizer to minimize question-path pairs training loss. We set the batch size to 64, set the learning rate to 1e-5, and set the max length of model input sequences to 100. What's more, we early stop when the F1 score on the validation set reaches the maximum. Differently, the γ in loss function of Implicit relation model is set to 0.9. In the results fusion step, we set the parameter k_1 to 0.7 and k_2 to 0.2. We feed top 5 paths from each similarity model into the fusion model.

4.2   Results

The Effectiveness of Candidate Generation Process. In order to evaluate the effectiveness of our candidate generation process named DPG, we set the hop threshold to two and compare it with a commonly used fixed candidate path generation process named FPG. In general method FPG, we first construct a candidate set in which all paths are within two hops and include only one entity, then rank paths to select the correct one, and finally restrict the correct answer. In our method, we dynamically generate candidate paths, which can have multiple entities, and then rank the candidate paths to obtain correct path and correct answer without any constrains. We can see the results based on Basic global model in Table 1. The experiment results demonstrate that our process in generating candidate paths is effective. It reduces the number of candidate paths from 4081 to 1203 and increases the F1 score from 61.80% to 67.65%.

The Effectiveness of Fusion Policy. We conduct multiple experiments to evaluate the effectiveness of our fusion policy. What's more, in order to reduce the impact of mistake during named entity recognition stage, we also create an

Table 1. The evaluation of different candidate generation process.

Process	Avg count	Macro precision	Macro recall	Avg F1
FPG	4081	62.31	63.94	61.80
DPG	**1203**	**68.07**	**69.11**	**67.65**

Table 2. The evaluation of results fusion policy.

Model	Linking method	Macro precision	Macro recall	Avg F1
Basic global model	Ent linking	68.07	69.11	67.65
Path diversity model	Ent linking	64.13	64.75	63.63
Implicit relation model	Ent linking	61.97	61.82	61.20
Path decomposition model	Ent linking	65.37	65.33	64.56
Basic global model	Path linking	60.02	60.49	59.56
Fusion model	-	**75.21**	**76.13**	**74.63**

elastic search (ES) index on triple which mask head entity or tail entity to find one-hop paths. When we search a question on the index, we can obtain one-hop candidate paths easily. The next steps for expanding candidate path set P are the same as described above. We evaluate the performance of proposed approaches by employing the macro precision, macro recall, and average F1 evaluation matrix. The results are shown in Table 2.

In the table, the ent linking means that we get one-hop candidate paths by topic entities, and the path linking means that we get one-hop paths by searching for questions on the triple-based ES index. To make up for the error of the named entity recognition, we build a candidate path set by linking question to path and rank paths by the simplest Basic global model. In addition, we also list results with a variety of hit rates for reference, as shown in Fig. 3. From the results, we observe that the F1 score has been significantly improved by fusing the five result files, which demonstrates our fusion policy can combine the advantages of each model.

The Effectiveness of Our Proposed Approach. We conduct experiments on CKBQA dataset which comes from the CCKS2109 KGQA competition. We compare the final results with the top four teams in the competition to illustrate the effectiveness of our approach. We can see the results in Table 3.

The top four teams in the competition all fine-tune the pre-trained language model Bert on the KGQA task and achieve promised results. Differently, the rank one uses a dictionary to segment the question to recall candidate entities, then they extracted 39 features in the similarity computation stage. In addition, in order to reduce the variance, 10-fold cross-validation was employed. The system of the first team has achieved good results but is too cumbersome. Compared with the rank one, our approach achieves better performance and it is more concise and interpretable.

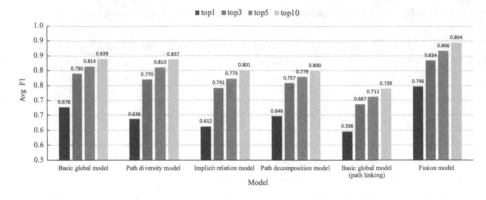

Fig. 3. The evaluation of model results with different statistical scope.

Table 3. The evaluation results on CKBQA dataset.

Teams	Avg F1
Rank 1	73.54
Rank 2	73.08
Rank 3	70.45
Rank 4	67.60
Our approach	**74.63**

5　Conclusion

In this paper, we propose a dynamic answering path based fusion model for the Knowledge Graph Question Answering task, and make a significant improvement on CKBQA dataset compared with representative ones. We propose a novel dynamic candidate path generation process which effectively reduces the number of candidate paths and makes the hop of answer paths not limited to a fixed number. Moreover, we design four independent models to learn the similarity representation of questions and candidate paths from different focuses. Lastly, the advantages of models are combined by an effective fusion policy which solves the problem of missing path scores caused by the dynamic path generation process. The experiment results demonstrate that our approach is effective at all stages of KGQA and we achieve better performance than the best one published in CCKS2019 competition.

Acknowledgment. This work was supported by project 2018AAA0102502.

References

1. Bordes, A., Chopra, S., Weston, J.: Question answering with subgraph embeddings. In: Proceedings of the 2014 Conference on Empirical Methods in Natural Language Processing (EMNLP), Doha, Qatar, pp. 615–620. Association for Computational Linguistics, October 2014
2. Bordes, A., Weston, J., Usunier, N.: Open question answering with weakly supervised embedding models. In: Proceedings of the 2014th European Conference on Machine Learning and Knowledge Discovery in Databases-Volume Part I, pp. 165–180 (2014)
3. Cai, Q., Yates, A.: Large-scale semantic parsing via schema matching and lexicon extension. In: Proceedings of the 51st Annual Meeting of the Association for Computational Linguistics (Volume 1: Long Papers), Sofia, Bulgaria, pp. 423–433. Association for Computational Linguistics, August 2013
4. Devlin, J., Chang, M.W., Lee, K., Toutanova, K.: Bert: pre-training of deep bidirectional transformers for language understanding. In: Proceedings of the 2019 Conference of the North American Chapter of the Association for Computational Linguistics: Human Language Technologies, Volume 1 (Long and Short Papers), pp. 4171–4186 (2019)
5. Dong, L., Wei, F., Zhou, M., Xu, K.: Question answering over Freebase with multicolumn convolutional neural networks. In: Proceedings of the 53rd Annual Meeting of the Association for Computational Linguistics and the 7th International Joint Conference on Natural Language Processing (Volume 1: Long Papers), Beijing, China, pp. 260–269. Association for Computational Linguistics, July 2015
6. Guo, D., Tang, D., Duan, N., Zhou, M., Yin, J.: Dialog-to-action: conversational question answering over a large-scale knowledge base. In: NeurIPS, pp. 2946–2955 (2018)
7. Gur, I., Yavuz, S., Su, Y., Yan, X.: DialSQL: dialogue based structured query generation. In: Proceedings of the 56th Annual Meeting of the Association for Computational Linguistics (Volume 1: Long Papers, Melbourne, Australia, pp. 1339–1349. Association for Computational Linguistics, July 2018
8. Hao, Y., et al.: An end-to-end model for question answering over knowledge base with cross-attention combining global knowledge. In: Proceedings of the 55th Annual Meeting of the Association for Computational Linguistics (Volume 1: Long Papers), Vancouver, Canada, pp. 221–231. Association for Computational Linguistics, July 2017
9. Hochreiter, S., Schmidhuber, J.: Long short-term memory. Neural Comput. 9(8), 1735–1780 (1997)
10. Hu, S., Zou, L., Yu, J.X., Wang, H., Zhao, D.: Answering natural language questions by subgraph matching over knowledge graphs. IEEE Trans. Knowl. Data Eng. 30(5), 824–837 (2018)
11. Lafferty, J., Mccallum, A., Pereira, F.: Conditional random fields: probabilistic models for segmenting and labeling sequence data. In: Proceedings of the Eighteenth International Conference on Machine Learning, pp. 282–289 (2001)
12. Liang, P., Jordan, M., Klein, D.: Learning dependency-based compositional semantics. In: Proceedings of the 49th Annual Meeting of the Association for Computational Linguistics: Human Language Technologies, Portland, Oregon, USA, pp. 590–599. Association for Computational Linguistics, June 2011
13. Mikolov, T., Chen, K., Corrado, G., Dean, J.: Efficient estimation of word representations in vector space. arXiv preprint arXiv:1301.3781 (2013)

14. Pennington, J., Socher, R., Manning, C.: Glove: global vectors for word representation. In: Proceedings of the 2014 Conference on Empirical Methods in Natural Language Processing (EMNLP), Doha, Qatar, pp. 1532–1543. Association for Computational Linguistics, October 2014

15. Radford, A., Wu, J., Child, R., Luan, D., Amodei, D., Sutskever, I.: Language models are unsupervised multitask learners (2019)

16. Reddy, S., Täckström, O., Collins, M., Kwiatkowski, T., Das, D., Steedman, M., Lapata, M.: Transforming dependency structures to logical forms for semantic parsing. Trans. Assoc. Comput. Linguist. **4**, 127–140 (2016)

17. Tai, K.S., Socher, R., Manning, C.D.: Improved semantic representations from tree-structured long short-term memory networks. In: Proceedings of the 53rd Annual Meeting of the Association for Computational Linguistics and the 7th International Joint Conference on Natural Language Processing (Volume 1: Long Papers), Beijing, China, pp. 1556–1566. Association for Computational Linguistics, July 2015

18. Wang, R., Wang, M., Liu, J., Chen, W., Cochez, M., Decker, S.: Leveraging knowledge graph embeddings for natural language question answering. In: Li, G., Yang, J., Gama, J., Natwichai, J., Tong, Y. (eds.) Database Systems for Advanced Applications, pp. 659–675. Springer International Publishing, Cham (2019)

19. Wiseman, S., Rush, A.M.: Sequence-to-sequence learning as beam-search optimization. In: Proceedings of the 2016 Conference on Empirical Methods in Natural Language Processing, pp. 1296–1306 (2016)

20. Xu, K., Feng, Y., Reddy, S., Huang, S., Zhao, D.: Enhancing freebase question answering using textual evidence. CoRR abs/1603.00957 (2016). http://arxiv.org/abs/1603.00957

21. Yang, Z., Dai, Z., Yang, Y., Carbonell, J., Salakhutdinov, R.R., Le, Q.V.: XLNET: generalized autoregressive pretraining for language understanding. In: Advances in Neural Information Processing Systems, pp. 5754–5764 (2019)

22. Yao, X., Van Durme, B.: Information extraction over structured data: question answering with Freebase. In: Proceedings of the 52nd Annual Meeting of the Association for Computational Linguistics (Volume 1: Long Papers), Baltimore, Maryland, pp. 956–966. Association for Computational Linguistics, June 2014

23. Zhang, Y., Dai, H., Kozareva, Z., Smola, A.J., Song, L.: Variational reasoning for question answering with knowledge graph. In: Thirty-Second AAAI Conference on Artificial Intelligence (2018)

Improving Deep Item-Based Collaborative Filtering with Bayesian Personalized Ranking for MOOC Course Recommendation

Xiao Li[1], Xiang Li[2(✉)], Jintao Tang[1], Ting Wang[1], Yang Zhang[1], and Hongyi Chen[1]

[1] National University of Defense Technology, Changsha, China
{xiaoli,tangjintao,tingwang,yangzhang15,chenhongyi}@nudt.edu.cn
[2] Academy of Military Science, Beijing, China
lixiang41@126.com

Abstract. With the advancement of big data and education technology, MOOCs (Massive Online Open Courses) has become a popular education model in online education community. A large number of online courses with diverse disciplinary background are offered freely to global learners. When a learner finishes a series of courses, it is very important to effectively and efficiently recommend the most relevant courses to study next. Traditional item-based recommendation methods are all pointwise approaches where models bias towards estimating the precise rating or relevance score of each item. It would be better to model this problem from a pairwise learning perspective which is more close to the ranking nature of course recommendation. In this paper, we combine item-based collaborative filtering and Bayesian Personalized Ranking for course recommendation problem. We theoretically derive the optimization schema based on Bayesian Personalized Ranking and develop a novel neural network model, called Bayesian Personalized Ranking Network (BPRN), which learns pairwise course preference for each user given her historically enrolled courses. With extensive experiments on a large-scale MOOCs enrollment dataset from XuetangX, we empirically demonstrate that our BPRN framework performs better than state-of-the-art item-based course recommendation methods.

Keywords: Course recommendation · Collaborative filtering · Bayesian analysis · Deep learning

1 Introduction

The rapid development of MOOCs (Massive Online Open Courses) has attracted widespread interests globally as worldwide students can learn high-quality online courses from famous universities with very low cost. According to Classcentral, more than 13,500 courses with diverse disciplinary background have been offered

© Springer Nature Switzerland AG 2020
G. Li et al. (Eds.): KSEM 2020, LNAI 12274, pp. 247–258, 2020.
https://doi.org/10.1007/978-3-030-55130-8_22

online in 2019.[1] When a student finishes a series of courses, it may be difficult for her to choose a proper course to study next. A careless course enrollment may not only waste time but also raise the attrition rate [17]. Therefore, how to effectively and efficiently recommend the most relevant courses for students to study next is very crucial. Typically, the recommender system needs to recommend a ranking list of n most relevant courses to a user based on her historically enrolled courses. Inspired by the music and movie recommendation domain, we can also model this problem as a collaborative filtering problem with implicit feedbacks. The term *implicit feedbacks* means that instead of knowing a rating or score about how much interest the user has for a course, we only know whether the user has enrolled the course or not.

In MOOCs datasets, usually the number of users is much larger than the number of courses. For example, in Coursera[2], there are 45 million registered users studying 3,800 courses in 2019. As the course size is much smaller than the user size, we study this problem from an item-based collaborative filtering perspective where user interests are implicitly modeled by learning her historically enrolled courses. Therefore, the computational burden of explicitly model a huge amount of users can be reduced.

Traditional item-based course recommendation methods [2,5] use feature engineering to mine useful signal from data. Recently, deep neural network based methods have attracted much research attention. For example, FISM [6] models items as latent vectors and uses the average of all historical item vectors to represent user interests. As different users may have different tastes on different items, NAIS [3] improves FISM by using attention mechanism to learn personalized weight for different users. To further remove noisy items (e.g., occasionally enrolling a course), HRL [15] proposes a hierarchical reinforcement learning algorithm to interactively revise the noisy item list.

Previous works have shown promising results. However, there still exists some limitation. From the learning to rank perspective, a major issue of previous works is that they are all pointwise approaches where models try to predict the precise relevance score or rating of each item. As item order is important in ranking based top-n recommendation problem, it would be better to model the course recommendation from a pairwise learning to ranking perspective where predicting relative preference order is more close to the ranking nature of top-n recommendation.

In this paper, we propose a pairwise learning model that could capture preference ordering information among courses for users. We theoretically derive the optimization schema based on the Bayesian Personalized Ranking [12] and develop a novel neural network model, called Bayesian Personalized Ranking Network (BPRN), which learns pairwise course preference for each user given her historically enrolled courses. With extensive experiments on a large-scale MOOCs enrollment dataset collected from XuetangX, we empirically demonstrate that our

---

[1] https://www.classcentral.com/report/mooc-stats-2019/.
[2] https://www.coursera.org.

BPRN framework performs better than state-of-the-art item-based course recommendation methods.

To summarize, the contributions of this paper include:

- We propose an item-based pairwise learning-to-rank model based on Bayesian Personalized Ranking.
- We develop the Bayesian Personalized Ranking Network (BPRN) and demonstrate its effectiveness using experiments.
- We release a large-scale course recommendation dataset with 647,381 course enrollment logs in https://github.com/go2school/BPRN.

2 Related Works

The problem of course recommendation requires the recommender system to rank a list of unenrolled courses at time $t + 1$ given historically enrolled courses at time t. Thus, users can make more cautious decision on what course to choose next. In this section, we review related literature development in the area of item-based collaborative filtering and ranking-based recommendation algorithms.

Item-Based Collaborative Filtering [13] refers to a class of collaborative filtering (CF) models that rely on measuring the similarity among items for generating the item recommendation list. Different to user-based CF methods [11,12], item-based CF does not need to explicitly model user embedding vector. Usually, the size of items is much smaller than the size of users. Therefore, the scalability of item-based CF is better than user-based CF. That's why item-based CF methods are more popular in industry [3,9].

Traditional item-based CF methods [9,13,14,16] use statistic metrics, such as Pearson correlation coefficient and cosine similarity to measure item similarity. Recently, machine learning based methods have shown superior performance over traditional heuristic methods. SLIM [10] uses simultaneous regression approach to directly learn the item-item similarity matrix. However, the high dimensionality of the item-item similarity matrix makes the optimization difficult. In addition, it has been shown that SLIM can fail to capture the dependencies between items that have never been co-rated [6]. To solve this problem, FISM [6] learns the latent item vector with much smaller dimensionality and uses average pooling of past items as user profile. A limitation of FISM is that the sequential dependencies between past items is lost. To capture such sequential dependencies, previous works use gated recurrent network [4] or convolutional neural networks [7] to learn the dependencies between past items. However, these methods assume that all historical items are of equal importance in estimating user interests. This may not be true in real-world scenario. NAIS [3] uses neural attention mechanism to learn the varying importance of past items. To further explore the hierarchical structure of items, ACF [1] propose a hierarchical attention network that learns both item-level and component-level attention importance. Finally, to deal with the noisy items in NAIS, HAL [15] develops a hierarchical reinforcement learning algorithm to interactively revise item list.

Bayesian Personalized Ranking. The above item-based collaborative filtering algorithms are all point-wise rating estimation methods, which means that they directly optimize for minimizing the prediction error on the point value of each user-item rating. Despite its great success in practical applications, a major drawback of pointwise methods is that they could not explicitly model the ranking information among items. To overcome this challenge, pairwise item recommendation methods provide a novel perspective. As far as we know, the most widely used pairwise recommendation framework is Bayesian Personalized Ranking [12]. The original work [12] provides a generic pairwise optimization criterion for top-n recommendation algorithm using implicit feedbacks. Authors derive the Bayesian formulation of the ranking of each pair of items given by a specific user, and uses ranking statistic AUC to measure the correctness of the ranking. Based on this formulation, the learning algorithm proposed for solving BPR essentially optimizes for correctly ranking item pairs using a stochastic gradient descent procedure. Since a user often only interacts with a tiny portion of the items, the learning problem is highly skewed in terms of positive/negative sample ratio. For this reason, each SGD iteration of BPR samples a user, a positive item and a negative item randomly with replacement, which is essentially a negative sampling protocol.

In this paper, we will combine item-based collaborative filtering and Bayesian Personalized Ranking as basis and provide a systematic study on course recommendation by contributing a novel learning framework with sound theoretical analysis.

3 Methods

3.1 Problem Formulation

Let $R \in \mathbb{R}^{M \times N}$ denotes a user-course interaction matrix where M and N denote the number of users and courses. Let U and C be the set of all users and courses, respectively. Let $h^u = (c_1^u, \cdots, c_t^u) \subseteq C$ be the historically enrolled courses of each user u, supposing that the user has already enrolled t courses. h^u is often called user profile in item-based collaborative filtering. The course recommendation dataset can be defined as $D = \{(u, h^u, c_{t+1}^u) | u \in U \wedge c_{t+1}^u \in C\}$ where each training example contains a user ID u, a user profile h^u and the recommended course c_{t+1}^u. To simplify notation, in the rest of this paper, we use c_u^* to represent c_{t+1}^u and we denote unenrolled course set $C \setminus h^u$ as $\bar{h}^u$. Thus, given a user profile h^u, the task of course recommendation is to provide a total order (ranking) $>_{\bar{h}^u} \subseteq C^2$ in the set of courses not enrolled by the user. It should be noted that the total order $>_{\bar{h}^u}$ is related to the user profile h^u while in original BPR [12] the total order is only related to the user ID.

3.2 Bayesian Derivation for Pairwise Ranking

In this subsection, we formulate the ranking information among items in a Bayesian manner to overcome the limitation of pointwise approach. Given a dataset $D = \{(u, h^u, c^u_{t+1})|u \in U \wedge c^u_{t+1} \in C\}$, we assume that each user is independently sampled, thus the total order of courses between users are independent. We aim to maximize the posterior probability $P(\Theta|D)$ at time t where Θ are the model parameters.

$$P(\Theta|D) \propto P(D|\Theta)P(\Theta) = \prod_{u \in U} P(>_{\bar{h}^u} |\Theta, h^u)P(\Theta) \tag{1}$$

To estimate $P(\Theta)$, following the BPR model, we can also assume all model parameters have Gaussian priors with variance λ.

$$P(\Theta) \sim \mathcal{N}(0, \lambda I) \tag{2}$$

To estimate the probability of total order given historically enrolled courses $P(>_{\bar{h}^u} |\Theta, h^u)$, usually we need to estimate $P(c_i > c_j|\Theta, h^u)$ for all pairs of (c_i, c_j) in $\bar{h}^u$. However, such a quadratic enumeration is too expensive. In addition, it is difficult to estimate probability of pairs of courses that are not observed in training data. In this paper, we simplify this by only estimating the preference likelihood between the positive course $c^*_u = c^u_{t_u+1}$ and a small set of randomly sampled negative courses in $\bar{h}^u$.[3] Therefore, we rewrite $P(>_{\bar{h}^u} |\Theta, h^u)$ as

$$P(>_{\bar{h}^u} |\Theta, h^u) \propto \prod_{c \in \bar{h}^u \setminus \{c^*_u\}} P(c^*_u > c|\Theta, h^u) \tag{3}$$

To estimate the probability of $P(c^*_u > c|\Theta, h^u)$, we can use the sigmoid function to transfer preference score $f(c^*_u, h^u; \Theta) - f(c, h^u; \Theta)$ between two courses into probability.

$$P(c^*_u > c|\Theta, h^u) = \sigma\big(f(c^*_u, h^u; \Theta) - f(c, h^u; \Theta)\big) = \frac{1}{1 + e^{-(f(c^*_u, h^u; \Theta) - f(c, h^u; \Theta))}} \tag{4}$$

where $f(c, h^u; \Theta)$ is the pointwise prediction function that estimates the preference of course enrollment c given profile h^u.

Finally, following the Maximum A Posterior principle, the learning objective can be defined as

BPR-OPT : $\ln P(\Theta|D)$

$$\propto \ln\Big(P(\Theta) \prod_{u \in U} \prod_{c \in \bar{h}^u \setminus \{c^*_u\}} \sigma\big(f(c^*_u, h^u; \Theta) - f(c, h^u; \Theta)\big)$$

$$\propto -\lambda \|\Theta\|^2 + \sum_{u \in U} \sum_{c \in \bar{h}^u \setminus \{c^*_u\}} \ln \sigma\big(f(c^*_u, h^u; \Theta) - f(c, h^u; \Theta)\big)) \tag{5}$$

3.3 Bayesian Personalized Ranking Network

In this subsection, we propose the Bayesian Personalized Ranking Network (BPRN) for course recommendation with BPR-OPT objective. Figure 1 shows the network architecture of BPRN.

---

[3] Sampling negative examples for implicit feedback dataset is a hot research topic. While the focus of this paper is to propose a new neural network for course recommendation, therefore we adopt a simple uniform sampling approach.

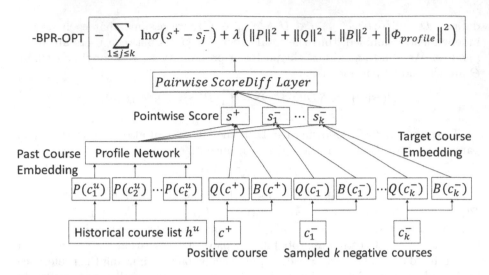

Fig. 1. Bayesian Personalized Ranking Network.

Input Pipeline. BPRN adopts a user-by-user input pipeline to optimize BPR-OPT. Specifically, the training example of each user fed into BPRN contains of a profile list h^u, a positive course c^+ and k sampled negative courses c_j^- $(1 \leq j \leq k)$. The profile of each user is evaluated only once. This is much faster than rating-by-rating input pipeline where the user profile has to be repeated evaluated for each course.

Profile Embedding. Given a historically enrolled course list h^u, BPRN first embeds each enrolled course c_t^u to a latent vector through an embedding matrix $P \in \mathbb{R}^{N \times d}$ where d is the latent dimension and yields vector $P(c_t^u)$ as the past course embedding. Those course embedding vectors will be sent into the profile network that learns user profile $Profile(h^u) \in \mathbb{R}^{N \times d}$ with various network models such as *average pooling* (FISM) and *attention network* (NAIS).

Target Course Embedding. When training BPRN, we input one positive course and k sampled negative course for learning pairwise preference. The IDs of these course will be embedded through an embedding matrix $Q \in \mathbb{R}^{N \times d}$ and yield $Q(c^+)$ and $Q(c_j^-)(1 \leq j \leq k)$. To capture the global properties of courses, BPRN also uses another embedding matrix $B \in \mathbb{R}^{N \times 1}$ to yield bias term.

Pairwise Loss Estimation. BPRN first evaluates pointwise ranking scores of all target courses by using the profile network, target course embedding and bias term (Pointwise Score in Fig. 1) as

$$f(h^u, c) = B(c) + \langle Profile(h^u) \cdot Q(c) \rangle \tag{6}$$

Second, the ranking scores s^+ and s_j^- $(1 \leq j \leq k)$ will be passed through the Pairwise ScoreDiff Layer which computes the score difference between s^+ and each s_j^- $(1 \le j \le k)$ as $\ln \sigma(s^+ - s_j^-)$. Third, to control the model complexity, BPRN performs regularization on P, Q, B and parameters of profile network $\Theta_{profile}$. As most deep learning library has automatic differentiation functionality, we will not manually derive the gradient update equation. We use ADAM [8] algorithm to optimize the negative of BPR-OPT loss.

4 Experiment

In this section, we conduct experiments to evaluate performance of BPRN. We aim at answering the following research questions:

RQ1 How does BPRN perform compared with state-of-the-art methods?
RQ2 How do hyper parameters affect BPRN performance?

4.1 Dataset

We collect a large-scale MOOCs enrollment dataset from XuetangX[4]. It contains $647,381$ course enrollment records with $35,661$ users on $3,127$ courses. Figure 2 show the frequency distribution of users by the number of enrolled courses, and the frequency distribution of courses by the number of enrolled learners. We can see that both frequency distribution approximately follow the Zipf distribution. A detailed analysis shows that although some users can enroll more than 100 courses, still more than 97% users enroll less than 50 courses.

4.2 Experimental Protocol

Basic Configuration. We split the dataset into a training set and a testing set with $60 : 40$ ratio by user-based splitting. As discussed in Sect. 3.1, each example in the dataset contains a user profile as well as a target course. Specifically, for each user, we use her previous $n - 1$ course interaction records as user profile and the last interaction as the positive target course.

To train BPRN in such an implicit feedback dataset (only positive interaction observed), we need to sample negative courses. In our experiments, we randomly sample four negative courses not enrolled by users during the training phase at each epoch.[5] In the testing phase, we adopt the evaluation method similar as [3] where 99 randomly sampled negative courses (not enrolled) and one true positive target course are mixed together for rating.

The experiments are repeated for five times. We plot curves of average NDCG score (the Normalized Discounted Cumulative Gain) at each epoch on the testing set to evaluate recommendation performance.

---

[4] http://www.xuetangx.com.
[5] In our pilot experiment, we find that sampling different negative course at each epoch can increase the variance of training set and reduce the risk of overfitting.

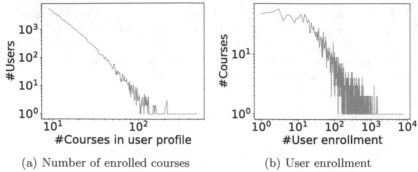

(a) Number of enrolled courses   (b) User enrollment

Fig. 2. The frequency distribution of users by the number of enrolled courses, and the frequency distribution of courses by the number of registered learners. The axis in both figures are logarithmic scaled.

Compared Methods. Three methods will be evaluated in our experiments.

- FISM [6]. This method applies an average pooling layer on the embeddings of historically enrolled courses for each user. The hyper parameter α that controls the strength of neighborhood similarity is set to 0.5 which is reported to have reasonable performance.
- NAIS [3]. This method improves FISM by using an attention network to assign weights to different historical courses. Both hyper parameters α and β (for smoothing softmax) are empirically set to 0.5. The NAIS-prod strategy is used as it is reported to perform better than NAIS-concat strategy [3].
- NAIS+BPRN. This is the BPRN version of NAIS where the profile network in BPRN is instantiated as NAIS which is a state-of-the-art item-based recommendation method.

Hyper-Parameter Tuning. The main hyper-parameters of BPRN include learning rate l, embedding size k, batch size b and regularization parameter λ. During performance comparison, we fix the learning rate, embedding size and batch size as $l = 0.01$, $k = 16$ and $b = 32$ for all methods. The regularization parameter λ is tuned in the range $[10^{-7}, 10^{-6}, 10^{-5}, \ldots, 0.1]$ during the first five epochs in the training phase. All network parameters are initialized in Gaussian distribution with zero mean and variance λ (see Eq. 2).

Implementation. We implement the BPRN framework as well as other competing methods in PyTorch. The experiments will run on a workstation with Intel i9-9820X 3.3 GHz CPU, 128 GB memory and GeForce RTX 2080 Graphics Card. To train network in batch, we need to deal with the dynamic length issue of course enrollment. In Sect 4.1, we show that more than 97% users enroll less than 50 courses. Thus, we set the maximum length of user profile as 50.

4.3   Performance Comparison (RQ1)

Table 1 shows the optimal NDCG@{1,3,5,10} score of each method with best hyper-parameters tuned on the training set[6]. We can see that NAIS+BPRN method is significantly (p-value < 0.01) better than the other two methods. Specifically, for NDCG@1, NAIS+BPRN has 3.3% and 6.2% relative improvement over NAIS and FISM respectively. When the k parameter of NDCG@k increases, the performance margin between NAIS+BPRN and the other two methods become small. This is under our expectation as the larger the k, more likely the positive recommended course will show in the ranking list. However, even for NDCG@10, our BPRN method still has 0.37% and 3.22% relative improvement over NAIS and FISM respectively.

Table 1. The optimal NDCG@{1, 3, 5, 10} score of each method.

Method	NDCG@1	NDCG@3	NDCG@5	NDCG@10
FISM	0.2204	0.3318	0.3742	0.4225
NAIS	0.2270	0.3405	0.3838	0.4349
NAIS+BPRN (ours)	**0.2345**	**0.3448**	**0.3877**	**0.4365**

We conduct a detailed analysis on the performance curves. Figure 3 shows the performance curves of NDCG@{1, 3, 5, 10} for different methods. We can see that our BPRN framework can significantly boost traditional pointwise recommendation methods. Specifically, during the initial epochs, the performance curves of FISM, NAIS and NAIS+BPRN have a large overlapping. After about 18 epochs, NAIS+BPRN starts to outperform the other two methods significantly and keep the margin space till the last epoch. This means that Bayesian pairwise learning is effective for improving the convergence rate of course recommendation models.

4.4   Hyper-Parameter Study (RQ2)

The BPRN framework relies on λ parameter to trade-off the MAP criteria and model complexity. How does this parameter affect the performance? As NAIS+BPRN performs the best in previous experiment, we plot NDCG@10 curves at different λ for NAIS+BPRN in Figure 4. We can see that the hyper-parameter λ has a great impact on BPRN framework. An improper choice of λ can degenerate performance. For example, for NAIS+BPRN with $\lambda = 0.0001$, although the performance curve raises quickly, it suffers the problem of overfitting after the 10th epcoh. Even its best performance still cannot surpass the optimal NDCG@10 score with $\lambda = 0.001$. Therefore, carefully tuning the hyper-parameter λ is very important in BPRN.

---

[6] The best λ for each method are 10^{-5} for FISM, 10^{-6} for NAIS and 0.001 for NAIS+BPRN.

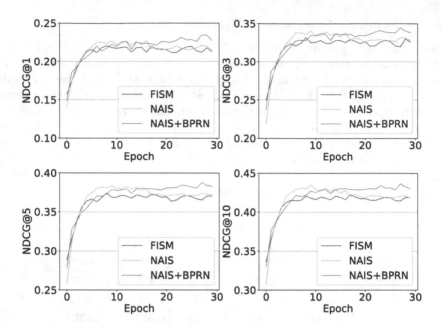

Fig. 3. Performance curves of NDCG@{1,3,5,10} for each method.

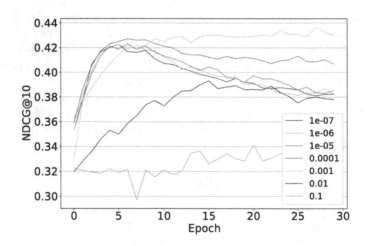

Fig. 4. Performance curves of NDCG@10 for NAIS+BPRN at different regularization parameter λ.

Table 2. Training time and NDCG@10 score at different number of negative courses.

#Negative	1	2	4	8	16
Time (Seconds)	206.3	226.9	273.9	351.6	579.7
NDCG@10	0.3762	0.4115	0.4365	0.4355	0.4425

Our BPRN framework needs to sample negative courses to form pairs of training examples. In our experiment, we sample four negative courses for each positive target course. Can we gain more performance boost with more negative training samples? To study this issue, we try different numbers of negative training courses (per positive course) in the set of $\{1, 2, 4, 8, 16\}$ for NAIS+BPRN. From Table 2, we can see that overall the more negative courses we sample, the more performance gain we can get. However, such performance gain is at the expense of longer training time. In practice, we believe that choosing four or eight negative courses is reasonably enough.

5   Conclusion and Future Works

In this paper, we study the course recommendation problem for MOOCs. We propose an item-based pairwise learning-to-rank model that could capture pairwise preference ordering information among courses. Based on our derivation on Bayesian Personalized Ranking, we develop a novel neural network, called Bayesian Personalized Ranking Network (BPRN), that can learn pairwise course preference. With extensive experiments on a large-scale MOOCs enrollment dataset from XuetangX, we empirically demonstrate that our BPRN framework performs better than state-of-the-art item-based recommendation methods.

In our future work, we plan to extend this work in two directions. First, we use the uniform sampling strategy to sample negative courses in training BPRN. However, the informativeness of those uniformly sampled negative courses may not be optimal due to the large sampling space. It is worthy to study more advanced negative sampling scheme where contextual information such as enrollment popularity and topic distribution are considered. Second, we only investigate the performance of BPRN by instantiating the profile network as one of the state-of-the-art item-based collaborative filtering method, i.e., NAIS. It is interesting to apply BPRN in a more broad family of network, such as convolution neural network, gated recurrent network and self-attention network.

Acknowledgement. We would like to thank the anonymous reviewers for their helpful comments. This work is supported by the National Key Research and Development Program of China (2018YFB1004502) and the National Natural Science Foundation of China (61702532, 61532001).

References

1. Chen, J., Zhang, H., He, X., Nie, L., Liu, W., Chua, T.S.: Attentive collaborative filtering: multimedia recommendation with item-and component-level attention. In: Proceedings of the 40th International ACM SIGIR conference on Research and Development in Information Retrieval, pp. 335–344. ACM (2017)
2. Elbadrawy, A., Karypis, G.: Domain-aware grade prediction and top-n course recommendation. In: Proceedings of the 10th ACM Conference on Recommender Systems, pp. 183–190. ACM (2016)

3. He, X., He, Z., Song, J., Liu, Z., Jiang, Y.G., Chua, T.S.: NAIS: neural attentive item similarity model for recommendation. IEEE Trans. Knowl. Data Eng. **30**(12), 2354–2366 (2018)
4. Hidasi, B., Karatzoglou, A., Baltrunas, L., Tikk, D.: Session-based recommendations with recurrent neural networks. In: The 4th International Conference on Learning Representations (2016)
5. Jing, X., Tang, J.: Guess you like: course recommendation in MOOCs. In: Proceedings of the International Conference on Web Intelligence, pp. 783–789. ACM (2017)
6. Kabbur, S., Ning, X., Karypis, G.: Fism: factored item similarity models for top-n recommender systems. In: Proceedings of the 19th ACM SIGKDD International Conference on Knowledge Discovery and Data Mining, pp. 659–667. ACM (2013)
7. Kim, Y.: Convolutional neural networks for sentence classification. In: Proceedings of the 2014 Conference on Empirical Methods in Natural Language Processing, pp. 1746–1751. ACL (2014)
8. Kingma, D.P., Ba, J.: Adam: a method for stochastic optimization. In: The 3rd International Conference on Learning Representations (2015)
9. Linden, G., Smith, B., York, J.: Amazon.com recommendations: item-to-item collaborative filtering. IEEE Internet Comput. **7**(1), 76–80 (2003)
10. Ning, X., Karypis, G.: SLIM: sparse linear methods for top-n recommender systems. In: IEEE 11th International Conference on Data Mining, pp. 497–506. IEEE (2011)
11. Rendle, S.: Factorization machines. In: Proceedings of the 2010 IEEE International Conference on Data Mining, pp. 995–1000. IEEE (2010)
12. Rendle, S., Freudenthaler, C., Gantner, Z., Schmidt-Thieme, L.: BPR: Bayesian personalized ranking from implicit feedback. In: Proceedings of the Twenty-Fifth Conference on Uncertainty in Artificial Intelligence, pp. 452–461. AUAI Press (2009)
13. Sarwar, B., Karypis, G., Konstan, J., Riedl, J.: Item-based collaborative filtering recommendation algorithms. In: Proceedings of the 10th International Conference on World Wide Web, pp. 285–295. ACM (2001)
14. Yang, C., et al.: RepoLike: a multi-feature-based personalized recommendation approach for open-source repositories. Front. Inform. Technol. Electron. Eng. **20**(2), 222–237 (2019)
15. Zhang, J., Hao, B., Chen, B., Li, C., Chen, H., Sun, J.: Hierarchical reinforcement learning for course recommendation in MOOCs, vol. 33, pp. 435–442. AAAI (2019)
16. Zhang, Y., Wu, Y., Wang, T., Wang, H.M.: A novel approach for recommending semantically linkable issues in Github projects. Sci. China Inf. Sci. **62**(9), 202–204 (2019)
17. Zheng, S., Rosson, M.B., Shih, P.C., Carroll, J.M.: Understanding student motivation, behaviors and perceptions in MOOCs. In: Proceedings of the 18th ACM Conference on Computer Supported Cooperative Work & Social Computing, pp. 1882–1895. ACM (2015)

Online Programming Education Modeling and Knowledge Tracing

Yuting Sun[1], Liping Wang[1(✉)], Qize Xie[1], Youbin Dong[1], and Xuemin Lin[1,2]

[1] Shanghai Key Laboratory of Trustworthy Computing,
East China Normal University, Shanghai, China
{51184501146,51194501074,10165101256}@stu.ecnu.edu.cn,
lipingwang@sei.ecnu.edu.cn
[2] The University of New South Wales, Sydney, Australia
lxue@cse.unsw.edu.au

Abstract. With the development of computer technology, more and more people begin to learn programming. And there are a lot of platforms for programmers to practice. It's often difficult for these platforms to customize the needs of users at different levels. In this paper, we address the above limitations and propose an intelligent tutoring model, to help programming platforms achieve better tutoring for different levels of users. We first devise a novel framework for programming education tutoring which is combined with programming education knowledge graph, crowdsourcing system and online knowledge tracing. Then, by ontology definition, information extraction and data fusion, we construct a knowledge graph to store the data in a more structured way. During the knowledge tracing stage, we extract behavior features and question knowledge features from a relational database and knowledge graph separately. Meanwhile, we improve the process for student ability evaluation and adapt the Knowledge Tracing algorithm to predict students' behavior on knowledge and questions. Experiment results on real-world user behavior data sets show that through the help of Knowledge Tracing algorithm, we can achieve considerably satisfied results on students' behavior prediction.

Keywords: Knowledge graph construction · Knowledge Tracing · Online programming

1 Introduction

Computer Programming is an important skill for software engineers and students majoring in computer science related fields. There have been many excellent platforms for programming practice exercise and examination, such as Leet-Code, Nowcoder, UsacoGate Online Judge, Zhejiang University Online Judge, etc. Recently, those online programming platforms with countless programming questions are the most common ways for programming exercise. However, with the popularity of the online platforms, there are new requirements from different

G. Li et al. (Eds.): KSEM 2020, LNAI 12274, pp. 259–270, 2020.
https://doi.org/10.1007/978-3-030-55130-8_23

users. For beginners, they need some help to diagnose bugs in their code. For Senior users, they usually need some guides for their study and avoid trying the same type of questions over and over again. And for teachers, they want to assign homework according to each student's knowledge level. In this paper, we address the above requirements and aim to provide an intelligent tutoring model for programming platforms, by which users with different programming levels will achieve effective one-to-one tutoring.

Intelligent Tutoring System (ITS) has been studied for several decades [10], which incorporate Artificial Intelligence (AI) techniques into education in order to achieve individualization of study process. For online education system, it is necessary to systematically investigate domain data mining and dynamic prediction model. In this paper, We devise a novel intelligent tutoring system for programming platform by intensive construction of the relative domain knowledge graph and introduction of effective knowledge tracing method. To the best of our knowledge, there is not a sophisticated approach for ITS in programming platform. The principle contributions of this paper are as follows:

- We devise a novel framework for programming education tutoring which is combined with programming education knowledge graph, crowdsourcing system and online knowledge tracing. Under the guidance of domain experts, the programming education knowledge graph is strictly constructed, which is utilized to serve for knowledge tracing algorithm and can be complemented by the crowdsourcing system.
- We extract behavior features and question knowledge features from a relation database and knowledge graph separately. Meanwhile, we improve the process for student ability evaluation and utilize the Dynamic Student Classification on Memory Networks [9] algorithm to make predictions on students' feedback on knowledge and questions.
- We conduct extensive experiments and evaluate knowledge tracing algorithms on real-world user behavior datasets to predict students' feedback on programming knowledge and questions separately. Comprehensive comparisons are made between algorithms. The experimental results demonstrate the effectiveness of our proposed method.

The rest of this paper is organized as follows: we briefly introduce the framework for programming education tutoring in Sect. 3. Domain ontology construction, information extraction and data fusion are represented in Sects. 3.1 and 3.2. Knowledge Tracing algorithms are devised in Sect. 4. We compare our method with other representative methods via some experiments, and depict a performance evaluation on student behavior prediction of our approach in Sect. 5. Finally, we conclude this paper in Sect. 6.

2   Related Work

Knowledge Graph Construction. There are two ways to construct Knowledge Graph [12], top-down and bottom-up. Top-down [18] construction refers to extracting ontology and pattern information from high-quality data and adding them to the knowledge graph with the help of structured data sources such as encyclopedia websites; the so-called bottom-up construction refers to extracting resource patterns from publicly collected data by certain technical means, selecting new patterns with high confidence and adding them to the knowledge base after manual review.

The bottom-up construction [15] usually takes three steps: information extraction, knowledge fusion, and knowledge processing. Information extraction [4] is to extract entities from various types of data sources. It's usually an automatic technology to extract structured information from structured or semi-structured data, involves entity extraction, relation extraction and attribute extraction. Some classical approaches for information extraction are based on (1) regularization, e.g. RoadRunner [5], (2) Template deduction [2], (3) Conditional random field [20], (4) Generalized hidden Markov model [19]. Some latest work for information extraction such as MGNER [17] for Multi-Grained Named Entity Recognition, DSGAN [13] using Generative Adversarial Networks for relation extraction. Knowledge fusion is to integrate the acquired new knowledge to eliminate contradictions and ambiguities. Knowledge processing is to obtain structured and organized knowledge, which mainly includes three aspects: ontology construction, knowledge reasoning and quality evaluation. Ontology construction can be divided into three main methods: manual ontology construction, automatic ontology construction and semi-automatic ontology construction. There are a lot of ontology construction methods been proposed, such as IDEF5, TOVE [1], MEHONTOLOGY, SKELETON, KACTUS, the Seven-Steps [8] method, etc. Among them, the Seven-Steps are proposed by Stanford University, which are to determine the scope of domain ontology, reuse existing ontology, determine domain terms, define hierarchical relationships, define attributes, define facets, and fill-in examples.

Knowledge Tracing. In the field of education, how to model students' mastery of knowledge is a crucial problem. Knowledge Tracing [3] aims to estimate students' level of mastery of a certain knowledge, based on his/her history interactions with questions. An accurate knowledge tracing enables us to grasp the needs of students and carry out accurate problem solving.

Traditional knowledge tracing are based on the first-order Markov Model, such as Bayesian Knowledge Tracing [6]. BKT models the knowledge state of a student as a set of binary variables, each variable represents if a student understands a knowledge. As a student keeps practice, the mastery of knowledge will also change dynamically. However, BKT can't handle the situation that one question might involve several knowledge.

Knowledge Tracing Machine [16] is a sequence prediction model. KTM uses the Factorization Machine [14] to solve the problem of sparse features, and capture the correlations between features. It can accurately and rapidly estimate students' performance and deal with questions with multiple knowledge. Also, KTM can handle questions with multiple skills.

With the popularity of deep learning [7] in recent years, it's also been applied to Knowledge Tracing. Deep Knowledge Tracing [11] uses the Long Short Term Memory to represent students' behavior. It can reflect the long-term knowledge dependencies. Unlike BKT assumed that once students master a certain knowledge, they will never forget it, DKT takes into account that students may forget what they have learned before after a period of time. It can capture students' recent performance to predict the results of their answers and make more use of students' recent performance.

3   Knowledge Graph for Programming Education

We investigate a programming education tutoring system which is constructed by a knowledge graph, crowdsourcing system and knowledge tracing algorithm with intelligence, real-time and dynamic. The framework of programming education tutoring system is shown in Fig. 1.

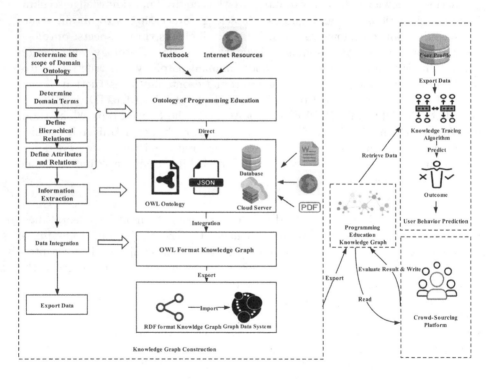

Fig. 1. Framework for knowledge graph construction and online knowledge tracing

For knowledge graph construction, during the preparation stage, textbooks, exercises, solutions for some difficult questions, related Internet resources, etc. are collected as the data sources. We carefully define the ontology of programming education using our domain knowledge and extract the information from different data sources. For unstructured data, regular expressions corresponding to textual content are devised. Then different formats of structural data are integrated and an RDF (Resource Description Framework) format knowledge graph is exported.

Moreover, in order to pursue a dynamical and updated knowledge graph, we devise a relevant crowdsourcing system which can publish tasks to related users and evaluate the answers. For the programming platform, in order to achieve more accurate results, both domain information in the knowledge graph and user behavior data in a relational database are served for programming tracing algorithms and online decision modules.

3.1   Domain Ontology Construction

The well-known knowledge graphs such as freebase, Wikidata, YaGo, etc. are widely utilized, but they haven't well covered the domain knowledge in programming education. To satisfy the domain requirement, we adapt the Seven-Steps method [8], and use four steps to define ontology: (1) determine the scope of domain ontology, (2) determine domain terms, (3) define hierarchical relationships, (4) define attributes and relationships.

Table 1. Modules and terms defined in programming education knowledge graph

Module	Terms
Data Structure (4 levels)	Linear Structure, List, Matrix, Tri Matrix, Queue, Stack, String, Tree Structure, Basic Tree Terms, Forest, Tree, Binary Tree, Multiway Tree, Tree Storage Structure, Graph, Basic Graph Term, Directed Graph, Undirected Graph, AOV, AOE, Graph Storage Structure, Graph Traversal, Minimal, Spanning Tree, Shortest Path, Searching, Basic Searching Term, Hash Searching, Linear Searching, Tree Searching, Sorting, Basic Sorting Term, External Sort, Internal Sort, Insert Sort, Merge Sort, Radix Sort, Select Sort, Swap Sort
Programming Language (4 levels)	Array, Branch Statement, Constant and Variable, Basic DataType, Constant, Variable Initialization, Variable Assignment, Variable Type, Custom Type, Dynamic Memory Management, Function, Customized Function, Library Function, IO Function, String Function, Recursive Function, Loop Statement, Operators and Expressions, Arithmetic Operator, Shift Operator, Expression Evaluation, Implicit Type Conversion, Positional Operator, Logical Operator, Relational Operator, Assignment Operator, Conditional Operator, Pointer, Preprocessing, Predefined Symbol, Conditional Compilation, File Contains, Macro
Dataset (2 levels)	Graph Structure Data, Linear Structure Data, Tree Structure Data, Others
Relevant Knowledge (3 levels)	Algorithm, Code Segment, Exercise, Extended Knowledge, Completion, Multiple Choice, Short Answer, True/False question

The ontology is constructed using the bottom-up approach. In the first step, for programming education domain scope determination, four major modules are identified and modules are guaranteed to be independent of each other as far as possible. Here, the defined four modules are data structure, programming language, datasets for experiment, and relevant knowledge(which can be reused as exercises separately). In the next two steps, terms belonging to each module are defined and arranged by hierarchical structure. The detailed definitions of terms in each module are listed in Table 1. Finally, based on the definition of modules and terms, the data property and object property are identified according to the domain knowledge.

3.2 Information Extraction and Data Fusion

After the domain ontology construction, we extract entities, attributes, and relations from different data sources which are combined with unstructured, semi-structured and structured data. We devise different extraction policies separately according to the structure of the data source. For the Data Structure and Programming modules defined in ontology, the data source mainly comes from Encyclopedia websites such as Baidu Encyclopedia and Wiki encyclopedia which are structured or semi-structured. In this case, the solution is straight-forward. An instance-class mapping table is constructed using the domain terms determined in Sect. 3.1 to accomplish the entity exaction. At the same time, a data property mapping table is created for attributes and relations extraction. These results are stored as JSON format temporarily.

For the relevant knowledge module defined in ontology, the source of the data is mainly from test questions set in the format of word and PDF. We devise the corresponding regular expressions to match the required data. In this process, both text and images embedded in the question are handled and stored into the relation database temporarily.

After information extraction, we achieve three independent data, that is domain ontology in OWL (Web Ontology Language) format (in Sect. 3.1), data that relates to Data Structure and Programming language in JSON format, and Relevant Knowledge data stored in relational databases. The JSON format data is converted to OWL format by instance-class mapping table (as shown in Algorithm 1) and the data stored in relation database is also converted to OWL format. Finally, the OWL file consists of instances of programming education Knowledge Graph, which can be exported to the graph database system.

Algorithm 1 is devised to determine the corresponding ontology of extracted information temporarily stored in JSON format. Given the clean data stored in JSON format, Algorithm 1 automatically completes the mapping from instance to ontology by data property-JSON mapping table and instance-class mapping table.

Algorithm 1. Ontology Instances generation

input: *JSONFile*: JSON Foramt File, *DDTable*: Data Property-JSON mapping table, *ECTable* :Instance class mapping table
output: *OWLFile*: Ontology Instances

```
 1: function CONVERTJSON(JSONFile, DDTable, ECTable)
 2:     construct an empty JSON object vector v
 3:     for each JSON object o in JSON file do
 4:         construct an empty object n
 5:         for each attribute of o do
 6:             if the key of attribute a can be found in DDTable then
 7:                 save a to object n
 8:         save object n to vector v
 9:     for each JSON object n in vector v do
10:         if attribute name can be found in ECTable then
11:             Find the class name c corresponding to name
12:             Construct corresponding OWLdata
13:             append OWL data to the end of OWL file
14:         else
15:             Find the class name c corresponding to Chinese name
16:             Construct corresponding OWLdata
17:             Append OWL data to the end of the OWL file
18:     return OWLFile
```

4   Student's Behavior Prediction

Problem Formulation. We formulate students' behavior prediction problem as follows. Given m users $\{u_1, u_2, \ldots, u_m\}$, n questions $\{q_1, q_2, \ldots, q_n\}$, l knowledge $\{k_1, k_2, \ldots, k_l\}$, behavior prediction on question aims to get the correct rate of student i on question s. Behavior prediction on knowledge aims to evaluate the ability of student i on knowledge j. Our goal is to predict students' behavior on questions or knowledge based on students' records on online programming platform.

We adopt the DSCMN [9] algorithm to help make predictions. The basic idea of DSCMN is to use time intervals to divide the sequence of questions that students have done (the number of time intervals is how many times a student attempted in the segment), and then calculate student ability based on segment. After this, cluster segments of students with similar abilities, assign clusters as an extra feature in the input. The features we were using are (1) knowledgeId, (2) question difficulty, (3) cluster, one-hot encoding is applied in each feature, and concat these one-hot encodings as input. And then bring the Recurrent Neural Network to trace student knowledge in each segment.

We first calculate the average score of student i towards a knowledge j in time interval z in Eq. 1, the weight w_j represents the important degree of the question that knowledge j belongs to, which is calculated by the correct rate of the question. $R(x_j)_{1:z}$ is the difference between how much student i performs on knowledge j (see Eq. 2), $d_{1:z}^i$ is a learning ability vector of student i on each knowledge for time interval 1 to z.

$$\text{Average score } (x_j)_{1:z} = \sum_{t=1}^{z} \frac{w_j x_{jt}}{|N_{jt}|} \tag{1}$$

$$R\,(x_j)_{1:z} = \frac{(\text{Average score } (x_j)_{1:z} - 50)}{50} \tag{2}$$

$$d_{1:z}^{i} = (R\,(x_1)_{1:z}, R\,(x_2)_{1:z}, \ldots, R\,(x_n)_{1:z}) \tag{3}$$

After calculating the learning ability vector, we use the k-means algorithm to assign segments of students to clusters. Student learning ability will change over time, so different time segments of a student can belong to different clusters (see Eq. 4).

We calculate students' abilities based on learning ability vectors $d_{1:z}^{i}$, and cluster students with similar abilities. We first choose k random centroids, and update them by the distance with each time interval, we calculate by iteration to get final centroids. After we got centroids, we can assign each time interval segment of student i into the nearest cluster by Eq. 4. We selected part of student records for visualization on Fig. 2. There are 6 clusters in the figure, the color Cyan means students have no interactions in the corresponding time intervals. Students with the same color means similar ability.

$$\text{Cluster } (Stu_i, Seg_z) = \arg\min_{C} \sum_{c=1}^{K} \sum_{d_{1:z-1}^{i} \in C_c} \left\| d_{1:z-1}^{i} - \mu_c \right\|^2 \tag{4}$$

We conduct experiments on DSCMN on different time intervals to find the best length of time intervals to separate student sequences. Experiment results show that when time intervals are set to 50, we got the best performance. (See Fig. 3).

And then the cluster for students at different time intervals will add as an extra feature in the input. A RNN model has been used to make predictions in each segment (Eq. 5). We can predict a question that involves multiple knowledge, by averaging the predictions of knowledge involved as the prediction to a question. In Eq. 5, v_t is the success and failure levels of knowledge k_t until time t − 1. v_t is calculated by $\frac{\sum_{1}^{t-1} \text{score}(u, k_t)}{|k_t|}$, W_{hx} represents the input weight matrix, W_{hh} represents the recurrent weight matrix, W_{yh} represents the readout weight matrix, b_h, b_y are biases for latent and readout units. In Eq. 6, y_t represents the probability of answering correctly on a question with knowledge.

$$h_t = \tanh\,(W_{hx}\,[x_{t-1}, k_t, v_t] + W_{hh} h_{t-1} + b_h) \tag{5}$$

$$y_t = \sigma\,(W_{yh} h_t + b_y) \tag{6}$$

Lastly, the sigmoid function is used to make predictions. After we get the predictions of the correct rate of knowledge i on students t, we can calculate the predictions of question q on the student, by averaging the correct rate of knowledge that belongs to q.

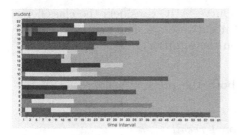

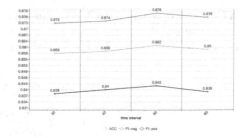

Fig. 2. Students' learning ability evolution over each time interval(50 attempts per time interval, each cluster is represented by different color)

Fig. 3. Performance at different time intervals

5   Experiment

5.1   Experiment Setups

Dataset. The dataset we used contains 202 users, 184 questions and 48 knowledge. The data are extracted from Mynereus programming Platform[1]. There are a total 86772 records, including records of students over a year of practice history on the website. These questions are C language and Data Structure programming oriented. And we have in total 105 terms, 156 inter-class attributes, 35 intra-class attributes, 8 inter modules relations, 2615 instances, and 8135 inter-instance relations in our Knowledge graph. Student records are extracted from a relational database, and the relations of questions and knowledge are from our knowledge graph.

Metrics. We choose (1) Accuracy, (2) AUC, (3) F1-score as metrics in our experiments.

Table 2. Confusion matrix

Actual/Predict values	Positive	Negative
Positive	True Positive	False Positive
Negative	False Negative	True Negative

Accuracy. The accuracy is calculated as follows:

$$accuracy = \frac{TP + TN}{TP + TN + FP + FN} \qquad (7)$$

---

[1] http://code.mynereus.com.

AUC. AUC is the Area Under the ROC Curve. AUC considers the classifier's classification ability for both positive and negative cases, and can still make a reasonable evaluation for the classifier when the samples are unbalanced.

F1-Score. F1-score takes into account both the precision and recall of the classification model.

$$F_1 = 2 \cdot \frac{precision \cdot recall}{precision + recall} \tag{8}$$

5.2   Results

We conduct two sets of experiments here. The first experiment is to predict students' behavior on questions, the second experiment is to predict students' behavior on knowledge. We select DKT-DSC and KTM as our comparison algorithms in the first experiment. Note that DSCMN is the derivation version of DKT-DSC.

Experiments for Predict Students' Behavior on Questions. The results for predicting students' behavior on questions are on Table 3. In DSCMN, we consider different questions that have different difficulty, and use the correct rate of each question as weight, and achieve the best result here. We also adapt KTM to make predictions on questions. KTM uses Factorization Machine to capture relevance between different features, the idea of Factorization Machine based on matrix decomposition, it captures relations between different features as in Eq. 9. Because the structure of KTM is very different from DSCMN and DKT-DSC, we select the best model in KTM to compare with the above two algorithms. In KTM, we use features (1) students, (2) questions, (3) knowledge, (4) attempts, (5) wins, (6) fails. Attempts are how many times a student answered a question, wins are how many times the student answered the question correctly.

$$\hat{y}(\mathbf{x}) := w_0 + \sum_{i=1}^{n} w_i x_i + \sum_{i=1}^{n} \sum_{j=i+1}^{n} \langle \mathbf{v}_i, \mathbf{v}_j \rangle x_i x_j \tag{9}$$

Table 3. Experiment results on predict students' behavior on questions

Model	ACC	AUC	F1-neg	F1-pos
DKT-DSC: student, question	0.748	0.808	0.753	0.744
DSCMN: student, question, knowledge	0.808	0.880	0.816	0.799
DSCMN: student, question, knowledge(weighted)	**0.818**	**0.882**	**0.825**	**0.809**
KTM: student, question, attempts, wins, fails	0.785	0.842	0.787	0.782

Experiments for Predict Students' Behavior on Knowledge. The experiment results of predict students' behavior on knowledge are presented on Table 4. Instead of predict success or failure on questions, DSCMN focus on students' performance on knowledge. In practice sometimes we care more about student mastery on knowledge rather than a single question.

It is noteworthy that we improved the performance of DSCMN a lot by adding weight to different questions. DSCMN outperfoms the other two algorithms in every model, achieving an accuracy of 0.818, 0.882 on AUC on predicted questions, 0.865 on accuracy and 0.942 on AUC on predicted knowledge.

Table 4. Experiment results on predict students' behavior on knowledge

Model	ACC	AUC	F1-neg	F1-pos
DKT-DSC: student, knowledge	0.853	0.934	0.869	0.834
DSCMN: student, question, knowledge	0.862	0.938	0.878	0.842
DSCMN: student, question, knowledge(weighted)	**0.865**	**0.942**	**0.879**	**0.849**

6   Conclusion and Future Work

In this paper, we devise a novel framework for programming education tutoring which is combined with programming education knowledge graph, crowdsourcing system and online knowledge tracing. The constructed knowledge graph can represent the programming data and better capture relations between students, questions and knowledge, etc. We successfully extract domain features and improve the Knowledge Tracing algorithm to help us predict students' behavior for knowledge and questions. And, we conduct extensive experiments and evaluate knowledge tracing algorithms on real-world user behavior data sets. Comprehensive comparisons are made between algorithms. The experimental results demonstrate the effectiveness of our proposed method. In the future, based on the predictions we got before, we can recommend students' questions and enhance the online algorithm dynamically.

Acknowledgment. This work was partially supported by NSFC 61401155 and NSFC 61502169. The first author thanks University Côte d'Azur, France and Inria Sophia Antipolis Méditerranée, France where she conducted her master's final year project and internship.

References

1. Research on document clustering based on BP neural net. Computer Science (2002)
2. Arasu, A., Garcia-Molina, H.: Extracting structured data from web pages. In: Proceedings of the 2003 ACM SIGMOD International Conference on Management of Data, pp. 337–348 (2003)

3. Corbett, A.T., Anderson, J.R.: Knowledge tracing: modeling the acquisition of procedural knowledge. User Model. User Adap. Inter. **4**(4), 253–278 (1994)
4. Cowie, J., Lehnert, W.: Information extraction. Commun. ACM **39**(1), 80–91 (1996)
5. Crescenzi, V., Mecca, G., Merialdo, P., et al.: Roadrunner: towards automatic data extraction from large web sites. VLDB. **1**, 109–118 (2001)
6. Freudenthaler, C., Schmidt-Thieme, L., Rendle, S.: Bayesian factorization machines (2011)
7. LeCun, Y., Bengio, Y., Hinton, G.: Deep learning. Nature **521**(7553), 436–444 (2015)
8. Li, W., Han, J., Pei, J.: CMAR: accurate and efficient classification based on multiple class-association rules. In: Proceedings 2001 IEEE International Conference on Data Mining, pp. 369–376. IEEE (2001)
9. Minn, S., Desmarais, M.C., Zhu, F., Xiao, J., Wang, J.: Dynamic student classification on memory networks for knowledge tracing. In: Yang, Q., Zhou, Z.-H., Gong, Z., Zhang, M.-L., Huang, S.-J. (eds.) Pacific-Asia Conference on Knowledge Discovery and Data Mining, pp. 163–174. Springer, Heidelberg (2019). https://doi.org/10.1007/978-3-030-16145-3_13
10. Nwana, H.: Intelligent tutoring systems: an overview. Artif. Intell. Rev. (1990)
11. Piech, C., et al.: Deep knowledge tracing. In: Advances in Neural Information Processing Systems, pp. 505–513 (2015)
12. Qiao, L., Yang, L., Hong, D., Yao, L., Zhiguang, Q.: Knowledge graph construction techniques. J. Comput. Res. Dev. **53**(3), 582–600 (2016)
13. Qin, P., Xu, W., Wang, W.Y.: Dsgan: Generative adversarial training for distant supervision relation extraction. arXiv preprint arXiv:1805.09929 (2018)
14. Rendle, S.: Factorization machines with libFM. ACM Trans. Intell. Syst. Technol. (TIST) **3**(3), 1–22 (2012)
15. Van Der Vet, P.E., Mars, N.J.: Bottom-up construction of ontologies. IEEE Trans. Knowl. Data Eng. **10**(4), 513–526 (1998)
16. Vie, J.J., Kashima, H.: Knowledge tracing machines: factorization machines for knowledge tracing. In: Proceedings of the AAAI Conference on Artificial Intelligence, vol. 33, pp. 750–757 (2019)
17. Xia, C., et al.: Multi-grained named entity recognition. arXiv preprint arXiv:1906.08449 (2019)
18. Zhao, Z., Han, S.K., So, I.M.: Architecture of knowledge graph construction techniques. Int. J. Pure Appl. Math. **118**(19), 1869–1883 (2018)
19. Zhong, P., Chen, J.: A generalized hidden Markov model approach for web information extraction. In: 2006 IEEE/WIC/ACM International Conference on Web Intelligence (WI 2006 Main Conference Proceedings) (WI 2006), pp. 709–718. IEEE (2006)
20. Zhu, J., Nie, Z., Wen, J.R., Zhang, B., Ma, W.Y.: 2d conditional random fields for web information extraction. In: Proceedings of the 22nd International Conference on Machine Learning, pp. 1044–1051 (2005)

Enhancing Pre-trained Language Models by Self-supervised Learning for Story Cloze Test

Yuqiang Xie[1,2], Yue Hu[1,2(✉)], Luxi Xing[1,2], Chunhui Wang[3], Yong Hu[3], Xiangpeng Wei[1,2], and Yajing Sun[1,2]

[1] Institute of Information Engineering, Chinese Academy of Sciences, Beijing, China
{xieyuqiang,huyue,xingluxi,weixiangpeng,sunyajing}@iie.ac.cn
[2] School of Cyber Security, University of Chinese Academy of Sciences, Beijing, China
[3] Effyic Intelligent Technology (Beijing) Co., Ltd., Beijing, China
{wangch,huyong}@effyic.com

Abstract. Story Cloze Test (SCT) gains increasing attention in evaluating the ability of story comprehension, which requires a story comprehension model to select the correct ending to a story context from two candidate endings. Recent advances, such as GPT and BERT, have shown success in incorporating a pre-trained transformer language model and fine-tuning operation to improve SCT. However, this framework still has some fundamental problems in effectively incorporating story-level knowledge from related corpus. In this paper, we introduce three self-supervised learning tasks (Drop, Replace and TOV) to transfer the story-level knowledge of ROCStories into the backbone model including vanilla BERT and Multi-Choice Head. We evaluate our approach on both SCT-v1.0 and SCT-v1.5 benchmarks. The experimental results demonstrate that our approach achieves state-of-the-art results compared with baseline models.

Keywords: Self-supervised learning · Story comprehension · Story Cloze Test

1 Introduction

Story comprehension is an extremely challenging task in natural language understanding with a long-running history in Artificial Intelligence [11]. Recently, an efficient open task, named *Story Cloze Test* (**SCT**) [8], has been introduced to evaluate the quality of the story comprehension. This task requires a story comprehension model to select the right ending from two candidate endings with the given story context.

To address the SCT challenge, both traditional machine learning approaches [2,12] and neural network models [1,14] have been used. Recently, Pre-trained Language Models (PLM), such as GPT [10] and BERT [3], have shown success in incorporating a pre-trained transformer language model and fine-tuning operation to

© Springer Nature Switzerland AG 2020
G. Li et al. (Eds.): KSEM 2020, LNAI 12274, pp. 271–279, 2020.
https://doi.org/10.1007/978-3-030-55130-8_24

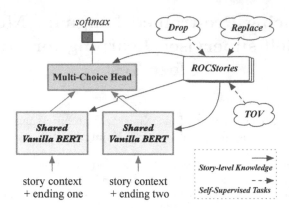

Fig. 1. The overview of our approach for Story Cloze Test. We introduce three self-supervised tasks (***Drop***, ***Replace*** and ***TOV***) to help the backbone model (shared vanilla BERT and Multi-Choice Head) learn the story-level knowledge from ROCStories.

improve SCT. Besides, TransBERT [5] achieves state-of-the-art results by transferring knowledge from semantically related tasks (like MNLI [17]) for SCT.

The major issue of the aforementioned methods is that they lack story-level knowledge for SCT. Compared to BooksCorpus [19] used in pre-training BERT, ROCStories is a large-scale in-domain unlabeled knowledge source for SCT. Through our observations, there are many closely-related connections between SCT and ROCStories. Thus, we argue that the story-level knowledge from the daily stories is efficient for reasoning the right ending.

In this paper, we enhance pre-trained language models by self-supervised learning, which aims to assist model in learning story-level knowledge of ROCStories for SCT task. As shown in Fig. 1, we introduce three self-supervised tasks (***Drop***, ***Replace*** and ***TOV***) to learn the story-level knowledge from daily stories of ROCStories, based on the backbone model including vanilla BERT [3] and Multi-Choice Head.

Our contributions are as the followings:

- We enhance pre-trained language models by self-supervised learning, which aims to assist model in learning story-level knowledge of ROCStories for SCT task.
- We design three self-supervised tasks (Drop, Replace and TOV) to further pre-train backbone model on daily stories of ROCStories.
- We conduct experiments on both SCT-v1.0 and SCT-v1.5 benchmarks. The results show that our approach achieves new state-of-the-art results (97.1% on SCT-v1.0 and 97.3% on SCT-v1.5), which are much closer to human performance.

2 Methodology

In this section, we will introduce the definition of the target SCT task, the backbone model and our designed self-supervised tasks.

2.1 Task Definition

Our task is formulated as follows: Given a story context $S = \{\text{sent.}_1, \text{sent.}_2, ..., \text{sent.}_L\}$ where L is 4 in SCT, and two candidate endings end.$_1$ and end.$_2$. The objective of SCT is to select the right ending. It can be formulated as a standard multi-choice task, requires a model to select the correct ending out of two choices so that the completed story is coherent and reasonable.

2.2 Backbone Model

In this paper, the backbone model consists two parts: a vanilla BERT [3] and a Multi-Choice Head. The vanilla BERT exploits transformer block [15] as the basic computational unit. In this paper, we employ the usual input format of the vanilla BERT: [CLS] sent. A [SEP] sent. B [SEP], where the first token is always a special [CLS] and two sent. s are separated by a special token [SEP]. Besides, the components of the two input sentences are quite different between SCT and our self-supervised tasks, which will be described in Sect. 2.3. The structure of input embeddings is constructed by summing the corresponding word, segment, and position embeddings.

Afterwards, we utilize the final aggregate [CLS] token representation $C_i \in \mathbb{R}^H$ to perform as the head of encoder, where H is the hidden size of the transformer output and i stands for the i-th choice. To adapt for the multiple choice task, we introduce task-specific parameters, a vector $V \in \mathbb{R}^H$, to dot product with [CLS] token representation as to the score for each choice i. The probability distribution is the softmax over all choices:

$$P_i^{\text{Task}} = \frac{e^{V \cdot C_i^{\text{Task}}}}{\sum_{j=1}^{N} e^{V \cdot C_j^{\text{Task}}}} \tag{1}$$

$$C_i^{\text{Task}} = \text{BERT}(\text{sent.}_A^{\text{Task}}, \text{sent.}_B^{\text{Task}}) \tag{2}$$

where N is the number of choices in tasks.

2.3 Self-supervised Methods

In this section, we will describe three self-supervised tasks. Through solving each self-supervised task, the resulting model is expected to learn the story-level knowledge from daily stories in ROCStories, which will then be transferred to solve the target SCT task. Note that we train the vanilla BERT and Multi-Choice Head together for the SCT task.

Drop. The *Drop* task is to randomly drop one sentence of a story and then predict whether a story is completed. As is illustrated in Fig. 2-(1), we fix the front sentences (sent. A) and drop a sentence from next sentences (sent. B) to generate a negative pseudo-sample. The model needs to select the completed story from two candidate sequences.

Replace. The *Replace* task randomly replaces one sentence with the other sentence of the story and predicts whether a story has repetitive events. Similar to the *Drop* task, we keep the front sentences (sent. A) unchanged which is shown in Fig. 2-(2) and replace a sentence in next sentences (sent. B) to create a negative pseudo-sample.

TOV. The *Temporal Order Verification (TOV)* task is to randomly choose two sentences of a story and change their position. Then, the system needs to predict whether a story is coherent. Sentence ordering tasks have been proved effective in the literature [7,16,18]. As shown in Fig. 2-(3), in the *TOV* task, we treat the front sentences (e.g., the first sentence of a story) as sent. A. And we the concatenate the original (or shuffled) next sentences as sent. B.

In this paper, we use the vanilla BERT and Multi-Choice Head to predict and minimize the cross-entropy loss. 50% of the time sent. B actually follows sent. A, and 50% of the time it is a pseudo-example by the operation of self-supervised tasks.

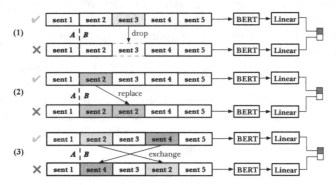

Fig. 2. To learn the story-level knowledge from ROCStories, we further train vanilla BERT and Multi-Choice Head with surpervised learning. Top down are our self-supervised tasks: (1) *Drop*, (2) *Replace* and (3) *TOV*.

2.4 Fine-Tuning on SCT

For the target SCT task, we treat the four-sentence story (context) as sent. A and each candidate ending as sent. B. The final probability distribution as follow:

$$P_i^{\text{SCT}} = \text{RM}([\text{sent.}_1; ...; \text{sent.}_4], \text{end.}_i) \tag{3}$$

where $\text{RM}(\cdot)$ stands for the resulting model (consists of vanilla BERT and Multi-Choice Head) which trained by each self-supervised learning task. In the SCT task, we use cross-entropy loss as our objective function.

3 Experiment

To evaluate the effectiveness of the self-supervised tasks (**Drop, Replace** and **TOV**), we conduct the experiments both on SCT-v1.0 and SCT-v1.5.

3.1 Dataset

We firstly train a vanilla BERT and a Multi-Choice Head with our self-supervised tasks on the unlabeled ROCStories for the SCT task. This corpus is a collection of 98,162 crowd-sourced complete five-sentence stories. Then, we evaluate our methods on the labeled data, SCT-v1.0 and SCT-v1.5[1]. SCT-v1.0 [8] consists of 3,742 stories which contain a four-sentence plot and two candidate endings. SCT-v1.5 [13] is a recently released revised version in order to overcome the human-authorship biases [12] discovered in SCT-v1.0.

In this paper, we randomly split 80% of examples in SCT-v1.0 evaluation set as our training set (1,479 cases), 20% of examples in SCT-v1.0 evaluation set as our validation set (374 cases). Besides, we utilize the SCT-v1.0 test set as our testing set (1,871 cases). For SCT-v1.5, we use the 1,871 SCT-v1.0 test set for training purpose and test on SCT-v1.5 validation set (1,571 cases).

3.2 Baselines

Model. We compare our approach with the following BERT based methods. Note that these models absolutely ignore ROCStories and directly fine-tune on SCT validation set.

- **BERT** [3] solves the SCT by employing a standard pre-train-then-fine-tune framework with MLM and NSP pre-training objectives.
- **RoBERTa** [6] is the upgraded BERT with a robustly optimized training procedure, which removes the NSP pre-training objective.
- **TransBERT** [5] further fine-tune BERT on semantically related supervised tasks (e.g., MNLI) and then fine-tune on the SCT validation set.

Pre-training Objectives. To evaluate the effectiveness of our self-supervised tasks, we also treat MLM+NSP (original pre-training task of BERT) and MLM (the task of RoBERTa) as our baselines.

- **Masked Language Model (MLM)** task aims to predict the masked tokens. In our work, we randomly mask 15% of all tokens in each story and model needs to predict the masked tokens.
- **Next Sentence Prediction (NSP)** task is to predict whether two segments follow each other in the original text. In this paper, we creat negative examples by pairing segments from different stories.

---

[1] https://competitions.codalab.org/competitions/15333.

3.3 Implementation

We use PyTorch [9] and open-source framework[2] to implement our technique and baselines. In our self-supervision part, the initial learning rate for Adam [4] is 2e-5 for BERT$_{BASE}$ and 1e-5 for BERT$_{LARGE}$. The maximum of the input sequence length is 128. The number of epochs for pre-training is 2 to 5. In the fine-tuning stage, learning rate is 2e-5 for BERT$_{BASE}$ and 9e-6 for BERT$_{LARGE}$. The other settings are the same as self-supervision module. We implement RoBERTa [6] and TransBERT [5] with the same settings described in their work.

3.4 Results on SCT-v1.0

We evaluate baselines and our model using accuracy as the metric on the SCT-v1.0. Results are summarized in Table 1.

BERT Based Methods. The vanilla BERT is a pre-trained bidirectional transformer via MLM and NSP objectives. All of our self-supervised learning tasks outperform vanilla BERT and BERT+MLM+NSP. Particularly, TOV improves vanilla BERT$_{BASE}$ to 90.7% (+2.1%). Besides, vanilla BERT$_{LARGE}$ with TOV achieves the best performance of 92.2% (+1.2%) between BERT based methods. The results demonstrate that our self-supervised tasks can bring excellent improvements for vanilla BERT.

RoBERTa Based Methods. Vanilla RoBERTa is the upgraded BERT with a robustly optimized training procedure, which removes the NSP objective. Results show that our self-supervised tasks outperform the baseline models and pre-training objectives, except the large model. It could be the human-authorship biases existed in SCT-v1.0.

Table 1. Accuracy on SCT-v1.0 test set and SCT-v1.5 validation set. Compared to vanilla BERT [3] and vanilla RoBERTa [6], our self-supervised approach brings considerable improvements. Note that $\star$ represents the pre-training task of the model is inconsistent with the current task.

Method	Accuracy (%) on SCT-v1.0				Accuracy (%) on SCT-v1.5			
	BERT		RoBERTa		BERT		RoBERTa	
	BASE	LARGE	BASE	LARGE	BASE	LARGE	BASE	LARGE
Vanilla	88.6	91.0	92.8	96.7	87.8	91.4	92.9	95.7
TransBERT	90.6	91.8	93.2	**97.4**	89.3	92.1	93.5	96.1
MLM	$\star$	$\star$	92.9	96.5	$\star$	$\star$	93.1	95.9
MLM+NSP	88.9	91.2	$\star$	$\star$	88.2	91.5	$\star .$	$\star$
Drop	89.9	91.8	93.4	96.7	88.9	91.9	93.5	96.6
Replace	90.2	92.0	93.4	96.6	89.3	91.7	93.9	96.3
TOV	**90.7**	**92.2**	**93.8**	97.1	**89.7**	**92.3**	**94.3**	**97.3**

---

[2] https://github.com/huggingface/transformers.

TransBERT Models. TransBERT further trains vanilla BERT on semantically related supervised tasks. Then the resulting model will be fine-tuned on the SCT validation set. Table 1 shows that TOV almost outperforms TransBERT except for RoBERTa$_{LARGE}$. The human-authorship bias in SCT-v1.0 could explain this phenomenon.

In summary, our self-supervised tasks all bring the vanilla BERT with improvements on the target SCT task and outperform original pre-training tasks in the procedure of exploiting ROCStories. Specifically, TOV obtains the best results. One explanation is that TOV could capture more generalized story-level knowledge from the daily stories.

3.5 Results on SCT-v1.5

Table 1 shows results of baselines and our methods on SCT-v1.5 which has less human-authorship biases. We can see that each of our self-supervised tasks brings improvements and TOV obtains a new state-of-the-art result of 97.3% (+1.6%) with RoBERTa$_{LARGE}$. Furthermore, all of our self-supervised tasks outperform the MLM task used in pre-training RoBERTa. These results further strengthen the point that our methods help improve BERT for the Story Cloze Test task.

In summary, the results in Table 1 demonstrate the best performance of TOV. And, we will further analyze the TOV task in the following context.

4   Analysis and Discussion

4.1   Analysis of Combination Forms

We investigate different combination forms of TOV which bring improvements on the target SCT task. In order to maintain the textual information of the story, we set the number of subsentences (sent. A and sent. B) to five or four. Then, we tag each form with the lower case letters: *a-g*. In Table 2, we shows all combination forms and summarize the best accuracy of BERT$_{BASE}$+TOV on the SCT-v1.0 test set.

Based on the results, we can conclude that the number of subsentences could influence the performance of TOV. When the number of subsentences is four (*d-g*), the model is required to finish TOV with a story segment. The explanation of the favorable results could be that the harder task form could learn more generalized story-level knowledge. Besides, it is interesting that our approach can obtain much more improvements if sent. B is longer, where the shuffled sentences are of greater randomness.

Table 2. Results of different combination forms in BERT$_{BASE}$+TOV on SCT-v1.0. We mark each combination form with *a-g*.

Tag	Sent. A	Sent. B	Acc. (%)
a	[1]	[2, 3, 4, 5]	90.1
b	[1, 2]	[3, 4, 5]	89.8
c	[1, 2, 3]	[4, 5]	89.7
d	[1]	[2, 3, 4]	**90.7**
e	[2]	[3, 4, 5]	90.3
f	[1, 2]	[3, 4]	90.1
g	[2, 3]	[4, 5]	90.2

Table 3. Ablation study of best model RoBERTa$_{LARGE}$+TOV on SCT-v1.5.

RoBERTa$_{LARGE}$ + TOV	Acc. (%)
Original (w/ two choices)	97.3
(1) w/ three choices	96.5
(2) w/ four choices	96.2
(3) w/o transfer Multi-Choice Head	96.8
(4) w/o fine-tuning on SCT-v1.0	76.1
(5) w/ Drop	97.1
(6) w/ Replace	96.9
(7) w/ Drop & Replace	96.8

4.2   Ablation Study

We perform an ablation study on our best model RoBERTa$_{LARGE}$+TOV on the SCT-v1.5 validation set. As shown in Table 3, the original model generates samples with two choices and achieves an accuracy of 97.2%. In ablation (1, 2), we modify the number of choices. The results indicate that TOV with two choices are more suitable for SCT. One reason is that SCT has two choices, and we should design a self-supervised task with two choices for ROCStories aiming at adapting for SCT. Ablation (3) demonstrates that the story-level knowledge is partly learned by the Multi-Choice Head and parts of story-level knowledge can be learned by RoBERTa. In ablation (4), we drop the procedure of fine-tuning on the target labeled data and directly test on the SCT-v1.5. This demonstrates that our self-supervised tasks are valid on the SCT task. Finally, we try to jointly train RoBERTa$_{LARGE}$ with Drop, Replace and TOV. Ablation (5, 6, 7) show that choosing only one self-supervised learning task once a time is more appropriate for the target SCT task. It is possible that different tasks could influence each other within the joint training.

5   Conclusion and Future Work

In this paper, we propose a simple and effective self-supervised learning approach to help vanilla BERT learn the story-level knowledge from daily stories in unlabeled ROCStories, which will benefit the target SCT task. In further work, it is meaningful to modify the Multi-Choice Head with other architecture like graph neural networks. Besides, our approach could be helpful for other low-source NLP applications.

Acknowledgement. We thank the reviewers for their insightful comments. We also thank Effyic Intelligent Technology (Beijing) for the computing resource support. This work was supported by in part by the National Key Research and Development Program of China under Grant No. 2016YFB0801003.

References

1. Cai, Z., Tu, L., Gimpel, K.: Pay attention to the ending: strong neural baselines for the ROC story cloze task. In: ACL, pp. 616–622 (2017)
2. Chaturvedi, S., Peng, H., Roth, D.: Story comprehension for predicting what happens next. In: EMNLP, pp. 1603–1614 (2017)
3. Devlin, J., Chang, M.W., Lee, K., Toutanova, K.: Bert: pre-training of deep bidirectional transformers for language understanding. arXiv preprint arXiv:1810.04805 (2018)
4. Kingma, D.P., Ba, J.: Adam: a method for stochastic optimization (2014)
5. Li, Z., Ding, X., Liu, T.: Story ending prediction by transferable Bert. IJCAI, August 2019
6. Liu, Y., Ott, M., Goyal, N., Du, J., Joshi, M., Chen, D., Levy, O., Lewis, M., Zettlemoyer, L.S., Stoyanov, V.: Roberta: A robustly optimized bert pretraining approach. ArXiv (2019)
7. Logeswaran, L., Lee, H., Radev, D.R.: Sentence ordering and coherence modeling using recurrent neural networks. In: AAAI (2016)
8. Mostafazadeh, N., et al.: A corpus and cloze evaluation for deeper understanding of commonsense stories. In: NAACL-HL, June 2016
9. Paszke, A., et al.: Automatic differentiation in PyTorch. In: NIPS (2017)
10. Radford, A., Narasimhan, K., Salimans, T., Sutskever, I.: Improving language understanding by generative pre-training (2018)
11. Schubert, L.K., Hwang, C.H.: Episodic logic meets little red riding hood: a comprehensive, natural representation for language understanding, pp. 111–174 (2000)
12. Schwartz, R., Sap, M., Konstas, I., Zilles, L., Choi, Y., Smith, N.A.: Story cloze task: UW NLP system. pp. 52–55 (2017)
13. Sharma, R., Allen, J., Bakhshandeh, O., Mostafazadeh, N.: Tackling the story ending biases in the story cloze test. In: ACL, pp. 752–757 (2018)
14. Srinivasan, S., Arora, R., Riedl, M.O.: A simple and effective approach to the story cloze test. In: NAACL-HLT (2018)
15. Vaswani, A., et al.: Attention is all you need. In: NIPS (2017)
16. Wang, H., et al.: Self-supervised learning for contextualized extractive summarization. In: ACL (2019)
17. Williams, A., Nangia, N., Bowman, S.R.: A broad-coverage challenge corpus for sentence understanding through inference. In: NAACL-HLT (2017)
18. Wu, J., Wang, X., Wang, W.Y.: Self-supervised dialogue learning. In: ACL (2019)
19. Zhu, Y., et al.: Aligning books and movies: towards story-like visual explanations by watching movies and reading books. In: ICCV, pp. 19–27 (2015)

MOOCRec: An Attention Meta-path Based Model for Top-K Recommendation in MOOC

Deming Sheng, Jingling Yuan$^{(\boxtimes)}$, Qing Xie, and Pei Luo

School of Computer Science and Technology, Wuhan University of Technology,
Wuhan 430070, China
{shengdeming,yjl,felixxq,wardoluo}@whut.edu.cn

Abstract. With the surge of the courses and users on Massive Open Online Courses (MOOC), MOOC has accumulated rich educational data. However, the utilization of MOOC resources is not high enough to satisfy the dynamic and diverse demands of different individuals. Meanwhile, the traditional recommendation model for MOOC dataset underperforms in both precision and recall. To address those issues, we collect and collate a MOOC dataset and then propose an attention meta-path based recommendation model named MOOCRec to jointly learn explicit and implicit relationships between students and courses. By extracting the knowledge points of the whole course information, we successfully construct different heterogeneous information networks (HINs) in MOOC and then we elaborately design multiple meta-paths based context to exploit the heterogeneity of other HINs in MOOC, which enables MOOCRec to offer abundant course resources. In particular, we leverage three attention mechanisms under MOOC to further enhance factors that effectively influence student preferences to improve the precision of our model. What's more, we adopt another classical dataset called Movielens, reconstruct HINs and redesign meta-paths to demonstrate that the extensive availability of MOOCRec.

Keywords: Educational recommendation model · MOOC · Heterogeneous information network · Meta path · Attention mechanism

1 Introduction

With the rapid development of the Internet and the continuous deepening of education informatization, the online education data is also expanding rapidly, forming a huge heterogeneous information network, which provides good conditions for intelligent auxiliary teaching and data-driven education. However, there are still many problems in the utilization of teaching resources and can not satisfy the individual needs of numerous students. The existing resource management mechanism is difficult to make reasonable resource allocation based on the operating status of teaching resources. In order to effectively improve the efficiency and performance of educational resource recommendation services, it is

© Springer Nature Switzerland AG 2020
G. Li et al. (Eds.): KSEM 2020, LNAI 12274, pp. 280–288, 2020.
https://doi.org/10.1007/978-3-030-55130-8_25

necessary to further extract and analyze the large-scale data of MOOC teaching resources and user groups.

Meanwhile, there has been relatively little research into online learning resource services. Therefore, we build a recommendation model in MOOC by analyzing existing recommendation systems in social networks. As a promising direction, HIN contains rich implicit information and multiple types of nodes. By extracting knowledge points in MOOC and basic information of users and courses, we construct different HINs in MOOC. With the assistance of HINs, we propose a novel attention recommendation model based on meta-path which not only extracts the implicit relationship between users, courses, teachers and knowledge points but also captures the rich features of each item. In summary, our main contributions in this paper are three folds:

- We collect and collate a MOOC dataset containing course details and user reviews of the course. By calculating the TF-IDF of each word in the whole course information, we successfully extract a Knowledge Points Set. And then we designed different meta-paths in the HINs to capture rich context information for offering abundant course resources.
- We propose an attention meta-path recommendation model named MOOCRec to accommodate students dynamics and diverse preferences. MOOCRec leverages both explicit and implicit relationships to better model the preference of students. In addition, after reconstructing HINs and redesigning meta-paths in Movielens, MOOCRec shows the potential for excellent versatility.
- We devise a collaborative attention mechanism under MOOC to better understand what factors affect student preferences effectively. With the enhancement of different embedding methods, the performance of MOOCRec has improved and exceeded the traditional model.

2 Related Work

Recently, the further development of graph neural network makes people explore the heterogeneous information network in recommendation. In the heterogeneous information network, a node represents an object and a side represents their rich relations. With the HIN, the implicit relationship can be dug out. Almost every HIN-based methods utilize paths to enhance the interaction data. Unfortunately, All of them ignore the explicit representation of path or meta-path. To solve this, meta-path based context model emerges [5]. Meanwhile, network embedding has shown great advantages in structure feature extraction and has many mature applications [8]. But it's worth noting that they learn network structure in homogeneous networks.

Our work is inspired by the recent progress of HIN-based model [5] and attention mechanism [1] in the recommendation. We draw on Deng et al. results [2]: a simple baseline with no graph topology and only aggregate node features can achieve similar performance as SoTA GNN. Thus we not only designed different meta-paths to consider the structure of HINs in MOOC, but also focused on analyzing the features of each network node, especially the knowledge points. At the same time, we designed the collaborative attention mechanism in MOOC to enhance the influencing factors of students' course selection.

3   The Proposed Model

3.1   Framework

The framework we propose approach is showed in Fig. 1. This framework includes three parts. The first part is course embedding. In this part, we take the information crawled by MOOC websites and the extracted knowledge points as explicit features. The second part is meta-path embedding. In this part, we devise a series of different meta-paths to capture implicit features, which enables our model to recommend abundant course resources. The last part is the attention mechanism. In the last part, we utilize the attention mechanisms to better model student preference and enhance the effects of explicit and implicit features on the final result.

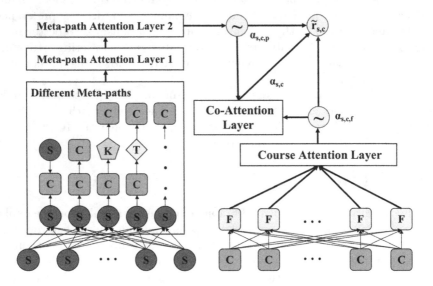

Fig. 1. An attention meta-path based recommendation model

3.2   Knowledge Points Extraction

Knowledge points refer to a set of key knowledge of each course. The process of extracting knowledge points is shown in Fig. 2. We take all the information crawled from the MOOC website as the extraction document of knowledge points. And then we filtered the results of document segmentation, added the weight parameters of each part of the document, calculated the weight of Term

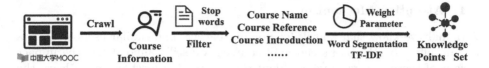

Fig. 2. The process of extracting knowledge points of MOOC courses

Frequency-Inverse Document Frequency (TF-IDF), and finally obtained the ten knowledge points of each course.

The set of knowledge points extracted from the course can be directly used as the features of the course, and the implicit interaction between students and the course can be further explored by analyzing the similarity between the two courses based on the meta-path.

3.3   Meta-paths in MOOC HINs

As shown in Table 1, we have elaborately designed several different meta-paths to mine the implicit connections between students and courses. The essence of a student choosing a course is to choose the knowledge point he needs, the recommendation of his classmates or some teachers he prefers. Therefore, we select four types of nodes: courses, users, knowledge points and teachers, respectively. For Movielens, meta-paths are UMMM, UUUM, UMTM, and UMUM. U represents user, M represents movie and T represents the type of movie.

Table 1. Different meta-paths for MOOC recommendation

Path	Schema	Description
P1	$S \xrightarrow{a^*} C \xleftarrow{a} S$	Students who have attended the same course
P2	$S \xrightarrow{a} C \xrightarrow{s^*} C$	Students who have attended the similar course
P3	$S \xrightarrow{a} C \xrightarrow{c^*} K \xleftarrow{c} C$	Students who have attended the courses that contain same knowledge point
P4	$S \xrightarrow{a} C \xrightarrow{t^*} T \xleftarrow{t} C$	Students who have attended the courses that are taught by the same teacher
P5	$S \longrightarrow ... \longrightarrow C$	Paths based Random Walks [8], we set the maximum path length L to 8

* $\xrightarrow{a}$ denotes the attend (review) relation; $\xrightarrow{s}$ denotes the similar relation; $\xrightarrow{c}$ denotes the contain relation; $\xrightarrow{t}$ denotes the teach relation.

3.4   Attention Mechanism

Although we know that knowledge points, teachers and similar users all have a certain effect on the user's choice of courses, we still cannot know the impact weight of these factors. Here we adopt three attention mechanisms to learn the impact of different meta-paths and explicit features on the final result.

Student Embedding and Course Embedding. Following [4], we adpot a embedding layer to transform the original one-hot representations of students and courses into low-dimensional dense vectors. Given a specified student-course pair $<s, c>$, the embedding layer can replace $p_s \in \mathbb{R}^{|s| \times 1}$ and $q_c \in \mathbb{R}^{|c| \times 1}$, the original features of students and courses, into $P \in \mathbb{R}^{|s| \times d}$ and $Q \in \mathbb{R}^{|c| \times d}$ respectively, which d is the dimension size of embeddings:

$$x_s = P^T \cdot p^s, \qquad y_c = Q^T \cdot q^c \tag{1}$$

Relation Embedding. Unlike student embedding and course embedding, here we exploit a Convolution Neural Network (CNN) to deal with sequences of variable lengths. Given a meta path or random walk path p_i, the CNN with a convolution layer and max pooling layer can learn relation embedding of this path instance p_i:

$$h_{p_i} = CNN(X^{p_i}; \theta) \tag{2}$$

where the $X^{p_i} \in \mathbb{R}^{L \times d}$ denotes the embedding matrix formed by concatenating nodes, L is the length of p_i, d is the dimension size of embedding, and θ denotes all the related parameters in CNN.

As is shown in Table 1, there are a series of meta-paths in the MOOC HIN, which have different semantics, here we adopt the max pooling method to get the final relation embedding r_p between a student-course pair $<s, c>$:

$$r_p = maxpooling(h_{p_i}{}_{i=1}^{N}) \tag{3}$$

where the N denotes the number of meta-paths, and the relation embedding r_p aims to capture implicit relationships.

Attention for Meta-path. To further enhance the implicit relationships, here we adopt two meta-path attention layers to weight student embedding x_s, course embedding y_c, and relation embedding r_p for a specified meta-path p_i in Table 1:

$$\alpha_{s,c,p_i}^{(1)} = f(W_s^{(1)} x_s + W_c^{(1)} y_c + W_{p_i}^{(1)} r_{p_i} + b_{s,c,p_i}^{(1)}) \tag{4}$$

$$\alpha_{s,c,p_i}^{(2)} = f(W_a^{(2)} \alpha_{s,c,p_i}^{(1)} + b_{s,c,p_i}^{(2)}) \tag{5}$$

Where the $W_*^{(*)}$ and $b^{(*)}$ denote the weight matrix and the bias vector, respectively. And for the $f(*)$, we adopt the ReLU function. At last, we utilize the softmax function to normalize the above $\alpha_{s,c,p}^{(2)}$:

$$\alpha_{s,c,p} = \frac{\exp^{\alpha_{s,c,p_i}^{(2)}}}{\sum_{i=1}^{N} \exp^{\alpha_{s,c,p_i}^{(2)}}} \tag{6}$$

Attention for Course Features. Due to the limitation of MOOC, we cannot collect more information of users, so we choose the characteristic information of the course, including the teachers and ten main knowledge points. Here we adopt a course attention layer to figure out which explicit feature matters most:

$$\alpha_{s,c,f} = f(W_s x_s + W_c y_c + b_{s,c,f}) \tag{7}$$

Co-attention. With the attention for meta-path and course features, both implicit and explicit features are enhanced. Naturally, we leverage a co-attention layer to capture them jointly:

$$\alpha_{s,c} = f(W_p \alpha_{s,c,p} + W_f \alpha_{s,c,f} + b_{s,c}) \tag{8}$$

For student x_s and course y_c, we trained three embeddings: student x_s, course y_c, and meta-path embedding r_{p_i}. And with the above attentional parameters, the rating $r_{s,c}$ can be expressed as:

$$r_{s,c} = x_s \times \alpha_{s,c,p} + y_c \times \alpha_{s,c,f} + \alpha_{s,c} \tag{9}$$

At last, we adopt the squared error loss to learn the parameters of our model.

4   Experiments

4.1   DataSets

As shown in Table 2, we collect and collate a MOOC dataset and adopt another classical dataset named Movielens to demonstrate the effectiveness and extensive availability of MOOCRec:

MOOC[1]. A popular E-learning platform, offering free courses online. We crawled the data of courses and users. MOOC dataset contains 85,564 reviews and ratings about 254 courses from 64,435 users. We construct our dataset by using reviews as evidence a student attends a class.

Movielens[2]. A dataset about movie ratings, which contains 943 user 100,000 rating information on 1,682 movies obtained from IMDB and Movie DataBase.

---

[1] https://www.icourse163.org.
[2] https://grouplens.org/datasets/movielens.

4.2 Baselines

To illustrate the effectiveness of our model, we consider four kinds of recommendation methods: CF-based (ItemCF and UserCF), Model-based (MF, GMF and NeuMF), Generic feature-based (FM, LFM and DeepFM), Path-based (Meta-path), as follows:

- **CF-based:** CF-based model exploits all the interactions of $<u, i>$ to make recommendations. For example, ItemCF, UserCF, etc.
- **Model-based:** Model-based methods define a parameter model to describe the relationship between users and items, users and users, items and items, and then using the existing user-item scoring matrix to optimize the solution to obtain parameters. For example, MF [7], GMF [4], NeuMF [4], etc.
- **Generic feature-based:** In the above methods, only the interaction data (rating data) between the user and the item is used, and a large amount of side information is not used, while Generic feature-based methods catch the diverse features to make recommendations. For example, FM [9], LFM [6], DeepFM [3], etc.
- **Path-based:** With different paths in the user-item graph, Path-based methods utilize the rich heterogeneous information to model user preference. For example, Meta-path [10], etc.

To evaluate our model, we adapt four widely used ranking-based metrics [4]: Precision (Prec @K), Recall@K (Rec@K), F1-Score (F1 @K) and Normalized Discounted Cumulative Gain (NDCG@K), and we set K = 10.

Table 2. Overall performance comparison. We use **bold** to mark the best performance and <u>underline</u> to indicate the best performance other than MOOCRec.

Model	MOOC (@10, %)				Movielens (@10, %)			
	Prec	Rec	F1	NDCG	Prec	Rec	F1	NDCG
ItemCF	0.22	21.16	0.44	0.22	31.01	14.65	19.90	52.09
UserCF	0.03	2.67	0.06	2.87	31.21	14.79	20.07	52.43
MF	21.81	8.03	11.74	29.78	31.57	20.41	24.79	65.07
GMF	25.35	21.29	23.14	35.21	31.34	20.21	24.57	64.58
NeuMF	<u>50.80</u>	4.02	7.45	<u>70.71</u>	32.43	20.84	25.37	66.92
FM	36.67	18.47	24.57	51.21	32.52	20.87	25.42	66.79
LFM	28.64	**38.55**	**32.86**	39.80	31.97	18.93	23.78	65.55
DeepFM	19.38	4.02	6.66	26.35	<u>33.18</u>	<u>21.46</u>	<u>26.06</u>	<u>67.22</u>
Meta-path	48.43	12.05	19.30	66.83	32.07	21.34	25.63	66.71
MOOCRec	**51.90**	13.25	21.11	**74,22**	**34.23**	**22.52**	**27.17**	**69.33**
%Improv.	2.17	–	–	4.96	3.16	4.94	4.26	3.14

4.3   Overall Comparison

The performance comparison results on two datasets are shown in Table 2. We can observe that:

- Meta-path is much better than the methods based on leveraging explicit relationships in cases where there are not many cross-links between users and items. Compared to the Meta-path based method, MOOCRec has successfully integrated the features of the course itself, performing even better with the attention mechanism we designed. We can draw the conclusion that a mutual enhancement way can better model student preference.
- We design a series of different meta-paths to try to provide students with a richer variety of courses, but we still found that MOOCRec did not perform well on the recall value. It is worth noting that the Meta-path based model also suffers the same problem of low recall value. We suspect that many students may tend to choose those popular courses with high reputation, and then MOOCRec will focus more on those courses under our attention mechanisms. In the end, the recall rate comes down while improving precision. At the same time, this is also in line with the characteristics of our MOOC dataset: the number of students is much larger than the number of courses, and on average each student only chooses a certain course.

5   Conclusion and Future Work

In this paper, we propose an attention meta-path based recommendation model for users' dynamic and diverse preferences. We adopt three attention mechanisms to enhance different embeddings, furthering improving the performance of MOOCRec. Unfortunately, due to the lack of information on MOOC users, we can't further analyze the impact of user information on the MOOC dataset recommendation effect. For future work, we would like to expand our MOOC dataset to collect multiple kinds of contextual information, build a Knowledge Graph and explore how to incorporate knowledge graph into the recommendation on our MOOC dataset. In addition, we also wanted to explore the performance impact of different embedding methods on recommendation.

References

1. Chen, J., Zhang, H., He, X., Nie, L., Liu, W., Chua, T.-S.: Attentive collaborative filtering: multimedia recommendation with item-and component-level attention. In: Proceedings of the 40th International ACM SIGIR Conference on Research and Development in Information Retrieval, pp. 335–344 (2017)
2. Deng, C., Zhao, Z., Wang, Y., Zhang, Z., Feng, Z.: GraphZoom: a multi-level spectral approach for accurate and scalable graph embedding. arXiv preprint arXiv:1910.02370 (2019)
3. Guo, H., Tang, R., Ye, Y., Li, Z., He, X.: DeepFM: a factorization-machine based neural network for CTR prediction. arXiv preprint arXiv:1703.04247 (2017)

4. He, X., Liao, L., Zhang, H., Nie, L., Hu, X., Chua, T.-S.: Neural collaborative filtering. In: Proceedings of the 26th International Conference on World Wide Web, pp. 173–182 (2017)
5. Hu, B., Shi, C., Zhao, W.X., Yu, P.S.: Leveraging meta-path based context for top-n recommendation with a neural co-attention model. In: Proceedings of the 24th ACM SIGKDD International Conference on Knowledge Discovery & Data Mining, pp. 1531–1540 (2018)
6. Jenatton, R., Roux, N.L., Bordes, A., Obozinski, G.R.: A latent factor model for highly multi-relational data. In: Advances in Neural Information Processing Systems, pp. 3167–3175 (2012)
7. Koren, Y., Bell, R., Volinsky, C.: Matrix factorization techniques for recommender systems. Computer **42**(8), 30–37 (2009)
8. Perozzi, B., Al-Rfou, R., Skiena, S.: DeepWalk: online learning of social representations. In: Proceedings of the 20th ACM SIGKDD International Conference on Knowledge Discovery and Data Mining, pp. 701–710 (2014)
9. Rendle, S., Gantner, Z., Freudenthaler, C., Schmidt-Thieme, L.: Fast context-aware recommendations with factorization machines. In: Proceedings of the 34th International ACM SIGIR Conference on Research and Development in Information Retrieval, pp. 635–644 (2011)
10. Sun, Y., Han, J.: Mining heterogeneous information networks: principles and methodologies. Synthesis Lect. Data Min. Knowl. Discov. **3**(2), 1–159 (2012)

Knowledge-Based Systems

PVFNet: Point-View Fusion Network for 3D Shape Recognition

Jun Yang and Jisheng Dang[✉]

School of Electronic and Information Engineering, Lanzhou Jiao Tong University, Lanzhou, China
1442342449@qq.com

Abstract. 3D object recognition has enjoyed much of research attention in the machine vision filed. Deep learning methods for 3D shape recognition such as the multi-view based methods and the point cloud based methods have achieved the state-of-the-art performance. However, little attention has been paid to the correlation between the high-level global single-point semantic features, the local geometric features of the point cloud data and the view features of the multi-view data, which is highly beneficial and can be taken advantage of in making the aforementioned features complementary to each other in our consideration. In this paper, we introduce a Point-View Fusion Network (PVFNet) which is an effective network used to integrate the point cloud data with the view data towards the goal of achieving a joint 3D shape recognition. More specifically, firstly, the high-level global single-point semantic features of the point cloud are extracted by our Global Single Point Network (GSPNet) and projected onto the subspace of the view features by a point-view attention fusion layer to describe the relationship between the point cloud data and the view data. Secondly, the global single-point features and the enhanced view features are projected onto the subspace of the local geometric features of the point cloud data by a point-view-point attention fusion layer to describe the relative correlation and significance of different local structures. Finally, all three types of the global features are further connected to obtain a unified global feature descriptor for the purpose of recognizing the 3D shapes involved. Our PVFNet so far has been evaluated on the ModelNet40 and ModelNet10 datasets to achieve 3D shape recognition and retrieval. Preliminary experimental results demonstrated that our framework can achieve a superior performance when compared to the state-of-the-art models.

Keywords: 3D shape recognition · Point cloud · Multi-view · Attention fusion · Neural network

1 Introduction

With the wide usage of 3D shape recognition applications such as autonomous vehicle [1, 2], robotic mapping and navigation [3] and AR (Augmented Reality)/VR (Virtual Reality), 3D shape recognition has become an important research topic in the machine vision field. With the significant success achieved by convolutional neural networks

© Springer Nature Switzerland AG 2020
G. Li et al. (Eds.): KSEM 2020, LNAI 12274, pp. 291–303, 2020.
https://doi.org/10.1007/978-3-030-55130-8_26

(CNN) [4] in computer vision, methods of 3D shape recognition have evolved from handcrafted descriptors SIFT [5], ORB [6], to deep learning methods. In order to leverage the advantages of CNNs. A multi-view convolutional neural network [7] that is able to extract features from 2D views of 3D shapes and aggregate them through a max pooling layer was proposed for the first time. Feng et al. [8] proposed group view features by aggregating view features with a group as a unit, then the group view features were aggregated by weights learnt during the training to obtain global feature descriptors used for 3D shape recognition. For 3D shape retrieval tasks, various framework models are presented. Guo et al. [9] presented a deep embedding framework supervised by both classification loss and triplet loss to settle the complex intra-class and inter-class variations. He et al. [10] presented Triplet-center loss as a well-designed loss function especially for retrieval. Han et al. [11] proposed sequential views to sequential Labels as a novel deep learning model with an encoder-decoder structure based on Recurrent Neural Networks (RNNs). However, these methods suffer from redundant view features and loss of position information of 3D models. However, limited by the camera angles, each view feature can only present partial local structures of an entire 3D shape, resulting in some view features might not being discriminative enough to represent 3D shapes. In short, multi-view based models may contain many redundant view features and tend to lose their geometric structure information.

VoxNet [12] transforms irregular point cloud into a standard volumetric grid that indicates the spatial occupancy for each voxel, and then a 3D CNN followed to extract the features of the volumetric grid and predict the categories of the 3D objects. Octree [13] subdivides the spatial structure and extracts the features hierarchically. However, the sparse structure of the volumetric data and the high computation cost constrain the performance. Point cloud is the simplest representation used to describe 3D objects. PointNet [14] proposes deep neural networks that can process raw point cloud directly, thus becoming the first effective method ever used for processing irregular 3D point cloud data. However, PointNet only focuses on the features of each point independently and lacks the ability to capture local structural features. Recent works have mostly concentrated on how to capture local features efficiently. For example, PointNet++ [15] extracts different scale local features by processing point cloud hierarchically. Kd-Network [16] extracts and aggregates features and presents them as Kd-trees by subdividing point cloud. DGCNN [17] proposes EConv (Edge Convolution) to capture edge features between points and their searched neighbors. X-Conv [18] is proposed for setting a potentially normative order to unorganized points, and then extracting local features of point cloud. However, the existing methods have so far still not developed the hidden correlations between different local structures and the discriminative fine-grained features of the point cloud, which results in restricting the development of point cloud analysis.

Through analyzing the pros and cons of multi-view based methods and the deep learning network focusing on point sets, we noticed that both multi-view and point cloud features are beneficial and complementary to each other. It is naturally thought that the fusion of multi-view features and point cloud features can be made use of in obtaining a better performance for 3D shape recognition. Recent works [19] for 3D shape recognition achieved the fusion between the global view features and the local geometric features. We found that this type of fusion ignores the implied correlation between different view features, global single-point features with a high-level semantic recognition capability and the local geometric features of the point cloud. This type of fusion also doesn't take into consideration of distinguishing the fine-grained features of both point cloud and view data. To solve this issue, we propose a Point-View Fusion Network (PVFNet) for 3D shape recognition that achieves a better joint between point cloud data and view data. First, we present a Global Single Point Network (GSPNet) to fully exploit the deep global single-point features of point cloud with a high-level semantic recognition capability, which are further exploited in a point-view attention fusion layer to help model the relationship between different views and point cloud. Then the single-point features and the enhanced view features are both projected onto the subspace of the local geometric features to exploit the relative correlations between different local structures and their discriminative information using a point-view-point attention fusion layer. The main contributions of our work are as follows: (1) The GSPNet we proposed captures high-level multi-semantic feature by splitting input features and convolution kernel parameter into groups and introducing dense connection. (2) The PVFNet we proposed utilizes a joint of the feature fusion of the high-level global single-point semantic features, the local geometric features of the point sets and the view features of the multi-view data towards the goal of achieving a better completion of the 3D shape recognition. Our PVFNet employs the high-level global single-point features extracted from the point cloud by the GSPNet to mine the relationship between the point cloud and the view data. Then the single-point features and the enhanced view features can be used to distinguish the significance of different local geometric features. (3) We designed a point-view attention fusion layer and a point-view-point attention fusion layer used as an adaptive feature selector to enhance distinguishing features and restrain useless features of the point cloud and the view data, and these layers are more efficient at representing discriminative information of point cloud and view data.

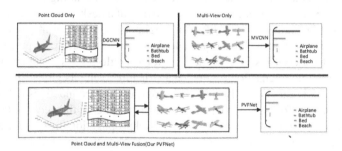

Fig. 1. Top: Illustration of 3D shape recognition framework DGCNN, which consumes point cloud data, and 3D shape recognition framework MVCNN, which is based on multi-view data. Bottom: Illustration of our proposed framework PVFNet, which is based on fusing of point cloud and multi-view data.

2 Point-View Fusion Network

In this section we will be introducing the details of our proposed PVFNet framework. Point cloud and multi-views are firstly fed into the framework individually to extract global single-point features, the local geometric features of the point cloud data and the view features of the view data, which are fused effectively to complete the 3D shape recognition task. The architecture of the PVFNet framework is shown in Fig. 1.

2.1 Feature Extraction

GEConv and GMLP. Massive efforts for point cloud analysis mainly employ MLP and graph-based convolution as a basic feature extraction module, which have made a great breakthrough. However, the previous works don't mention how to reduce redundancy and have inherent shortcomings. For example, MLP convolution based methods loss local information and graph convolution based methods are space consuming. Thus, we get the Group Edge Convolution (GEConv) and Group MLP (GMLP) by dividing the channels of the graph convolution [17] and MLP into groups respectively, which are shown in Fig. 2. For input point cloud and convolution kernel, the number of parameters is reduced to $1/g$ for a single operation when we split it to groups, which reduces the computational complexity and allows to encode more useful information. Inspired by [20], in order for each group to include information of all other groups, we stack group convolutions together and then shuffle channel to sufficiently fuse features to capture fine-grained features for all groups. Compared to previous methods, group convolution can ease the training of deep neural networks by reducing redundancy and increase the width of the neural networks to allow more feature channels which is beneficial to encoding more discriminating information on each group.

Global Single-Point Feature Branch. Applying the graph convolution to only calculate the edge features of the local point cloud, while regarding the original point cloud as a set of small blocks, results in losing the global structure information and absolute position information of the point cloud, which are inevitable to be ambiguous in some cases. For example, for some points located on a flat surface of a desk, the local representations distance tends to be similar among different neighbors, so these points cannot be accurately recognized based on the local representation alone. Therefore, we present a Global Single Point Network (GSPNet) to extract the global single-point semantic features from the point cloud as a compensation for the local structure losing its global shape structure information by applying GMLP. Besides, we also use dense connections [21] to transfer hierarchical feature information from each layer to the other in the network (See Fig. 3). These connections also lead to a better gradient backpropagation through the network as well as is beneficial to learn multi-semantic features. Note that this branch is compatible with the other models used for extracting local geometric features.

2.2 Point-View Attention Fusion and Point-View-Point Attention Fusion

Inspired by the recent advances of the attention mechanisms used for different tasks in the machine vision field [22], we propose two novel feature fusion strategies named point-view attention fusion and point-view-point attention fusion. Our strategy focuses mainly on utilizing the high-level global single-point semantic features as shown in

Fig. 3 and the details of the structures of the proposed point-view attention fusion and point-view-point attention layers are shown in Fig. 4 and Fig. 5 respectively.

Point-View Attention Fusion Layer.

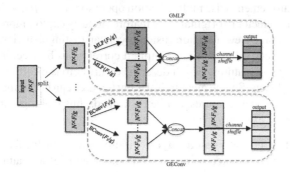

Fig. 2. GEConv and GMLP layer.

Point-View Attention Fusion Layer. Figure 3 shows the GSPNet are used to transform the original coordinates of single point to global semantic feature descriptor p_g of shape $1 \times k$, which is concatenated to each view feature v by an operation that first repeats the single-point features of the point cloud M times, where we define feature fusion as $\partial(\cdot) = \mathbb{R}^{M \times C1} \times \mathbb{R}^{1 \times k} \to \mathbb{R}^{M \times (C1+k)}$. This operation follows MLP layer, the global single-point features of the point cloud and each of the view features are fused together to form a view feature selector that is normalized by a normalization function,

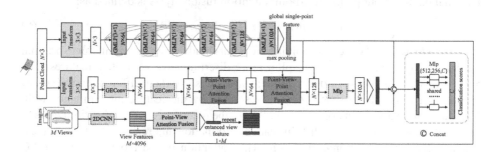

Fig. 3. Architecture of the proposed PVFNet.

$$\ell(\cdot) = sigmoid\,(\exp(\partial(\cdot))) \tag{1}$$

which can normalize the output of the descriptor to range between [0, 1]. Due to the large input to the sigmoid function, we also add the exp function to avoid the output closing to 0 and 1 before the sigmoid. Then the view feature selector $R(p_g, v)$ can be defined as:

$$R(p_g, \mathrm{v}) = \ell(MLP(\partial(p_g, \mathrm{v}))) \tag{2}$$

with the range of [0, 1] representing the significance of different view features. The final output of our point-view attention layers is defined as:

$$T(p_g, v) = v * R(p_g, v) \tag{3}$$

where $*$ indicates an element-wise multiplication operation, $v * R(p_g, v)$ indicates applying the view feature selector to original view features to fully exploit more discriminative view features. Our point-view attention fusion layer can distinguish view features to act as feature selectors very well by mining implied correlations between point cloud features and different view features, which describe contributions of different views to the recognition of 3D objects. So, the meaningful view features can be augmented adaptively and useless redundant view features can be restrained, thus making our network more robust and efficient at extracting view features.

Point-View-Point Attention Fusion Layer. Figure 3 shows the local geometric feature p_l of the point cloud is extracted using GEConv in the geometric feature branch. After being repeated N times, both of the enhanced view features and the single-point semantic features are concatenated to the local geometric features, where we define feature fusion as $\partial(\cdot) = \mathbb{R}^{N \times C1} \times \mathbb{R}^k \times \mathbb{R}^{N \times C2} \to \mathbb{R}^{N \times (C1 + k + C2)}$. This fusion operation is followed by the MLP layer, and both the high-level single-point features and the enhanced view features are fused together with geometric features to serve as a geometric structure discriminator. The point cloud feature discriminator $R(p_g, v, p_l)$ can be defined as:

$$R(p_g, v, p_l) = \ell(MLP(\partial(p_g, v, p_l))) \tag{4}$$

with the range of [0, 1] representing the significance of different local structures. We also adopt the residual connection for the better utilization of the feature discriminator. The final output of our point-view-point attention fusion layers is defined as:

$$T(p_g, v, p_l) = p_l * (1 + R(p_g, v, p_l)) \tag{5}$$

where $pl * R(p_g, v, p_l)$ indicates applying the point cloud feature discriminator to original local geometric features to fully exploit more discriminative geometric features. The relative relationship between different local structures and their different contributions to the recognition of 3D shapes can be described by the feature discriminator well. Then the feature discriminator can well identify the features as feature discriminators to highlight local structures with discriminative information and restrain the useless local structure features. We arrange two point-view attention fusion blocks before the FC layer because as the network deepens, the features become clearer and more suitable for attention mechanisms. The first attention fusion layer tends to capture mid-level geometric features, and the second attention fusion layer is more sensitive to high-level geometric features.

3 Experiments and Discussion

3.1 3D Shape Recognition and Retrieval

The performance of our PVFNet on ModelNet [23] dataset is compared with that of various models based on different representations. As shown in Table 1, our proposed

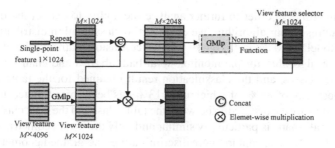

Fig. 4. Point-view attention fusion layer.

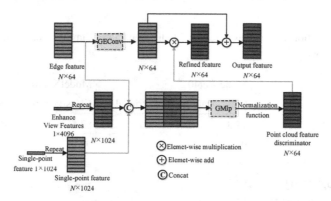

Fig. 5. Point-view-point attention fusion layer.

framework PVFNet achieves the best performance with a classification accuracy of 94.1% and a retrieval mAP of 90.8% on the ModelNet40 dataset. Compared to the state-of-the-art multi view based model GVCNN, PVFNet improves the performance by 1.5% and 6.3% for classification and retrieval respectively. Compared to the state-of-the-art point cloud based model DGCNN, our PVFNet improves the performance by 1.9% for classification on the ModelNet40 dataset. Compared to multimodal models, our PVFNet also outperforms them by a noticeable margin.

Using experimental results, we analyze both the qualitative results and the pros and cons of our method. We first present the visual results of several typical misclassifications of the models used on the ModelNet40 dataset. As shown in Fig. 6, the rightmost column shows both the true labels assigned to the 3D point clouds and the recognition results obtained by our algorithm. The similarity of the structure of the mispredicted labels to the true labels is obvious. For example, Fig. 6(c) shows that the true label is the plant and the predicted label of our algorithm is the flower pot, and the flower pot is mispredicted as the plant in Fig. 6(d). The reason is that some of the flower pots have plants in them while the other flower pots don't. So, the question of how to ignore the interference features and focus on only the significant features will be the motivation for improving on our PVFNet.

At the same time, in order to further analyze the ability to extract the fine-grained geometric features of our algorithm, we counted the classification distribution of each class in the ModelNet10 dataset. Figure 7 shows confusion matrix of the classification accuracy of our algorithm for the monitor, sofa, bathtub, bed and chair class is 100% for all of these classes, and the classification result obtained for the table class has a satisfactory accuracy of 85%, of which only 13% has been mispredicted for the desk class. The reason why the distinction between the table class and the desk class is difficult is that their overall shape is particularly similar and only the local geometric structure is different. Our PVFNet can capture more discriminating fine-grained geometric features of the point clouds after the point-point attention fusion layer distinguishes between different local structures using high-level global view features and single-point features.

Table 1. Classification and retrieval results of different algorithms on the ModelNet40 and ModelNet10 datasets (%).

Model	ModelNet40		ModelNet10	
	Classification	Retrieval	Classification	Retrieval
3D ShapeNet [23]	77.3	49.2	83.5	68.3
VoxNet [12]	93.0	–	92.0	–
MVCNN (AlexNet) [7]	89.9	80.2	–	–
MVCNN (GoogleNet) [7]	92.2	83.0	–	–
GVCNN (GoogleNet) [17]	92.6	84.5	–	–
PointNet [14]	89.2	–	–	–
PointNet++ [15]	90.7	–	–	–
Kd-Net [16]	91.8	–	94.0	–
DGCNN [17]	92.2	–	–	–
SeqViews [11]	93.3	89.1	94.8	91.4
PVNet [19]	93.2	89.5	–	–
PVFNet (AlexNet)	**94.1**	**90.8**	**95.0**	**91.7**

For the retrieval task, the 256-dimensional features are used after two fully-connected layers as the 3D shape descriptor in our PVFNet to complement retrieval task. MVCNN and GVCNN further apply a low-rank Mahalanobis metric learning to boost the retrieval performance and the triplet-center loss is used as the loss function. Our PVFNet, trained with the general softmax loss and without the low-rank Mahalanobis metric, achieved the state-of-the-art performance for shape retrieval with a 90.8% mAP, which strongly demonstrates the effectiveness of our strategy of using feature fusion for the shown 3D shape representations.

3.2 Ablation Studies

In order to explore the effectiveness of the GSPNet, point-view attention fusion layer and point-view-point attention fusion layer within our PVFNet framework, we evaluate the performance of the proposed method using different combinations of the components. The experimental results are shown in Table 2. The overall recognition accuracy of GSPNet gains 0.8% over PointNet by grouping MLP and EConv and dense connection, because group convolution can encode more useful information and dense connection can strengthen the interactivity of features of different levels, mining deep high-level semantic features. "Late Fusion" denotes that the point cloud global feature and the multi-view global feature are directly concatenated in a late fusion way. It is obvious that our PVFNet not only outperforms the point cloud and multi-view models by a large margin, but also gains 1.0% and 1.6% over the late fusion model with regard to both the mean class accuracy and the overall accuracy of the classification, respectively. It can also be seen that both point-view attention fusion and point-view-point attention fusion are beneficial to the performance. With more meaningful view features and discriminative local geometric features included in the point-view attention fusion and point-view-point attention fusion layers respectively, the performance of our framework improves from 92.5% to 94.1% progressively, which validates the effectiveness of usage of two attention fusion layers.

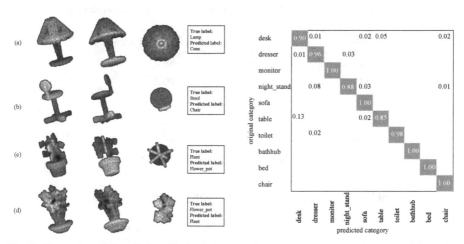

Fig. 6. Examples of the misclassified point cloud models.

Fig. 7. Confusion matrix on ModelNet10

3.3 Influence of Missing Views and Points

In this section, the robustness of our PVFNet regarding missing input data is evaluated. First, we explore the impact of missing views in favor of network performance. 12 views and 1024 points as input data are used to train our PVFNet. During testing, the number of points is fixed at 1,024 but the number of views is set to vary from 4 to 8, 10, 12. Table 3 shows the comparison between results obtained by MVCNN and GVCNN and between those obtained by our PVFNet. As seen from the table, compared with MVCNN and GVCNN, the better performance of PVFNet can still be retained regardless of the reduced views. It can also be seen that, with their number of views ranging from 12 to 4, the overall accuracy of MVCNN and GVCNN drops by 5.3% and 2.3% respectively and the performance of our PVFNet drops by only 2.0%.

At the same time observing the effect of missing points while keeping the number of views fixed at 12 during testing, the numbers of points selected for the experiment are 1,024, 768, 512, 384, 256, 128. Figure 8 shows the performance of state-of-the-art DGCNN degrades greatly because of the missing data when compared with our PVFNet. When the number of points is reduced from 1024 to 256, the recognition accuracy of DGCNN drops to about 50%, while our PVFNet can keep up a satisfactory performance of around 85%. The satisfactory performance of our framework benefits from the supplement of the corresponding view information which can help to compensate for the effect of the missing points. The verification of the robustness of our PVFNet to the problem of missing data is shown by the detailed results obtained from the experiments. Our framework can compensate for the two multimodal features. It is obvious that the strategy adopting point cloud global single-point semantic features, local geometric features and view features as complements of each other during the fusion process, is highly effective.

Table 2. Effectiveness of different components of our framework (%).

Models	Mean accuracy	Overall accuracy
PoinNet	86.2	89.2
GSPNet	87.1	90.0
Multi-view model	87.6	89.9
Late fusion	90.8	92.5
PVFNet (Point-View)	91.2	93.2
PVFNet (P-V + P-V-P)	91.8	94.1

Table 3. The accuracy comparison of different numbers of views.

Models	Number of views	mA (%)	OA (%)
MVCNN	4	82.5	84.6
	8	87.2	89.0
	12	87.6	89.9
GVCNN	4	88.1	90.3
	8	90.2	92.1
	12	90.7	92.6
PVFNet	4	89.8	92.1
	8	91.6	93.7
	12	91.8	94.1

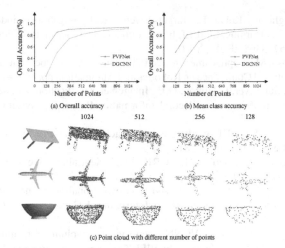

(a) Overall accuracy   (b) Mean class accuracy

1024   512   256   128

(c) Point cloud with different number of points

Fig. 8. Accuracy comparison of different numbers of input points.

4   Conclusion

In this paper, we propose a point-view fusion convolutional network called PVFNet for the tasks of 3D shape recognition and retrieval. PVFNet is a novel multimodal fusion network that takes full advantage of the correlation between global single-point semantic features, local geometric features of point cloud data and view features of view data. In our framework, we present GSPNet to capture global fine-grained single-point features with a high-level multi-semantic recognition capability and then fuse view features. The fusion in turn provides guidance for the point-view attention fusion layer to adaptively discard redundant view features. Then comes the introduction of our point-view-point attention fusion layer which uses both the enhanced view features and the global single-point features to describe the relative correlation and significance of different local structures. Finally, three types of global features are fused to get the unified feature descriptor which completes the task of 3D shape representation. We evaluate the effectiveness of our PVFNet by carrying out comprehensive experiments on the ModelNet for the tasks of 3D shape recognition and retrieval. The results of ablation studies and the influence of missing input data indicate that our point-view and point-view-point based feature fusion strategy has contributed significantly to the proposed PVFNet.

Acknowledgments. The authors would like to thank the reviewers for their valuable comments. This work is supported by National Natural Science Foundation of China under grant No. 61862039.

References

1. Liu, Z., Chen, H., Di, H.: Real-time 6D lidar slam in large scale natural terrains for UGV. In: IEEE Intelligent Vehicles Symposium (IV), pp. 662–667. IEEE (2018)

2. Zhu, Y., Mottaghi, R., Kolve, E.: Target-driven visual navigation in indoor scenes using deep reinforcement learning. In: IEEE International Conference on Robotics and Automation (ICRA), pp. 3357–3364. IEEE (2017)
3. Biswas, J., Veloso, M.: Depth camera based indoor mobile robot localization and navigation. In: IEEE International Conference on Robotics and Automation, pp. 1697–1702. IEEE (2012)
4. Krizhevsky, A., Sutskever, I., Hinton, G.E.: ImageNet classification with deep convolutional neural networks. In: Advances in Neural Information Processing Systems, pp. 1097–1105 (2012)
5. Ng, P.C., Henikoff, S.: SIFT: predicting amino acid changes that affect protein function. Nucleic Acids Res. **31**(13), 3812–3814 (2003)
6. Rublee, E., Rabaud, V., Konolige, K.: ORB: an efficient alternative to SIFT or SURF. In: ICCV, pp. 2564–2571. IEEE (2011)
7. Su, H., Maji, S., Kalogerakis, E.: Multi-view convolutional neural networks for 3D object recognition. In: Proceedings of the IEEE International Conference on Computer Vision, pp. 945–953. IEEE (2015)
8. Feng, Y., Zhang, Z., Zhao, X.: GVCNN: group-view convolutional neural networks for 3D object recognition. In: Proceedings of the IEEE Conference on Computer Vision and Pattern Recognition, pp. 264–272. IEEE (2018)
9. Guo, H., Wang, J., Gao, Y.: Multi-view 3D object retrieval with deep embedding network. IEEE Trans. Image Process. **25**(12), 5526–5537 (2016)
10. He, X., Zhou, Y., Zhou, Z.: Triplet-center loss for multi-view 3D object retrieval. In: Proceedings of the IEEE Conference on Computer Vision and Pattern Recognition, pp: 1945–1954. IEEE (2018)
11. Han, Z., Shang, M., Liu, Z.: Seqviews2seqlabels: learning 3D global features via aggregating sequential views by RNN with attention. IEEE Trans. Image Process. **28**(2), 658–672 (2018)
12. D. Maturana, S. Scherer. VoxNet: a 3D convolutional neural network for real-time object recognition, in: RSJ International Conference on Intelligent Robots and Systems (IROS), pp. 922–928. IEEE (2015)
13. Riegler, G., Osman, A., Geiger, A.: OctNet: learning deep 3D representations at high resolutions. In: Proceedings of the IEEE Conference on Computer Vision and Pattern Recognition, pp. 3577–3586. IEEE (2017)
14. Qi, C.R., Su, H., Mo, K.: PointNet: deep learning on point sets for 3D classification and segmentation. In: Proceedings of the IEEE Conference on Computer Vision and Pattern Recognition, pp. 652–660. IEEE (2017)
15. Qi, C.R., Yi, L., Su, H.: PointNet++: deep hierarchical feature learning on point sets in a metric space. In: Advances in Neural Information Processing Systems, pp. 5099–5108 (2017)
16. Klokov, R., Lempitsky, V.: Escape from cells: deep Kd-networks for the recognition of 3D point cloud models. In: Proceedings of the IEEE International Conference on Computer Vision, pp. 863–872. IEEE (2017)
17. Wang, Y., Sun, Y., Liu, Z.: Dynamic graph cnn for learning on point clouds. ACM Trans. Graph. (TOG) **38**(5), 1–12 (2019)
18. Li, Y., Bu, R., Sun, M.: PointCNN: convolution on x-transformed points. In: Advances in Neural Information Processing Systems, pp. 820–830 (2018)
19. You, H., Feng, Y., Ji, R.: PVNet: a joint convolutional network of point cloud and multi-view for 3D object recognition. In: 2018 ACM Multimedia Conference on Multimedia Conference, pp. 1310–1318. ACM (2018)
20. Zhang, X., Zhou, X., Lin, M.: ShuffleNet: an extremely efficient convolutional neural network for mobile devices. In: Proceedings of the IEEE Conference on Computer Vision and Pattern Recognition, pp. 6848–6856. IEEE (2018)
21. Iandola, F., Moskewicz, M., Karayev, S.: DenseNet: implementing efficient ConvNet descriptor pyramids (2014). arXiv:1404.1869

22. Wang, F., Jiang, M., Qian, C.: Residual attention network for image classification. In: Proceedings of the IEEE Conference on Computer Vision and Pattern Recognition, pp. 3156–3164. IEEE (2017)
23. Wu, Z., Song, S., Khosla, A.: 3D ShapeNets: a deep representation for volumetric shapes. In: Proceedings of the IEEE Conference on Computer Vision and Pattern Recognition, pp. 1912–1920. IEEE (2015)

HEAM: Heterogeneous Network Embedding with Automatic Meta-path Construction

Ruicong Shi[1,2], Tao Liang[1,2], Huailiang Peng[1(✉)], Lei Jiang[1], and Qiong Dai[1]

[1] Institute of Information Engineering, Chinese Academy of Sciences, Beijing, China
{shiruicong,liangtao0305,penghuailiang,jianglei,daiqiong}@iie.ac.cn
[2] School of Cyber Security, University of Chinese Academy of Sciences, Beijing, China

Abstract. Heterogeneous information network (HIN) embedding is widely used in many real-world applications. Meta-path used in HINs can effectively extract semantic information among objects. However, the meta-path faces challenges on the construction and selection. Most of the current works construct dataset-specific meta-paths manually, which rely on the prior knowledge from domain experts. In addition, existing approaches select a few explicit meta-paths, which lack of much subtle semantic information among objects. To tackle the problems, we propose a model with automatic meta-path construction. We develop a hierarchical aggregation to learn effective heterogeneous embeddings with meta-path based proximity. We employ a multi-layer network framework to mine long meta-paths based information implicitly. To demonstrate the effectiveness of our model, we apply it to two real-world datasets and show the performance improvements over state-of-the-art methods.

Keywords: Heterogeneous information network · Heterogeneous embedding · Meta-path

1 Introduction

Many real-world data, including social networks, citation networks, and biological networks, can be naturally modelled as graph data to enhance various application. Therefore, network embeddings (NE), also known as graph embedding, gain popularity in recent years. Network embedding maps nodes of a network into a low-dimensional continuous vector space, simultaneously preserving the structure information of the original network. Then taking the embedding representation as input of various downstream graph analysis tasks, such as node classification, clustering, link prediction, etc. Different from homogeneous networks [1–5], where all vertices belong to the same class, heterogeneous networks [6–11] involves multiple node types and/or multiple edge types. For complex heterogeneous networks, the network schema [12] of a graph specifies type constraints on objects and relationship constraints among objects. Network schema

G. Li et al. (Eds.): KSEM 2020, LNAI 12274, pp. 304–315, 2020.
https://doi.org/10.1007/978-3-030-55130-8_27

makes heterogeneous networks semi-structured, which guides the semantic explorations of meta-paths [13] in the network. Meta-path [13] is defined as a relationship sequence connecting two objects of a heterogeneous network. Meta-path is widely adopted to learn the semantic information in various heterogeneous graph tasks, such as heterogeneous network embedding [6,9–11], similarity search [13–16], recommendation [18–20] and so on. Generally, meta-path based approaches are dependent on meta-path to extract node pairs that have no direct edge and preserve the proximity of the node pairs in embeddings. Different meta-paths express different semantic proximity, taking a citation network DBLP as an example, the meta-path A-P-A means a co-author relation, and A-P-C-P-A expresses a relationship of two authors who's papers are accepted in the same conference.

Although meta-path is widely adopted in the heterogeneous information networks, how to construct appropriate meta-paths for a given HIN remains indeterminate. Existing meta-path based works construct dataset-specific meta-paths manually [6,9–11,13,15,17–20], which strongly depend on dataset and prior knowledge from domain experts. More importantly, lots of experiments prove that meta-paths can critically affect the quality of the learned network embeddings. For example, in many algorithms, using the meta-path "A-P-C-P-A" always performs better than using meta-path "A-P-T-P-A" on both node classification and clustering tasks on dataset DBLP. When facing complex HIN, relying on prior knowledge is hard to find the optimal meta-path for the downstream analysis.

Based on the above intuition, we propose an end-to-end **H**eterogeneous network **E**mbedding with **A**utomatic **M**eta-path construction (**HEAM**) model. It enumerates all possible meta-paths within K-order neighborhoods which avoids manually construct meta-paths and prevents the effective meta-paths from being missed. Then we develop a hierarchical aggregation, including node aggregation and path aggregation, to aggregate meta-path guiding neighbors and different meta-paths respectively. Moreover, we use a multi-layer network framework to implicitly mine long meta-paths, which avoids the number of meta-paths exponentially growing produced by high-order neighborhoods.

Our contributions are as follows: (i) Our model can automatically construct meta-paths which don't relies on specific datasets and any prior knowledge. (ii) In our model, each node aggregates the features of the meta-path guiding neighbors with meta-path based proximity, which preserves the semantic information and structure information in the embedding space simultaneously. (iii) We assess the effectiveness of our algorithm on two datasets: ACM, DBLP. The extensive experiments prove that our model improves the performance on node classification tasks and clustering tasks by comparing with the state-of-the-art models.

2   Related Work

Recently, many pieces of research begin to focus on the representation learning of graph data. However, different from the complex HINs targeted in our paper,

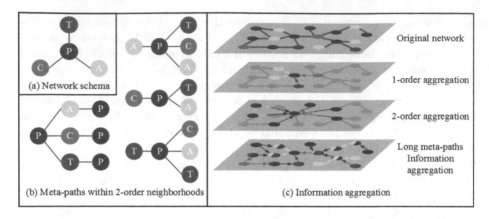

Fig. 1. (a) shows the network schema of DBLP dataset. (b) shows all existing meta-paths within the 2-order neighborhoods, including "AP", "APC", "APA", "APT" for author, "PA", "PC", "PT", "PAP", "PCP", "PTP" for paper, "CP", "CPA", "CPT" for conference, "TP", "TPC", "TPA", "TPT" for term. (c) shows the aggregation of the 1-order, 2-order and more long meta-paths based information.

several prior research focus on homogeneous information networks [1–5], which have a single node type and edge type. Random walk based approaches, such as Deepwalk [1] and Node2vec [3], perform random walks over the network to generate corpus and then input node sequences into a SkipGram [21] model to train node embeddings. Graph neural network based approaches, such as GraphSAGE [4] and GAT [5], employ deep neural network to train network embeddings based on the network structure and node attribute features simultaneously.

Heterogeneous information networks embedding learn the node representations for the networks which have various node/edge types. In the existing works, heterogeneous graph embedding techniques mostly adopt meta-paths to mine the structure information and semantic information. Metapath2vec [9] employs meta-path to guide random walk and then training the embedding with the SkipGram model. However, metapath2vec only utilizes a meta-path scheme which is hard to mine comprehensive semantic information. To overcome this issue caused by using a single meta-path in prior works, some other methods [6,11,15,19,22] combine multiple meta-paths to mine different semantic information in heterogeneous. ESim [15] learns node embedding in a specified embedding space according to the predefined meta-paths . However, it needs to conduct a search to find the optimal coefficients of meta-paths. HERec [19] uses meta-paths to guide random walks and employs the node sequences generated to learn network embedding. Then it designs different strategies to integrate multiple meta-paths embeddings. HIN2Vec [11] treats the predefined meta-paths as the target relations, then it implements multiple prediction tasks to learn the node embeddings and meta-paths simultaneously. HAN [6] utilizes a hierarchical attention mechanism, including both of node and semantic attention, to determine the weights of neighbors and different meta-paths simultaneously. These models incorporating more meta-paths have better performance than the model with a single meta-path.

However, all of the aforementioned methods require users to predefine meta-paths, which strongly depends on the prior knowledge and specific dataset. Due to the limitation, many methods are proposed to solve this problem. Meng et al. [23] utilizes the similarity function and GreedyTree framework to find out optimal meta-paths that express the relationship between given node pairs. Cao et al. [24] employs a greedy algorithm to find out meta-paths from schema-rich heterogeneous network. After that they employ a likelihood function to calculate the coefficients of meta-paths. However, those algorithms are not end-to-end methods and hard to learn meta-paths combined with specific application tasks. Yun et al. [25] transforms a heterogeneous graph into new graph structures for performing graph convolutions. It can discover meta-paths from heterogeneous graph automatically. Unlike the approach based on graph convolutions network, our model employs graph neural network, which can be used to large-scale networks.

3   Preliminary

In this chapter, we briefly state some definitions and notations used in this paper.

Definition 1. *Heterogeneous Information Network [13]. A heterogeneous information network, also known as heterogeneous graph, is a special graph as $\mathcal{G} = (\mathcal{V}, \mathcal{E})$ where $\mathcal{V}$ and $\mathcal{E}$ represent the set of nodes and edges, respectively. For any heterogeneous graph, it exists $|\mathcal{A}| + |\mathcal{R}| > 2$, where $\mathcal{A}$ and $\mathcal{R}$ denote the set of the node types and the edge types, respectively. We can obtain $\mathcal{A}$ and $\mathcal{R}$ via a node-type mapping function $\phi : \mathcal{V} \to \mathcal{A}$ and a edge-type mapping function $\psi : \mathcal{E} \to \mathcal{R}$.*

Taking citation network DBLP for example, the network contains four types of nodes (paper, author, conference, as well as term) and three types of edges (P-A, P-C, P-T). So this graph belongs to heterogeneous graph, and we can create a network schema for DBLP as shown in Fig. 1(a).

Definition 2. *Meta-path [13]. A meta-path ρ is defined as a object sequence in the form of $O_1 \xrightarrow{R_1} O_2 \xrightarrow{R_2} \cdots \xrightarrow{R_l} O_{l+1}$ (abbreviated as $O_1 O_2 \cdots O_{l+1}$) which expresses a combination of relation between objects O_1 and O_{l+1}.*

When a network schema is given, we can use different meta-paths to connect two nodes. Taking Fig. 1(b) for example, two papers can be connected via several meta-paths, for instance, "PAP", "PCP" and "PTP". Different meta-paths reveal different semantics, the meta-path "P-A-P" means the relationship of papers written by the same author, while meta-path "P-C-P" means the relationship of papers published in the same conference.

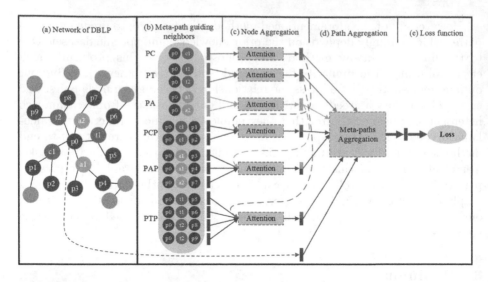

Fig. 2. The overall process of HEAM. We automatically construct all meta-paths within 2-order neighborhoods. In node aggregation, we use attentions to aggregate the meta-paths guiding neighbors. After that, we use path aggregation to fuse different mate-paths. We through a task-special loss function to calculate the loss.

4   The Proposed Model

In this chapter, we display our model in detail. Figure 2 presents the whole framework of HEAM. We decompose our model into three parts: automatic meta-path construction, node aggregation and path aggregation. We employ automatic meta-path construction to avoid hand-crafted meta-paths in previous methods. Then we develop a hierarchical aggregation mechanism, including node aggregation and path aggregation. Node aggregation apply an attention mechanism to aggregate the information from their neighbors on a meta-path, and path aggregation can aggregate different meta-paths information with a self-attention.

4.1   Automatic Meta-path Construction

We automatically construct meta-paths for all types objects without any prior knowledge. For a given network schema, as Fig. 1 shown, we use a simple traverse algorithm to discover all possible meta-paths within K-order neighborhoods. Take "P" as an example, we traverse the 1st-order neighborhood to generate meta-paths "P-A", "P-C", and "P-T", and traverse the 2th-order neighborhood to construct meta-paths "P-A-P", "P-C-P", and "P-T-P". In the same way, we traverse K-order neighborhoods to discover meta-paths automatically. Then, we divide the neighbors into different subsets guiding by different meta-paths, as Fig. 2(b) shown, which will be used in hierarchical attention mechanism later.

4.2 Node-Level Aggregation

Because the neighbors show different importance in the specific task, we use an attention to learn the weights flexibly. Due to different objects with different initial feature spaces, we need to project the features into the same feature space first, through a type-specific transformation matrix $\mathbf{W}_t$, where t represents the object type.

$$\mathbf{H}' = \mathbf{W}_t \cdot \mathbf{H}_t \tag{1}$$

For node i, it's neighbor j belongs to meta-path ρ guiding neighbors, denoted as $j \in \mathcal{N}_i^\rho$. We use a self-attention mechanism to calculate the coefficient of j, that shows the importance of neighbor j to the core i. Then we normalize it across all neighbors on meta-path ρ with the softmax function:

$$\alpha_{ij}^\rho = \frac{\exp\left(\mathbf{A}_\rho^{\mathrm{T}} \cdot \mathbf{E}_{ij}^\rho\right)}{\sum_{k \in \mathcal{N}_i^\rho} \exp\left(\mathbf{A}_\rho^{\mathrm{T}} \cdot \mathbf{E}_{ik}^\rho\right)}, \tag{2}$$

where $\mathbf{A}_\rho$ is a parametrized attention vector for meta-path ρ, $\mathbf{E}_{ij}^\rho$ is a path encoding representation. We suppose that $\rho = O_1 O_2 \cdots O_{l+1}$ and $\rho_1 = O_1 O_2 \cdots O_l$, then $\mathbf{E}_{ij}^\rho = [\mathbf{h}_i' \| \mathbf{z}_i^{\rho_1} \| \mathbf{h}_j']$, where $\|$ represents the concatenate operation, $\mathbf{z}_i^{\rho_1}$ means meta-path ρ_1 based representation for node i. Instead of using the node pairs representation, we employ path encoding representation to preserve the meta-path based proximity in the embeddings. Moreover, the comprehensive meta-path based representation $\mathbf{z}_i^{\rho_1}$ avoids the error propagation caused by the specific nodes between node i and j on meta-path ρ. After that, we perform a node-level aggregation to fuse the features of all neighbors on the meta-path ρ, we get a meta-path based representation:

$$\mathbf{z}_i^\rho = \sigma\left(\sum_{j \in \mathcal{N}_i^\rho} \alpha_{ij}^\rho \cdot \mathbf{h}_j'\right), \mathbf{h}_j' \in \mathbf{H}' \tag{3}$$

where $\sigma\left(\cdot\right)$ denotes the LeakyRelu activation function.

4.3 Path-Level Aggregation

From the previous step, we get all meta-path based representations of node i: $Z_i = \left\{\mathbf{z}_i, \mathbf{z}_i^1, \mathbf{z}_i^2, \ldots, \mathbf{z}_i^\rho\right\}$, where $\mathbf{z}_i^\rho$ denotes the meta-path ρ based representation, $\mathbf{z}_i$ denotes the representation of node i itself. Furthermore, to distinguish the importance of them, we use the self-attention mechanism to learn the weight for each meta-path.

$$\mu_\rho = softmax\left(\mathbf{q}^{\mathrm{T}} \cdot MLP\left(\mathbf{z}_i^\rho\right)\right). \tag{4}$$

Before we calculate the coefficients of each meta-path, we firstly input the meta-path based representations into a one-layer MLP with activation function tanh.

After that, we use an self-attention to calculate the coefficients of those meta-paths and normalize them across all meta-paths with the softmax function. Following that we fuse different meta-path based representations:

$$\mathbf{z}_i = \mathop{\Big\|}_{k=1}^{K} \left(\sigma \left(\sum_{\mathbf{z}_i^\rho \in Z} \mu_\rho \cdot \mathbf{z}_i^\rho \right) \right), \tag{5}$$

where the $\sigma(\cdot)$ denotes the activation function elu. We employ a multi-head attention and concatenate the output of each head attention.

Due to the number of meta-paths exponentially growing produced by the increase of neighborhoods, we set the k to 2 to simplify the calculation. Simultaneously, in order to mine long meta-path information, we use a multi-layer network framework. For example, instead of using the 5-order neighborhoods to mine meta-path "A-P-C-P-A" with length 5, we use 2-order neighborhoods to mine long meta-paths information implicitly, such as "A" gets the information from "C" through meta-path "A-P-C" and "C" gets the information from A through meta-path "C-P-A", through the information dissemination, "A" can aggregates the information of "A-P-C-P-A" in the second layer. Simultaneously, considering the similarity of nodes connected by long meta-paths is weak, we mine the long meta-paths information implicitly which avoids over-smoothing of the nodes embeddings.

We apply the final embeddings to specific tasks with different loss functions. For node classification, we apply the Cross-Entropy as the loss function and train the model by minimizing the objective function over all training data. Moreover, we use different classifiers for the prediction of different types of nodes in heterogeneous graph,

$$L(z) = \sum_{i=1}^{\ell} \Delta \left(f^{t_i}(z_i), y_i \right), \tag{6}$$

where $f^{t_i}(z_i)$ denotes the prediction of node i using a type-specified classification function, y_i denotes the ground-truth.

Table 1. Statistics of the datasets

Dataset	A-B	A	B	A-B	Feature	Training	Validation	Test
DBLP	Paper-Author	14328	14475	41794	1018	800	400	2857
	Paper-Conf	14328	20	14328				
	Paper-Term	14328	8920	114624				
ACM	Paper-Author	3025	6028	10055	1524	600	300	2125
	Paper-Subject	3025	56	3025				

5   Experiments

In this chapter, we assess our algorithm on two datasets (ACM and DBLP). We chose some state-of-art baselines for comparison, including both homogeneous and heterogeneous graph embedding approaches. We prove that our algorithm has excellent performance in both node classification task and clustering tasks.

5.1   Datasets

To evaluate the effectiveness of our algorithm, we perform node classification and clustering on two citation datasets: DBLP and ACM. The detailed descriptions are shown in the Table 1. For dataset **DBLP**, the set of node types is: $\{papers\,(P), authors\,(A), terms\,(T), conferences\,(C)\}$. Authors are labelled into four categories. **ACM** includes three node types: papers (P), authors (A), subject (S). Here we label the papers into three classes. The features of nodes in the two datasets is represented as bag-of-words of abstract.

5.2   Baselines

We introduce our model HEAM and several recent graph embedding methods as following: **DeepWalk** [1]: A model with random walks designed for homogeneous graphs embedding. **GAT** [5]: A deep neural network with node attention mechanism used for homogeneous graph embedding. **metapath2vec** [9]: A heterogeneous network embedding model which utilizes single meta-path guiding random walks. **HAN** [6]: A heterogeneous graph embedding learning with hierarchical attention mechanism. **HEAM**: A heterogeneous graph embedding proposed in this paper, which use all meta-paths within 2-order neighborhoods, and a 2-layer network framework. **HEAM-N**: A variant of HEAM which utilizes N-layer network framework, N equals 1 or 3.

5.3   Implementation Details

In the model training, we use the following hyperparameters settings: the learning rate is 0.005, the dropout of attention is 0.6, the multi-head attention mechanism K is 4, the patience of early stopping is 50. The dimension of the final embedding is set to 64 for our model and all baselines. For random walk based approaches, we use the context window with 5, each walk with length 40, the negative sample with 5. We select meta-paths "PAP" and "PSP" for ACM, "APA", "APCPA" and "APTPA" for DBLP. For GAT, metapath2vec, we perform experiments on all selected meta-paths and record the best result. We treat the heterogeneous as a homogeneous and perform DeepWalk on the graph.

5.4   Classification Task

In the classification task, we used KNN (k = 5) as the node classifier. To prove the reliability of the experimental data, we perform the experiment ten times

Table 2. Results (%) of the node classification task.

Datasets	Metrics	Training	Deepwalk	Metapath2vec	GAT	HAN	HEAM-1	HEAM-3	HEAM
ACM	Macro-F1	20%	75.13	66.36	88.16	90.96	91.96	89.68	**92.11**
		40%	79.39	69.59	88.38	90.71	91.74	90.07	**92.42**
		60%	80.79	70.85	88.29	90.87	92.37	90.47	**92.50**
		80%	83.57	72.84	88.15	91.10	92.10	90.69	**93.09**
	Micro-F1	20%	77.42	66.28	88.09	90.92	91.92	89.80	**92.00**
		40%	80.76	69.51	88.36	90.63	91.69	90.01	**92.37**
		60%	81.84	70.80	88.39	90.79	92.36	90.47	**92.45**
		80%	83.46	72.80	88.14	91.02	92.06	90.52	**92.98**
DBLP	Macro-F1	20%	86.09	91.31	88.88	90.68	86.39	**91.74**	91.43
		40%	87.91	92.07	89.52	91.40	87.12	92.11	**92.50**
		60%	88.79	91.96	89.80	91.02	86.96	92.21	**92.46**
		80%	89.11	91.86	90.15	91.24	87.08	92.39	**92.57**
	Micro-F1	20%	87.39	91.90	89.66	91.21	89.08	92.28	**92.36**
		40%	88.89	92.57	90.15	91.89	87.87	**92.65**	92.57
		60%	89.75	92.50	90.53	92.49	87.73	92.76	**92.95**
		80%	89.98	92.45	90.73	92.68	87.78	92.90	**93.45**

to calculate the average of *Macro-F1* and *Micro-F1* in Table 2. The result shows that our algorithm achieves the best performance. Generally, heterogeneous graph based approaches are better than the homogeneous graph based approaches, meta-path based approaches are better than the approaches without meta-path. Compare HEAM with its two variants, HEAM-3 is better than HEAM-1 in DBLP which relys on long meta-paths "APCPA" and "APTPA", but it is worse than HEAM-1 in ACM which relys on short meta-paths "PAP" and "PSP". It proves that the multi-layer framework can effectively mine the long path information, at the same time, it also leads to information redundancy in some situation. The 2-layer framework can balance those states and achieve the best performance.

5.5   Clustering Task

We choose the KMeans clustering algorithm for node Clustering task. We adopt the quantity of initial clusters as the same with the quantity of categories in the node classification task. We use two metrics (NMI and ARI) to evaluate the results. We perform the experiment ten times and show the average results in Table 3. As shown in the results, our algorithm achieves the best performance. The multiple meta-paths based algorithms, HAN and HEAM, are better than the single meta-path based algorithms, e.g. metapath2vec and GAT. We can find that HEAM performs better than its two variants.

5.6   Model Analysis

In this section, we analyze the availability of our algorithm. Firstly, existing approaches employ a few meta-paths, which lack of much subtle semantic

Table 3. Results (%) of the node clustering task.

Datasets	Metrics	DeepWalk	Metapath2vec	GAT	HAN	HEAM-1	HEAM-3	HEAM
ACM	NMI	48.01	23.37	64.87	71.10	68.91	65.89	**72.99**
	ARI	44.48	16.86	70.79	75.82	73.07	70.10	**78.59**
DBLP	NMI	75.25	73.34	66.85	73.48	61.67	72.53	**75.79**
	ARI	79.95	78.43	72.01	79.04	68.51	77.93	**81.56**

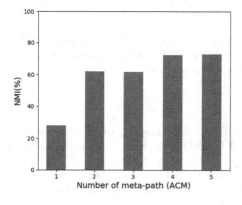
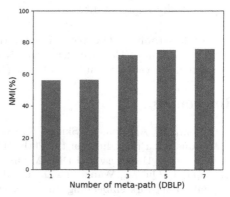

Fig. 3. The effect of different number of meta-paths on ACM clustering.

Fig. 4. The effect of different number of meta-paths on DBLP clustering.

information among objects. In our model, we use more rich meta-paths. In order to prove the effectiveness of these meta-paths, we evaluate the effect of the number of meta-paths on clustering task and show the results of ACM dataset and DBLP dataset in Fig. 3 and Fig. 4 respectively. As the result shown, we find that combining with more meta-paths, the performance of the learned embedding is improved. It proves that the subtle semantic information in heterogeneous graphs is useful.

Furthermore, we use the attention coefficient of different meta-paths to verify the validity of our model. The coefficient of meta-paths for paper in ACM is: "PAP" > "PA" > "PSP" > "PS". The papers published by the same author are more similar to the papers published in the same subject. The author's influence on the paper is greater than that of the subject. For author and subject, the most important meta-paths are "AP" and "SP" respectively. Through information dissemination in the network, we can mine some meta-paths implicitly at the second layer aggregation, such as "PAP", "PSP", "PAPAP" and so on. In DBLP, the coefficient of meta-paths for author is: "APA" > "APC" > "AP" > "APT". The co-authors are more similar than other relationships. For paper, conference, and term, the most important meta-paths are "PA", "CPA", and "TPA" respectively. At the second layer, we can get long meta-path "APAPA", "APTPA", and "APCPA" implicitly.

6   Conclusion

In our paper, we introduce a heterogeneous network embedding algorithm HEAM which resolves the problem of using predefine meta-paths in current works. Then we use node aggregation and path aggregation to fuse the meta-path guiding neighbors and different meta-paths respectively. The embedding vectors in our model contain structure information and semantic information simultaneously. By comprehensive analysis, the proposed HEAM also has proven its potentially expansibility.

Acknowledgments. This paper is supported by National Key Research and Development Program of China under Grant No. 2017YFB0803003 and National Science Foundation for Young Scientists of China (Grant No. 61702507).

References

1. Perozzi, B., Al-Rfou, R., Skiena, S.: DeepWalk: online learning of social representations. In: Proceedings of the 20th ACM SIGKDD International Conference on Knowledge Discovery and Data Mining, pp. 701–710. ACM (2014)
2. Tang, J., Qu, M., Wang, M., Zhang, M., Yan, J., Mei, Q.: Line: large-scale information network embedding. In: Proceedings of the 24th International Conference on World Wide Web, pp. 1067–1077. WWW (2015)
3. Grover, A., Leskovec, J.: Node2vec: scalable feature learning for networks. In: Proceedings of the 22nd ACM SIGKDD International Conference on Knowledge Discovery and Data Mining, pp. 855–864. ACM (2016)
4. Hamilton, W.L., Ying, R., Leskovec, J.: Inductive representation learning on large graphs. In: Proceedings of the 31st International Conference on Neural Information Processing Systems, pp. 1025–1035. NIPS (2017)
5. Veličković, P., Cucurull, G., Casanova, A., Romero, A., Liò, P., Bengio, Y.: Graph attention networks. In: International Conference on Learning Representations (2018)
6. Xiao, W., et al.: Heterogeneous graph attention network. WWW (2019)
7. Chang, S., Han, W., Tang, J., Qi, G.J., Aggarwal, C.C., Huang, T.S.: Heterogeneous network embedding via deep architectures. In: Proceedings of the 21st ACM SIGKDD International Conference on Knowledge Discovery and Data Mining, pp. 119–128. ACM (2015)
8. Tang, J., Qu, M., Mei, Q.: PTE: predictive text embedding through large-scale heterogeneous text networks. In: Proceedings of the 21st ACM SIGKDD International Conference on Knowledge Discovery and Data Mining, pp. 1165–1174. ACM (2015)
9. Dong, Y., Chawla, N.V., Swami, A.: Metapath2vec: scalable representation learning for heterogeneous networks. In: Proceedings of the 23rd ACM SIGKDD International Conference on Knowledge Discovery and Data Mining, pp. 135–144. ACM (2017)
10. Huang, Z., Mamoulis, N.: Heterogeneous information network embedding for meta path based proximity. ArXiv abs/1701.05291 (2017)
11. Fu, T.y., Lee, W.C., Lei, Z.: HIN2Vec: explore meta-paths in heterogeneous information networks for representation learning. In: Proceedings of the 2017 ACM on Conference on Information and Knowledge Management, pp. 1797–1806. ACM (2017)

12. Sun, Y., Han, J.: Mining heterogeneous information networks: a structural analysis approach. SIGKDD Explor. **14**, 20–28 (2012)
13. Sun, Y., Han, J., Yan, X., Yu, P.S., Wu, T.: PathSim: meta path-based top-k similarity search in heterogeneous information networks. VLDB **411**, 992–1003 (2011)
14. Yu, X., Sun, Y., Norick, B., Mao, T., Han, J.: User guided entity similarity search using meta-path selection in heterogeneous information networks, pp. 2025–2029, October 2012
15. Shang, J., Qu, M., Liu, J., Kaplan, L.M., Han, J., Peng, J.: Meta-path guided embedding for similarity search in large-scale heterogeneous information networks. CoRR abs/1610.09769 (2016)
16. Wang, C., Song, Y., Li, H., Sun, Y., Zhang, M., Han, J.: Distant meta-path similarities for text-based heterogeneous information networks. In: Proceedings of the 2017 ACM on Conference on Information and Knowledge Management, pp. 1629–1638. ACM (2017)
17. Sun, Y., Barber, R., Gupta, M., Aggarwal, C.C., Han, J.: Co-author relationship prediction in heterogeneous bibliographic networks. In: Proceedings of the 2011 International Conference on Advances in Social Networks Analysis and Mining, ASONAM 2011, pp. 121–128. IEEE Computer Society (2011)
18. Fan, S., et al.: Metapath-guided heterogeneous graph neural network for intent recommendation. In: Proceedings of the 25th ACM SIGKDD International Conference on Knowledge Discovery and Data Mining, pp. 2478–2486. ACM (2019)
19. Shi, C., Hu, B., Zhao, W., Yu, P.S.: Heterogeneous information network embedding for recommendation. IEEE Trans. Knowl. Data Eng. **31**(02), 357–370 (2019)
20. Hu, B., Shi, C., W.X.Z., Yu., P.S.: Leverage meta-path based context for top-n recommendation with a neural co-attention model. In: Proceedings of the 24th ACM SIGKDD International Conference on Knowledge Discovery and Data Mining. ACM (2018)
21. Mikolov, T., Chen, K., Corrado, G.S., Dean, J.: Efficient estimation of word representations in vector space. CoRR abs/1301.3781 (2013)
22. Chen, T., Sun, Y.: Task-guided and path-augmented heterogeneous network embedding for author identification. In: Proceedings of the Tenth ACM International Conference on Web Search and Data Mining, pp. 295–304. ACM (2017)
23. Meng, C., Cheng, R., Maniu, S., Senellart, P., Zhang, W.: Discovering meta-paths in large heterogeneous information networks. In: Proceedings of the 24th International Conference on World Wide Web, pp. 754–764. WWW (2015)
24. Cao, X., Zheng, Y., Shi, C., Li, J., Wu, B.: Link prediction in schema-rich heterogeneous information network. In: Bailey, J., Khan, L., Washio, T., Dobbie, G., Huang, J.Z., Wang, R. (eds.) PAKDD 2016. LNCS (LNAI), vol. 9651, pp. 449–460. Springer, Cham (2016). https://doi.org/10.1007/978-3-319-31753-3_36
25. Yun, S., Jeong, M., Kim. R., Kang, J., Kim, H.: Graph Transformer Networks. In: Proceedings of the 32th Advances in Neural Information Processing Systems, pp. 11983–11993 (2019)

A Graph Attentive Network Model for P2P Lending Fraud Detection

Qiyi Wang[1](✉), Hongyan Liu[2](✉), Jun He[1](✉), and Xiaoyong Du[1]

[1] Key Labs of Data Engineering and Knowledge Engineering, Ministry of Education, China
School of Information, Renmin University of China, Beijing 100872, China
{wqy1994202,hejun,duyong}@ruc.edu.cn
[2] Department of Management Science and Engineering, Tsinghua University, Beijing 100084,
China
liuhy@sem.tsinghua.edu.cn

Abstract. Fraud detection for peer-to-peer (P2P) lending is an important and challenging problem in both real application and research area. Different from existing methods which are mainly based on user demographic information, in this paper we study if other information such as user relationship represented by graph and transaction description information are helpful for the fraud detection problem. Meanwhile, attention mechanism is widely employed to explain how deep learning model works. However, existing studies don't discriminate the importance of neighbors and different edge features in graph. In this paper, we propose a new graph attentive network model called 'FDNE' for P2P lending fraud detection based on text information and/or user relationship information. We design a novel attention method called edge-feature attention and use a global normalization operation to identify influential edge feature. Experiments conducted on a real dataset demonstrate that our model significantly outperforms other baselines and can make reasonable explanations simultaneously.

1 Introduction

Online peer-to-peer (P2P) lending, is a way to provide unsecured microloans from individual investors to individual borrowers without the intermediation of financial institutions [18]. With the development of Web 3.0 applications, P2P platforms provide better services of introducing borrowers to investors. Through P2P platforms, borrowers could get small loans with less time comparing to traditional lending institutions such as banks, while investors and platforms could earn profits. However, lacking sophisticated risk assessment, online P2P companies face more risks than traditional lending institutions. Cheating borrowers often pretend to be normal ones, and they often repay loans for the first repayment periods and stop repayment later, making it difficult to be detected. Therefore, fraud detection for P2P lending is an important and challenging problem. P2P companies often regard overdue users as fraud users.

Most existing fraud detection or credibility assessment methods are mainly based on user demographic information that users provide when they apply for loan, which are most structured information. In this paper we study if other information such as user

G. Li et al. (Eds.): KSEM 2020, LNAI 12274, pp. 316–327, 2020.
https://doi.org/10.1007/978-3-030-55130-8_28

relationship represented by graph and transaction description text information is helpful for the fraud detection problem. Graph data contain lots of information and play an important role in prediction tasks. As far as we know, few works use graph information in peer-to-peer lending fraud detection problems. Besides, most of existing works neglect the interpretability of models, which cannot give hints about characteristics of fraud users. Attention mechanism is widely employed to explain how the model works in deep neural network. Some works adapt attention mechanism to graph learning and propose various graph attentive network models. However, existing studies don't discriminate the importance of neighbors and different edge features.

To solve this issue, we propose a new graph attentive network model for P2P lending fraud detection called 'FDNE (Fraud Detection considering both Node and Edge feature importance)'. In this model, we design a novel attention called edge-feature attention to deal with graph with heterogeneous edges, adding the influence from different kinds (features) of edges into the attention, and devise a global normalization operation to identify important features of each edge. At the same time, text information is analyzed to find important words for the fraud detection task, enhancing interpretability of the model.

The main contributions of the paper can be summarized as follows:

- We propose a novel graph attentive network model, FDNE, for P2P lending fraud detection, and design a new attention called edge-feature attention to deal with heterogeneous edges in the graph, taking the influence from different features of edges into consideration.
- We use global normalization operation to identify influential features of each edge. Meanwhile, influential neighbors and words are found through node attention and word attention in the model.
- Experiments conducted on a real dataset demonstrate that our model significantly outperforms other baselines and give useful information about fraud users simultaneously.

2 Related Work

There are some prior research works for fraud detection on P2P platforms [18–23]. Some proposed approaches focus on statistical analysis through analyzing the attributes of users, and prediction is based on rules [18–20]. Others propose machine learning methods [21–23]. For example, SVM, decision tree and simple neural network models are used to find overdue users. As far as we know, few works use graph data in peer-to-peer lending fraud detection problems. Graph data are involved for fraud detection on other scenario such as e-commerce in several works [1–7]. Several methods use graph neural network to learn node or graph representations [2–5]. Hu et al. [2] proposes a feature engineering model based on graph convolutional network [8]. Zhang et al. [3] maps the attributed heterogeneous information network to multiple single-view attributed graphs and employs graph convolutional network [8] to learn embeddings of each single-view attributed graph. Li et al. [5] detects spam advertisements based on GCN [8], while a heterogeneous graph and a homogeneous graph are integrated to capture local context and global context of comments. However, most of the existing works neglect the interpretability of models, and cannot explain the reason why users are overdue if they are used to solve the P2P lending fraud detection problem studied in this paper.

Recently, attention mechanism has been widely used in graph neural network to enhance the performance and add interpretation to the models [9–16]. Model GAT [9] introduces attention to graph neural network model, achieving good performance on most of graph classification problems. Some works pay attention to using edge attributes to learn node or graph representations [11–16]. Most of them don't use original edge features and edge features are represented by neighbor nodes. Based on GAT [9], Gong and Cheng [11] design a graph attentive network EGNN(A) using original edge features. For each edge feature, EGNN(A) aggregates the representation of edge features to calculate node attention. However, EGNN(A) doesn't discriminate the importance of neighbors and different edge features. The influence of different edge features on node representation is ignored.

In this paper, we want to solve the above-mentioned issues, and improve graph neural network for fraud detection in P2P lending, through designing new attention mechanism to distinguish the importance of neighbors and the importance of different edge features.

3   Problem Definition

In this section, we first introduce the dataset on which we study the fraud detection problem and data pre-processing methods. Then we define the problem of fraud detection studied in this paper.

3.1   Data

In existing financial fraud detection studies, we usually perform prediction based on user's structured demo graphic information. In this paper, we do not use this kind of information. Instead, we want to study how to use user's behavior information or social information for the detection task and if these kinds of information are helpful for the task. For each user, we collect two kinds of information. One is about user's transaction description information in e-commerce websites, a short text describing the purchased item in the past six months. The other is about user's communication information including phone call or text messaging information.

To deal with the text information, we regard all text information of each user as a document. The document is then preprocessed through word segmentation and is finally represented as an $1 \times M$ multi-hot vector, where M is the number of distinct words in all documents. If a word is appeared in the document, the figure in the corresponding dimension of the vector is set to 1, otherwise 0. Moreover, we get the word vector ω_i using Word2Vec [24] tool for each word w_i based on the document dataset.

Based on the communication information, we derive relationship between users. We classify user relationship into two types: direct relation and indirect relation. Two users have direct relationship if they ever communicated with each other through phone call or text messaging. Two users have indirect relationship if they both communicated with someone who is not a user through phone call or text messaging, or they left the same phone number as contact phone number for transactions in e-commerce platforms. We build a graph G to reflect the user's relationship. In the graph, nodes represent users and edges between users mean they have certain kinds of relation: direct and/or indirect relation. Each edge has two binary features: direct relation and indirect relation, as an edge may simultaneously has two kinds of relationship. For each node, we use the multi-hot vector derived from the user's document as the node's original feature.

3.2 Problem Definition

Let N be the number of users and N_{net} be the number of users who are in the relationship graph G. Let M be the number of distinct words in all documents and let ω be an $M \times F$ embedding matrix of words, where each row $\omega_i \in \mathbb{R}^F$ ($i = 1, 2, \ldots, M$) represents the embedding vector of word w_i. We use an $N \times M$ matrix d to represent document corpus, where d_i ($i = 1, 2, \ldots, N$) denotes document i, which is a multi-hot vector. Let P be the number of edge features in the graph and $E \in R^{N \times N \times P}$ be a tensor representing the edge feature values of the relationship graph G, where E_{ijp} ($i = 1, 2, \ldots, N_{net}, j = 1, 2, \ldots, N_{net}$) denotes the value of the pth edge feature between node i and node j. The subscript $\cdot$ is used to select the whole range (slice) of a dimension. Then, $E_{ij.} \in \mathbb{R}^P$ means the P-dimensional feature vector of the edge between node i and node j, where each dimension is the value of corresponding edge feature. Moreover, we use $\mathcal{N}_i$ ($i = 1, 2, \ldots, N_{net}$) to represent the set of neighbors of node i in graph G and make $\mathcal{N}_{ip}$ ($i = 1, 2, \ldots, N_{net}, p = 1, 2, \ldots, P$) be the set of neighbors of node i with edge feature p in graph G.

In this paper, we want to use both graph data and text data to get the prediction $\hat{y}_i \in \{0, 1\}$ for each user i. Let $y_i \in \{0, 1\}$ represents the true value, i.e., whether user i has overdue behavior in loan repayment. Meanwhile, we aim to devise a framework, in which users who don't have graph data can be classified by the text information.

4 Graph Attentive Network Model for Fraud Detection

4.1 Architecture Overview

Our proposed model is illustrated in Fig. 1, where different users are analyzed through different paths as shown by the orange solid line and the green dashed line. For user j who only has text data (the orange solid line in Fig. 1), document vector d_j is inputted and analyzed through a text processing module to get a new representation vector X_j, and then classification is performed through the classification module to get the prediction y_j.

Meanwhile, for user i who has both graph data and document vector d_i, analysis is performed by two paths as shown by the green dashed lines in Fig. 1. Document vector d_i is transformed to vector X_i by the text processing module. Then, graph information such as $\mathcal{N}_i$, E_i and X_i are inputted to the graph attentive network with edge features and node features and processed with L layers. In the end, node representation vector $h_i^{(L)}$ is produced. Finally, the final prediction result y_i is obtained through classification module. To improve interpretability, attention mechanisms are designed for both text processing module and the graph attentive network model.

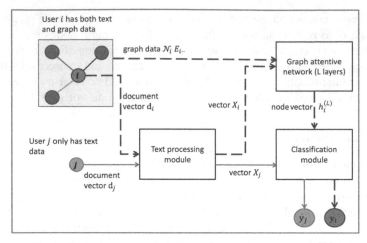

Fig. 1. The framework of the FDNE model (Color figure online)

4.2 EGNN(A): Attention Based EGNN Layer

In model EGNN(A) [11], edge features are used in attention mechanism. For each layer $l + 1$, user i's node representation vector $h_i^{(l+1)}$ is obtained from aggregation on its neighbor's node representation vectors, i.e. $\{h_j^{(l)}, j \in \mathcal{N}_i\}$. Therefore, for each neighboring node j and each feature p, node representation vector $h_j^{(l)}$ and edge feature $E_{ijp}^{(l)}$ are incorporated. First, attention vector $\hat{a}_{ijp}^{(l+1)}$, which represents the influence neighboring node j makes on node i is calculated:

$$\hat{a}_{ijp}^{(l+1)} = \exp\left\{\sigma\left[E_{ijp}^{(l)} v^T\left(W^{(l+1)}h_i^{(l)} \| W^{(l+1)}h_j^{(l)}\right)\right]\right\} \tag{1}$$

where σ is a non-linear activation, v is a parameter vector, $W^{(l+1)}$ is a parameter matrix, and $\|$ means concatenation operation. Then, the attention is normalized by a doubly stochastic normalization function $DS(\cdot)$ and node feature and edge feature are both updated as shown in Eq. (2) and Eq. (3) respectively.

However, EGNN(A) doesn't discriminate the importance of neighbors and different edge features. In this model, edge feature representation is initialized by the original edge feature value and is updated through each layer. It is designed on the level of each edge, neglecting influence on the level of different edge features.

$$h_i^{(l+1)} = \sigma\left\{\|_{p=1}^P \sum_{j \in N} DS(\hat{a}_{ijp}^{(l+1)})W^{(l+1)}h_j^{(l)}\right\} \tag{2}$$

$$E_{ijp}^{(l+1)} = DS(\hat{a}_{ijp}^{(l+1)}) \tag{3}$$

To solve this issue, we design a novel attention called edge-feature attention to take the influence on the level of different edge features into consideration and use a global normalization operation to know which feature of the edge is more influential.

4.3 Graph Attentive Network with Edge-Feature Attention

The edge-feature attention is designed for graph attentive network model to learn node representations in this paper. For each node i, we first use the representation of i's neighboring nodes to embed each edge feature, and then we calculate the influence each edge feature makes on the representation of node i. The result we get is the edge-feature attention and we use it for updating edge feature representation (as shown in Fig. 2).

For the graph attentive network model, given the user relationship graph G with node feature representation vector X and edge feature representation E, through L layer's convolution operation, we get node representation vector $h^{(L)}$.

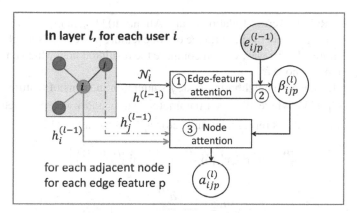

Fig. 2. Illustration of graph attentive network module

In layer l, we want to obtain node representation $h_i^{(l)}$ for each node i. For each edge feature p, we calculate edge-feature attention to update edge feature representation. Let $N_p(i)$ be the set of i's neighboring nodes with which node i has edge feature p with value 1. First, we represent the embedding of edge feature p, denoted by $h_{ip}^{(l)}$, as the sum of i's neighboring node representations $h_j^{(l-1)}(j \in N_p(i))$ as shown in Eq. (4). A is adjacency matrix, D is degree matrix and I is unit matrix in Eq. (5) and Eq. (6).

$$h_{ip}^{(l)} = \sum_{j \in N_p(i)} \bar{A}_{ij} h_j^{(l-1)} \tag{4}$$

$$\bar{A} = D^{-\frac{1}{2}} \tilde{A} D^{-\frac{1}{2}} \tag{5}$$

$$\tilde{A} = A + I \tag{6}$$

Then, edge-feature attention $\alpha_{ip}^{(l)}$ is calculated based on node representation vector $h_i^{(l-1)}$ and the embedding of edge feature p $h_{ip}^{(l)}$ in Eq. (7),

$$\alpha_{ip}^{(l)} = \sigma \left\{ \mathbf{u}_p^T \left[W_1^{(l)} h_i^{(l-1)} \big\| W_1^{(l)} h_{ip}^{(l-1)} \right] \right\} \tag{7}$$

where $W_1^{(l)}$ is parameter matrix, u_p is a parameter vector, $||$ means concatenate operation, and σ denotes the activation function.

For the edge between neighboring node j and node i, we obtain new edge feature representation $\beta_{ijp}^{(l)}$ using $\alpha_{ip}^{(l)}$ and $e_{ijp}^{(l-1)}$ $(e^{(0)} = E)$ below.

$$\hat{\beta}_{ijp}^{(l)} = e_{ijp}^{(l-1)} \alpha_{ip}^{(l)} \tag{8}$$

$$\beta_{ijp}^{(l)} = \frac{\exp\left(\hat{\beta}_{ijp}^{(l)}\right)}{\sum_{r=1}^{P} \sum_{v=1}^{N} \exp\left(\hat{\beta}_{ivr}^{(l)}\right)}, \forall i \sum_{p=1}^{P} \sum_{j=1}^{N} \beta_{ijp}^{(l)} = 1 \tag{9}$$

As used in ARGAT (Across-Relation Graph Attention) [13], in Eq. (9) we use global softmax function to measure the importance of edge feature p over all i's neighbor nodes and all edge features. In this way, we can compare the result to know which edge feature is more important for node i.

We then calculate each neighboring node j's attention in terms of feature p through Eq. (10) and Eq. (11), where $W_2^{(l)}$ is a parameter matrix and v is a parameter vector. Finally, we get the node i's representation vector $h_i^{(l)}$ based on Eq. (12).

$$\hat{a}_{ijp}^{(l)} = \sigma\left[\beta_{ijp}^{(l)} v^T \left(W_2^{((l))} h_i^{(l-1)} || W_2^{(l)} h_j^{(l-1)}\right)\right] \tag{10}$$

$$a_{ijp}^{(l)} = \frac{\hat{a}_{ijp}^{(l)}}{\sum_{k \in N_p(i)} \hat{a}_{ikp}^{(l)}} \tag{11}$$

$$h_i^{(l)} = \sigma\left\{ ||_{p=1}^{P} \sum_{j \in N} a_{ijp}^{(l)} W_2^{(l)} h_j^{(l-1)} \right\} \tag{12}$$

4.4 Text Processing Module

To extract information from documents, we design a simple neural network with attention model, which is illustrated in Fig. 3. First, we use topic model LDA [17] to get topic word distribution. For each topic, we choose the top 10 words with highest probability and average these words' vector obtained through word embedding model to represent the topic. For each word w_k in document i, based on topic word distribution, we choose the topic for which w_k has the highest probability among all the topics and use this topic's vector to represent the word, denoted as $t_k \in \mathbb{R}^F$.

For each word w_k, we concatenate its topic vector t_k with word vector ω_k as its final representation. Let V be the matrix with the kth row being the final representation of word w_k. For each word k, we calculate $\hat{a}_k'$ via a single fully connected layer in Eq. (13), where u_v is a parameter vector and b_v is bias. For document i, we then normalize $\hat{a}_k'$ to get the word attention a_{ik}' in Eq. (14). d_{ij} is the value in j^{th} dimension of d_i. We get

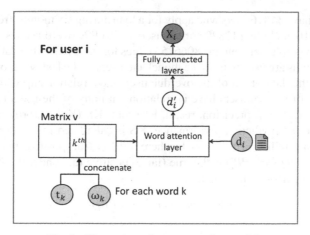

Fig. 3. Illustration of text processing module

the new representation d_i' of document i in Eq. (15), and d_i' is further processed by two fully-connected layers to get the final representation X_i for each user i.

$$\hat{a}_k' = u_v^T v_k^T + b_v \tag{13}$$

$$a_k' = \frac{d_{ik}\hat{a}_k'}{\sum_{j=1}^{|M|} d_{ij}\hat{a}_k'} \tag{14}$$

$$d_i' = \sum_{j=1}^{|M|} a_{ij}'v_j \tag{15}$$

4.5 Classification Module

In the classification module, we use two fully connected layers to get the prediction for each user. Tanh and Sigmoid activation function are used for the two layers respectively. Loss function is defined below.

$$Loss = -\sum_{i=1}^{N} y_i log\hat{y}_i + (1 - y_i)log(1 - \hat{y}_i) \tag{16}$$

5 Experiments

5.1 Datasets and Evaluation Measure

We conduct experiments on a real dataset from a peer-to-peer lending company and evaluate our proposed model's performance on finding out users who will be overdue in six months. We have 34816 users who apply for a loan during six months from June 2016 to November 2016, including 17820 normal users and 16996 overdue ones. We construct user relationship graph that contains 8066153 edges for the users in the dataset. Users in the first four months are used to train the model and users in the last two month forms the test data set. In the dataset, all of the overdue users have relationship with other users, but around 60% of normal users have no relation with any of the users in the training dataset. We use accuracy, precision, recall, F1-score, KS (Kolmogorov-Smirnov) and AUC to evaluate the prediction performance. KS is calculated in Eq. (16) to measure the distinguishing ability of the model, where TPR is true positive rate and FPR is false positive rate. TPR_j (FPR_j) is the true (false) positive rate under the jth probability threshold used to perform classification.

$$KS = max_j(TPR_j - FPR_j) \tag{17}$$

5.2 Performance Comparison

For all the experiments, we implement the algorithms in Python within Tensorflow framework. Following the experiment settings of [8, 9, 11], we set the number of layers in graph attentive network as 2 in all of experiments for fair comparison. Throughout the experiments, we use the SGD optimizer with learning rate 0.1 to optimize the loss function. An early stopping strategy with window size of 100 is adopted. We fix the dimension to 32 for all hidden layers in graph attentive network. We use GCN [8], GAT [9], EGNN(A) [11] as baseline models. To further investigate the effectiveness of each component of our proposed model, we conduct ablation study.

The name of the baseline model with '-a' means that the text processing method is the neural network with attention designed in this paper, while the name of the model which contains '-f' means that we process text using fully connected layers without word attention and topic information. Meanwhile, model 'FDNE-noedatt' is a simplified model by removing edge-feature attention from our model and 'FDNE-noglobal' uses softmax normalization for edge-feature attention instead of global normalization. The experimental results to compare different models are given in Table 1.

From Table 1, serval interesting phenomena can be observed. Overall, our approach 'FDNE' outperforms three baselines in all metrics. Especially for Recall, it reaches to 0.951, much higher than any of the baseline model. 'EGNN(A)-a' and 'FDNE' have better performance than 'GCN-a' and 'GAT-a', which means that aggregating edge features to calculate node attention has strong influence on the prediction. Meanwhile, comparison of 'FDNE' with 'EGNN(A)-a' demonstrates the effectiveness of edge-feature attention. Moreover, almost all models using designed neural network with attention to process text information outperform their corresponding baselines. For example, 'FDNE' gets better performance than 'FDNE-f' in terms of all evaluation measures except precision. It proves that word attention plays an important role in the prediction task. Meanwhile, comparing 'FDNE' with 'FDNE-noglobal', we can see that global normalization for edge feature representation also contributes to the prediction. We further evaluate our proposed model based on top-n precision, which is shown in Table 2. We can see that the top 10 users detected by our model are all true fraud users. Overall, our model has the best performance among the compared models.

Table 1. Classification performance comparison

Models	Precision	Recall	F1-score	Accuracy	KS	AUC
GCN-f	0.652	0.832	0.731	0.698	0.405	0.757
GCN-a	0.727	0.856	0.786	0.770	0.578	0.823
GAT-f	0.655	0.817	0.727	0.696	0.403	0.776
GAT-a	0.734	0.885	0.802	0.784	0.590	0.825
EGNN(A)-f	0.737	0.899	0.810	0.791	0.601	0.824
EGNN(A)-a	0.736	0.916	0.816	0.796	0.608	0.829
FDNE-f	**0.740**	0.919	0.820	0.800	0.622	0.830
FDNE-noedatt	0.721	0.889	0.797	0.776	0.557	0.823
FDNE-noglobal	0.734	0.882	0.801	0.784	0.588	0.825
FDNE	**0.740**	**0.951**	**0.832**	**0.811**	**0.634**	**0.835**

Table 2. Top-n precision comparison

Models	Top10	Top20	Top30	Top50	Top100	Top200
GCN-f	0.6	0.7	0.7	0.7	0.62	0.655
GCN-a	0.6	0.75	0.633	0.7	0.71	0.74
GAT-f	0.7	0.75	0.7	0.66	0.68	0.62
GAT-a	0.6	0.75	0.733	0.72	0.71	0.72
EGNN(A)-f	0.8	0.75	0.733	0.68	0.68	0.685
EGNN(A)_a	0.8	0.77	0.75	0.80	0.73	0.735
FDNE-f	0.9	0.7	0.733	0.8	0.76	0.735
FDNE-noedatt	0.8	**0.8**	0.8	0.82	0.73	0.735
FDNE-noglobal	0.8	0.75	0.8	0.74	0.72	0.75
FDNE	1	**0.8**	**0.833**	**0.825**	**0.79**	**0.76**

5.3 Interpretability

Based on node attention, we could know which neighboring nodes are more important to a node's representation learning. For example, from the experimental results we find that a user who is classified as an overdue user by 'FDNE', has a neighbor user who is also predicted as an overdue user and has 319 overdue neighbors in the graph. This neighbor is the most important one among all the neighbors. Regarding this neighbor as a core user, we further analyze its neighbors. According to edge feature representation $e^{(L)}$, we find that direct relations are more influential than indirect ones among relations between the core user and 319 overdue neighbors. Meanwhile, we can analyze which words contribute more to predict results based on word attention in the text processing module. Through the experiment, we find that some words about online games and peer-to-peer lending apps have high attention values. It is easily to understand that users who often use peer-to-peer lending apps might have financial problems, and they may have less ability to repay. Another interesting phenomenon is that users who often recharge for their online games account are easily overdue. These information gives useful hints for the P2P lending platform to evaluate a user's credibility.

6 Conclusions

Fraud detection for peer-to-peer (P2P) lending is an important and challenging problem in both real application and research area. As far as we know, existing works seldom use graph information or neglect the interpretability of models. Meanwhile, existing studies on graph attentive network don't discriminate the importance of neighbors and different edge features. So, in this paper, we propose a new graph attentive network model for P2P lending fraud detection. We design edge-feature attention to deal with heterogeneous edges in the graph and use global normalization operation to identify influential features of each edge. Meanwhile, important neighbors and words are found through node attention and word attention in the model. Experiments on a real dataset show that our mode improve the performances over the existing approaches significantly and provide insightful explanations in the mean time.

References

1. Li, X., et al.: FlowScope: spotting money laundering based on graphs. In: National Conference on Artificial Intelligence (2020)
2. Hu, B., Zhang, Z., Shi, C., Zhou, J., Li, X., Qi, Y.: Cash-out user detection based on attributed heterogeneous information network with a hierarchical attention mechanism. In: Proceedings of the AAAI Conference on Artificial Intelligence, vol. 33, pp. 946–953, July 2019
3. Zhang, Y., Fan, Y., Ye, Y., Zhao, L., Shi, C.: Key player identification in underground forums over attributed heterogeneous information network embedding framework. In: Proceedings of the 28th ACM International Conference on Information and Knowledge Management, pp. 549–558, November 2019
4. Wang, D., et al.: A semi-supervised graph attentive network for financial fraud detection. In: 2019 IEEE International Conference on Data Mining (ICDM), pp. 598–607. IEEE, November 2019
5. Li, A., Qin, Z., Liu, R., Yang, Y., Li, D.: Spam review detection with graph convolutional networks. In: Proceedings of the 28th ACM International Conference on Information and Knowledge Management, pp. 2703–2711, November 2019
6. Ki, Y., Yoon, J.W.: PD-FDS: purchase density based online credit card fraud detection system. In: KDD 2017 Workshop on Anomaly Detection in Finance, pp. 76–84, January 2018
7. Nilforoshan, H., Shah, N.: SilceNDice: mining suspicious multi-attribute entity groups with multi-view graphs. arXiv preprint arXiv:1908.07087 (2019)
8. Kipf, T.N., Welling, M.: Semi-supervised classification with graph convolutional networks. arXiv preprint arXiv:1609.02907 (2016)
9. Veličković, P., Cucurull, G., Casanova, A., Romero, A., Lio, P., Bengio, Y.: Graph attention networks. arXiv preprint arXiv:1710.10903 (2017)

10. Thekumparampil, K.K., Wang, C., Oh, S., Li, L.J.: Attention-based graph neural network for semi-supervised learning. arXiv preprint arXiv:1803.03735 (2018)
11. Gong, L., Cheng, Q.: Exploiting edge features for graph neural networks. In: Proceedings of the IEEE Conference on Computer Vision and Pattern Recognition, pp. 9211–9219 (2019)
12. Kim, J., Kim, T., Kim, S., Yoo, C.D.: Edge-labeling graph neural network for few-shot learning. In: Proceedings of the IEEE Conference on Computer Vision and Pattern Recognition, pp. 11–20 (2019)
13. Busbridge, D., Sherburn, D., Cavallo, P., Hammerla, N.Y.: Relational graph attention networks. arXiv preprint arXiv:1904.05811 (2019)
14. Shang, C., et al.: Edge attention-based multi-relational graph convolutional networks. arXiv preprint arXiv:1802.04944 (2018)
15. Linmei, H., Yang, T., Shi, C., Ji, H., Li, X.: Heterogeneous graph attention networks for semi-supervised short text classification. In: Proceedings of the 2019 Conference on Empirical Methods in Natural Language Processing and the 9th International Joint Conference on Natural Language Processing (EMNLP-IJCNLP), pp. 4823–4832, November 2019
16. Zhou, H., Young, T., Huang, M., Zhao, H., Xu, J., Zhu, X.: Commonsense knowledge aware conversation generation with graph attention. In: IJCAI, pp. 4623–4629, July 2018
17. Blei, D.M., Ng, A.Y., Jordan, M.I.: Latent dirichlet allocation. J. Mach. Learn. Res. **3**, 993–1022 (2003)
18. Emekter, R., Tu, Y., Jirasakuldech, B., et al.: Evaluating credit risk and loan performance in online Peer-to-Peer (P2P) lending. Appl. Econ. **47**(1), 54–70 (2015)
19. Freedman, S., Jin, G.Z.: The information value of online social networks: lessons from peer-to-peer lending. Int. J. Ind. Organ. **51**, 185–222 (2017)
20. Peng, H., Zhao, H., Zhou, Y.: Will borrowing statements affect the cost of borrowing and the success rate of borrowing? A text analysis based on online lending statements. Financ. Res. **04**, 158–173 (2016)
21. Blanco, A., Pino-Mejías, R., Lara, J., et al.: Credit scoring models for the microfinance industry using neural networks: evidence from Peru. Expert Syst. Appl. **40**(1), 356–364 (2013)
22. Kang, S., Cho, S.: Approximating support vector machine with artificial neural network for fast prediction. Expert Syst. Appl. **41**(10), 4989–4995 (2014)
23. Malekipirbazari, M, Aksakalli, V.: Risk assessment in social lending via random forests Pergamon Press, Inc. (2015)
24. Mikolov, T., Chen, K., Corrado, G., Dean, J.: Efficient estimation of word representations in vector space. Computer Science (2013)

An Empirical Study on Recent Graph Database Systems

Ran Wang[1][✉], Zhengyi Yang[2], Wenjie Zhang[2], and Xuemin Lin[2]

[1] East China Normal University, Shanghai, China
rwang@stu.ecnu.edu.cn
[2] The University of New South Wales, Sydney, Australia
{zyang,zhangw,lxue}@cse.unsw.edu.au

Abstract. Graphs are widely used to model the intricate relationships among objects in a wide range of applications. The advance in graph data has brought significant value to artificial intelligence technologies. Recently, a number of graph database systems have been developed. In this paper, we present a comprehensive overview and empirical investigation on existing property graph database systems such as Neo4j, Agens-Graph, TigerGraph and LightGraph (LightGraph has recently renamed to TuGraph.). These systems support declarative graph query languages. Our empirical studies are conducted in a single-machine environment against on the LDBC social network benchmark, consisting of three different large-scale datasets and a set of benchmark queries. This is the first empirical study to compare these graph database systems by evaluating data bulk importing and processing simple and complex queries. Experimental results provide insightful observations of various graph data systems and indicate that AgensGraph works well on SQL based workload and simple update queries, TigerGraph is powerful on complex business intelligence queries, Neo4j is user-friendly and suitable for small queries, and LightGraph is a more balanced product achieving good performance on different queries. The related code, scripts and data of this paper are available online (https://github.com/UNSW-database/GraphDB-Benchmark).

Keywords: Graph database systems · Labeled property graph · LDBC benchmark

1 Introduction

Graphs are widely used to model the intricate relationships among entities in real world. The recent blossoming of modern graph technologies has brought tremendous value to artificial intelligence (AI) and machine learning (ML) applications [25]. The related context in graph data enhances AI in many ways, ranging from graph based acceleration of ML, connected feature extraction, explainable AI,

G. Li et al. (Eds.): KSEM 2020, LNAI 12274, pp. 328–340, 2020.
https://doi.org/10.1007/978-3-030-55130-8_29

and especially, to the area of knowledge engineering. For example, using context-rich knowledge graphs for recommendation at eBay [27], extracting knowledge from the Lessons Learned database at NASA [23], and combining graph features and ML for fraud detection in security and finance [12].

Graph management systems, as one of the most fundamental infrastructures in graph-based AI and ML applications, have received lots of attention both in industry and academia. Two types of graph management systems are developed to manage graph data efficiently, effectively collecting explicit and implicit information. The first one is *graph analytics systems*, performing batch computations on large clusters, such as GraphLab [22], Giraph [2] and GraphX [16], etc. The latter is *graph databases*, fulfilling primitive requirements of storing and fast querying graph data, such as Neo4j [8], JanusGraph [5] and ArangoDB [3], etc. Specifically, graph databases primarily adopt labeled property graph model or resource description framework (RDF). As RDF databases have received lots of attention in literature [11] and the recent explosive popularity of property graph databases [9] such as Neo4j, in this paper, we aim to comprehensively evaluate graph databases that support labeled property graph model. Note that our focus in this paper is on the four popular enterprise graph database systems Neo4j, AgensGraph [1], TigerGraph [10] and LightGraph [6], though there are also graph databases proposed in the literature [17,18].

A number of graph databases have been used in enterprises to handle fast-evolving graph data in AI domains. Even though some popular graph databases have already taken a large share of the market and also been studied in academic communities, in recent years, a few new graph database systems, such as Agens-Graph, TigerGraph and LightGraph, are developed to satisfy the needs of higher level business demands, lower resource consumption, and enhanced performance. Inspired by these observations, in this paper, we empirically investigate major enterprise graph databases, and take the most popular Neo4j and several young graph databases AgensGraph, TigerGraph, LightGraph as examples, to evaluate their performance in detail under various experimental settings.

In the literature, various benchmarks [15,21,26] have been proposed to evaluate the performance of graph database systems. Additionally, social network is penetrating into every aspect of daily lives and marketing activities, producing enormous information. Thus, inspired by and following existing work [24], we employ Linked Data Benchmark Council Social Network Benchmark (LDBC-SNB) [15,26] as our evaluation tool, under which our experiments evaluate four graph databases by loading three different scale datasets and processing interactive and business intelligence query workloads.

The main contributions of this paper are as follows:

- We investigate the market of recent enterprise graph database systems, and present a study of prevalent products. Due to the rapid advance in graph databases, existing works [21,24] are either outdated or incomplete.
- Take four products Neo4j, AgensGraph, TigerGraph, and LightGraph as examples, we make further investigations on graph databases. To the best

of our knowledge, this is the first work comprehensively survey and compare popular choices of modern industrial-strength graph database systems.

– We adapt all queries in the LDBC_SNB benchmark and make significant adjustments to work with all four aforementioned graph databases. Based on this, we systematically evaluate them including the performance of data loading and querying under three different data scales.

– We thoughtfully analyse and conclude the results of our unified benchmark. This provides researchers and companies with insightful advice on how to select a proper graph database system in different use cases.

The rest of this paper is organized as follows. Section 2 states necessary preliminaries. Section 3 reviews related works. Section 4 presents our investigation. Section 5 shows experimental results. Finally, we conclude the paper and discuss future research directions in Sect. 6.

2 Preliminaries

2.1 Labeled Property Graph Model

A labeled property graph (LPG) is a directed graph with *labels* and *properties*, in which labels define different subsets (or classes) of vertices and edges and properties are arbitrary number of key-value pairs representing real-world attributes. The LPG model can easily represent many complex relations in real applications and has the advantages of high expressiveness. In addition, it is nature to understand making it increasingly popular in the database community.

2.2 LDBC Benchmark

LDBC_SNB [15] models a social network akin to Facebook, consisting of persons and their activities. It is rich in entities, relationships and attribute types, simulating the characteristics of information architecture in a real-world social network. It proposes three query workloads: *Interactive, Business Intelligence*, and *Graph Algorithms*. Since the last one is applicable for graph analytics systems which is out of the scope of this paper, we only take the first two as our testing objects, covering the aspects of evaluating the ability of commercial graph databases in processing actual business requirements.

Interactive workload [15] defines user-centric transactional-like interactive queries, consisting of 8 transactional update (IU), 7 simple read-only (IS) and 14 complex read-only (IC) queries. Among them, IU queries are simple vertex or edge insertions. Both of IS and IC queries are information retrieval operations, starting with a specific vertex. The difference is that IS targets at simple pattern matching, accessing at most 2-hop vertices and collecting basic data, while IC defines complex path traversal and returns computed and aggregated results. Business intelligence (BI) workload [26] defines 25 analytic queries to respond to business-critical questions. Different from interactive workload, BI queries start with multiple vertices and contains more complex graph analysis and result statistics operations, aiming at collecting valuable business information.

The structure of LDBC-SNB dataset and detailed definitions of workloads can be found in the official specification [20].

3 Related Work

With the development of graph databases, many evaluation and benchmark endeavors are conducted to compare these products. For existing evaluation works, some [13,19] only focus on listing the characteristics of graph databases without any empirical exploration, some [14,21] restrict evaluation on micro operations, small scale data and a limited set of products, and some [24] aim at present the superiority of their own products. Meanwhile, many existing benchmarks cannot evaluate the ability of graph databases in processing actual and business requirements. For instance, micro-benchmark [21] is limited in micro operations, and RDF-benchmark [11] is only suitable for finding RDF structures.

The experimental study in our paper fills the gap. We not only conclude the characteristics of prevalent and young enterprise graph databases, but also evaluate the performance of representative products on large-scale datasets, employing benchmark LDBC-SNB which simulates a realistic social network and defines both of micro and macro query workloads.

4 Graph Database

Section 4.1 reviews the features of enterprise graph database systems. Then, a detailed introduction of evaluated databases is presented in Sect. 4.2.

4.1 An Overview of Graph Databases

Graph database is a type of NoSQL database emerged as a response to the limitations of traditional relational databases. Graph databases are highly optimised for handling connected data. Popularity of graph databases dramatically increases in recent years because of their high performance and the ease of use. We give an overview of graph database systems in the current commercial market, including well-developed graph and multi-model database systems, such as Neo4j, JanusGraph (successor of Titan) and ArangoDB, and other relatively young systems, such as AgensGraph, TigerGraph, LightGraph and Nebula [7]. Their basic features are concluded in Table 1, including database type, storage structure, open source or not (Op.), supporting distributed processing or not (Ds.), transactional or not (Ta.), schema-free or not (Sf.), implementation languages (Impl.) and query languages (Lang.).

From Table 1, we can see that, as enterprise products, these graph database systems are almost all transactional with ACID guarantees. Moreover, most native graph databases prefer self-designed storage structures and graph query languages, while hybrid databases are more flexible and support multiple storage engines or query languages. Besides, with the increasing requirements for processing large data, many products target at high scalability, supporting distributed storage of data and can be deployed in a cluster of machines. Among all studied graph databases, only TigerGraph and LightGraph are not open source projects, so the studies of their storage technologies are limited.

Table 1. Basic information of graph database systems.

System	Type	Storage	Op.	Ds.	Ta.	Sf.	Impl.	Lang.
Neo4j	Native	Linked lists	Yes	No	Yes	Yes	Java	Cypher
JanusGraph	Hybrid	Cassandra/HBase	Yes	Yes	Yes[a]	No	Java	Gremlin
ArangoDB	Hybrid	MMFiles/RocksDB	Yes	Yes	Yes	Yes	C++	AQL
AgensGraph	Hybrid	PostgreSQL	Yes	No	Yes	Yes	C	Cypher,SQL
TigerGraph	Native	Native engine	No	Yes	Yes	No	C++	GSQL
LightGraph	Native	Native engine	No	No	Yes	No	C++	Cypher
Nebula	Native	RocksDB	Yes	Yes	No[b]	No	C++	nGQL

[a]The transaction isolation levels in JanusGraph is determined by the choice of storage.
[b]The transactional support in Nebula is still under development.

Then we choose several systems to conduct further research and performance evaluation based on the benchmark LDBC_SNB. To summary, the selection criteria are: 1) support the labeled property graph model, 2) support declarative graph query languages, 3) support OLTP, 4) could fully implement query workloads in LDBC_SNB, and 5) full licence available. Therefore, we take Neo4j, AgensGraph, TigerGraph and LightGraph to investigate and benchmark. Especially the last three, as young graph databases, get little attention in existing work even though they demonstrate promising performance in their own reports.

4.2   Graph Databases Details

Neo4j. Neo4j [8] is the most popular graph database system. Compared with other products, mature community is one of its biggest advantages. It provides user-friendly web interface and APIs and supports many third-party applications, frameworks and programming language drivers. Neo4j also developed the graph query language Cypher, which is a widely-used and easy-reading language. However, Neo4j typically cannot scale to large graphs as it dose not support data sharing in distributed environment. Even in the enterprise version, the distributed mode just replicate the data in all machines to achieve high availability. With the storage structure of linked lists, Neo4j stores vertices, edges and attributes natively and separately. This design usually causes high memory consumption and low efficiency, even if indexes are created [24].

AgensGraph. AgensGraph [1] is proposed as a new generation multi-model graph database. As a multi-model database, AgensGraph supports graph, relational, document and key-value data models at the same time, adopting SQL and Cypher as query languages. Thus, it can integrate the two languages in one single query, supporting more flexible use cases and very friendly to experienced SQL-users. AgensGraph adopts the PostgreSQL RDMS as storage engine. However, when processing large data, it will take a very long time to load and consume large memory to store data. In the process of querying complex requirements, a large temporary space will be occupied. AgensGraph is a developing product and cannot support all grammars in Cypher. In this work, we implement queries with high complexity in LDBC_SNB by integrating Cypher and SQL.

TigerGraph. TigerGraph [10] is one of the rising stars in distributed graph database in recent years. Its core system is developed from scratch using C++. By combining native graph storage with MapReduce, massively parallel processing and fast data compression/decompression, TigerGraph shows strong scalability and great performance, especially in handling complex queries. More importantly, TigerGraph can be easily deployed to a large scale of clusters and process queries distributedly, allowing it to answer queries on extremely large graphs that are likely to fail in a single machine setting. It also develops its own query language GSQL, which is a powerful and procedure-like language. However, as a non open source commercial product, TigerGraph is not freely available.

LightGraph. LightGraph [6] is a high-performance graph database product, implemented in C++. As an enterprise-focused product, LightGraph is still under development, but has already demonstrated impressive strength. LightGraph supports storing and querying billion-scale data in single machine. Meanwhile, it adopts a lockless design, greatly improving the throughput under high loads and enabling queries to be processed with high parallelism. Although LightGraph provides Cypher interface, it still cannot support most grammars. Therefore, for complex querying requirements, it suggests to implement stored procedures by Python or C++ APIs. Please note that, we only implement IU and IS micro queries with Cypher, and for IC and BI queries in the benchmark, we directly use the plug-in offered by the authors to run.

Other Databases. JanusGraph [5], previously known as Titan, is a massively scalable graph database and natively integrates with the Apache TinkerPop framework. As a hybrid system, it can integrate with third-party systems like HBase, Cassandra, ElasticSearch for storing and indexing the graph structures. ArangoDB [3] is a native multi-model database, supporting graph, document and key-value models, and provides a unified database query language to coverer three models in a single query. Nebula [7] is a newly released high-performance and scalable graph database capable of hosting very large scale graphs with low latency, getting more attention recently.

However, JanusGraph adopts an imperative query language Gremlin[1]. Meanwhile, both query languages in Nebula and ArangoDB are still developing and not fully compatible with all LDBC_SNB query workloads at this moment. Since only Cypher and GSQL are in the process towards an upcoming international standard language for property graph querying [4], we only consider Cypher- and GSQL-based systems, that is Neo4j, AgensGraph, TigerGraph and LightGraph.

5   Empirical Studies

Section 5.1 presents the experimental setup. Then, we discuss experimental results in Sect. 5.2. After that, an overall evaluation is given in Sect. 5.3.

5.1   Setup

We generated three datasets with different scales by using LDBC data generator [15]. All datasets have the same structure [20]. The statistics of datasets are summarized in Table 2. We can find that the number of vertices and edges and the raw size of datasets increase almost linearly with scale factor.

Table 2. Statistics of datasets.

Dataset	Scale factor	$|V|$ (Million)	$|E|$ (Million)	Size (GB)
DG1	1	3.182	17.256	0.798
DG10	10	29.988	176.623	8.257
DG100	100	282.638	1775.514	85.238

For benchmark LDBC_SNB, there are overall 54 queries in interactive and business intelligence workloads. Their implementations in four graph databases are acquired from their official staff or implemented by ourselves, like BI queries in AgensGraph, IU and IS queries in LightGraph, guaranteeing the optimality of query statements in each database as much as possible. To conduct a fair comparison, we take the same parameters for the same query in all systems. Note that, for LightGraph, as its Cypher interface has not been fully implemented and optimized, we only implemented IU and IS queries with Cypher, and used C++ stored procedures provided by the authors to do other experiments. The other three databases were operated by query languages.

We conducted all experiments on a machine with two 20-core processors Intel Xeon E5-2680 v2 2.80 GHz, 96 GB main memory, and 960G NVMe SSD, running Ubuntu 16.04.5 operating system. We tested performance on fast NVMe SSD storage, as it is already a default option for database systems. To perform experiments, we ran each micro query (IU and IS) 100 times and each macro query (IC and BI) 3 times, setting mean running time as result. By default, we set timeout as 1 h for each query processing, and use symbols 'TO' and 'OOM' to represent timeout and out of memory respectively.

---

[1] Gremlin has limited degree of declarative support.

5.2   Experimental Results

Data Importing. Due to the large size of raw data, datasets were bulk imported into graph database systems. Data importing is also a common operation in production. We present the data importing time of four databases and the storage size of each loaded dataset respectively in Fig. 1.

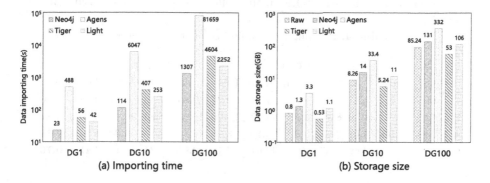

(a) Importing time   (b) Storage size

Fig. 1. The bulk importing time and storage size of datasets.

Figure 1-a shows that, for all three scale datasets, the data importing time in Neo4j is the shortest, followed by LightGraph, TigerGraph and AgensGraph. Actually, even if we consider the indexes creating time in Neo4j and preprocessing time of stored procedures in LightGraph, both of them still present better bulk importing performance than TigerGraph and AgensGraph. AgensGraph, limited by its third-party relational storage engine PostgreSQL, the speed of data bulk importing is much slower than other systems. For DG100, AgensGraph will take about one day to load data, unreasonable in common usage.

Figure 1-b presents the raw and loaded storage sizes for three datasets in four databases, showing that TigerGraph requires the least space to store data, even less than raw data, followed by LightGraph, Neo4j and AgensGraph. The storage consumption in LightGraph is about twice as much as that in TigerGraph, Neo4j is about three times large. AgensGraph requires the largest storage space.

Based on the above experiments, Neo4j performs best in the efficiency of large data importing, but its storage cost is very large. TigerGraph costs least space to store data, but take a little long time to load data. LightGraph presents good performance overall. AgensGraph shows the worst performance.

Processing Interactive Queries. Interactive workload consists of IU, IS and IC queries. Among them, both of IU and IS are micro operations. IC are macro queries under more complex interactive scenarios. We select several representative results to present and full data are available online. Since the running time of IU and IS operations is not more than one second, we record them in milliseconds, and record the running time of IC queries in seconds.

Table 3. Running time (millisecond) for IU queries.

IU	DG1				DG10				DG100			
	Neo4j	Agens	Tiger	Light	Neo4j	Agens	Tiger	Light	Neo4j	Agens	Tiger	Light
2	4.62	**0.46**	8.14	0.56	4.45	**0.52**	9.81	0.63	4.13	**0.54**	10.36	0.98
4	30.50	8.47	8.55	**1.02**	40.56	7.94	8.52	**1.07**	38.41	9.75	8.90	**1.07**
6	5.02	2.08	8.38	**1.39**	16.10	3.53	8.99	**1.58**	12.01	2.99	9.72	**1.52**
8	1.14	**0.42**	8.18	0.69	1.13	**0.50**	8.45	0.56	1.28	**0.50**	9.40	0.61

Table 4. Running time (millisecond) for IS queries.

IS	DG1				DG10				DG100			
	Neo4j	Agens	Tiger	Light	Neo4j	Agens	Tiger	Light	Neo4j	Agens	Tiger	Light
1	1.05	1.83	3.47	**0.60**	1.45	1.70	3.27	**0.61**	1.60	1.85	3.65	**0.66**
2	18.49	10.01	10.90	**8.90**	33.11	21.40	**11.09**	15.96	25.80	26.20	**9.91**	16.00
4	1.33	**0.33**	4.01	0.53	1.11	**0.34**	3.74	2.34	0.57	**0.34**	4.06	0.61
6	1.33	13.11	4.66	**0.67**	2.37	14.61	4.7	**0.65**	1.34	15.40	4.41	**0.70**

Table 3 lists the processing time of IU queries, showing that LightGraph and AgensGraph are more efficient than other two databases. LightGraph is good at inserting vertices, such as IU_4 and IU_6, and AgensGraph is good at inserting edges, like IU_2 and IU_8. Neo4j is faster than TigerGraph in updating edges, while slower in updating vertices.

The experimental results of IS queries are listed in Table 4. We find that LightGraph is the most efficient in most cases. TigerGraph and AgensGraph only show good performance in several cases, like IS_2 and IS_4, and worse than Neo4j in most cases according to the full data.

Table 5. Running time (second) for IC queries.

IC	DG1				DG10				DG100			
	Neo4j	Agens	Tiger	Light	Neo4j	Agens	Tiger	Light	Neo4j	Agens	Tiger	Light
1	0.60	0.32	**0.03**	0.06	2.36	1.17	**0.12**	0.36	10.22	8.26	**0.56**	2.48
3	2.01	0.35	0.06	**0.01**	24.06	7.95	0.37	**0.05**	616.54	63.82	1.32	**0.37**
6	2.58	0.17	0.09	**0.01**	113.92	0.36	0.31	**0.03**	TO	5.66	0.97	**0.11**
10	0.50	0.71	**0.03**	0.04	2.32	2.93	**0.06**	0.12	9.33	12.41	**0.15**	0.37
12	0.19	2.58	**0.02**	0.04	0.66	4.96	**0.06**	0.15	0.60	55.97	**0.13**	0.16
13	0.01	**0.01**	0.01	0.01	0.03	0.02	**0.01**	0.01	**0.00**	0.03	0.02	0.01
14	340.55	43.34	0.20	**0.01**	424.05	488.61	0.27	**0.02**	63.83	TO	0.31	**0.03**

Table 5 shows the running time of IC queries, there are big differences in performance between graph databases. Overall, both of TigerGraph and Light-Graph present efficient performance with little difference, while AgensGraph and Neo4j perform much worse than other two systems. In the case of executing IC_14 for DG1 and DG10, Neo4j is even four orders of magnitude slower than LightGraph. However, for IC_13, which requires to return the length of the shortest path between two given vertices, the results show little difference among all databases across all datasets. This is because both Neo4j and AgensGraph support the keyword shortestPath and optimize internal computing process.

For interactive workload, all databases work well in micro queries, but show big differences in macro queries. LightGraph performs best in most cases. Tiger-Graph is only efficient in processing complex queries, while bad at micro operations. AgensGraph and Neo4j are suitable for micro operations, wile work much worse than other two databases when processing IC queries.

Processing Business Intelligence Queries. Business intelligence workload, considering more actual business applications, requires higher performance for databases to response to complex scenarios. Table 6 lists 12 representative queries as examples, recording results in seconds.

Actually, all graph databases cannot successfully execute all BI queries across all datasets. Overall, TigerGraph performs best, followed by LightGraph. Neo4j and AgensGraph still perform bad, the results are timeout in many cases. In the case of executing BI_7 for DG1, Neo4j is four orders of magnitude and Agens-Graph is five orders of magnitude slower than LightGraph, showing big performance difference. Although there may be some reasons in query statements optimization, the biggest problem lies in the technologies of data storing and query processing. For example, for BI_16, AgensGraph is timeout even under the DG1. By splitting its original query statement, we found that AgensGraph cannot match pattern (Person)-[:KNOWS*3..5]-(Person) in a reasonable time, which maybe related to its storage structure or query execution mechanism.

Table 6. Running time (second) for BI queries.

BI	DG1				DG10				DG100			
	Neo4j	Agens	Tiger	Light	Neo4j	Agens	Tiger	Light	Neo4j	Agens	Tiger	Light
2	3.36	6.67	0.59	**0.44**	29.16	102.2	**5.27**	9.19	237.6	1388.2	**42.67**	221.8
4	1.58	0.21	**0.02**	0.03	14.50	0.62	0.12	**0.09**	173.3	4.76	**1.19**	3.57
7	375.8	1647.8	0.88	**0.04**	TO	TO	0.88	**0.74**	TO	TO	OOM	**38.63**
8	0.41	1.65	**0.03**	0.08	3.89	6.46	**0.16**	1.00	43.31	69.90	**1.32**	60.07
10	941.2	48.15	0.05	**0.03**	TO	692.2	**0.31**	0.40	TO	TO	**4.11**	33.91
13	0.77	1.39	0.18	**0.12**	5.56	14.36	1.63	**1.07**	46.35	415.5	**13.21**	82.66
16	3.36	TO	**0.49**	0.58	39.97	TO	**4.59**	6.62	429.9	TO	**50.64**	426.1
17	0.41	0.27	0.01	**0.01**	34.65	3.55	**0.03**	0.06	TO	999.1	**0.20**	0.88
18	6.44	352.6	**0.37**	2.95	76.00	TO	**4.71**	58.02	700.1	TO	**52.57**	1742.2
20	4.73	36.53	1.21	**0.84**	44.53	255.1	**14.69**	30.94	732.7	TO	OOM	**290.8**
23	0.18	0.24	**0.02**	0.05	1.55	1.20	**0.06**	0.37	13.44	11.22	**0.51**	50.10
24	16.25	29.96	0.78	**0.66**	191.5	373.2	**7.57**	14.73	TO	TO	**75.71**	1198.7

In the comparison of TigerGraph and LightGraph, for DG1, their performances show little difference. But with the increasing scale of datasets, the processing time in LightGraph increases faster than that in TigerGraph. Thus, TigerGraph is much more efficient under DG10 and DG100, in part, thanks to is build-in parallel processing mechanism and stored procedure-like query statements implemented by GSQL. However, for BI_7 and BI_20 in DG100, TigerGraph is out of memory, but LightGraph finishes them successfully, showing that TigerGraph requires more memory when processing large datasets.

5.3   Overall Evaluation

We only list part of experimental results, full data is available online. According to full results, the four databases present different performance and no one can perform best in all scenarios. The findings can be summarized as follows:

- Neo4j is user-friendly and the most efficient one in data importing. However, it is only suitable for micro queries and small-scale datasets, and shows bad performance in running complex business intelligence queries.
- AgensGraph works well on SQL accompanied workload and simple update and query operations, while performs very bad in processing complex queries and managing large datasets.
- TigerGraph is powerful on complex business intelligence queries, such as IC and BI queries, especially on large datasets like DG100. It need the least memory to store data, although takes a little long time to load data.
- LightGraph is a more balanced product, achieving good performance on all types of queries. However, it cannot fully support Cypher grammars, and complex queries only can be implemented by stored procedures.

Although we cannot get a closer look at the implementation details of closed source products TigerGraph and LightGraph, we list three potential reasons of their performance advantages over community-driven graph databases, namely, Neo4j and AgensGraph. The first reason is the language differences in implementation as TigerGraph and LightGraph are implemented in C++, which generally shows a greater performance than the language implements Neo4j, namely, Java. The second reason is the other two systems are schema-free, which TigerGraph and LightGraph both have fixed schema, which allows more optimizations to be done. The third reason is commercial products tend to use advanced algorithms and optimizations to improve their efficiency. Finally, as for the hybrid system, AgensGraph, the extra layer to its underlaying relational database encounters significant extra costs to the graph database overall.

These four graph database systems are all business products, stable, professional and suitable for processing actual and business query requirements. According to our experience in usage, Neo4j and AgensGraph are more user-friendly, supporting more complete query language grammars. The query statements of TigerGraph and LightGraph are more like stored procedures, requiring some technologies in expressing queries. Most importantly, the two products are not free available, users cannot change or optimize by themselves.

6 Conclusion and Future Work

In this paper, we investigate and conclude some graph database systems. Moreover, we present a further research and evaluation on Neo4j, AgensGraph, TigerGraph, LightGraph. Based on the benchmark LDBC_SNB and single machine environment, experiments across three different scale datasets show that LightGraph and TigerGraph have significantly better performance in managing large data and processing high complexity queries. Neo4j and AgensGraph are suitable for simple micro operations and give friendly use experience. In the future work, we will extend our study to more graph database products, and evaluate their performance under distributed environment.

Acknowledgments. Xuemin Lin is supported by 2019B1515120048, 2018AAA-0102502 and 2018YFB1003504.

References

1. Agensgraph. https://bitnine.net/
2. Apache giraph. http://giraph.apache.org
3. Arangodb. https://www.arangodb.com/
4. Gql. https://www.gqlstandards.org/
5. Janusgraph. https://janusgraph.org/
6. Lightgraph. https://fma-ai.cn/
7. Nebula. https://nebula-graph.io/cn/
8. Neo4j. https://neo4j.com/
9. Ranking of graph dbms. https://db-engines.com/en/ranking/graph+dbms
10. Tigergraph. https://www.tigergraph.com/
11. Abdelaziz, I., Harbi, R., Khayyat, Z., Kalnis, P.: A survey and experimental comparison of distributed sparql engines for very large RDF data. Proc. VLDB Endow. **10**(13), 2049–2060 (2017)
12. Akoglu, L., Tong, H., Koutra, D.: Graph based anomaly detection and description: a survey. Data Min. Knowl. Disc. **29**(3), 626–688 (2014). https://doi.org/10.1007/s10618-014-0365-y
13. Besta, M., et al.: Demystifying graph databases: analysis and taxonomy of data organization, system designs, and graph queries. arXiv preprint arXiv:1910.09017 (2019)
14. Ding, P., Cheng, Y., Lu, W., Huang, H., Du, X.: Which category is better: benchmarking the RDBMSs and GDBMSs. In: Shao, J., Yiu, M.L., Toyoda, M., Zhang, D., Wang, W., Cui, B. (eds.) APWeb-WAIM 2019. LNCS, vol. 11642, pp. 207–215. Springer, Cham (2019). https://doi.org/10.1007/978-3-030-26075-0_16
15. Erling, O., et al.: The LDBC social network benchmark: interactive workload. In: Proceedings of the 2015 ACM SIGMOD, pp. 619–630 (2015)
16. Gonzalez, J.E., Xin, R.S., Dave, A., Crankshaw, D., Franklin, M.J., Stoica, I.: Graphx: graph processing in a distributed dataflow framework. In: Proceedings of OSDI 2014, pp. 599–613 (2014)
17. Hao, K., Yang, Z., Lai, L., Lai, Z., Jin, X., Lin, X.: PatMat: a distributed pattern matching engine with cypher. In: Proceedings of the 28th ACM International Conference on Information and Knowledge Management, pp. 2921–2924 (2019)

18. Kankanamge, C., Sahu, S., Mhedbhi, A., Chen, J., Salihoglu, S.: Graphflow: an active graph database. In: Proceedings of SIGMOD 2017, pp. 1695–1698 (2017)
19. Kolomičenko, V., Svoboda, M., Mlỳnková, I.H.: Experimental comparison of graph databases. In: Proceedings of 2013 iiWAS, pp. 115–124 (2013)
20. LDBC SNB task force: The LDBC social network benchmark. Technical report, LDBC (2019). https://ldbc.github.io/ldbc_snb_docs/ldbc-snb-specification.pdf
21. Lissandrini, M., Brugnara, M., Velegrakis, Y.: Beyond macrobenchmarks: microbenchmark-based graph database evaluation. Proc. VLDB Endow. **12**(4), 390–403 (2018)
22. Low, Y., Bickson, D., Gonzalez, J., Guestrin, C., Kyrola, A., Hellerstein, J.M.: Distributed GraphLab: a framework for machine learning and data mining in the cloud. Proc. VLDB Endow. **5**(8), 716–727 (2012)
23. Meza, D.: How NASA finds critical data through a knowledge graph. https://neo4j.com/blog/nasa-critical-data-knowledge-graph/
24. Rusu, F., Huang, Z.: In-depth benchmarking of graph database systems with the linked data benchmark council (LDBC) social network benchmark (SNB). arXiv preprint arXiv:1907.07405 (2019)
25. Sahu, S., Mhedhbi, A., Salihoglu, S., Lin, J., Özsu, M.T.: The ubiquity of large graphs and surprising challenges of graph processing. Proc. VLDB Endow. **11**(4), 420–431 (2017)
26. Szárnyas, G., et al.: An early look at the LDBC social network benchmark's business intelligence workload. In: Proceedings of the 1st ACM SIGMOD Joint Workshop on GRADES-NDA, pp. 1–11 (2018)
27. Wang, X., Wang, D., Xu, C., He, X., Cao, Y., Chua, T.S.: Explainable reasoning over knowledge graphs for recommendation. In: Proceedings of the AAAI Conference on Artificial Intelligence, vol. 33, pp. 5329–5336 (2019)

Bibliometric Analysis of Twitter Knowledge Management Publications Related to Health Promotion

Saleha Noor[1] , Yi Guo[1,2,3](✉) , Syed Hamad Hassan Shah[4] ,
and Habiba Halepoto[5]

[1] School of Information Science and Engineering, East China University of Science and
Technology, Shanghai 200237, People's Republic of China
guoyi@ecust.edu.cn

[2] National Engineering Laboratory for Big Data Distribution and Exchange Technologies,
Business Intelligence and Visualization Research Center, Shanghai 200436,
People's Republic of China

[3] Shanghai Engineering Research Center of Big Data and Internet Audience, Shanghai 200072,
People's Republic of China

[4] Glorious Sun School of Business and Management, Donghua University, Shanghai, China

[5] College of Information Science and Technology, Donghua University, Shanghai, China

Abstract. This study enriches our understanding by systematically reviewing Knowledge management twitter health (KMTH) articles extracted from Web of Science (WoS) using two types of bibliometric analysis (a) citation through HistCite and (b) co-authorship analysis through VOSviewer–for the last 11 years. A total of 798 KMTH articles were found from 2009 to 2019, which were analyzed based on the most influential authors, articles, journals, institutions, and countries. Furthermore, they were analyzed base on co-authorship among different countries of KMTH articles. This study found USA and UK the most influential countries according to KMTH publications. Harris JK was found most prominent author while *Journal of internet health research* was on top ranking. This study opens new avenues for all health care providers to utilize Twitter as a KM platform to promote health care. This is the first bibliometric analysis of KMTH publications according to our best knowledge.

Keywords: Twitter · Bibliometric analysis · Health promotion · Knowledge management

1 Introduction

The word "Web 2.0" was coined first in 2004, started a new era of communication to exchange information and knowledge. Social networking Site (SNS) e.g. twitter, Facebook and YouTube are birth of web 2.0. SNS as a group of virtual communities helps to share information in form of text, pictures and video with openness and in interactive manner [15]. Use of social media (SM) is emerging as a popular way to seek health information for the general public [32, 40]. For adult Internet users, the Internet is a primary

G. Li et al. (Eds.): KSEM 2020, LNAI 12274, pp. 341–354, 2020.
https://doi.org/10.1007/978-3-030-55130-8_30

source of health information, second only to health care providers; 80% of US Internet users have looked online for health information [6, 8]. 60% of state health departments (SHDs) reported using a SM site and many are using their account(s) daily to disseminate information and knowledge on healthy behaviors and health conditions [30].

Twitter has been used in different health contexts at both individual and organizational levels platform where people can quickly and easily contribute and engage in new forms of knowledge seeking, knowledge acquisition and knowledge dissemination behavior [13, 14]. They also support traditional models of engagement such as communities of practice – CoP [10]. For instance, health-minded individuals discuss health problems with their peers and seek support from experts [28]. The participative nature of Web 2.0 levels off the playing field for knowledge sharing in CoPs. Individuals with varying degrees of expertise from medical experts to lay-persons can seek and disseminate health information [5]. For example, twitter-based community *Health Care Social Media Canada* (#hcsmca), founded in 2010, uses twitter platform for health related knowledge sharing and health promotion.

Prior studies have exhibited huge contribution of twitter as a knowledge acquisition [37], knowledge seeking [17] and knowledge dissemination [23] platform and also its undeniable role in health promotion as eHealth and m-health. Moreover, bibliometric studies of SM have been conducted with respect to different research areas e.g. event detection in SM [3], bibliometric of sentiment analysis [16] and bibliometric of SM in psychology [39] whereas some studies conducted bibliometric analysis of different health problems using SM as platform e.g. Muller *et al.* [20] provide an overview of the eHealth and mHealth research field related to physical activity, sedentary behavior, and diet. Li *et al.* [18] study purpose was to examine publication trends and explore research hot spots of Internet health information seeking behavior through Co-Word Biclustering Analysis. But there are scarcest studies that shed light on overall contribution of twitter publications in heath sector except one systematic review (but not bibliometric analysis) that highlighted the role of twitter in health care academic research [27]. However, there was no comprehensive study found to integrate twitter role in health construct at a glance especially what are significant research dynamics (e.g. most prominent authors, articles, journals, countries and institutions). To cover this gap, this study reviewed twitter articles to highlight research dynamics of health research under umbrella of twitter and allow us to answer following question. This study will serve at one-stop source to provide what role has twitter offered as so far and what is still to come.

- What is the growth output of KMTH publications?
- Which are the most productive institutions and countries of KMTH publications?
- What are the most influential articles, authors and journals of KMTH publications?
- What are the co-authorship links among countries of KMTH publications?

2 Methodology

Bibliometric analysis can help researchers to identify the origins and the current significance of a given concept [7]. This method has been applied to specific journals [29, 36] and diverse disciplines, such as Twitter [22], social media and knowledge management [21], social care data [32], prosumption [24, 25], technology assisting people with dementia [2] and sedentary behavior of diet-related eHealth and mHealth Research [20].

2.1 Inclusion Criteria for Related Research Articles

We started by query string for: topic: "twitter" AND topic: "health" in the WoS Core Collection database and got 1300 articles. This database covered timespan of all years (1985 - march 2019) and consisted of Science Citation Index Expanded (SCI-EXPANDED), Social Science Citation Index (SSCI) and Conference Proceedings Citation Index - Science (CPCI-S). As we were more interested to explore the KM contribution of twitter in health sector, therefore we refined the above query string with key worlds "information", "knowledge" and "awareness" and consequently got 695, 144 and 119 research articles respectively. From the previous literature we also found that E-health and M-health are emerging research concepts that are dramatically involving SM especially twitter in health promotion and knowledge seeking and disseminating activities [9]. Therefore we searched for: topic: "e-health" OR topic: "ehealth" and refined by: topic: "twitter" and got 30 research articles. Finally, we also searched for: topic: "m-health" OR topic: "mhealth" and refined by: topic: "twitter" and got 12 research articles.

2.2 Exclusion Criteria for Related Research Articles

After this we extracted these five data files comprising of 695, 144, 119, 30 and 12 research articles from WoS in the form of plain text file and saved it into text document file. Now we have total 1000 articles but there might be some articles that may duplicate in different data set. To delete the duplication of the research articles, we imported these all file into Hiscite software and finally got 798 articles comprising all necessary information to conduct bibliometric analysis (title, abstract, authors' name, language used in writing paper, type of document, and cited reference used in the paper etc.). Using Hiscite software, we also got the most influential authors, articles, journals, research document, language used in paper, the most influential institutions, countries and Citation Mapping as an output information related to our study (Garfield, Pudovkin and Istomin [10]). Furthermore, we imported these files in VOSviewer software to get a bibliographic mapping analysis. Form these outputs, authors explored the related research streams of twitter articles related to health in KM discipline.

2.3 Analytical Tool Used for Data Analysis

In this study, HistCite and VOSviewer software have been used for data analysis. HistCite is frequently used as for analyzing systematic literature reviews as well as bibliometric analysis [38]. Apriliyanti and Alon [1] used HistCite as a bibliometric co-citation tool for 336 articles and VOSviewer software as bibliometric cartography tool for 2088 articles to isolate five research streams for absorptive capacity [1]. VOSviewer is widely used in bibliometric analysis, especially in thematic analysis, cartography, and cluster analysis [21, 24, 25]. Examination of the literature shows that both HistCite and VOSviewer are extensively used. Therefore, this study uses both software in its analysis. We applied HistCite to conduct a bibliometric descriptive analysis to find out the most significant authors, journals, cited reference, institutions, and countries (Thelwall [31]). Five types of bibliometric mapping analysis can be used: named as co-author, co-occurrence of keywords, citation, bibliographic coupling, and co-citation through VOSviewer (van Eck and Waltman [34]; Eck and Waltman [35]) but we applied only co-authorship as type of analysis and countries as unit of analysis to explore co-authorship among 30 most productive countries. In keyword analysis, VOSviewer utilizes a text-mining technique to analyze the content of titles, keywords, and abstracts. Furthermore, we applied co-occurrence as type of analysis and all keywords as unit of analysis to explore emerging themes in KMTH publications. As a consequence, we found different clusters of closely associated items (countries in our study), which are denoted by the same color. The larger the item, the greater its significance and popularity with respect to the other items (Perianes-Rodriguez, Waltman and van Eck 2016).

3   Results and Discussion

3.1   Document Type

There were 798 KMTH publications in total from 2009 up until now (5 May 2019) as shown in Table 1. The total sources published under umbrella of KMTH research field got 11060 total global citation score (TGCS) and 860 Total local citation score (TLCS) as shown in Table 1. TGCS represents number of times KMTH articles were cited by other articles around the globe while TLCS refer to the number of citations of the paper within the collection data retrieved from WoS (798 articles in this study), which represents the importance and significance of article within that particular research field [1]. 568 KMTH articles received 9066 TGCS and 685 TLCS whereas 146 proceeding papers got only 445 TGCS and 66 TLCS. Despite of having only 56 Review papers as compared to 146 proceeding papers, their TGCS was approximate double (914) than proceeding papers TGCS (445) which indicate that review papers got more attention from KMTH academia. Furthermore, editorial material (15) were ten time less than proceedings papers (146) but have almost same TGCS of proceedings papers (see Table 1) which as well indicates more significance of editorial material than proceedings papers. All document types along total publications (Recs), TLCS, Avg. TLCS (number of local citations per document), TGCS and Avg. TGCS (number of global citations per document) were listed in Table 1.

3.2   Yearly Output of KMTH Publications

The yearly growth of the academic research literature production in the KMTH publications has been shown in Table 2. The production of academic research sources published in 2009 was only 2 articles. From 2010 to 2012 the publication growth was increased gradually, adding almost 10 papers every year, from 10 to 27 articles. After that period, the publications per year got dramatic rise from 63 to 93 publications from 2013 to 2015. This showed the significant attention of academia in KMTH research. In recent years, an

Table 1. Types of documentations

#	Document type	Recs	TLCS	Avg. TLCS	TGCS	Avg. TGCS
1	Article	568	685	1.2	9066	15.96
2	Proceedings Paper	146	66	0.45	445	3.04
3	Review	56	55	0.98	914	16.32
4	Editorial Material	15	27	1.8	441	29.4
5	Article; Proceedings Paper	6	26	4.33	190	31.66
6	Meeting Abstract	2	0	–	0	–
7	Review; Early Access	2	0	–	1	0.5
8	Article; Data Paper	1	1	1	3	3.00
9	Article; Early Access	1	0	–	0	–
10	Letter	1	0	–	0	–
	Total	798	860		11060	

Recs = Records; TGCT = total global citation score; TLCS = Total local citation score

exponential growth has been observed from 2016 to 2018 from 148 to 175 publications. The year 2018 was a peak year for the KMTH publications. In 2018, 175 maximum numbers of articles were published. This analysis clearly narrated the emergence and prominence of KMTH research field and still has a long way to go.

Table 2. Yearly growth output of KMTH publications

#	Publication year	Recs	Percent	TLCS	TGCS
1	2009	2	0.3	33	460
2	2010	10	1.3	91	907
3	2011	20	2.5	43	919
4	2012	27	3.4	52	1959
5	2013	63	7.9	163	1615
6	2014	77	9.6	120	1480
7	2015	93	11.7	124	1463
8	2016	148	18.5	148	1382
9	2017	145	18.2	61	596
10	2018	175	21.9	22	263
11	2019	38	4.7	3	15
	Total	798	100	860	11059

Recs = Records; TGCT = total global citation score; TLCS = Total local citation score

3.3 Most Influential Countries of KMTH Publications

The KMTH publications were published in 53 different countries around the globe and top ten most productive countries are listed here in Table 3. USA published five time more articles (427) with 7390 TGCS than UK (97 publications and ranked second highest among top ten countries), showing the greater influence and unbeatable publication share in KTMH research. It was interesting to know that UK number of publications were high than Canada (81) but Canada TGCS (1630) was much higher than UK TGCS (780) which indicates that KMTH publications produced by Canada were more influential than UK. Similarly, Switzerland TGCS was higher than India and Saudi Arabia despite of having more publications. This showed Switzerland articles were more influential as compared to India and Saudi Arabia. It is also interesting to note that only USA authors published almost as many articles as all other authors from the rest of 9 countries in the top ten most productive countries. This analysis also revealed e-health and m-health as emerging fields in developed countries, like USA, UK and Canada, and played vital role in the promotion of health care through twitter as KM platform and now twitter platform for health promotion taking roots in developing countries gradually like India and Saudi Arabia.

Table 3. Most influential countries in KMTH publications

#	Country	Recs	Percent	TLCS	% age TLCS	Avg. TLCS	TGCS	Avg. TGCS	% age TGCS
1	USA	427	53.5	622	70.05	1.46	7390	17.31	63.08
2	UK	97	12.2	63	7.09	0.65	786	8.10	6.71
3	Canada	81	10.2	72	8.11	0.89	1630	20.12	13.91
4	Australia	70	8.8	71	8.00	1.01	875	12.50	7.47
5	China	38	4.8	23	2.59	0.61	401	10.55	3.42
6	Spain	31	3.9	6	0.68	0.19	180	5.81	1.54
7	Italy	26	3.3	12	1.35	0.46	194	7.46	1.66
8	Saudi Arabia	22	2.8	18	2.03	0.82	99	4.50	0.84
9	India	22	2.8	0	0.00	0.00	48	2.18	0.41
10	Switzerland	17	2.1	1	0.11	0.06	113	6.65	0.96
	Total	**831**	**104.4**	**888**			**11716**		

Avg. = number of citations per document; % age TLCS = percentage of Total local citations % age TGCS = percentage of Total global citations

3.4 Co-authorship Between Most Influential Countries of KMTH Publications

One interesting fact was also observed during analysis of most influential countries of KMTH publications that cumulative percentage was 104.4% (see Table 3). This indicates that the publications of each country are not mutually exclusive and might have overlapping with each other where authors played significant role to contribute in writing same publication through co-authorship. This evoked us to conduct co- authorship analysis of most influential countries of KMTH publications through VOSviewer. We chose type of analysis as "co-authorship" and unit of analysis as "countries" in VOSviewer. Furthermore, we chose 5 minimum numbers of publications of a country as a threshold value and consequently got 30 most noteworthy countries in co-authorship network as shown in Fig. 1. Through this analysis five clusters were emerged to highlight the co-authorship significance of 30 countries as shown in Table 4. In first cluster (Red), all the countries are from Europe except Brazil which indicates that European countries' authors cited each other work in KMTH publication. Similarly in cluster 2 (green), most of countries are from Europe except Iran. Cluster 3 (Blue) was observed most diverse cluster in term of geographic location as it was comprised of three different continents (USA and Canada from North America; China, and Saudi Arabia from Asia and Australia). Forth cluster (Yellow) exhibited all countries list from Asia except Norway while fifth cluster (purple) was comprised of only two Asian countries (see Table 4). Furthermore, despite of having great variation of number of publication USA (427) and Australia (70), they have same links with other countries (24-LWOC) in 30 top productive countries as shown in Table 4.

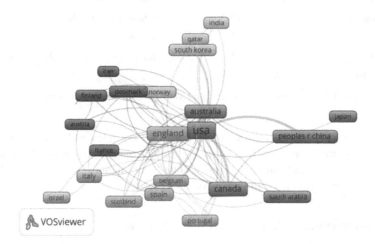

Fig. 1. CO-authorship Network of 30 countries in KMTH publications. (Color figure online)

3.5 Most Influential Institutions of KMTH Publications

Through Hiscite software, authors found that the 798 KMTH research publications were published by 1092 institutes around the globe. Top 10 institutes were sorted according to the number of publications as shown in Table 5. It was interesting to know that all top institutes were from USA and except one from Canada and one from Australia. It revealed huge contribution of USA institutes in health promotion by using twitter.

Table 4. Co-authorship mapping clusters of KMTH publications

#	Label	Recs	LWOC	#	Label	Recs	LWOC
	Cluster 1: Red				**Cluster 3: Blue**		
1	Belgium	10	10	1	Australia	70	24
2	Brazil	5	5	2	Canada	81	14
3	England (UK)	97	22	3	Ireland	8	7
4	Netherlands	16	11	4	Japan	7	3
5	Portugal	11	7	5	China	38	10
6	Scotland	12	10	6	Saudi Arabia	22	6
7	Spain	31	9	7	USA	427	24
8	Switzerland	17	16		**Cluster 4: Yellow**		
	Cluster 2: Green			1	India	22	6
1	Austria	5	5	2	Malaysia	6	3
2	Denmark	7	10	3	Norway	6	8
3	Finland	7	10	5	Qatar	7	4
4	France	12	12	5	South KOREA	15	6
5	Germany	15	17		**Cluster 5: Purple**		
6	Greece	7	2	1	Israel	7	2
7	Iran	5	5	2	Italy	26	12
8	Sweden	7	6				

Recs = Records (Total number of publications); LWOC = link with other countries

University of Toronto gave contribution in form of 15 articles and these 15 article got highest TGCS among all top 10 institutes. This fact made Canada publications more influential. Similarly, Colombia University with 10 publications and 460 TGCS came forward to create impact in health promotion through twitter.

3.6 Most Influential Authors of KMTH Publications

This study found 2930 authors contribution in writing 798 publications. We sorted ten most influential authors according to their total publications. We found that Harris JK was on top level with 11 publications and 184 TGCS. Harris JK authored consecutively publications for health promotion every year, in 2013 (3), 2014 (2), 2015 (1), 2016 (1), 2017 (2), 2018 (2). Fu KW, Fung ICH and Tse ZTH were found with 9 publications and 21 TGCS from 2013 to 2019. Dredze M, Househ M, Merchant RM, Thompson MA wrote 8 publications with TGCS 344, 142, 50 and 33 respectively. Dredze (King Saud Bin Abdul-Aziz University) TGCS was highest among all top 10 authors and reflected Dredze publications have been influential in health promotion. His one publication "National and Local Influenza Surveillance through Twitter: An Analysis of the 2012–2013 Influenza Epidemic" published in 2013 got 148 TGSC.

Table 5. Most Influential institution of KMTH publications Top 10

#	Institution	Country	Recs	Percent	TLCS	TGCS
1	University Washington	USA	24	3.0	43	334
2	University Georgia	USA	22	2.8	84	286
3	University of Pennsylvania	USA	16	2	27	166
4	University Sydney	Australia	16	2	8	152
5	University Kentucky	USA	15	1.9	24	118
6	University Toronto	Canada	15	1.9	2	916
7	Johns Hopkins University	USA	14	1.8	3	378
8	University Calif San Francisco	USA	11	1.4	47	236
9	Harvard University	USA	11	1.4	1	232
10	Columbia University	USA	10	1.3	121	460

Table 6. Most influential authors of KMTH publications

#	Author	Country	Recs	TLCS	TLCS/t	TGCS	TGCS/t
1	Harris JK	USA	11	37	6.85	184	32.12
2	Fu KW	China	9	21	5.42	91	19.03
3	Fung ICH	USA	9	21	5.42	91	19.03
4	Tse ZTH	USA	9	21	5.42	91	19.03
5	Dredze M	USA	8	2	1	344	59.48
6	Househ M	Saudi Arabia	8	21	3.96	142	26.12
7	Merchant RM	USA	8	22	6.17	50	14.33
8	Thompson MA	USA	8	15	3.3	33	7.53
9	Asch DA	USA	7	21	3.96	141	25.79
10	Bragazzi NL	Italy	7	2	0.67	69	19.1

Recs = Records; TGCT = total global citation score; TLCS = Total local citation score.
TLCS/t = average total local citations; TGCS/t = average total global citations.

3.7 Most Influential Articles of KMTH Studies

Next, we present the top 10 highly cited papers of KMTH studies (see Table 4). Chew and Eysenbach [4] article ranked first with 407 TGCS and 40.7 average TGCS per year (TGCS/t). This study illustrated the potential of using SM to conduct "Infodemiology" studies for public health. Signorini et al. [26] article was second highly cited paper with 402 GCS but its Avg. TGCS was 44.67, slightly more than first article. Similarly, Grajales III *et al.* [12] article was ranked 6th, but it was interesting to know that its 30.17 Avg. TGCS was more than articles ranked 3rd, 4th, and 5th which indicates that comparatively this paper is gaining more attention of KMTH academia. In this study, they examined the use of information embedded in the tweets to track rapidly evolving public sentiment with respect to H1N1 (swine flu). The 3rd ranked paper in the list (Table 7) got highest TLCS (84) with 8.4 TLCS/t which indicate that in pool of 798 KMTH retrieved articles from WoS, 84 studies cited this paper with average 8.4 citation score per year.

Table 7. Most influential KMTH articles

#	Title	TLCS	LCS/t	TGCS	GCS/t
1	Pandemics in the Age of Twitter: Content Analysis of Tweets during the 2009 H1N1 Outbreak	0	0	407	40.7
2	The Use of Twitter to Track Levels of Disease Activity and Public Concern in the US during the Influenza A H1N1 Pandemic	0	0	402	44.67
3	Dissemination of health information through social networks: Twitter and antibiotics	84	8.4	274	27.4
4	Infodemiology and Infoveillance: Framework for an Emerging Set of Public Health Informatics Methods to Analyze Search, Communication and Publication Behavior on the Interne	0	0	252	22.91
5	Social Internet Sites as a Source of Public Health Information	33	3	208	18.91
6	Social Media: A Review and Tutorial of Applications in Medicine and Health Care	0	0	181	30.17
7	Patients' and health professionals' use of social media in health care: Motives, barriers and expectations	26	3.71	163	23.29
8	Crowdsourcing, citizen sensing and sensor web technologies for public and environmental health surveillance and crisis management: trends, OGC standards and application examples	0	0	151	16.78
9	National and Local Influenza Surveillance through Twitter: An Analysis of the 2012–2013 Influenza Epidemic	0	0	148	21.14
10	Adoption and use of social media among public health departments	0	0	144	18

3.8 Most Influential Journals of KMTH Studies

In Hiscite bibliometric analysis, authors found that 798 KMTH articles were published in 459 academic sources (mostly journals and proceedings conferences). Journal of Medical Internet Research was ranked first among top 10 journal with maximum no of records (80) and TGCS 1750. Its importance and contribution in the field of KMTH is undeniable, e-health is not only hotspot in journal articles but also average TGCS indicated how remarkably journal has been successful to attract research community attraction. Similarly, Plos One role is incredibly strong. Though Plos One records were 25 but its TGCS 1308 was almost near to Journal of Medical Internet Research. This also depicted Plos One articles were also playing role as nuclei of KMTH publication. The rest of top ten journals for KMTH publications are listed in Table 7. American Journal of Infection Control was found with 8 publication and 396 TGCS at rank no 7. Its TGCG were more than journal raked at no 3, 4, 5 and 6. This showed American Journal of Infection Control was most influential journal after Journal of Medical Internet Research and Plos One (Table 8).

Table 8. Most influential Journals of KMTH publications

#	Journal	Recs	Percent	TGCS	Avg. TGCS	TLCS
1	Journal of Medical Internet Research	80	10	1705	298.28	116
2	Plos One	25	3.1	1308	177.12	27
3	Journal of Health Communication	19	2.4	209	39.81	41
4	Computers In Human Behavior	15	1.9	151	38.27	27
5	Health Communication	14	1.8	162	35.17	23
6	BMJ Open	10	1.3	56	12.36	8
7	American Journal of Infection Control	8	1	396	51.91	18
8	International Journal of Medical Informatics	8	1	35	10.75	14
9	Current Hematologic Malignancy Reports	7	0.9	52	13.83	15
10	International Journal of Environmental Research And Public Health	7	0.9	14	7	4

4   Conclusion

Twitter as KM platform for health promotion was emerged in 2009. After that, it gained histrionic popularity as knowledge seeking, knowledge acquisition and knowledge disseminating platform for public health promotion by the time but still it has a long way to go. This study provided an overall contribution of the twitter research dynamics in KMTH academia at a glance. The results were obtained by applying two type of bibliometric analysis: first, a performance analysis that includes both influence (citation score) and second, productivity (total publications) indicators to highlight most noteworthy journals, articles, countries, institutions and authors of KMTH articles.

4.1 Practical Implications

Our study opened new avenues for all health care provider to utilize twitter as a KM platform to promote health care. It's not only cost effective and user-friendly but also provides ease to spread the useful information rapidly. In our study three clusters provided clear evidences that twitter is effetely compliable as KM platform for health promotion. A report entitled "enabling health care decision-making through clinical decision support and knowledge management," published by the agency for healthcare research and quality (AHRQ) provided strong evidence that clinical KM IT systems are inevitable in promoting and improving health care process and will become a key part of all healthcare organizations over the next few years. As this transformation unfolds, the effective health KM through SM will become an essential skill that all healthcare providers will need to master [19].

4.2 Limitations and Future Research

Twitter role in health is an emerging research, and it is a significantly offering its services for health promotion and for health seeking behavior platform. Our main limitation in this paper was that during co-authorship and co-occurrence mapping, we only considered countries with five minimum number of publications and publications with five minimum number of keywords respectively. Due to this reason several countries and documents were excluded from analysis and their inclusion could have been proposed different results. Second, HistCite cannot accommodate non-WoS source databases. This study is only focused on articles extracted from WoS, so the articles in our study are from prestigious journals. A bias therefore might exist for high-quality publications, and non-WoS journal articles information is not shown in our analysis that might impact differently twitter role in health promotion. In a country analysis an important problem might be that many non-English speaking countries may publish also research in local languages and most of this research is not included in WoS. Therefore, these publications are not considered and usually not cited. This issue could also produce deviations in the results. However, considering the current world standards for research, the material published in WoS is sufficiently representative to be considered as a general sample in order to identify important results and conclusions.

Conflict of Interest. The authors declare no conflict of interest.

Funding. This research is financially supported by The National Key Research and Development Program of China (grant number 2018YFC0807105), National Natural Science Foundation of China (grant number 61462073) and Science and Technology Committee of Shanghai Municipality (STCSM) (under grant numbers 17DZ1101003, 18511106602 and 18DZ2252300).

References

1. Apriliyanti, I.D., Alon, I.: Bibliometric analysis of absorptive capacity. Int. Bus. Rev. **26**(5), 896–907 (2017). https://doi.org/10.1016/j.ibusrev.2017.02.007
2. Asghar, I., et al.: Assistive technology for people with dementia: an overview and bibliometric study. Health Inf. Libr. J. **34**(1), 5–19 (2017). https://doi.org/10.1111/hir.12173

3. Chen, X., et al.: A bibliometric analysis of event detection in social media. Online Inf. Rev. **43**(1), 29–52 (2019). https://doi.org/10.1108/OIR-03-2018-0068
4. Chew, C., Eysenbach, G.: Pandemics in the age of Twitter: content analysis of tweets during the 2009 H1N1 outbreak. PLOS ONE **5**(11) (2010). https://doi.org/10.1371/journal.pone.001 4118
5. Choo, E.K., et al.: Twitter as a tool for communication and knowledge exchange in academic medicine: a guide for skeptics and novices. Med. Teach. **37**(5), 411–416 (2015). https://doi. org/10.3109/0142159X.2014.993371
6. Couper, M.P., et al.: Use of the Internet and ratings of information sources for medical decisions: results from the DECISIONS survey. Med. Decis. Making **30**(5), 106S–114S (2010). https://doi.org/10.1177/0272989X10377661
7. Fellnhofer, K.: Toward a taxonomy of entrepreneurship education research literature: a bibliometric mapping and visualization. Educ. Res. Rev. **27**, 28–55 (2019). https://doi.org/10. 1016/j.edurev.2018.10.002
8. Fox, S.: The social life of health information (2011)
9. Gan, C., Wang, W.: A bibliometric analysis of social media research from the perspective of library and information science. Digit. Serv. Inf. Intell. **2014**, 23–32 (2014)
10. Garfield, E., Pudovkin, A.I., Istomin, V.S.: Mapping the output of topical searches in the web of knowledge and the case of watson-crick. Inf. Technol. Libr. **22**(4), 183–187 (2003). https:// doi.org/ISI>://000188258600008
11. Gilbert, S.: Learning in a Twitter-based community of practice: an exploration of knowledge exchange as a motivation for participation in #hcsmca. Inf. Commun. Soc. **19**(9), 1214–1232 (2016). https://doi.org/10.1080/1369118X.2016.1186715
12. Grajales III, F.J. et al.: Social media: a review and tutorial of applications in medicine and health care. J. Med. Internet Res. **16**(2) (2014). https://doi.org/10.2196/jmir.2912
13. Harris, J.K., et al.: Social media adoption in local health departments nationwide. Am. J. Publ. Health. **103**(9), 1700–1707 (2013). https://doi.org/10.2105/AJPH.2012.301166
14. Harris, J.K.: The network of web 2.0 connections among state health departments. J. Pub. Health Manag. Pract. **19**(3), E20–E24 (2013). https://doi.org/10.1097/phh.0b013e318268 ae36
15. Kaplan, A.M., Haenlein, M.: Users of the world, unite! The challenges and opportunities of Social Media. Bus. Horizons **53**(1), 59–68 (2010). https://doi.org/10.1016/j.bushor.2009. 09.003
16. Keramatfar, A., Amirkhani, H.: Bibliometrics of sentiment analysis literature. J. Inf. Sci. **45**(1), 3–15 (2019). https://doi.org/10.1177/0165551518761013
17. Lagos-ortiz, K.: Technologies and Innovation. **749**(2017), 63–76 (2017). https://doi.org/10. 1007/978-3-319-67283-0
18. Li, F., et al.: Mapping publication trends and identifying hot spots of research on internet health information seeking behavior: a quantitative and co-word biclustering analysis. J. Med. Internet Res. **17**(3) (2015). https://doi.org/10.2196/jmir.3326
19. Lobach, D., et al.: Enabling health care decision making through clinical decision support and knowledge management. Evidence Report No. 203. AHRQ Publication No. 12-E001-EF. Evidence Report/Technology Assessment 203 (2012)
20. Muller, A.M. et al.: Physical activity, sedentary behavior, and diet-related eHealth and mHealth research: bibliometric analysis. **20**(4) (2018). https://doi.org/10.2196/jmir.8954
21. Noor, S., et al.: Bibliometric analysis of social media as a platform for knowledge management. Int. J. Knowl. Manag. **16**(3) (2020)
22. Noor, S. et al.: Research synthesis and thematic analysis of Twitter through bibliometric analysis. Int. J. Semant. Web Inf. Syst. **16**(3) (2020). 22 pages
23. Okazaki, S., et al.: Using Twitter to engage with customers: a data mining approach. Internet Res. **25**(3), 416–434 (2015). https://doi.org/10.1108/IntR-11-2013-0249

24. Shah, S.H.H., et al.: Prosumption: bibliometric analysis using HistCite and VOSviewer. Kybernetes **49**(3), 1–24 (2019). https://doi.org/10.1108/K-12-2018-0696
25. Shah, S.H.H., et al.: Research synthesis and new directions of prosumption: a bibliometric analysis. Int. J. Inf. Manag. Sci. **31**(1), 79–98 (2020). https://doi.org/10.6186/IJIMS.202003 31(1).0005
26. Signorini, A., et al.: The use of Twitter to track levels of disease activity and public concern in the U.S. during the influenza A H1N1 pandemic. PLoS ONE **6**(5) (2011). https://doi.org/10.1371/journal.pone.0019467
27. Sinnenberg, L., et al.: Twitter as a tool for health research: a systematic review. Am. J. Pub. Health **107**(1), E1–E8 (2017). DOI:https://doi.org/10.2105/AJPH.2016.303512
28. Sugawara, Y. et al.: Cancer patients on Twitter: a novel patient community on social media. BMC Res. Notes (2012). https://doi.org/10.1186/1756-0500-5-699
29. Tang, M., et al.: 2018. ten years of sustainability (2009 to 2018): a bibliometric overview. Sustainability **10**(5), 1–21 (2018). https://doi.org/10.3390/su10051655
30. Thackeray, R., et al.: Adoption and use of social media among public health departments. BMC Public Health **242**(12), 1–6 (2012). https://doi.org/10.1186/1471-2458-12-242
31. Thelwall, M.: Bibliometrics to webometrics. J. Inform. Sci. **34**(4), 605–621 (2008). https://doi.org/10.1177/0165551507087238
32. Urquhart, C., Dunn, S.: A bibliometric approach demonstrates the impact of a social care data set on research and policy. Health Inf. Libr. J. **30**(4), 294–302 (2013). https://doi.org/10.1111/hir.12040
33. Vance, K., et al.: Social internet sites as a source of public health information. Dermatol. Clin. **27**(2), 133–136 (2009). https://doi.org/10.1016/j.det.2008.11.010
34. Van Eck, N.J., Waltman, L.: Software survey: VOSviewer, a computer program for bibliometric mapping. Scientometrics **84**(2), 523–538 (2010). https://doi.org/10.1007/s11192-009-0146-3
35. Van Eck, N.J., Waltman, L.: VOSviewer Manual. Universitteit Leiden (2018). http://www.vosviewer.com/documentation/Manual_VOSviewer_1.5.4.pdf
36. Vošner, H.B., et al.: A bibliometric retrospective of the journal computers in human behavior (1991–2015). Comput. Hum. Behav. **65**(2016), 46–58 (2016). https://doi.org/10.1016/j.chb.2016.08.026
37. Wang, W., et al.: Harnessing twitter "big data" for automatic emotion identification. In: Proceedings - 2012 ASE/IEEE International Conference on Privacy, Security, Risk and Trust and 2012 ASE/IEEE International Conference on Social Computing, SocialCom/PASSAT 2012, pp. 587–592 (2012)
38. Zupic, I., Čater, T.: Bibliometric methods in management and organization. Organ. Res. Methods. **18**(3), 429–472 (2015). https://doi.org/10.1177/1094428114562629
39. Zyoud, S.H. et al.: Global trends in research related to social media in psychology: mapping and bibliometric analysis. Int. J. Mental Health Syst. **12**(1), 4 (2018). https://doi.org/10.1186/s13033-018-0182-6
40. Pew Internet & American Life Project. Choice Rev. Online. **51**(05), 51-2434 (2013). https://doi.org/10.5860/choice.51-2434

Automatic Cerebral Artery System Labeling Using Registration and Key Points Tracking

Mengjun Shen[1,2], Jianyong Wei[3], Jitao Fan[3], Jianlong Tan[1,2],
Zhenchang Wang[4], Zhenghan Yang[4], Penggang Qiao[4], and Fangzhou Liao[1,2(✉)]

[1] Institute of Information Engineering, Chinese Academy of Sciences, Beijing, China
{shenmengjun,tanjianlong,liaofangzhou}@iie.ac.cn
[2] School of Cyber Security, University of Chinese Academy of Sciences,
Beijing, China
[3] ShuKun (Beijing) Technology Co., Ltd., Jinhui Bd, Qiyang Rd, Beijing, China
{weijy,fanjt}@shukun.net
[4] Department of Radiology, Beijing Friendship Hospital, Capital Medical University,
No. 95 YongAn Road, Beijing 100050, People's Republic of China

Abstract. The cerebral artery network is a complex system. Each cerebral hemisphere has three main arteries: anterior/middle/posterior cerebral artery (ACA/MCA/PCA). In the middle of the brain, these arteries form the Willis circle. The vessel diseases around the Willis circle are generally a warning signal of stroke. Computed tomography angiography (CTA) is an important tool to visualize the cerebral artery system, but due to the complexity of the artery network, it is difficult to generate straightened lumen of the arteries. Hence, radiologists need to spend a long time inspecting every corner on the original CTA image and detecting the stenosis. In this paper, we propose an automatic pipeline that can identify the main arteries, dissect them into segments, and generate their straightened lumen. The main idea is to utilize the registration to identify key points of the cerebral artery system and apply the shortest path finding algorithm to extract the main routes of the artery system. Our approach is robust to all kinds of brain/neck scanning protocols and most topological variations of the artery system. Furthermore, our approach is very fast (15.7 s per case on average). Consequently, our method is more useful in real scenarios compared with previous methods.

Keywords: Cerebral artery · Image registration · Vessel labeling

1 Introduction

Computed tomography angiography (CTA) is a minimally invasive and cost-efficient imaging modality. It is widely used in routine clinical diagnoses of the

Supported by the National Key Research and Development Program of China under Grant No. 2017YFB0803003.

head and neck vessels [9], especially in cerebrovascular disease which represents one of the leading causes of severe disability and mortality worldwide [11]. Automated anatomical labeling [5] of cerebral arteries can be very helpful. It can not only help the diagnostic process for radiologists and physicians, but also provide essential information for many subsequent procedures, such as generating medical imaging reports, calculating statistics for important anatomical branches, and visualizing regions of interest. The main challenge for cerebral artery labeling is the large individual variations among populations [10]. Length, tortuosity, the radius of vessels and the topology of vascular networks are highly variable. Therefore, the robustness of handling individual variations is the focus of the automated anatomical labeling problem.

Towards this problem, Bilgel et al. [1] use a random forest classifier as the initial vessel segment classifier and a belief propagation on the Bayesian network representation of the vessel tree as the label correcting step. However, it only handles the anterior circulation. Bogunović et al. [2] model the whole Willis circle as a graph with topological and spacial attributes, and detect the bifurcation points to dissect the graph. They try various standard reference graphs to match the query case for handling different individual variations. Robben et al. [7] propose to simultaneously perform vessel segmentation and labeling by formulating the vessel labeling as an integer program problem. Nevertheless, these studies do not fully fulfill the clinical request: firstly, these methods do not cover segment A2 (i.e., segment II of ACA) which is very hard to identify but crucial for diagnosis. Secondly, these studies assume that the image contains the whole brain, which is not always satisfied. Thirdly, all these methods use machine learning algorithms, which require sophisticated manual labeling, and their decision process can hardly be interpreted. Finally, all these approaches involve iteration or searching, which makes their running speed unbearable in clinical use.

In this paper, we propose a simple, effective, robust and fast method for cerebral artery system labeling problem utilizing registration and key points tracking. We firstly design a loss function for the registration to solve the partial data registration problem. Then we find robust key points on the vessel graph as start/end points for the vessel tracking. Finally, we use simple rules to detect fork points and dissect the graph into segments. To track the A2 path with higher precision, we adopt a deep learning method as a correction. With the proposed approach, we achieve 100% registration accuracy, 98.1% segment identification accuracy (A2 not included) and 64.3% A2 identification accuracy. Besides, our approach is fast enough (15.7 s) to be utilized in real clinical scenarios.

2   Methods

2.1   Terminology and Dataset

Here we list all abbreviations used in this paper:

- L-/R-: prefix for left/right that indicates the cerebral hemisphere.
- ACA/MCA/PCA: anterior/middle/posterior cerebral artery.

- A1/A2, M1/M2, P1/P2: segment I/II of ACA, MCA, PCA.
- ACoA/PCoA: anterior/posterior communicating artery.
- ACAx/MCAx/PCAx/ACoAx/PCoAx: candidate path of its corresponding vessel.
- ICA, BA: internal carotid artery, basilar artery.

A sketch of the cerebral artery system is illustrated in Fig. 1.

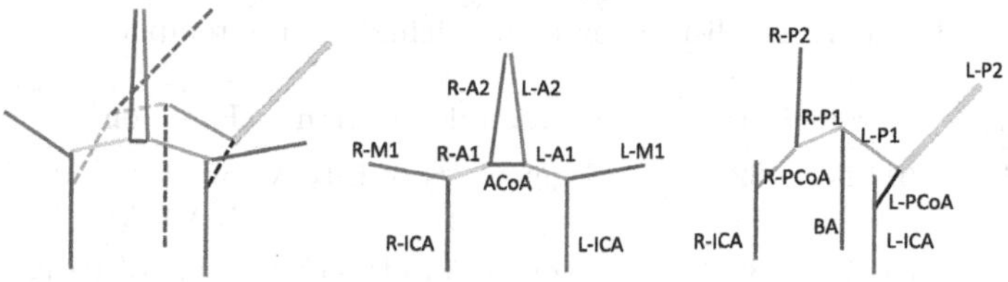

Fig. 1. Left: a sketch of the cerebral artery system (front view). Vessels in the back are shown in dash lines. Middle: the anterior circulation and its segment labeling. Right: the posterior circulation and its segment labeling.

We collect 194 head-neck CTA images from Beijing Friendship Hospital, Capital Medical University. The scanning target is inhomogeneous: 12 of them only contain the brain, 7 of them only contain the neck, and 175 of them include both the brain and the neck. 132 of them are males, and 62 of them are females. Their age ranges from 15 to 91, with a mean of 61.8 and a standard deviation of 11.7. Along with the raw images, we also collect their artery segmentation results.

2.2   CT Template Generation

We choose a patient with the standard skull shape, and we manually select rotation angles and shift parameters so that the hemisphere fissure co-aligns with the image mid-line and the canthomeatal line becomes parallel to the X-Y plane. Then we crop the image so that only the skull is in the field of view, and a small blank margin is left in the X/Y axis. Finally, we resize the image to the size of $224 \times 320 \times 320$ using the linear interpolation. The three-view of the template is shown in Fig. 2.

Prior Key Points on Template. Since the images are registered, we can define key points directly based on the coordinate. Unlike the previous study [2], we do not directly look for fork points in the Willis circle because they lie in a complex vessel network and merely using the coordinate is not robust enough. Instead, we define several key points that stand for the start and end points of the arteries. In most cases, these key points are very close to the real vessels, and there are no distractors nearby. Their names, coordinates and other descriptions are listed in Table 1. The relationship between the key points and

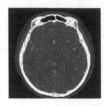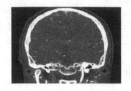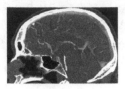

Fig. 2. The template used in our registration.

Table 1. The list of key points defined in the template.

Name	Coordinate (z, y, x)	Position description	Function
L-c6	161, 142, 182	Segment VI of L-ICA	Start of L-ACAx, L-MCAx
R-c6	161, 142, 139	Segment VI of R-ICA	Start of R-ACAx, R-MCAx
a2	84, 98, 160	End of L-A2, R-A2	End of L/R-ACAx
L-m1	134, 120, 230	End of L-M1	End of L-MCAx
R-m1	134, 120, 90	End of R-M1	End of R-MCAx
L-m1$_m$	134, 128, 208	Middle of L-M1	End of L-MCAx when L-m1 is missing
R-m1$_m$	134, 128, 112	Middle of R-M1	End of R-MCAx when R-m1 is missing
ba	161, 161, 160	Middle of BA	Start of L/R-PCAx
L-p2	139, 190, 195	Middle of L-P2	End of L-PCAx
R-p2	139, 190, 125	Middle of R-P2	End of R-PCAx

the segmentation of one case is shown in Fig. 3. Note that the key points do not lie on the vessels precisely, and that is why we use these key points with tolerance in the following.

2.3 Registration

We utilize the affine registration to register other cases to the template image. The registration process is done in a custom-designed registration toolbox, which is inspired by the Airlab [8].

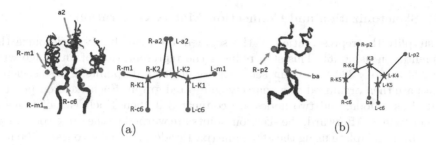

Fig. 3. Demonstration of the positions of the key points in one sample case registered to the template. (a) Left: the anterior circulation and its related key points. Right: a sketch for the key points and fork points for the anterior circulation. (b) The same illustration for the posterior circulation.

Registration Loss. The registration loss is based on mean squared error (MSE), with a slight modification. For CTA images that only contain the neck, only a small part of them can be aligned with the template. Consequently, if the raw MSE loss is applied directly between the warped image and the template, no matter how to choose the padding value, the loss would be very high. To solve this problem, when calculating the loss function, the padding area is omitted. However, this setting can also lead to another extreme situation: the warped image only has its background aligned with the background of the template. Under such circumstance, the loss can easily become 0. Hence, if the loss is too low (smaller than 20 in our experiments), we manually set the loss to positive infinity and restart the optimization with another initial parameter.

Optimization. We choose Adam [4] as the optimization method. The initial learning rate is 0.05 and decay by a factor of 0.1 every 50 iterations. The maximum number of iterations is set to 150. The registration would be considered successful if the final loss is smaller than 1800. The optimization would terminate before the 150th iteration if the convergence condition is met.

Multi-start Registration. Since the scan range varies a lot in this study, one fixed initial affine matrix θ^0 cannot guarantee the successful convergence. So we did a simple grid search for θ^0 in the shifting and zooming space:

- $s_z \in \{0, 0.4, 0.8\}$, $\sigma \in \{0.7, 1, 1.3\}$,

where s_z denotes the shift in z-axis. A positive value means an upward shift, and a negative value represents a downward shift. In our experiments, these three values correspond with three types of scanning: the brain scanning (no shift), the brain-neck joint scanning (moderate upward shift) and the neck scanning (large upward shift). σ denotes the scaling factor, which accounts for different brain sizes.

2.4 Skeletonization and Connection Matrix Generation

To simplify the representation of the segmentation, we firstly skeletonize the segmentation using [6]. This step reduces the representation of an image with size $224 \times 320 \times 320$ to a point set with about 3000 3D points. These skeleton points are then organized as a sparsely connected graph. Each skeleton point is treated as a node, and two nodes are connected if their Euclidean distance is smaller than 2. Afterward, the skeleton points are warped to the same coordinate as the brain template using the affine matrix. Besides, the points outside the new image are ruled out. Thus, the shape of the connection matrix is also changed. We denote the new skeleton point set as $\mathcal{V}$ and its connection matrix as $\mathcal{G}$. All graphs mentioned in this study are the undirected graph.

As mentioned above, we define some prior key points on the template. For each key point, if the distance between its nearest skeleton point and itself is small enough, we treat the nearest skeleton point as its surrogate. The distance threshold is 20 for L-m1$_m$ and R-m1$_m$, and 40 for other key points. Otherwise, the key point is considered as "missing".

2.5 Predefined Path-Finding

Path Finding with Blockage. With a graph and some key points, we are ready to find the path between these points on the graph. The Dijkstra algorithm is applied to find the shortest path. Nevertheless, in our experiments, because of the Willis circle, an alternative path might be found when the main road is missing, which misleads us to the conclusion that the main road exists. Therefore, we define an algorithm that finds the path between two points with blocking all other alternative ways:

For simplicity, in the following formulas, except for ACoA, only the candidate path of each vessel in the left hemisphere is introduced since the right one is symmetrical.

Find MCAx. MCA is the first vessel to be found. Because the path is straight forward, there is no need to block other locations, i.e., the blocking node set $\mathcal{B}$ is set to $\emptyset$. When L-m1 is missing, L-m1$_m$ is used as the end node instead.

$$\text{L-MCAx} = \begin{cases} \mathcal{F}(\mathcal{G}, \text{L-c6}, \text{L-m1}, \emptyset), & \text{if } \exists \text{L-m1}, \\ \mathcal{F}(\mathcal{G}, \text{L-c6}, \text{L-m1}_m, \emptyset), & \text{otherwise}. \end{cases} \tag{1}$$

Find ACAx. The L-A2 and R-A2 are very close (see Fig. 1), so it is very hard to determine two a2 key points. Besides, L-A2 and R-A2 are generally tangled, and the skeletonization algorithm would give one common skeleton instead of two parallel skeletons (see Fig. 4b left).

We design a pipeline to detect both L-ACAx and R-ACAx for solving the above problem. Firstly, as we already locate the position of one a2, we need to identify the other a2. Secondly, we try to re-track the skeleton points of L-ACAx and R-ACAx so that they are disentangled.

Algorithm 1. Path finding with blockage

Input: Graph $\mathcal{G}$, start node n_1, end node n_2, blocking node set $\mathcal{B}$
Output: *Path* (a list of node, standing for shortest path from n_1 to n_2 on $\mathcal{G}$, without passing any node in $\mathcal{B}$). *None* if path finding failed
1: **function** $\mathcal{F}(\mathcal{G}, n_1, n_2, \mathcal{B})$
2:     **if** $n_1 == None$ or $n_2 == None$ **then**
3:         **return** *None*
4:     **end if**
5:     copy $\mathcal{G}$ to $\mathcal{G}'$
6:     **for** $n \in \mathcal{B}$ **do**   ▷ cut all connections of $\mathcal{B}$ in $\mathcal{G}'$
7:         $\mathcal{G}'[n, :] = 0$
8:         $\mathcal{G}'[:, n] = 0$
9:     **end for**
10:     $Dist, Path = \text{DIJKSTRA}(\mathcal{G}', n_1, n_2)$
11:     **if** $Dist > 0$ **then**   ▷ if n_1 and n_2 are connected
12:         **return** *Path*
13:     **else**
14:         **return** *None*
15:     **end if**
16: **end function**

The method to find another a2 point is shown in Fig. 4a. Firstly, by comparing the path length, we determine which side the existing a2 point belongs to. Secondly, we cut all incoming paths of A2 and find the point furthest to a2 (endpoint). Thirdly, we elevate the energy of the existing c6-a2 path and try to connect c6 with the endpoint, so that the new path would find another way to the endpoint and bypass the existing a2. Finally, we cut the new path at the position nearest to the existing a2. The cut point is the other a2 point.

To solve the problem of the entangled skeleton, we need to re-find a path between a2 and c6. Inspired by [3], we train a model that can predict the "centerness" of the vessel. Namely, at the center of the vessel, the output of the centerness model is expected to be 1, and decay to 0 at the vessel's border. The detail of the network design and training can be found in the original article [3]. Then we use the Dijkstra algorithm to find the shortest path between a2 and c6 in this "centerness" map (see Fig. 4b).

Find PCAx. Finding L-PCAx is very straight forward in normal cases. We just need to find the shortest path between ba and L-p2. In Fetal-type PCA cases where L-p1 is missing, although ba and L-p2 are not directly connected, they may be connected by the Willis circle. So we need to block the pre-found anterior circulation to prevent finding the wrong path:

$$L\text{-}PCAx = \mathcal{F}(\mathcal{G}, ba, L\text{-}p2, \mathcal{B}^p), \qquad (2)$$

$$\mathcal{B}^p = L\text{-}ACAx + R\text{-}ACAx + L\text{-}MCAx + R\text{-}MCAx. \qquad (3)$$

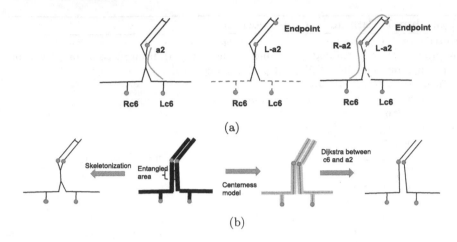

Fig. 4. The pipeline to find two ACAx. (a) Find the other a2 point. (b)Re-track the path between c6 and a2.

Find PCoAx

$$L\text{-PCoAx} = \mathcal{F}(\mathcal{G}, L\text{-c6}, L\text{-p2}, \mathcal{B}^C), \tag{4}$$

$$\mathcal{B}^C = R\text{-ACAx} + R\text{-MCAx} + R\text{-PCAx}. \tag{5}$$

Find ACoAx. Before finding ACoAx, we need to find the L-K1 and R-K1 points (fork points for C7/A1/M1, where C7 refers to the segment VII of the ICA, see Fig. 3). The method to find L-K1 and R-K1 is described in the next section.

$$ACoAx = \mathcal{F}(\mathcal{G}, L\text{-K1}, R\text{-K1}, \mathcal{B}^A), \tag{6}$$

$$\mathcal{B}^A = L\text{-PCAx} + R\text{-PCAx}. \tag{7}$$

To prevent the ACoAx goes through the entangling area of A2, we set some limits to it: it should be straight enough (measured by the second-order derivative), and its coordinate on the Z-axis should not be smaller than 110. Otherwise, the ACoAx is set to *None*.

2.6   Dissecting Vessel Graph by Detecting Fork Point

To parse the predefined path to segments, we need to find the fork points (K1 to K5 in Fig. 3). Firstly, we define a fork point detection function $\mathcal{K}$, which receives two paths and returns their last common point, i.e., fork point.

Then we find the first important fork points:

- K1: C7/A1/M1 fork point: L-K1 = $\mathcal{K}$(L-ACAx, L-MCAx).
- K2: A1/A2/ACoA fork point: L-K2 = $\mathcal{K}$(L-ACAx, ACoAx).
- K3: BA/L-P1/R-P1 fork point: K3 = $\mathcal{K}$(L-PCAx, R-PCAx).
- K4: P1/P2/PCoA fork point: L-K4 = $\mathcal{K}$(L-PCAx, L-PCoAx).
- K5: ICA/PCoA fork point: L-K5 = $\mathcal{K}$(L-PCAx, L-ACAx/L-MCAx).

Under an ideal condition, we can find all these fork points, thus dissecting the graph accordingly. However, people with all the target vessels and key points are rare. Most people lack one or several ACoA, PCoA and P1. Extreme cerebral ischemic stroke cases may lack M1 or A2. Hence, we design a backup algorithm to detect the fork points under these circumstances. Fortunately, these fork points usually lie on the turning of the vessel, so we apply the simple linear programming to find them (See Fig. 5). One exception is that when P1 is missing, the border between PCoA and P2 is not obvious. We choose to cut it at the point that is closest to BA (See Fig. 5(f)).

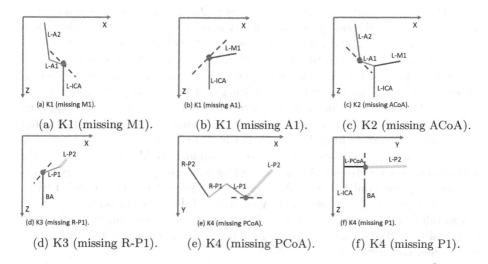

(a) K1 (missing M1).   (b) K1 (missing A1).   (c) K2 (missing ACoA).

(d) K3 (missing R-P1).   (e) K4 (missing PCoA).   (f) K4 (missing P1).

Fig. 5. Finding fork points using linear programming.

3   Result

3.1   Registration and Running Speed

The greatest challenge of the registration procedure is to deal with those neck scanning. In those images, only the lower part of the brain presents, yet we still need to identify their MCA and PCA. With the proposed registration loss function and the multi-start registration method, the registration procedure achieves 100% success rate. Figure 6 shows some of the demo results.

Thanks to the GPU support for the registration and the model inference, the total running time of our approach is 15.7 s (including registration, centerness model and graph dissecting) on average, which is much faster than [7] (510 s, registration time not included). Bogunović et al. [2] and Bilgel et al. [1] did not report the running time, but we infer that their methods should be much slower than ours according to their descriptions (finding maximal cliques, graph matching [2] and belief propagation on the graph [1]).

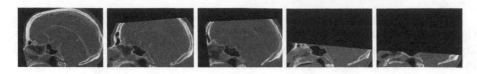

Fig. 6. The saggital view of some of the registration results. Note that even when the image only contains a small part of the brain, the registration is successful.

3.2 Manual Score

Since the research target of previous studies [1,2,7] is different from ours (magnetic resonance angiography vs. computed tomography angiography, ideal clean data vs. real clinical data, A2 not included vs. A2 included), we cannot provide a comparison with them. Besides, it is difficult to use a quantified score to measure the performance of the skeleton extraction procedure and vessel labeling. Accordingly, we choose to use manual scoring to measure the performance of our method. We invite one doctor to inspect our results and give a correct/wrong label to each segment the algorithm found, a missing label to the segment that the algorithm failed to find. The result is shown in Table 2.

It is shown that our method is good at finding MCA and PCA, but the performance on identifying A2 is not satisfactory enough. The main reason is that our approach did not solve ACA entangling perfectly. The ground truth of the vessel centerness is constructed by normalized distance transform [3], which is also influenced by the vessel entangling. Hence, the prediction of the model is not ideal when the entangling is severe. The overall accuracy (excluding A2) is 98.1%, and the accuracy of A2 is 64.3%.

3.3 Demonstration of Results

Firstly, we show the vessel labeling results of some cases whose scanning includes the brain (Fig. 7). Note that our method successfully identifies all existing segments and correctly reports the missing segments.

Table 2. Scoring by the doctor for the proposed vessel labeling system.

	LA1	RA1	LA2	RA2	LICA	RICA	LM1	RM1	ACoA	BA	LPCoA	RPCoA	LP1	RP1	LP2	RP2
#Found	177	163	177	164	184	190	184	190	81	173	61	70	173	169	190	187
#Missing	0	0	2	10	1	1	0	0	0	3	0	0	0	0	0	0
#Wrong	0	0	57	57	0	2	0	3	5	0	1	4	10	7	1	4

As mentioned above, we aim to tackle every scanning situation in clinical conditions. Figure 8 shows the results of some special cases. The first to third cases are real neck scanning, which lacks upper vessel segmentation results. The fourth and fifth cases are artificial hard cases: we cropped ACA for the fourth case

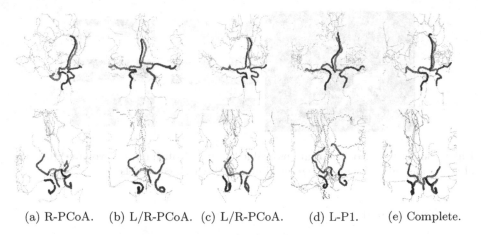

(a) R-PCoA.   (b) L/R-PCoA.   (c) L/R-PCoA.   (d) L-P1.   (e) Complete.

Fig. 7. The vessel labeling results in five normal cases. The results are separated as the anterior circulation (upper) and the posterior circulation (lower). The small red dots represent the whole vessel skeleton graph. Other colorful lines stand for the target vessels identified in the graph. Sub-captions (a–d) indicate the missing segments.

Fig. 8. The vessel labeling results in five special cases. All segments are shown in one figure.

and ACA+R-MCA for the fifth case. Our approach can still correctly identify all existing vessels and report the missing segments.

With the extracted vessel skeletons, we can do some advanced visualizations. For example, we can locate the Willis circle, which is usually hidden under the dense vessel network. The Willis circle plays a vital role in the collateral circulation (when one part of the circle is blocked, the influence on blood supply would be compensated by other parts). With this method, we can do volume rendering for the Willis circle at any view angle without occlusion (See Fig. 9 left). The most common disease in the cerebral artery system is stenosis. The best way to visualize stenosis is the straightened lumen, which enables radiologists to inspect the vessel radius at any position in one image (See Fig. 9 right).

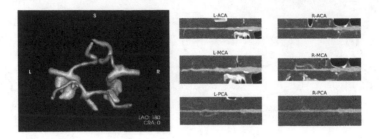

Fig. 9. Some advanced visualizations enabled by vessel labeling. Left: volume rendering of the Willis circle. Right: straightened lumen of extracted vessels.

4 Conclusion

In this paper, we propose an automatic vessel labeling method for the cerebral artery system based on registration and key points tracking. This method firstly utilizes the registration to recognize key points in the cerebral artery system, then identifies main arteries by finding the shortest path between key points and dissects main arteries into segments. Further, it generates the straightened lumen of vessel segments. Our method is robust and effective to execute the cerebral vessel labeling in most cases. It proves the feasibility of using a simple graph tracking method to solve this complex graph parsing problem. Nonetheless, its overall performance still needs to be further optimized for clinical use. Especially due to the A2 entangling problem, many A2 straightened lumens are distorted, which may mislead radiologists to a wrong diagnosis. A possible way is to perform centerline labeling and train centerness model with the labeled centerline.

References

1. Bilgel, M., Roy, S., Carass, A., Nyquist, P.A., Prince, J.L.: Automated anatomical labeling of the cerebral arteries using belief propagation. In: Medical Imaging 2013: Image Processing, vol. 8669, p. 866918 (2013)
2. Bogunović, H., Pozo, J.M., Cárdenes, R., San Román, L., Frangi, A.F.: Anatomical labeling of the circle of willis using maximum a posteriori probability estimation. IEEE Trans. Med. Imaging **32**(9), 1587–1599 (2013)
3. Guo, Z., et al.: DeepCenterline: a multi-task fully convolutional network for centerline extraction. In: Chung, A.C.S., Gee, J.C., Yushkevich, P.A., Bao, S. (eds.) IPMI 2019. LNCS, vol. 11492, pp. 441–453. Springer, Cham (2019). https://doi.org/10.1007/978-3-030-20351-1_34
4. Kingma, D.P., Ba, J.: Adam: a method for stochastic optimization. In: Proceedings of the 3rd International Conference on Learning Representations (ICLR) (2015)
5. Kitasaka, T., et al.: Automatic anatomical labeling of arteries and veins using conditional random fields. Int. J. Comput. Assist. Radiol. Surg. **12**(6), 1041–1048 (2017). https://doi.org/10.1007/s11548-017-1549-x
6. Lee, T.C., Kashyap, R.L., Chu, C.N.: Building skeleton models via 3-D medial surface axis thinning algorithms. CVGIP Graph. Models Image Process. **56**(6), 462–478 (1994)

7. Robben, D., et al.: Simultaneous segmentation and anatomical labeling of the cerebral vasculature. Med. Image Anal. **32**, 201–215 (2016)
8. Sandkühler, R., Jud, C., Andermatt, S., Cattin, P.C.: AirLab: autograd image registration laboratory. arXiv preprint arXiv:1806.09907 (2018)
9. Saxena, A., Ng, E.Y.K., Lim, S.T.: Imaging modalities to diagnose carotid artery stenosis: progress and prospect. Biomed. Eng. Online **18**(1), 66 (2019)
10. Wu, D., et al.: Automated anatomical labeling of coronary arteries via bidirectional tree lstms. Int. J. Comput. Assist. Radiol. Sur. **14**(2), 271–280 (2019)
11. Xu, G., Ma, M., Liu, X., Hankey, G.J.: Is there a stroke belt in China and why? Stroke **44**(7), 1775–1783 (2013)

Page-Level Handwritten Word Spotting via Discriminative Feature Learning

Jie Gao, Xiaopeng Guo, Mingyu Shang, and Jun Sun[✉]

Wangxuan Institute of Computer Technology, Peking University,
Beijing 100871, People's Republic of China
{gaojie2018,mingyu.shang,jsun}@pku.edu.cn, xpguo123@gmail.com

Abstract. Handwritten word spotting (HWS) is a task of retrieving word instances within handwritten documents, which is typically assisted by word annotations (word-level HWS). Previous methods following this paradigm are always in the manual feature modelling fashion, failing to capture sufficient discriminative information of the original input; also, they are always quite time-consuming in the artificial segmentation phase, limiting their applications in practice. To address these problems, we revisit HWS and model it on page-level via discriminative feature learning. Two distinct components modelled as neural networks are combined: word discriminative representation learning by Siamese Feature Network (SFNet) and the word discriminative spotting by Word Discriminative Spotting Network (WdsNet). Even without annotation of boxes, our WdsNet reaches impressive results on the IAM benchmark dataset with 76.8% mAP for the full page word spotting, revealing its superiority over other competitors.

Keywords: Handwritten word spotting · Discriminative feature learning · Page-level word feature extraction

1   Introduction

Handwritten text broadly exists in daily life such as historical manuscripts, offline exam and homework, blackboard text and so on. The automatic handwritten document indexing is crucial since it's the basis of document comprehension. Therefore, Handwritten word spotting (HWS) [6,10] arises, which is the task of retrieving the word from a given image (Query-by-Example, QbE) or a given string (Query-by-String, QbS) within the handwritten documents (see Fig. 1).

According to the existence of annotations, HWS is modelled in three ways: word-level, line-level and page-level. Both the Word-level and line-level HWS can be considered as a segmentation-based method, which assumes that all the documents are preceded into lines or words. The goal of Page-level HWS, by contrast, is to greatly reduce the amount of annotation work that has to be performed. A different taxonomy of HWS is depending on the shape of the query word, which can be a word image (Query-by-Example, QbE) or a word

© Springer Nature Switzerland AG 2020
G. Li et al. (Eds.): KSEM 2020, LNAI 12274, pp. 368–379, 2020.
https://doi.org/10.1007/978-3-030-55130-8_32

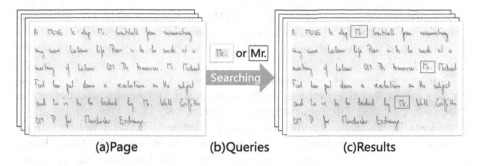

(a)Page   (b)Queries   (c)Results

Fig. 1. Illustration of the page-level HWS. We search "Mr." in pages by the sample image or the word string.

string (Query-by-String, QbS). Page-level HWS is still a challenging task due to the combination of QbE and QbS with no annotations of words. Moreover, page-level HWS is limited to the number of annotated page datasets.

In this work, we propose a page-level HWS method, which allows both QbE and QbS under the same HWS system. Most methods regard the word spotting as a classification task, which leads to the zero-shot or one-shot [5] problem. Instead, we formulate the word spotting as an encoding task. The key idea is that we encode the word images and texts into a common space and keep same words be close to each other. Then we search the queried word in this word discriminative space. Specifically, to address the zero-shot problem, we propose a siamese convolutional network for discriminative feature learning, called Siamese Feature Network (SFNet). For full-page word spotting, we propose a word discriminative spotting network (WdsNet) based on the faster-rcnn [15].

The main contributions of our work are four-fold: (i) We propose a SFNet for discriminative feature learning. (ii) We propose a WdsNet which embedded the SFNet. The WdsNet can query by example and query by string simultaneously on page-level. (iii) Our model is trained by segmented words and fine-tunes by full page, can finally spotting words on page level. In other words, we can feed our model in words rather than a large number of annotated pages. (iv) In the WdsNet, the multi-stage word segmentation architecture is proposed to improve the quality of word proposals. Our model has been proved to get impressive results on the IAM dataset with mAP 76.8% for the full page word spotting without any manual annotation.

2   Related Work

Concerning word-level HWS, the methods are classified into Optical Character Recognition (OCR) based method [7,12] and word embedding method [14,19,20], respectively. Methods based on OCR recognize all the scanned manuscripts to digital texts by OCR system firstly, then search the word through digital texts. The key point of the OCR-free method is word representations, or word embedding. The work of Fisher et al. [13] and the PHOC [1,17] embedding method enhanced the OCR-free word spotting networks. However, the two methods above have the same drawback: the zero-shot problem which exists widely could not be disposed.

To solve the zero-shot problem, researchers begin to regard the word spotting task as a metric learning problem. Initially proposed by Bromley and LeCun for signature verification [3], Siamese networks become a widely-known method for metric learning which can handle zero-shot problem well. Siamese networks have shown much potential in word spotting, which is further verified by Berat Kurar Barakat et al. [2]. However, the Siamese networks need two word images as the input that couldn't be employed for page-level word spotting directly.

Without annotations of boxes, line-level or page-level HWS will be more challenging. Since the word segmentation is a time-consuming work for people, spotting on line-level or page-level is indispensable for historical manuscripts indexing. Inventively, Retsinas et al. [16] proposed a word spotting method working on line-level. Retsinas's method learns PHOC embedding by a convolutional neural networks with word images. Then they use the same network for line level embedding. After all, the sliding window has been employed to match the feature of the query word and the lines. However, this method fails to handle the zero-shot problem and couldn't fulfill page-level word spotting.

Since the page-level HWS can be formulated to an object detection problem, the popular network faster-rcnn [15] proposed a Region Proposal Network (RPN) for predicting object bounds and objectness scores simultaneously. However, the results of the searching are determined by the annotation quality, the cascade rcnn [4] is proposed for region optimization. The cascade rcnn adjust bonding boxes by two or three stage of regression layers, aims to find a good set of proposal boxes for training the next stage. In this way, it can improve the recall of word boxes for the encoding training.

3   Methodology

3.1   Overview

In this work, we propose a page-level HWS method by combining two main networks: Firstly, we propose a siamese feature network (SFNet) for discriminative feature learning, which enables to encode the words into a word discriminative space. Secondly, we use the proposed WdsNet for page-level word encoding. The WdsNet employs the segmentation part and the SFNet contributes to fine-tune the embedding weights for page-level HWS.

The whole word spotting pipeline includes three parts as shown in Fig. 2: (a) The word discriminative feature learning. We propose a Siamese Feature network (SFNet), trained to learn the deep discriminative features in metric learning style. (b) The word discriminative feature extraction. We propose the Word Discriminative Spotting Network (WdsNet) to extract the word region features from full pages. We combine the Projection Text Proposals (PTP) and Dilated Text Proposals (DTP) for word pre-segmentation, then we employ the cascade

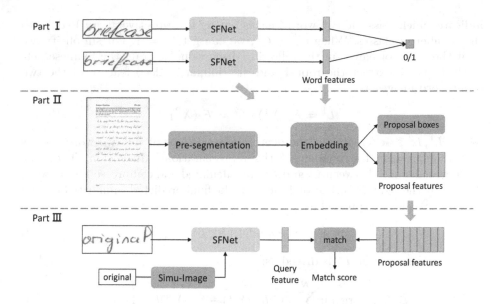

Fig. 2. Workflow of the word spotting networks. Part I is the SFNet, the word image pairs are the input and the distances of each pair are the labels. The word image pairs going through a convolutional siamese network for the discriminative feature learning. Part II is the WdsNet for end-to-end word spotting, consisting of pre-segmentation and feature extraction. Particularly, the feature extraction part reuses the SFNet, and the labels are the word features extracted by the SFNet. Part III is the process of matching the word feature of input query with proposal features extracted by part II.

regression for region fine-tuning. Finally, we embed the SFNet into WdsNet for feature extraction. (c) The word searching. We match the feature of query words extracted by the SFNet with the feature instances of the pages extracted by the WdsNet for word spotting.

3.2   Word Discriminative Representation Learning

As mentioned above, manual word representation methods couldn't handle the one-shot or zero-shot problem. We propose the Siamese Feature Network (SFNet) which consists of two convolutional neural networks that share weights with each other. The input of SFNet is a pair of word images and the output is the distance of the deep representation of the two word images. After embedding by the SFNet, the calculated distance between same words is shorter and at the same time, the calculated distance between different words is longer.

For the SFNet $\mathcal{F}_\beta$ parameterized by β , we construct a training dataset $D = \left\{ \left(\left(X_n^1, X_n^2 \right), Y_n, C_n^1, C_n^2 \right) \right\}$, where $X_n^1 \in \mathbb{R}^{W_n^1 \times H_n^1}$ and $X_n^2 \in \mathbb{R}^{W_n^2 \times H_n^2}$ are a pair of word images, $Y_n \in \{0, 1\}$ denotes whether the word on the two images are same, W_n^1, H_n^1 and W_n^2, H_n^2 represent the wide and high of the two word images, respectively. We consider different words as different classes, $C_n^1, C_n^2 \in \mathbb{R}^V$

indicate which class the two word images belong to respectively and V denotes the number of classes. When $C_n^1 = C_n^2$, we can get $Y_n = 1$. For simplicity, we omit the index of pair n in the following discussion. To train the Siamese network, first, we extract the robust and discriminative deep feature of the two input word images as:

$$L^1 = \mathcal{F}_\beta(X^1), \; L^2 = \mathcal{F}_\beta(X^2) \tag{1}$$

where $L^1, L^2 \in \mathbb{R}^D$ represent the deep representations of the two word images and D represents the number of the representations' dimension. Then the Euclidean distance between L^1 and L^2 are calculated and compressed to between 0 and 1 by the sigmoid activation function. The final predicted similarity between L^1 and L^2 can be expressed as:

$$\hat{Y} = Sigmoid(E(L^1, L^2)) \tag{2}$$

The frist loss function $\mathcal{L}_1$ is drived as:

$$\mathcal{L}_1(\beta) = \arg\min_\beta \sum_{n=1}^{N} -Y^n log(\hat{Y}^n) - (1-Y^n)log(1-\hat{Y}^n) \tag{3}$$

Besides, to make the latent representation more general, we add an additional dense layer as the classifier which takes the latent representation L^1 (or L^2) as the input and output the distributions $\hat{C}_1$ (or $\hat{C}_2$) . So the second loss $\mathcal{L}_2$ can be expressed as:

$$\mathcal{L}_2(\beta) = \arg\min_\beta \sum_{i=1}^{N} \sum_{j=1}^{V} -C_1^{ij} log(\hat{C}_1^{ij}) - C_2^{ij} log(\hat{C}_2^{ij}) \tag{4}$$

During training, the above two losses are jointly optimized, the final loss function is defined as:

$$\mathcal{L}(\beta) = \mathcal{L}_1(\beta) + \lambda \mathcal{L}_2(\beta) \tag{5}$$

where λ is the balanced parameter to be set.

3.3 WdsNet

We propose the WdsNet to get word embedding pool by inputting a handwritten page. It consists of four parts below: (i) The pre-segmentation part for calculating the potential word bounding boxes by Projection Text Proposals (PTP) and Dilated Text Proposals (DTP) before training. (ii) The cascade regression part for adjusting the bounding boxes by cascade regression before sending the related regions to the SFNet. (iii) The pre-trained SFNet. (iv) The parallel fully connected (FC) layer branches: One is a classification block. It takes the deep representation of the regions as input and outputs the classification results; the other calculates wordness scores for each proposed region. By retraining the SFNet with two additional branches, we can get more robust and discriminative representations for the proposed regions (Fig. 3).

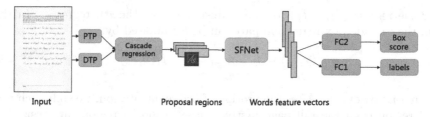

Proposal regions   Words feature vectors

Fig. 3. The WdsNet. Given handwritten pages, first going through a Projection Text Proposals (PTP) and Dilated Text Proposals (DTP) layer. With the cascade regression and ROI pooling, the region proposals all be adjusted to a fixed size. Then all the feature of proposal regions are fed into the box score branch and the embedding branch. The label of the embed feature is the word discriminative feature learned by the SFNet above.

Pre-segmentation. We choose Projection Text Proposals (PTP) joint with the Dilated Text Proposals (DTP) as an external segmentation component rather than the Region Proposal Networks (RPN) applied in faster R-CNN [15], enables to get more word regions with keeping WdsNet succinct. Given a grayscale image and a threshold, PTP first calculates the horizontal projection and segments the image to j lines. For each line, PTP calculates the vertical projection and segments them to l words. Then the PTP changes the threshold to segment again. The DTP is based on connected components recently introduced from [21]. The two methods all rely on the threshold defined before the training. On this basis, we joint this two methods for picking up proposals as more as possible.

Given a full page image denoted by $G \in \mathbb{R}^{W \times H}$, where the W, H are the width and height of the page. The PTP is denoted by $\mathcal{P}$ and the DTP is denoted by $\mathcal{D}$ the proposed potential regions are formulated as:

$$\{p_n\} = \mathcal{P}(G) \cup \mathcal{D}(G) \tag{6}$$

where $p_n = \{x_n, y_n, w_n, h_n\}$, $n = 1, 2, 3, ..., N$, N represents the number of the proposed potential regions. The x_n, y_n denote the coordinates of the proposed potential regions center point and w_n, h_n is the width and height of the proposed potential region.

Box Regression. After the pre-segmentation, we use a Non Maximum Suppression (NMS) component for de-duplication. The [4] introduced that the IOU is an important parameter for object detection backbone. Increasing IOU can keep the most inputs of the classification part be right, however, it can also lead to over-fitting due to less inputs. Based on this observation, we proposed muti-stage box regression before input proposals to SFNet.

The bounding box of proposal region can be formulated by $b_p = \{x, y, w, h\}$ containing the four coordinates of a page patch. The task of bounding box regression is to regress a candidate bounding box b_p into a target bounding box b_g, using the regressor $f_i(x, b_p)$ for box adjust, i means the ith regressor. In this work, the cascade box regression part can be formulated by:

$$f(x, b_p) = f_n \circ f_{n-1} \circ \cdots \circ f_1(x, b_p) \tag{7}$$

Feature Extraction. After the process of pre-segmentation, we spilt each proposed region from the full page firstly. Those proposed regions are resized to the same size and fed to pre-trained SFNet for word deep representation. Then the features are send to two branches. The first branch denoted by $\mathcal{F}'_\gamma$, we use a fully connected (FC) layer to get a confidential score about the wordness of each box; the second branch denoted by $\mathcal{F}''_\lambda$, it is also a FC layer. Considering different words as different classes, the second branch is utilized to classify the deep representation to V classes, where V represents the number of the word class. The output of the deep representation of P_n can be expressed as:

$$L_n = \mathcal{F}_\beta(P_n) \tag{8}$$

where $P_n \in \mathbb{R}^{W_n \times H_n}$ represents one of proposed regions on the full page. The confidential score $\hat{W}_n$ about wordness of P_n can be formulated as:

$$\hat{W}_n = \mathcal{F}'_\gamma(L_n) \tag{9}$$

And the distributions $\hat{C}_n$ indicates which class P_n belongs to can be formulated as:

$$\hat{C}_n = \mathcal{F}''_\lambda(L_n) \tag{10}$$

Losses of WdsNet. Noted that the SFNet $\mathcal{F}_\beta$ and the two parallel branches $\mathcal{F}'_\gamma$, $\mathcal{F}''_\lambda$ after it are joint trained. To optimize the first branch $\mathcal{F}'_\gamma$, we derive the loss $\mathcal{L}_1$ as:

$$\mathcal{L}_w(\beta, \gamma) = \arg\min_{\beta, \gamma} \sum_{n=1}^{N} -W^n log(\hat{W}^n) - (1-W^n) log(1-\hat{W}^n) \tag{11}$$

where the W is the ground truth label and $\hat{W}$ is the output confidential score of the first branch, n represents the number of the proposed potential regions.

To optimize the second branch $\mathcal{F}''_\lambda$ and the SFNet $\mathcal{F}_\beta$, we derive the loss $\mathcal{L}_c$ as:

$$\mathcal{L}_s(\beta, \lambda) = \arg\min_{\beta, \lambda} \sum_{i=1}^{N} \sum_{j=1}^{V} -C_1^{ij} log(\hat{C}_1^{ij})) - C_2^{ij} log(\hat{C}_2^{ij}) \tag{12}$$

where the $C \in R^V$ is the ground truth label which represents the class of the proposed region, $\hat{C} \in R^V$ is the output of $\mathcal{F}''_\lambda$, V represents the number of the word class.

The total loss is the combination of two losses:

$$\mathcal{L}_{all}(\beta, \gamma, \lambda) = \mathcal{L}_w(\beta, \gamma) + \alpha \mathcal{L}_c(\beta, \lambda) \tag{13}$$

3.4 Word Searching

In this section, we search the query word through pages. Before matching, we should get the representation of the query word. Due to the input of the words can be string or image, we firstly transform the string to image. Then we use the same method for feature extraction.

Given a word string, we first simulate an image with the same word. We first normalize all the word images from the train split and crop them to characters, then joint the handwritten characters to make a word image.

For the word image, we extract the word discriminative feature by the SFNet. For the handwritten pages, we extract the word discriminative feature by the WdsNet. Then we use the cosine distance to rank the similarity between the word image feature and the proposal region feature pool. The nearest region is most likely to be the same word with the query.

4 Experiments

4.1 Dataset

IAM Offline Handwriting Dataset [11]: The IAM Offline Handwriting Dataset is a well-known dataset consisting of 1539 pages, or 115320 words, written by 657 writers. We use the official train/val/test split for writer independent text line recognition, where there is no writer overlap between the different splits. Following official protocol, we get rid of all the stop words. Respectively, in the SFNet, we only use words appeared twice or more for learning the word discriminative feature. In the WdsNet, we remove ground truth boxes that are too small to down-sampled by a factor 8.

George Washington (GW) Dataset: The George Washington (GW) dataset [23] written in English consists of 20 pages, or 4860 words. We follow the evaluation procedure used in [33], by splitting the pages into a training and validation set of 15 pages, setting aside 5 for testing, and also doing a 5–15 split of train/val and testing. In both cases, we use 1 page as a validation set. The results reported are the average of four cross validations.

4.2 Training

The full networks are implemented on pytorch, and we train the SFNet about 3 h and the WdsNet about 2 h on an NVIDIA Titan GTX. As we mentioned before, the inputs of SFNet are image pairs. We establish a triple list consisting of anchor image, positive image, negative image for SFNet training. We use the resnet34 [8] as the backbone of SFNet. We used the train split word image of IAM and GW for learning and the test split word image for the evaluation. The WdsNet adopts pages as input, and the word classes as the embedding label for fine-tuning the SFNet included.

4.3   Ablation Experiments

Stages of Cascade Regression. The number of stages impacts the results, it is shown in Tabel 1. Adding one or two stages can reach obvious improvement. However, the performance will be decreased when it comes to three stages. Too many stages still can filter too much proposals, which leads to over-fitting easier.

Table 1. The impact of the number stages of cascade regression and the parameter IOU. We set IOU to be 25%, 50%, 60%, respectively.

	IOU 25%	IOU 50%	IOU 60%
Stage 1	74. 3	72. 2	71. 1
Stage 2	76. 8	74. 6	72. 1
Stage 3	70. 3	65. 4	63. 1

4.4   Performance

Segmentation Comparison. In this paper, we chose the jointed PTP-DTP for segmentation before training. In the respect of handwritten word segmentation, it can be observed in Table 2 that this method will get higher recall than RPN. The RPN method based on the sliding window, may be not so efficient for word segmentation due to the variety of the length-width ratio.

Table 2. Recall (%) of the PTP-DTP method and the Region Proposal Network (RPN).

Dataset	IAM		GW	
IOU	25%	50%	25%	50%
RPN	58. 1	39. 1	87. 0	82. 3
PTP-DTP	88. 7	87. 1	95. 0	94. 1

Comparison with the Segmentation-Based Methods. We evaluated our model by the protocol metric for word spotting, Mean Average Precision (MAP), where the Average Precision (AP) is defined as:

$$MAP = \frac{\sum_q^Q AP(q)}{|Q|}, AP = \frac{\sum_k^N P(k) \times r(k)}{|r|} \tag{14}$$

where $P(k)$ is the precision measured at cut-off k in the returned list and $r(k)$ is an indicator function that is 1 if a returned result at rank k is relevant, and

Table 3. mAP (%) performance evaluation on the segmentation based methods. Note that methods marked with * only work on the segmented word image.

Method	IAM	
	QbE	QbS
Embed attributes [1]*	88. 4	73. 7
DCToW [22]*	77. 0	85. 3
TPP-PHOCNet (CPS) [18]*	82. 7	93. 4
DeepEmbed [9]*	90. 4	94. 0
Our	76. 8	74. 6

0 otherwise. A retrieved word is considered relevant if its IoU overlap with a ground truth box is greater than a threshold $0.25, 0.5$ and the label matches the query. The MAP score is the mean of the AP over the queries (Table 3).

In this section, we compare our method with the state-of-the-art segmentation-based word spotting method and line-level method. All the data are for the IAM dataset. Considering the line-level method, our QbE result is higher even we query words from full page. We observe that we have competitive results for the IAM dataset in the QbE, it is near with the results of segmentation-based methods, even we didn't use any ground truth boxes information.

Table 4. mAP (%) performance evaluation comparison with the line-level method.

	IAM		GW	
	QbE	QbS	QbE	QbS
George [16]	71. 6	83. 2	-	-
Our	**76. 8**	74. 6	85. 7	75. 4

Comparison with the Line-Level Methods. We compared our method with the George's in Table 4. George matches the words by a sliding window style method on line level. However, we still can get better performance at QbE. The spotting performance of our proposed method may be improved by the encoding of the word discriminative feature. The word discriminative feature let our system be robust in example matching section.

5   Conclusion

We proposed a page-level word spotting framework via the word discriminative feature in this paper. By leveraging the metric learning merit in original Siamese network, we established the SFNet to obtain the discriminative feature for a better representation of input query. Then, we proposed the Word discriminative spotting network (WdsNet) and further embedded the proposed SFNet into it to achieve the page-level word spotting. This work is an initial exploration of page-level word spotting and we believe our observation of modelling page-level word spotting in a discriminative feature learning fashion will encourage future research.

Acknowledgements. This work is supported by National Natural Science Foundation of China under contract No. 61671025.

References

1. Almazan, J., Gordo, A., Fornes, A., Valveny, E.: Word spotting and recognition with embedded attributes. IEEE Trans. Pattern Anal. Mach. Intell. **36**(12), 2552–2566 (2014). https://doi.org/10.1109/tpami.2014.2339814
2. Barakat, B.K., Alasam, R., El-Sana, J.: Word spotting using convolutional siamese network. In: 2018 13th IAPR International Workshop on Document Analysis Systems (DAS). IEEE (2018). https://doi.org/10.1109/das.2018.67
3. Bromley, J., et al.: Signature verification using a "siamese" time delay neural network. In: Advances in Neural Information Processing Systems, pp. 737–744 (1994). https://doi.org/10.1142/9789812797926_0003
4. Cai, Z., Vasconcelos, N.: Cascade r-CNN: Delving into high quality object detection. In: 2018 IEEE/CVF Conference on Computer Vision and Pattern Recognition. IEEE (2018). https://doi.org/10.1109/cvpr.2018.00644
5. Fei-Fei, L., Fergus, R., Perona, P.: One-shot learning of object categories. IEEE Trans. Pattern Anal. Mach. Intell. **28**(4), 594–611 (2006). https://doi.org/10.1109/tpami.2006.79
6. Giotis, A.P., Sfikas, G., Gatos, B., Nikou, C.: A survey of document image word spotting techniques. Pattern Recogn. **68**, 310–332 (2017). https://doi.org/10.1016/j.patcog.2017.02.023
7. Graves, A., Liwicki, M., Fernández, S., Bertolami, R., Schmidhuber, J.: A novel connectionist system for unconstrained handwriting recognition **31**, 855–868 (2009). https://doi.org/10.1109/TPAMI.2008.137
8. He, K., Zhang, X., Ren, S., Sun, J.: Identity mappings in deep residual networks. In: Leibe, B., Matas, J., Sebe, N., Welling, M. (eds.) ECCV 2016. LNCS, vol. 9908, pp. 630–645. Springer, Cham (2016). https://doi.org/10.1007/978-3-319-46493-0_38
9. Krishnan, P., Dutta, K., Jawahar, C.: Word spotting and recognition using deep embedding. In: 2018 13th IAPR International Workshop on Document Analysis Systems (DAS). IEEE (2018). https://doi.org/10.1109/das.2018.70
10. Manmatha, R., Han, C., Riseman, E.: Word spotting: a new approach to indexing handwriting. In: Proceedings CVPR IEEE Computer Society Conference on Computer Vision and Pattern Recognition. IEEE (1996). https://doi.org/10.1109/cvpr.1996.517139
11. Marti, U.V., Bunke, H.: The IAM-database: an English sentence database for offline handwriting recognition. Int. J. Doc. Anal. Recogn. **5**(1), 39–46 (2002). https://doi.org/10.1007/s100320200071

12. Marti, U.V., Bunke, H.: Using a statistical language model to improve the performance of an HMM-based cursive handwriting recognition system. In: Hidden Markov Models: Applications in Computer Vision, pp. 65–90. World Scientific (2001). https://doi.org/10.1142/s0218001401000848
13. Perronnin, F., Dance, C.: Fisher kernels on visual vocabularies for image categorization. In: 2007 IEEE Conference on Computer Vision and Pattern Recognition. IEEE (2007). https://doi.org/10.1109/cvpr.2007.383266
14. Rath, T., Manmatha, R.: Word image matching using dynamic time warping. In: 2003 IEEE Computer Society Conference on Computer Vision and Pattern Recognition, 2003. Proceedings. IEEE Computer Society (2003). https://doi.org/10.1109/cvpr.2003.1211511
15. Ren, S., He, K., Girshick, R., Sun, J.: Faster r-CNN: towards real-time object detection with region proposal networks. IEEE Trans. Pattern Anal. Mach. Intell. **39**(6), 1137–1149 (2017). https://doi.org/10.1109/tpami.2016.2577031
16. Retsinas, G., Louloudis, G., Stamatopoulos, N., Sfikas, G., Gatos, B.: An alternative deep feature approach to line level keyword spotting. In: 2019 IEEE/CVF Conference on Computer Vision and Pattern Recognition (CVPR). IEEE (2019). https://doi.org/10.1109/cvpr.2019.01294
17. Sudholt, S., Fink, G.A.: PHOCNet: A deep convolutional neural network for word spotting in handwritten documents. In: 2016 15th International Conference on Frontiers in Handwriting Recognition (ICFHR). IEEE (2013). https://doi.org/10.1109/icfhr.2016.0060
18. Sudholt, S., Fink, G.A.: Attribute CNNs for word spotting in handwritten documents. Int. J. Doc. Anal. Recogn. **21**(3), 199–218 (2018). https://doi.org/10.1007/s10032-018-0295-0
19. Wahlberg, F., Dahllöf, M., Mårtensson, L., Brun, A.: Data mining medieval documents by word spotting. In: Proceedings of the 2011 Workshop on Historical Document Imaging and Processing - HIP. ACM Press (2011). https://doi.org/10.1145/2037342.2037355
20. Wei, H., Zhang, H., Gao, G.: Representing word image using visual word embeddings and RNN for keyword spotting on historical document images. In: 2017 IEEE International Conference on Multimedia and Expo (ICME). IEEE (2017). https://doi.org/10.1109/icme.2017.8019403
21. Wilkinson, T., Brun, A.: A novel word segmentation method based on object detection and deep learning. In: Bebis, G., et al. (eds.) ISVC 2015. LNCS, vol. 9474, pp. 231–240. Springer, Cham (2015). https://doi.org/10.1007/978-3-319-27857-5_21
22. Wilkinson, T., Brun, A.: Semantic and verbatim word spotting using deep neural networks. In: 2016 15th International Conference on Frontiers in Handwriting Recognition (ICFHR). IEEE (2016). https://doi.org/10.1109/icfhr.2016.0065

NADSR: A Network Anomaly Detection Scheme Based on Representation

Xu Liu[1,2] , Xiaoqiang Di[1,2,3(✉)], Weiyou Liu[1], Xingxu Zhang[1], Hui Qi[1,2],
Jinqing Li[1,2], Jianping Zhao[1,2], and Huamin Yang[1,2]

[1] School of Computer Science and Technology, Changchun University of Science and
Technology, Changchun 130022, China
dixiaoqiang@cust.edu.cn
[2] Jilin Province Key Laboratory of Network and Information Security,
Changchun 130022, China
[3] Information Center, Changchun University of Science and Technology,
Changchun 130022, China

Abstract. Deep learning has been widely used for identifying anomaly
network traffic. It trains supervised classifiers on a pre-screened numerical traffic feature dataset in the most cases, so the classification effectiveness depends heavily on feature representation. There is no unified
feature representation method, and the current feature representation
methods cannot profile traffic precisely. Therefore, how to design a traffic feature representation method to profile traffic is challenging. We
propose a Network Anomaly Detection Scheme based on data Representation (NADSR). Data representation method converts raw network
traffic into images by treating every numerical feature value as an image
pixel and then creating a circulant pixel matrix for a traffic sample. It
retains the traffic feature's spatial structure instead of padding empty
pixels with constant values while directly reshaping a long feature vector into a pixel matrix. Experimental results verify the effectiveness of
the proposed NADSR. It improves the overall detection accuracy compared with state-of-the-art methods, and also provides reference to solve
security-related classification problems.

Keywords: Anomaly detection · Traffic feature · Data representation

1 Introduction

With the fast development of the Internet, network security has become increasingly challenging. The Report of National Computer Network Emergency Technical Response Center in 2018 points out that malicious network attack is still

Supported by the Department of Science and Technology of Jilin Province grant
NO. 20190302070GX, the Education Department of Jilin Province grant NO.
JJKH20190598KJ, Jilin Education Science Planning Project (GH180148) and Jilin
Province College and University "Golden Course" Plan Project (Network protocol and
network virus virtual simulation experiment).

© Springer Nature Switzerland AG 2020
G. Li et al. (Eds.): KSEM 2020, LNAI 12274, pp. 380–387, 2020.
https://doi.org/10.1007/978-3-030-55130-8_33

one of the most important network security issues and deserves public attentions [1]. Anomaly detection, as a necessary part of security defense, is used to identify anomalies [5,8] for further defense. Though network anomaly detection has obtained some achievements with the help of deep learning methods, it has occurred with some difficulties. How to represent traffic feature is critical [7]. If extracted features cannot be learned to profile traffic efficiently, the classification accuracy will be influenced.

Therefore, we propose a Network Anomaly Detection Scheme based on data Representation (NADSR). The main challenge to be addressed in this paper is the network traffic feature representation in the data preprocessing stage.

To solve this challenge, we design a traffic-to-image conversion method for representation learning in the anomaly detection. Generally, one raw traffic sample is represented as a long feature vector [14]. Some researches have explored to reshape a long vector into a pixel gray value matrix directly to obtain an image [2,10], but it might disrupt the spatial structure and further decrease classification accuracy [16]. Some symbolic network traffic features [6,7,15] are encoded by One-Hot encoder and word embedding technique, this will produce massive zeroes that influence layer-by-layer processing effect of deep learning [11]. Therefore, to avoid the spatial feature loss and reduce the number of zeros within the pixel matrix, we bundle the sparse discrete features that are encoded by One-Hot scheme first, and then design a novel representation learning method – re-circulation pixel permutation strategy (RPP) by creating circulant pixel matrix, which retains spatial structure of raw network traffic. Finally, the data representation method coupled with Convolutional Neural Network (CNN) is constructed in the proposed NADSR.

The experiments are carried on two public network traffic datasets: NSL-KDD [12] and UNSW-NB15 [9]. Experimental results verify the effectiveness of NADSR. The contributions of this study are summarized as follows:

(1) The discrete features bundling method reduces the number of zeros within the pixel matrix, which is helpful for CNN to learn the traffic feature.
(2) The proposed NADSR improves the overall classification accuracy to 81.4% and 94.9% on NSL-KDD and UNSW-NB15, respectively.
(3) The proposed representation learning method RPP is not only suitable for CNN but also for other detection algorithms.

The remaining of this paper is structured as follows. Section 2 states the main problem and Sect. 3 illustrates the main methodology. The experimental results are described in Sect. 4. Section 5 concludes the full paper and the future work.

2   Problem Description

In a general network anomaly detection problem, the detection algorithm aims to identify anomalies deviated from the normal network traffic, and then report to the security operator for further analysis. This paper focuses on identifying anomalies, which can be abstracted as a binary-classification problem. Benefit from deep learning technique, it can be solved by training an effective classifier on the labeled data.

In this paper, network packets are captured first by the tcpdump, and then the features are extracted. After that, they are encoded into the image pixel matrices to represent as images. The classification model, CNN is used in this paper and it is trained on the image dataset. Finally, CNN will be evaluated on the new coming test data.

The core work focuses on how to retain the spatial characteristics of traffic feature in the data representation stage and then ensure a high accuracy of the detection model that is trained on the represented images. In this paper, we design an image conversion method which assumes traffic features have structural relationship and then represents them into images by the designed Re-circulation Pixel Permutation (RPP) strategy.

3   NADSR: A Scheme to Network Anomaly Detection Based on Representation

In the proposed NADSR scheme, we first encode one record of traffic feature into a long vector, then discard useless features and bundle the discrete features, and subsequently perform normalization. With the normalized features, data representation method is designed to convert traffic features into images. The images are used for training classification model. This section outlines the main work. The data pre-processing is introduced first, data representation method is detailed subsequently.

3.1   Data Pre-processing

There are both numerical and symbolic features in the traffic feature. The symbolic features can also be seen as the discrete features, and a discrete feature can be seen as a binary feature. To eliminate the influence of symbolic features on traffic representation, we bundle the discrete features. The discrete features are encoded into 0–1 vector by One-Hot encoder first. Assume all the discrete features are the same weight, the 0–1 vectors are bundled further to obtain a decimal feature. For example, there are four discrete features, after the bundling process, four discrete features are reduced into one numerical feature. This bundling operation not only maintains unity of each feature, but also keeps combination among each discrete feature.

3.2   Feature Reduction

To optimize the remaining features, a feature filter is designed to remove useless features. As the dimensions of features are different, using standard deviation to compare discreteness of features is inappropriate, so the coefficient of variance C_v is introduced and defined as (1).

$$C_{vi} = \frac{\sigma_i}{\mu_i} * 100\% \tag{1}$$

where σ_i and μ_i are standard deviation and mean of ith feature. Generally, a higher C_v indicates a higher discreteness, and the feature of higher C_v plays a more important role. Specially, when the mean μ_i is equal to zero, the corresponding feature will be seen as unimportant relatively.

3.3 Data Normalization

Data normalization can eliminate differences among different dimensional data, so it is therefore widely used in machine learning. Because features of different scales will result in the unreliability of training model, we normalize them in the same distribution. Rescale-Min-max normalization is designed in this paper as (2).

$$x_i' = \frac{x_i - x_{min}}{x_{max} - x_{min}} * (1 - a) + a \tag{2}$$

where x_{max} and x_{min} represent the maximum and minimum value of feature x_i, x_i and x_i' represent the raw feature and the normalized feature. To change the minimum value of the normalized, we re-scale the range of the normalized feature from $[0, 1]$ into $[a, 1]$ where indicator $a \in (0, 1)$.

3.4 Image Representation

To learn the deep characteristics of traffic feature automatically, we convert the feature vector of every traffic sample into a pixel matrix and then feed them into CNN as images. A Recirculation Pixel Permutation (RPP) strategy is designed. The RPP function is defined in (3) which is used to convert a long vector into a circulant matrix, where x_i is i^{th} sample of the dataset, and it is an original long vector with n elements. x_i' is obtained by moving every element $x_{ij}(j = 1, 2, \cdots, n)$ of x_i one unit forward every time, then x_i' is used to represent pixel values of the transformed image whose dimension is $n * n$. RPP not only retains the original spatial structure of sample, but also facilitates the detection algorithm to mine relationships among the adjacent features deeply.

$$x_i = [x_{i1}, x_{i2}, ..., x_{in}] \rightarrow \begin{bmatrix} x_{i1} & x_{i2} & \cdots & x_{in} \\ x_{i2} & \cdots & x_{in} & x_{i1} \\ \cdots & x_{in} & \cdots & \cdots \\ x_{in} & x_{i1} & x_{i2} & \cdots \end{bmatrix}_{n*n} = x_i' \tag{3}$$

3.5 Classification Model

After the data representation, classification model is used to verify their effectiveness and adaptability. CNN, as a typical deep learning algorithm, has achievements in the area of image processing. We use CNN to evaluate the performance and effectiveness of the proposed NADSR. The workflow of the proposed NADSR is detailed in Algorithm 1.

Algorithm 1. NADSR workflow

1: Input: Dataset D = training data, validation data, test data;
2: **Preprocess:** Feature Reduction, Data Normalization;
3: **Representation:**
4: **while** $x \in D$ **do**
5:    $x'(1,:) = x = [x_1, x_2, \cdots, x_n]$
6:    **for** $1 \le i \le n$ **do**
7:       $x'(i+1,:) = [x_{i+1}, x_{i+2}, \cdots, x_n, x_1, \cdots, x_i]$
8:    **end for**
9: **end while**
10: **Train** $\rightarrow$ **Validation** $\rightarrow$ **Test**
11: Output: acc_test, loss_test, and test report

---

4   Experiment

All experiments are conducted on an Ubuntu 16.04 LTS machine with Intel Xeon (R)W-2123, 3.6 GHz CPU, GeForce GTX TITAN Xp COLLECTORS EDITION GPU and 12 GB VRAM. Two public datasets, NSL-KDD [12] and UNSW-NB15 [9] are used for evaluation.

4.1   Data Representation Method Comparison

Figure 1 shows the comparison results with other researches that use different data representation methods. CN1 and CN2 represent the results obtained from [6] and [13], respectively on NSL-KDD test[+], and CU1 and CU2 indicate the results obtained from [13] and [3], respectively on UNSW-NB15. It can be found that, method used in previous research [13] obtains a lower FNR and a higher Recall than the others. In contrast, our result is better than the others in the perspectives of Accuracy, Precision and F1. Additionally, the accuracy rate in this paper is larger than [4] by nearly three percent. Though the results in research [6] are close to us, there is an over-fitting in its work by analyzing its confusion matrix.

In all, compared with other representation methods, the proposed data representation method is useful and competitive. It represents almost no distortion on the raw data. As a consequence, it can be regarded as supplying almost complete knowledge during representation. Its well performance can be contributed to the effective representation method and also the detection algorithm CNN, therefore, it provides the positive influence of data representation on the anomaly detection.

4.2   Detection Algorithm Comparison

The proposed representation method RPP works in data preprocessing stage, so its function is to help the further detection algorithm to improve classification result. Therefore, we measure the effectiveness of the proposed data representation on other detection algorithms. Figure 2 shows the comparison results conducted on NSL-KDD. We compare six approaches, support vector machine (SVM), k-Nearest Neighbor (KNN), Decision Tree (DT), Random Forest (RF), Naive Bayesian (NB) and Logistic Regression (LR), they are applied to test on the test$^+$ and test^{-21} of NSL-KDD.

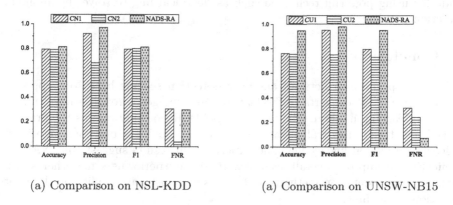

(a) Comparison on NSL-KDD   (a) Comparison on UNSW-NB15

Fig. 1. Comparison result with other representation methods.

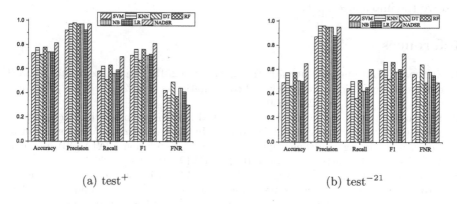

(a) test$^+$ (b) test^{-21}

Fig. 2. Comparison with other detection algorithms on NSL-KDD.

The obvious finding is that all evaluation metrics are close, and it suggests that the proposed data representation method is not only helpful for the CNN but also helpful for the other detection algorithms. Take a deep comparison, our method performs better than other detection algorithms. When performing data fitting, deep learning models can extract complex features than traditional machine learning models and mine hidden characteristics of the samples. Hence, deep learning models have a better representation learning ability than the shallow machine learning models [7].

4.3   Discussion

The classification based on image representation is still at the beginning stage and not mature. Though it might not be the best solution compared with other state-of-the-art approaches in some perspectives, it explores a new appliance mode for using powerful technique such as deep learning to solve the anomaly detection problem.

5   Conclusion

This paper proposes a network anomaly detection scheme based on representation, NADSR. It converts traffic features into images through a novel representation learning method. The proposed NADSR not only retains the original spatial structure of raw traffic features, but also propels the detection algorithm to learn the hidden knowledge. The experimental results suggest that NADSR is effective to improve overall accuracy. It also outperforms some other state-of-the-art representation methods and shows a robust adaptability on different detection algorithms.

The proposed representation method has improved classification effectiveness, but there still exists performance imbalance such as the rare anomaly is hard to detect. In the future, we will tend to study the data re-sampling method to solve the imbalance issue.

References

1. Summary of internet security situation in china in 2018, national computer network emergency technology processing and coordination center (2019). http://www.cac.gov.cn/2019-04/17/c_1124379080.htm
2. Blanco, R., Malagón, P., Cilla, J.J., Moya, J.M.: Multiclass network attack classifier using CNN tuned with genetic algorithms. In: 28th International Symposium on Power and Timing Modeling, Optimization and Simulation (PATMOS), pp. 177–182. IEEE (2018). https://doi.org/10.1109/PATMOS.2018.8463997
3. Khan, N.M., Madhav C, N., Negi, A., Thaseen, I.S.: Analysis on improving the performance of machine learning models using feature selection technique. In: Abraham, A., Cherukuri, A.K., Melin, P., Gandhi, N. (eds.) ISDA 2018 2018. AISC, vol. 941, pp. 69–77. Springer, Cham (2020). https://doi.org/10.1007/978-3-030-16660-1_7
4. Kwon, D., Natarajan, K., Suh, S.C., Kim, H., Kim, J.: An empirical study on network anomaly detection using convolutional neural networks. In: IEEE 38th International Conference on Distributed Computing Systems (ICDCS), pp. 1595–1598 (2018). https://doi.org/10.1109/ICDCS.2018.00178
5. Kwon, D., Kim, H., Kim, J., Suh, S.C., Kim, I., Kim, K.J.: A survey of deep learning-based network anomaly detection. Cluster Comput. 22(1), 949–961 (2017). https://doi.org/10.1007/s10586-017-1117-8
6. Li, Z., Qin, Z., Huang, K., Yang, X., Ye, S.: Intrusion detection using convolutional neural networks for representation learning. In: Liu, D., Xie, S., Li, Y., Zhao, D., El-Alfy, E.-S.M. (eds.) ICONIP 2017. LNCS, vol. 10638, pp. 858–866. Springer, Cham (2017). https://doi.org/10.1007/978-3-319-70139-4_87

7. Liu, H., Lang, B., Liu, M., Yan, H.: CNN and RNN based payload classification methods for attack detection. Knowl. Based Syst. **163**, 1–10 (2018). https://doi.org/10.1016/j.knosys.2018.08.036

8. Luo, X., Di, X., Liu, X., Qi, H., Li, J., Cong, L., Yang, H.: Anomaly detection for application layer user browsing behavior based on attributes and features, vol. 1069, pp. 1–9. Elsevier, Suzhou (2018). https://doi.org/10.1088/1742-6596/1069/1/012072

9. Moustafa, N., Slay, J.: UNSW-NB15: a comprehensive data set for network intrusion detection systems (UNSW-NB15 network data set). In: Military Communications and Information Systems Conference (2015). https://doi.org/10.1109/MilCIS.2015.7348942

10. Nsunza, W.W., Tetteh, A.Q.R., Hei, X.: Accelerating a secure programmable edge network system for smart classroom. In: IEEE SmartWorld, Ubiquitous Intelligence and Computing, Advanced and Trusted Computing, Scalable Computing and Communications, Cloud and Big Data Computing, Internet of People and Smart City Innovation, pp. 1384–1389. IEEE (2018). https://doi.org/10.1109/SmartWorld.2018.00240

11. Potluri, S., Ahmed, S., Diedrich, C.: Convolutional neural networks for multi-class intrusion detection system. In: Groza, A., Prasath, R. (eds.) MIKE 2018. LNCS (LNAI), vol. 11308, pp. 225–238. Springer, Cham (2018). https://doi.org/10.1007/978-3-030-05918-7_20

12. Tavallaee, M., Bagheri, E., Lu, W., Ghorbani, A.A.: A detailed analysis of the KDD cup 99 data set. In: IEEE International Conference on Computational Intelligence for Security and Defense Applications (2009). https://doi.org/10.1109/CISDA.2009.5356528

13. Vinayakumar, R., Alazab, M., Soman, K.P., Poornachandran, P., Al-Nemrat, A., Venkatraman, S.: Deep learning approach for intelligent intrusion detection system. IEEE Access **7**, 41525–41550 (2019). https://doi.org/10.1109/ACCESS.2019.2895334

14. Vinayakumar, R., Soman, K., Poornachandran, P.: Applying convolutional neural network for network intrusion detection. In: International Conference on Advances in Computing, Communications and Informatics (ICACCI), pp. 1222–1228. IEEE (2017). https://doi.org/10.1109/ICACCI.2017.8126009

15. Wu, K., Chen, Z., Li, W.: A novel intrusion detection model for a massive network using convolutional neural networks. IEEE Access **6**, 50850–50859 (2018). https://doi.org/10.1109/ACCESS.2018.2868993

16. Xie, K., Li, X., Xin, W., Cao, J., Zheng, Q.: On-line anomaly detection with high accuracy. IEEE/ACM Trans. Netw. **26**(3), 1222–1235 (2018). https://doi.org/10.1109/TNET.2018.2819507

A Knowledge-Based Scheduling Method for Multi-satellite Range System

Yingguo Chen[1], Yanjie Song[1(✉)] [iD], Yonghao Du[1], Mengyuan Wang[2], Ran Zong[3], and Cheng Gong[4]

[1] College of Systems Engineering, National University of Defense Technology, Changsha 410073, Hunan, China
ygchen@nudt.edu.cn, songyj_2017@163.com, duyonghao15@163.com
[2] CBT Nuggets, Eugene, OR 97401, USA
mengyuanwang93@gmail.com
[3] Unit 96625, Zhangjia Kou, China
zongran630@126.com
[4] Hunan Military Area Command, Changsha 41010, Hunan, China
gcclg195@126.com

Abstract. Satellite range systems (SRSs) play an important role in managing satellite resources and ensuring the smooth measurement and control of satellites. The increasing demand for satellite applications has made it more difficult to manage measurement and control tasks. Therefore, an efficient measurement and control system is needed to fulfil these task requests. Satellite range system, selects execution orders and locations for tasks according to the improved adaptive large neighbourhood search algorithm (ALNS-I). ALNS-I introduces heuristic rules into the initial population generation by considering the characteristics of the SRSP problem in the destruction and repair methods. A neighborhood search method is used to perform a local search when there is no clear optimization effect. The efficiencies of the framework and the algorithm are verified by experiments. The experimental results show that ALNS-I can achieve a higher task reward and task completion rate than comparison algorithms.

Keywords: Satellite range scheduling · Evolutionary · Neighborhood search · Knowledge

1 Introduction

A satellite range system (SRS) is a complex and core system in satellite management that controls the measurement and control processes for a satellite to ensure normal operation. This system can perform multiple tasks such as satellite resources management and satellite operations management. The most critical task of an SRS is the development of a measurement and control task execution plan for each satellite within a prescribed time range. The measurement and control plan requires that the measurement and control tasks are successfully completed to ensure normal satellite operation. Thus, there are

© Springer Nature Switzerland AG 2020
G. Li et al. (Eds.): KSEM 2020, LNAI 12274, pp. 388–396, 2020.
https://doi.org/10.1007/978-3-030-55130-8_34

strict timeliness and reliability requirements for an SRS. Designing an efficient SRS is of great significance for the continued development of the aerospace industry.

The SRSP has been analysed in related studies [1]. This problem involves multiple satellites and multiple ground stations and can thus also be called a multi-satellite multi-station scheduling problem. The SRSP solution is the development of a reasonable task execution plan for measurement and control tasks.

The satellite range scheduling problem is a research focus in the satellite field. Heuristic algorithms [2, 3] and meta-heuristic algorithms [4] have become the main methods for dealing with an increasing number of tasks.

The paper is structured as follows: in the second section, an optimization model for the satellite range scheduling problem is developed; in the third section, an improved adaptive large neighbourhood search algorithm (ALNS-I) is developed for the SRSP; in the fourth section, the results obtained using the developed algorithm are compared with those obtained using other methods; in the final section, the research results of this paper are summarized, and the direction of development of relevant research is analysed.

2   Optimization Model

2.1   Definition of Symbols

The symbols in this paper are defined to correspond with those in related studies [3].

Input Data:

- $Task = \{task_1, task_2, task_3, \ldots, task_n\}$ denotes the set of satellite measurement and control tasks. For a task $task_i$, the following attributes are defined:
- eat_i is the earliest available time for $task_i$ and is determined by satellite operation;
- lat_i is the latest available time for $task_i$ and is determined by users' requests;
- r_i is the reward for $task_i$ and depends on the duration, the importance and the urgency of the task;
- dur_i is the duration of $task_i$ and is determined by satellite operation; and
- $TW = \{tw_1, tw_2, tw_3, \ldots, tw_m\}$ is the set of ground stations time windows. The following attributes are defined for a time window tw_j:
- evt_j is the earliest visible time of the time window j and
- lvt_j is the latest visible time of the time window j.

Decision Variables:

- $x_{i,j}$ is a decision variable. If the time window j is chosen for the satellite measurement and control task i, then $x_{i,j}$ is 1 and is 0 otherwise.
- $st_{i,j}$ is the start time for the satellite measurement and control task i.

2.2 Assumptions

An SRS is a complex system involving multiple equipment fields. To theoretically model this practical application, the following assumptions are made about the SRSP.

1. The planning horizon for the problem is known in advance.
2. Each task is deterministic and will therefore not change the deadline and not be cancelled.
3. An antenna can only execute one measurement and one control task at a time.
4. A task cannot be suspended or pre-empted after it has been started.
5. Each task can only be completed once at most, irrespective of its periodicity.
6. Each task can only be completed once at most, irrespective of the destruction of task execution.

2.3 Mathematical Model

In this section, a model is constructed for the SRSP using mixed integer programming to facilitate subsequent task scheduling. The objective function of the model is to maximize the reward of the task sequence.

Objective Function:

$$\max \sum_{i=1}^{n} \sum_{j=1}^{m} r_i x_{ij}, \tag{1}$$

Constraints:

$$st_i + dur_i \leq st_{i+1}, \ i \in \{1, 2, 3, \ldots, n-1\}, \tag{2}$$

$$st_i < st_{i+1}, i \in \{1, 2, 3, \ldots, n-1\}, \tag{3}$$

$$\sum_{i=1}^{n} \sum_{j=1}^{m} x_{ij} \leq 1, \tag{4}$$

$$evt_j \leq st_i \leq lvt_j, \ i \in Task, \ j \in TW, \tag{5}$$

$$eat_i \leq st_i \leq lat_i, \ i \in Task, \tag{6}$$

$$evt_j \leq st_i + dur_i \leq lvt_j, \ i \in Task, \ j \in TW, \tag{7}$$

$$st_i + dur_i \leq lat_i, \ i \in Task, \tag{8}$$

$$x_{ij} \in \{0, 1\} \tag{9}$$

Constraints (2) and (3) ensure that there is an interval between two measurement and control tasks such that these two tasks do not overlap. A subsequent task can only be executed after the previous task has been completed. Constraint (4) defines the number of task executions in the planning horizon. Each measurement and control task may or may not be executed, and if executed, it is executed at most once. Constraints (5) and (6) ensure that the start time of the tasks is within the time allowed for the execution of the task and the available time windows of the ground stations. Constraint (7) ensures that the task completion time is within the visible time window of the selected ground station. Constraint (8) ensures that the tasks are completed before their respective deadlines. Constraint (9) allows the decision variable to take two values: 0 or 1.

3 Improved ALNS (ALNS-I)

3.1 Structure of ALNS-I

The improved adaptive large neighbourhood search algorithm that is developed here is described below.

Step 1: Initialization of parameters and data. Initialize all of the parameters in the adaptive large neighbourhood search algorithm and the jump parameters for the local search. Input the set of measurement and control tasks and the set of ground station time windows.

Step 2: Generate an initial plan. Heuristic rules are used to guide the generation of the solution, and the reward for the solution is calculated. At this time, the reward value is r_0, and the current solution is recorded as a local optimal solution s^l and a global optimal solution s^g.

Step 3: Destroy the current solution. The method for destroying the current solution uses roulette selection rules.

Step 4: Repair the broken solution. After the current solution is destroyed, a portion of the task sequence fragments are reinserted into the task sequence, where the task insertion is performed using the rules for the repair methods.

Step 5: Calculate the reward value of the newly generated solution. The reward value of the new task sequence is calculated and compared with the local optimal solution s^l and the global optimal solution s^g. If the current solution is better than the global solution, the current solution becomes the new global optimal solution. The comparative results are used to update the scores of the destruction and repair methods and obtain new weight values. After that, a newly obtained solution is updated to a new local optimal solution s^l.

Step 6: Reset the weight of the destruction and repair rules according to a condition. A weight reset condition is specified so that if the optimization iteration reaches its threshold condition, all of the weights of the destruction method and the repair method are reset.

Step 7: Carry out a local search. If the reward for the new solution is continuously lower than the reward for the solution for the previous generation, a local search optimization is performed.

Step 8: Determine if the algorithm termination condition is reached. If the algorithm termination condition is not reached, return to step 3; otherwise, output the optimal solution as the final solution.

3.2 Initial Solution Generation

The original adaptive large neighbourhood search algorithm exhibits good performance for traditional optimization problems. The heuristic rule comes from a related study and has been verified to obtain a good reward for the task sequence. This rule is defined as a CONSTRUCT RULE and is described below.

CONSTRUCT RULE: Sort according to the reward of measurement and control tasks, selecting the task execution positions for the measurement and control tasks with larger task rewards first.

3.3 Destruction Method

Destruction and repair methods are the key components of the ALNS-I and play an important role in improving the solution quality and obtaining a higher reward solution. A destruction method is the selection and removal of a partial sequence segment from the original task sequence. The goal of a destruction method is to incorporate randomness and diversity. Three task sequence destruction rules are used in this paper: DESTROY 1, DESTROY 2 and DESTROY 3. The specific treatment methods for the three destruction rules are given below.

DESTROY 1: Select a position from the task sequence randomly as the starting point of the destruction operation and specify the task segment from the starting point to the fixed interval and delete this task segment.
DESTROY 2: Select a certain length of the task segment from the foremost position of the task sequence for deletion.
DESTROY 3: Select a certain length of the task segment from the middle position of the task sequence for deletion.

3.4 Repair Method

The repair method is another important component of ALNS-I. This method is used to add the deleted task sequence segment back into the deleted sequence after the original segment has been destroyed. The choice of repair method directly affects the performance of the new task sequence. Three repair methods are used in this paper, REPAIR 1, REPAIR 2 and REPAIR 3, which are defined below.

REPAIR 1: Insert the deleted task segment into the beginning of the task sequence, and move the other positions of the original task sequence backwards.
REPAIR 2: Insert the deleted task fragment into the last position of the task sequence, and move the other positions of the original task sequence forwards.
REPAIR 3: Insert the deleted task segment into a random location as the insertion start point of the task segment, and move the tasks after the insertion point in the original task sequence backward.

4 Computational Results

4.1 Experimental Settings

4.1.1 Experimental Scene Setting

To verify the effect of the ALNS-I algorithm, the number of test data sets of tasks in this study is varied from 50 to 500. The data set is divided into 50, 100, 150, 200, 250, 300, 400, and 500 tasks, corresponding to a total of 8 scales. There are two task scene distribution types for measurement and control tasks: in Asia and in the global region. This experimental design shows how the developed algorithm can be adapted to different scenes and task scales, thereby reflecting the comprehensive effect of the algorithm.

4.1.2 Experimental Environment Setting

ALNS-I is implemented by Matlab2017a on a desktop with a Core I7-7700 3.6-GHz CPU, an 8-GB memory, and a Windows 7 operating system. Each algorithm runs 30 times under the same conditions in this experimental environment.

A neighbourhood search algorithm (NSA) is chosen as one of the comparison algorithms to serve as a commonly used algorithm in scheduling problems. Another comparison algorithm is a constructive heuristic algorithm, DSA, which sorts by the duration of the task and prioritizes long-duration measurement and control tasks.

4.2 Results and Discussion

In this paper, the experimental results for the SRSP in Asia are verified first. The task scale is selected from 50 to 300 tasks, and the results are shown in Table 2. In this table, ave represents the average reward that the algorithm can reach, max represents the maximum reward value, min represents the minimum reward value, and std. represents the standard deviation of the algorithm operation.

The results that are presented in Table 1 show that the ALNS-I algorithm that is developed in this paper exhibits the best performance for 50 to 300 tasks. However, in terms of the optimal reward, ALNS-I produces the maximum reward value for all of the scenes. In terms of the minimum reward value, for scales of 100, 200, or 300, the reward value obtained using the DSA algorithm is higher than the worst cases of the other two comparison algorithms. The results for another satellite range scheduling experiment in the global region are shown in Table 2.

Both ALNS-I and DSA produce the optimal solution, and ALNS-I reaches the optimal solution for every algorithmic operation for 50 tasks in the global region. This result reflects that the 50 global measurement and control tasks pose no difficulty in solving the SRSP of the SRS. The measurement and control process can easily be completed for of all of the tasks in the planning horizon. ALNS-I exhibits more volatility for optimizing tasks in the global region than for the range scheduling in Asia because each available time window can execute more tasks than in Asia. For a scale of 300, the minimum reward value optimized by the DSA algorithm is better than for the developed algorithm. Next, a large-scale range scheduling problem is used to verify the effect of

Table 1. Results of 50–300 tasks in Asia

Task set	ALNS-I				NSA				DSA
	Ave	Max	Min	Std.	Ave	Max	Min	Std.	
50	609	609	609	0.00	502.4	609	416	4.84	492
100	1030.2	1060	1006	1.69	830	911	736	5.37	879
150	1043.7	1124	1002	3.72	728.9	881	594	7.92	988
200	1183.3	1260	1112	5.59	864.6	1000	765	7.16	1013
250	1324.2	1423	1247	5.09	894.2	1057	754	7.74	1274
300	1063.6	1175	1005	5.13	733	896	631	8.20	866

Table 2. Results of 50–300 tasks in global region

Task set	ALNS-I				NSA				DSA
	Ave	Max	Min	Std.	Ave	Max	Min	Std.	
50	665	665	665	0.00	661.6	665	654	0.36	665
100	1209.9	1210	1209	0.03	1159.5	1169	1147	0.79	1166
150	1934.8	1957	1911	1.44	1769.1	1838	1718	3.96	1795
200	2157.4	2197	2124	2.06	1981.9	2036	1892	4.79	2107
250	2772.3	2854	2716	3.78	2457	2546	2344	7.43	2652
300	2852.5	2933	2803	3.63	2523.7	2687	2369	9.38	2892

the developed algorithm. The results of the different algorithms for solving a large-scale SRSP are visually compared in the box-line diagram in Fig. 1.

The results of the abovementioned experiments show that the ALNS-I algorithm that has been developed here can be used to solve large-scale SRSPs. In the developed algorithm, the heuristic rules play an important role in the initial solution generation.

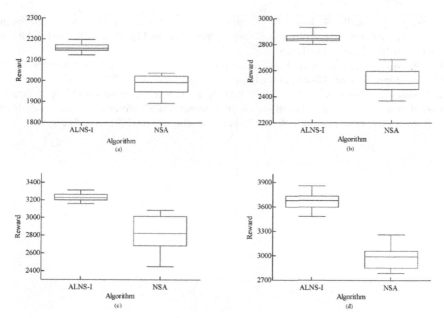

Fig. 1. Box plot of SRSP results for 200–500 mission scales in global region: Figures (a)–(d) represent 200 tasks, 300 tasks, 400 tasks, and 500 tasks, respectively.

5   Conclusion

A satellite range system is a key ground system that can guarantee normal operation for a satellite and the successful completion of various tasks. Among these tasks, the main task of an SRS is reasonable scheduling of satellite resources. Satellite range scheduling becomes challenging in large-scale scenes, and only efficient design algorithms can achieve high-reward solutions in a limited amount of time. Satellite range system uses a modified adaptive large neighbourhood search algorithm to allocate a time window for each measurement and control task. Experimental results show that the ALNS-I algorithm can be well applied to SRSs.

In future research, we will study more complex heuristic rules to improve the algorithmic stability. The uncertainty range scheduling problem will also be a primary direction in our next research study.

Acknowledgment. Yingguo Chen and Yanjie Song contribute equally to the article.

Funding. This work was supported by the National Natural Science Foundation of China (61473301, 71690233).

References

1. He, Y., Chen, Y., Lu, J., Chen, C., Wu, G.: Scheduling multiple agile earth observation satellites with an edge computing framework and a constructive heuristic algorithm. J. Syst. Archit. **95**, 55–66 (2019). https://doi.org/10.1016/j.sysarc.2019.03.005

2. Marinelli, F., Nocella, S., Rossi, F., Smriglio, S.: A lagrangian heuristic for satellite range scheduling with resource constraints. Comput. Oper. Res. **38**(11), 1572–1583 (2011). https://doi.org/10.1016/j.cor.2011.01.016
3. Luo, K., Wang, H., Li, Y., Qiang, L.: High-performance technique for satellite range scheduling. Comput. Oper. Res. **85**, 12–21 (2017). https://doi.org/10.1016/j.cor.2017.03.012
4. Barbulescu, L., Howe, A.E., Watson, J.P., Whitley, L.D.: Satellite range scheduling: a comparison of genetic, heuristic and local search. In: International Conference on Parallel Problem Solving from Nature, pp. 611–620 (2002). https://doi.org/10.1007/3-540-45712-7_59

IM-Net: Semantic Segmentation Algorithm for Medical Images Based on Mutual Information Maximization

Yi Sun[1(✉)] and Peisen Yuan[2]

[1] School of Computer Science, Fudan University, Shanghai 200433, China
ysun@fudan.edu.cn
[2] College of Information Science and Technology, Nanjing Agricultural University, Nanjing 210095, China

Abstract. The medical image is often noisy, which makes it difficult to extract the image features from the medical image segmentation model. Because the noise is often generated randomly, it is difficult to use supervised information for denoising. In this paper, we focus on this challenging problem and propose an IM-Net algorithm for medical image segmentation based on mutual information maximization. The IM-Net can remove the noise and therefore improve the quality of the extracted feature by maximizing the mutual information between the extracted feature and the input image. IM-Net uses the Binary Cross Entropy with Logits Estimation to approach the true value of mutual information and uses a bilinear interpolation function as a discriminator to maximize the mutual information estimator. Extensive experiments are conducted and the IM-Net is compared with different methods to demonstrate the effectiveness of our model. Experimental results show that the training efficiency and segmentation precision are greatly improved.

Keywords: Medical image segmentation · Mutual information maximization · Image feature extraction

1 Introduction

Extracting image features with strong representation ability from the original image has been a research focus in the field of image processing. At present, the widely used method to extract image features can be broadly divided into the following categories: (1) statistical feature extraction of color or gray level [1]; (2) texture and edge feature extraction [2]; (3) algebraic feature extraction[3]; (4) characteristics of transformation coefficients or filter coefficients [4]; (5) feature

---

This work was supported in part by National Natural Science Foundation of China (No. 61806097); College Students Entrepreneurship Training Program 2019 (No. S20190025). The study was funded by Shanghai Science and Technology Committee (17511104200, 18411952100, 17411953500).

G. Li et al. (Eds.): KSEM 2020, LNAI 12274, pp. 397–405, 2020.
https://doi.org/10.1007/978-3-030-55130-8_35

extraction based on deep learning [5–8]. The traditional feature extraction methods have strong explanatory power and can provide inspiration for the design of deep learning methods.

At present, the method based on deep learning has made remarkable achievements in the field of image feature extraction. There are many methods based on the convolution neural network [6,7,9]. But we find that the feature extraction of medical images based on a convolution neural network is still a challenge. It is difficult to ensure that the feature vectors extracted from the convolution layer are useful for subsequent tasks. For example, in the image segmentation algorithm based on U-shaped network[10–13], the process of feature extraction by down-sampling is mostly unsupervised, which needs to get the feedback of annotated data after up-sampling and forecasting. In this case, the low quality of deep feature coding reduces the training efficiency and segmentation accuracy of the network.

To solve this problem, we derive inspiration from the information theory. If a medical image x is treated as a sample generated by a random variable X, the encoder with the parameter ψ encodes the image x as $y = E_\psi(x)$, as a sample generated by random variable Y, we can capture useful features and filter noise by maximizing the mutual information between the original image and the feature encoding.

In facing of the challenges of low resolution, high noise and difficult feature extraction of medical images, we propose an image semantic segmentation algorithm IM-Net (**InfoMax Network**) based on mutual information maximization for medical images. It consists of two subnets and a discriminator. One subnet divides the image, while the other subnet and discriminator estimate and maximize the mutual information to improve the quality of deep features.

Our main contributions can be summarized as follows:

1. Binary Cross Entropy with Logits Estimation is defined according to the specific image segmentation task, which can remain in a high gradient state when approaching the true value of the mutual information, thus speeding up the convergence rate of the model;
2. We use the bilinear interpolation function as the discriminator of the IM-Net, and use the discriminator to identify whether the image and the sample are a match or not. The discriminator is also used to filter the noise of the image under the unsupervised condition so that the feature coding can capture the helpful information in the image skillfully;
3. The IM-Net model combines the algorithm of maximizing feature encoding to extract image features with the algorithm of image segmentation, which greatly improves the training efficiency and accuracy of the segmentation network.

2 Image Segmentation Algorithm Based on Mutual Information Maximization

2.1 Measurement and Lower Bound of Mutual Information

The mutual information between the random variables X and Y can be represented as the Kullback Leibler (KL) divergence of the product of their joint distributions and their respective marginal distributions, namely

$$I(X, Y) = D_{KL}(\mathbb{P}_{XY} || \mathbb{P}_X \otimes \mathbb{P}_Y), \tag{1}$$

From the Donsker-Varadhan form of KL-divergence, we can obtain a lower bound for the KL-divergence of the distributions $\mathbb{P}$ and $\mathbb{Q}$

$$D_{KL}(\mathbb{P} || \mathbb{Q}) \geq \sup_{T \in \mathcal{F}} \mathbb{E}_{\mathbb{P}}[T] - \log(\mathbb{E}_{\mathbb{Q}}[e^T]). \tag{2}$$

Our specially designed mutual information estimator for IM-Net is called the Binary Cross Entropy with Logits Estimation and is defined as follows

$$\hat{I}_{\Theta}^{(BL)}(X, E_{\psi}(X)) := \sup_{\theta \in \Theta} \mathbb{E}_{\mathbb{P}}[\log \sigma(T_{\theta}(x, E_{\psi}(x)))] + \tag{3}$$

$$\mathbb{E}_{\mathbb{P} \times \tilde{\mathbb{P}}}[\log(1 - \sigma(T_{\theta}(x', E_{\psi}(x))))], \tag{4}$$

where $\sigma(z)$ is a sigmoid function.

2.2 Overall Framework for IM-Net

The framework of IM-Net is shown in Fig. 1. IM-Net consists of two subnets and a bilinear interpolation function discriminator. One subnet performs feature extraction and image segmentation, and the other subnet and discriminator simultaneously estimates and maximizes the mutual information between image and feature coding to improve the quality of image deep features. The IM-Net model can filter the noise of medical images without supervision, capture the information which is helpful for image classification, and improve the accuracy of image segmentation. The details of IM-Net is as follows.

The contraction path follows the typical architecture of convolution networks, consisting of three down-sampled modules, each of which contains two 3×3 convolutions followed by a Rectified Linear Unit (ReLU), and a 2×2 pooling layer. With each down-sampling step, the number of feature channels is doubled, and finally the feature coding is obtained. In each step of the expansion path, the characteristic spectrum is sampled first, then 2×2 deconvolved, and then the number of the characteristic channels is halved, connected with the characteristic spectrum obtained from the corresponding levels in the contraction path, followed by a linear rectification function through two 3×3 convolutions. Finally, the 1×1 convolution maps each pixel to its own category.

The number of categories (including background) is L, the number of pixels is N, the p_{ic} indicates the probability of predicting that the i pixels belong to the c category, and the g_{ic} represents the 0–1 value of whether the i pixels actually

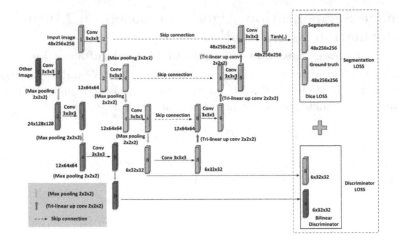

Fig. 1. The framework of IM-Net

belong to the c category, then the split loss function for the first subnet can be expressed as Eq. 5.

$$\text{DiceLoss} = \frac{1}{L} \sum_{c=1}^{L} \left(1 - \frac{2\sum_{i=1}^{N} p_{ic} g_{ic}}{\sum_{i=1}^{N} p_{ic}^2 + \sum_{i=1}^{N} g_{ic}^2 + \epsilon}\right), \tag{5}$$

where ϵ is a small positive number to ensure that the top denominator is not 0.

Another subnetwork of the IM-Net is the mutual-information encoder, which still uses the U-Net. Input images from the first subnetwork and other random images are input into the mutual information encoder, and the last layer of the mutual information encoder is used as their characteristic spectra.

The parameter θ is used to train a discriminator that distinguishes the matching between the feature map and the feature code in the sample. We need to find the discriminant function parameter θ that maximizes the BL-estimator to ensure that the estimator is close to the true value of the mutual information. The base encoder parameter ψ is used to maximize the BL-estimator to ensure that the learned image encoding is optimal. That is, the parameters θ and ψ are determined by the following formula

$$(\theta^*, \psi^*) = \arg\max_{\theta, \psi} \mathbb{E}_{\mathbb{P}}[\log \sigma(T_\theta(x, E_\psi(x)))] + \mathbb{E}_{\mathbb{P} \times \tilde{\mathbb{P}}}[\log(1 - \sigma(T_\theta(x', E_\psi(x))))]. \tag{6}$$

In addition, in order to train both the subnetworks and the discriminator parameters at the same time, we maximize the discriminator's target function, the BL mutual information estimator (Eq. 3), and minimize the discriminant loss function. Let the number of training samples be M, $T_\theta(x^{(i)}, y^{(i)})$ mean the discriminator's score on the i, and h_i represents whether or not the i sample is actually a value of 0–1 for a positive sample, then the loss function $\mathcal{L}_{\text{BCE}}$ of the

IM-Net discriminator can be expressed as Eq. 7.

$$\mathcal{L}_{\text{BCE}} = -\frac{1}{M}\sum_{i=1}^{M}\left[h_i\log\sigma(T_\theta(x^{(i)},y^{(i)})) + (1-h_i)\log(1-\sigma(T_\theta(x^{(i)},y^{(i)})))\right].$$

(7)

Thus, the loss function of the entire IM-Net is an addition of the split loss function and the discriminant loss function, and when its gradient drops to 0, the two subnetworks and the discriminator are trained.

2.3 Selection of Discriminant Function

We use the bilinear interpolation function as the discriminator of IM-Net, the result of bilinear interpolation is not linear, but the product of two linear functions. Assuming that the number of pixels in the feature map is N, the number of components in the feature code is K, the i-th pixel of the feature map is x_i, and the j-th of the feature code is y_j, then the bilinear function is expressed as Eq. 8.

$$T_\theta(x,y) = \sum_{i=1}^{N}\sum_{j=1}^{K}\theta_{ij}x_iy_j + \sum_{i=1}^{N}\theta_{i,0}x_i + \sum_{j=1}^{K}\theta_{0,j}y_j + \theta_{00},$$

(8)

where $\theta_{ij}(i \geq 1, j \geq 1)$ is a quadratic coefficient, $\theta_{i,0}$ and $\theta_{0,j}$ are a single coefficients, and θ_{00} is a constant coefficient.

3 Experiments

3.1 Datasets and Experiment Setup

The liver tumor image segmentation of LIVER 100 dataset was used to quantitatively evaluate the segmentation accuracy of the IM-Net. Our goal was to segment the medical images in the LIVER 100 dataset correctly into three categories: background, liver, and tumor. In the experiment, 80% of the LIVER 100 dataset were used for training and 20% for testing.

3.2 Performance Comparison

Results of the LIVER 100 data are shown in Fig. 2 with the IM-Net. The result of the test set image segmentation part of the data, the first medical imaging for the LIVER 100 original image, we use it as the IM-Net input images, the second for the prediction of the IM-Net output image segmentation, the third for real annotation data graph, which is marked red while the liver tumor is marked in blue. It can be seen from the qualitative analysis of Fig. 2 that the position, shape and size of the liver in the test set image of IM-Net are very close to the actual label. For identifying the tumor in the original image, the IM-Net is able to accurately determine whether there is a tumor in the medical image. In particular, the prediction accuracy of tumor location is promising. However, in

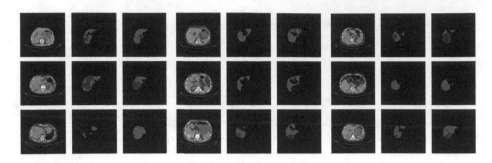

Fig. 2. Partial results of liver segmentation and tumor detection by the IM-Net algorithm on LIVER 100 dataset are presented.

the determination of tumor size, the prediction results of the IM-Net still have errors compared with real labeling.

In order to verify the effectiveness of the IM-Net algorithm, we compared the segmentation effect of IM-Net with some other recently proposed models that can be used for liver image segmentation. We trained U-Net [14], U-Net++ [15], DialResUnet [16], AgNet [17] and ra-unet [18] on LIVER 100 dataset. The tumor dice coefficients of the IM-Net were increased by 20.7%, 25.8%, 7.5%, 24.3%, 20.6%, compared respectively. After the analysis, it is believed that the IM-Net can accurately identify the image of the tumor because being compared with the traditional U-Net and its variants, the IM-Net can maximize the mutual information and capture the deep features to help the classification tasks, which can distinguish between different image tumor under noises, and determine the tumor's location and size (Table 1).

Table 1. Dice coefficients comparisons on LIVER 100 dataset

Algorithms	Tumor dice coefficient
U-Net	0.2899
U-Net++	0.2782
DialResUnet	0.3245
AgNet	0.2814
RA-Unet	0.2902
IM-Net (Our method)	**0.3499**

3.3   The Influence of Different Discriminant Functions

The discriminator with the bilinear function is used to identify whether the feature map and feature coding is a match. However, in related work [5,19], the neural network is used as discriminator. It is reasonable to use CNN as discriminant, the function expression ability of the neural network makes the mutual information estimator $\hat{I}_\theta(X, Y)$ approximate to the true value of mutual information $I(X, Y)$ with arbitrary accuracy. To test whether double linear function or neural network is more suitable as a criterion in medical image segmentation tasks, we have designed the following experiments: the IM-Net is replaced by a convolution bilinear function neural network. Then we conducted experiments and compared the results for segmentation effect.

Table 2. Dice coefficients of liver segmentation and tumor detection comparing on LIVER 100 dataset with different discriminant functions

Algorithms	Tumor dice coefficient
IM-Net with bilinear	0.3499
IM-Net with CNN	0.2815

In order to compare the effectiveness of the bilinear function discriminator and the convolutional neural network discriminator on the image segmentation, the results on the LIVER 100 dataset with different discriminators are shown in Table 2. Table 2 shows the dice coefficients of liver and tumor on the whole test set of the IM-Net with two different discriminant functions. After the replacement of discriminator by the convolutional neural network, the dice coefficient of tumor decreased by 19.5% compared with the original model.

3.4   The Influence of Different Loss Functions

The loss function in the IM-Net algorithm is composed of split dice loss function of U-Net and cross-entropy loss function of a bilinear discriminator. To explore the influence on the result of the loss function of the IM-Net training, we have designed the following experiments: modify the original IM-Net loss function as three parts, U-Net segmentation dice loss function for image segmentation and feature extraction, U-Net segmentation dice loss function of the mutual information of the encoder for extracting feature mappings, and the bilinear function BCEwithLogits discriminant criterion of the loss function. IM-nets with different loss functions were trained and tested on the same LIVER 100 dataset to quantitatively compare the effectiveness of loss functions during training.

Table 3 is the dice coefficient of tumor on the whole test set of IM-Net with two different loss functions. After the loss function was changed to three parts, the dice coefficient of tumor detection decreased by 24.1% compared with the original model. The experimental results in this section proved the correctness of our model in the selection of loss functions.

Table 3. Dice coefficients of liver segmentation and tumor detection were compared on LIVER 100 dataset with different loss functions

Algorithms	Tumor dice coefficient
The original IM-Net	0.3499
Changing the IM-Net loss function	0.2654

4 Conclusions

In this paper, the IM-Net model for medical image segmentation is proposed. Experimental results show that the bilinear discriminant function of our proposed can provide U-Net with low-dimensional denoizing image feature coding, which is helpful for classification tasks by discriminating positive and negative samples, thereby improving the performance and the accuracy of medical image segmentation.

References

1. Shen, D., Ip, H.H.S.: Discriminative wavelet shape descriptors for recognition of 2-D patterns. Pattern Recogn. **32**(2), 151–165 (1999)
2. Tamura, H., Mori, S., Yamawaki, T.: Textural features corresponding to visual perception. IEEE Trans. Syst. Man Cybern. **8**(6), 460–473 (1978)
3. Hong, Z.-Q.: Algebraic feature extraction of image for recognition. Pattern Recogn. **24**(3), 211–219 (1991)
4. Goupillaud, P., Grossmann, A., Morlet, J.: Cycle-octave and related transforms in seismic signal analysis. Geoexploration **23**(1), 85–102 (1984)
5. Hjelm, R.D., et al.: Learning deep representations by mutual information estimation and maximization. arXiv preprint arXiv:1808.06670 (2018)
6. Yang, S., Ramanan, D.: Multi-scale recognition with DAG-CNNs. In: Proceedings of the IEEE International Conference on Computer Vision, pp. 1215–1223 (2015)
7. Herranz, L., Jiang, S., Li, X.: Scene recognition with CNNs: objects, scales and dataset bias. In: Proceedings of the IEEE Conference on Computer Vision and Pattern Recognition, pp. 571–579 (2016)
8. Gao, H., Ji, S.: Graph u-nets. In: Proceedings of the 36th International Conference on Machine Learning, ICML 2019, vol. 97, pp. 2083–2092. PMLR (2019)
9. Yoo, D., Park, S., Lee, J.-Y., Kweon, I.S.: Multi-scale pyramid pooling for deep convolutional representation. In: Proceedings of the IEEE Conference on Computer Vision and Pattern Recognition Workshops, pp. 71–80 (2015)
10. Chen, C., Liu, X., Ding, M., Zheng, J., Li, J.: 3D dilated multi-fiber network for real-time brain tumor segmentation in MRI. In: Shen, D., et al. (eds.) MICCAI 2019. LNCS, vol. 11766, pp. 184–192. Springer, Cham (2019). https://doi.org/10.1007/978-3-030-32248-9_21
11. Huang, C., Han, H., Yao, Q., Zhu, S., Zhou, S.K.: 3D U²-net: a 3D universal U-net for multi-domain medical image segmentation. In: Shen, D., Liu, T., Peters, T.M., Staib, L.H., Essert, C., Zhou, S., Yap, P.-T., Khan, A. (eds.) MICCAI 2019. LNCS, vol. 11765, pp. 291–299. Springer, Cham (2019). https://doi.org/10.1007/978-3-030-32245-8_33

12. Brügger, R., Baumgartner, C.F., Konukoglu, E.: A partially reversible U-net for memory-efficient volumetric image segmentation. In: Shen, D., Liu, T., Peters, T.M., Staib, L.H., Essert, C., Zhou, S., Yap, P.-T., Khan, A. (eds.) MICCAI 2019. LNCS, vol. 11766, pp. 429–437. Springer, Cham (2019). https://doi.org/10.1007/978-3-030-32248-9_48

13. Orlando, J.I., et al.: U2-net: a Bayesian U-net model with epistemic uncertainty feedback for photoreceptor layer segmentation in pathological OCT scans. In: 16th IEEE International Symposium on Biomedical Imaging, ISBI 2019, pp. 1441–1445. IEEE (2019)

14. Ronneberger, O., Fischer, P., Brox, T.: U-net: convolutional networks for biomedical image segmentation. In: Navab, N., Hornegger, J., Wells, W.M., Frangi, A.F. (eds.) MICCAI 2015. LNCS, vol. 9351, pp. 234–241. Springer, Cham (2015). https://doi.org/10.1007/978-3-319-24574-4_28

15. Zhou, Z., Rahman Siddiquee, M.M., Tajbakhsh, N., Liang, J.: UNet++: a nested U-net architecture for medical image segmentation. In: Stoyanov, D., et al. (eds.) DLMIA/ML-CDS -2018. LNCS, vol. 11045, pp. 3–11. Springer, Cham (2018). https://doi.org/10.1007/978-3-030-00889-5_1

16. Milletari, F., Navab, N., Ahmadi, S.: V-net: fully convolutional neural networks for volumetric medical image segmentation. In: 2016 Fourth International Conference on 3D Vision (3DV), pp. 565–571, October 2016

17. Oktay, O., et al.: Attention U-net: learning where to look for the pancreas. arXiv preprint arXiv:1804.03999, 2018

18. Jin, Q., Meng, Z., Sun, C., Wei, L., Su, R.: RA-UNET: a hybrid deep attention-aware network to extract liver and tumor in CT scans. arXiv preprint arXiv:1811.01328 (2018)

19. Belghazi, M.I., et al.: Mutual information neural estimation. International Conference on Machine Learning, pp. 531–540 (2018)

Data Processing and Mining

Fast Backward Iterative Laplacian Score for Unsupervised Feature Selection

Qing-Qing Pang[1] and Li Zhang[1,2]([✉])

[1] School of Computer Science and Technology and Joint International Research
Laboratory of Machine Learning and Neuromorphic Computing,
Soochow University, Suzhou 215006, Jiangsu, China
20184227025@stu.suda.edu.cn, zhangliml@suda.edu.cn
[2] Provincial Key Laboratory for Computer Information Processing Technology,
Soochow University, Suzhou 215006, Jiangsu, China

Abstract. Iterative Laplacian Score (IterativeLS), an extension of
Laplacian score (LS) for unsupervised feature selection, iteratively
updates the nearest neighborhood graph for evaluating the importance of
a feature by its local preserving ability. However, LS and IterativeLS separately measure the importance of each feature and do not consider the
association of features. To remedy it, this paper proposes an enhanced
version of IterativeLS, called fast backward iterative Laplacian score
(FBILS). The goal of FBILS is to fast remove some unimportant features
by taking into account the association of features. The proposed FBILS
evaluates the feature importance according to the joint local preserving
ability that reflects the association of features. In addition, FBILS deletes
more than one feature in an iteration, which would speed up the process
of feature selection. Extensive experiments are conducted on UCI and
microarray gene datasets. Experimental results confirm that FBILS can
achieve a good performance.

Keywords: Unsupervised learning · Feature selection · Laplacian
score · Local preserving · Iteration algorithm

1 Introduction

With the development of technology and storage, the dimensionality of data
could be very high in many applications, such as image annotation [1], object
tracking [5] and image classification [15]. Usually, data may contain irrelevant
and redundant information, which would have a negative effect on learning algorithms owing to the curse of dimensionality. As a technique of dimensionality
reduction, feature selection has attracted a lot of attention in pattern recognition,

Supported by the Natural Science Foundation of the Jiangsu Higher Education Institu-
tions of China under Grant No. 19KJA550002, the Six Talent Peak Project of Jiangsu
Province of China under Grant No. XYDXX-054, and the Priority Academic Program
Development of Jiangsu Higher Education Institutions.

© Springer Nature Switzerland AG 2020
G. Li et al. (Eds.): KSEM 2020, LNAI 12274, pp. 409–420, 2020.
https://doi.org/10.1007/978-3-030-55130-8_36

machine learning and data mining. Feature selection can eliminate irrelevant and redundant features, which promotes the computational efficiency and keeps the interpretation of reduced description [13,15].

According to the situations of data labels, feature selection methods can be divided into three types: supervised, unsupervised and semi-supervised ones [2]. Both supervised and semi-supervised methods for feature selection, to some extent, depend on the label information to guide the feature evaluation by encoding features' discriminative information contained in labels [10]. For unsupervised methods, feature importance is assessed by the ability to maintain specific attributes of data, such as the variance value [4], and Laplacian score (LS) [9]. LS was proposed based on the spectral graph theory and uses a neighborhood graph to determine optimal features. Zhu et al. [17] proposed an iterative Laplacian score (IterativeLS), which progressively changes the neighborhood graph by discarding the least important features in each iteration and assesses the importance of the feature by its Laplacian score value. In each iteration, IterativeLS would reconstruct a neighborhood graph using the rest features. In doing so, higher time complexity is required for IterativeLS. Moreover, both LS and IterativeLS separately measure the importance of each feature and ignore the association of features.

To enhance both LS and IterativeLS, this paper presents a fast backward iterative Laplacian score (FBILS) method for unsupervised feature selection. Inspired by IterativeLS, FBILS also adopts a recursive scheme to select features. FBILS differs from IterativeLS in that it deletes more than one feature at each iteration, greatly reducing the number of iterations. The criterion of evaluating the feature importance in FBILS is to calculate the joint local preserving ability of features, which is totally different from those in both LS and IterativeLS. In summary, FBILS could speed up the process of iterative feature selection and take into account the association of features. The validity and stability of FBILS is confirmed by experimental results.

The remainder of this paper is organized as follows. In Sect. 2, we review two unsupervised methods for feature selection. Section 3 proposes the FBILS method and discusses its properties. In Sect. 4, we conduct experiments on UCI and gene datasets to compare the proposed method with the existing unsupervised methods. This paper is summarized in Sect. 5.

2   Related Methods

This section briefly reviews two unsupervised feature selection methods: LS and IterativeLS, which are very related to our work.

Assume that there is a set of unlabeled data $X = \{\mathbf{x}_1, \cdots, \mathbf{x}_u\}$, where $\mathbf{x}_i \in \mathbb{R}^n$, n is the number of features, and u is the number of samples. Let $F = \{f_1, \cdots, f_n\}$ be the feature set with features f_k, $k = 1, \cdots, n$ and $\mathbf{X} \in \mathbb{R}^{u \times n}$ be the sample matrix with row sample vectors $\mathbf{x}_i$, $i = 1, \cdots, u$.

2.1 LS

He et al. [9] proposed LS based on manifold learning. The goal of LS is to select features which can keep the local structure of the original data. That is to say that LS concerns the local structure rather than the global structure of data.

For the given X, LS first constructs the neighborhood graph that can be represented by a weight matrix $\mathbf{S}$:

$$S_{ij} = \begin{cases} \exp\left\{-\gamma\|\mathbf{x}_i - \mathbf{x}_j\|^2\right\}, & if \ (\mathbf{x}_i \in KNN(\mathbf{x}_j) \ \vee \ \mathbf{x}_j \in KNN(\mathbf{x}_i)) \\ 0, & otherwise \end{cases} \tag{1}$$

where $\gamma > 0$ is a constant to be tuned, and $KNN(\mathbf{x}_i)$ denotes the set of K nearest neighbors of $\mathbf{x}_i$.

LS measures the importance of feature f_k by calculating its Laplacian score:

$$J_{LS}(f_k) = \frac{\sum_{i=1}^{u} \sum_{j=1}^{u} (x_{ik} - x_{jk})^2 S_{ij}}{\sum_{i=1}^{u} (x_{ik} - \mu_k)^2 D_{ii}} \tag{2}$$

where x_{ik} denotes the kth feature of the ith sample, $\mu_k = \frac{1}{u}\sum_{i=1}^{u} x_{ik}$ is the mean of all samples on feature f_k, and $\mathbf{D}$ is a diagonal matrix with $D_{ii} = \sum_j S_{ij}$.

The smaller the score $J_{LS}(f_k)$ is, the more important the kth feature is for keeping the local structure of data. The computational complexity of constructing $\mathbf{S}$ is $O(nu^2)$, and the computational complexity of calculating scores for n features is $O(nu^2)$. Hence, the overall computational complexity of LS is $O(nu^2)$.

2.2 IterativeLS

IterativeLS was presented by introducing the iterative idea into LS [17]. Experimental results in [17] indicated that IterativeLS outperforms LS on both classification and clustering tasks.

The key idea of IterativeLS is to gradually improve the neighborhood graph based on the remaining features. In each iteration, IterativeLS discards the least relevant feature with the greatest score among the remaining features. Similar to LS, IterativeLS evaluates the importance of a feature by its Laplacian score that is calculated according to (2).

Assume that A is the remaining feature subset in the current iteration. Let $|A| = n'$. In the current iteration, the computational complexity of constructing $\mathbf{S}$ is $O(n'u^2)$, and computational complexity of calculating scores for n' features is $O(n'u^2)$. Note that $n' < n$. Hence, the overall computational complexity of IterativeLS is $O(n^2u^2)$ for n iterations.

3 Fast Backward Iterative Laplacian Score

This section presents the novel method for unsupervised feature selection: FBILS, which is an extension of LS. Both LS and IterativeLS measure the importance of features separately using the ability to maintain the local structure that is measured by the Laplacian score. Similar to both LS and IterativeLS, FBILS also considers maintaining the local structure. However, FBILS measures the importance of feature f_k by considering its joint local preserving ability. Similar to IterativeLS, FBILS adopts a recursive scheme. The difference is that IterativeLS discards the least relevant feature with the greatest Laplacian score in each iteration, and FBILS uses a new criterion to remove more than one feature in each iteration. Therefore, FBILS greatly reduces the running time with less iterations.

3.1   Criterion

FBILS iteratively removes features according to a novel criterion, the joint local preserving ability. We first discuss how to measure the joint Laplacian score for feature subsets before we introduce the criterion to measure the importance of feature f_k.

Both LS and IterativeLS calculate the Laplacian score of only single feature. Because we want to consider the association between features, we have to calculate the Laplacian score of feature subsets, or joint Laplacian score that is defined as follows.

Definition 1. *Given the weight matrix* $\mathbf{S}$, *the centered sample matrix* $\widetilde{\mathbf{X}}_A$ *and its corresponding feature subset* A, *the joint Laplacian score of* A *is defined as*

$$J(A) = \frac{trace\left(\widetilde{\mathbf{X}}_A^T \mathbf{L} \widetilde{\mathbf{X}}_A\right)}{trace\left(\widetilde{\mathbf{X}}_A^T \mathbf{D} \widetilde{\mathbf{X}}_A\right)} \tag{3}$$

where $trace(\cdot)$ *is the sum of the diagonal elements of a matrix,* $\mathbf{L} = \mathbf{D} - \mathbf{S}$ *is the Laplacian matrix, and* $\mathbf{D}$ *is a diagonal matrix with* $D_{ii} = \sum_j S_{ij}$.

According to Definition 1, we need to construct the weight matrix $\mathbf{S}$ to represent the neighborhood graph. For the given subset of data X_A that is a part of X and consists of features in A, we need to make it centered, which results in a new set with zero mean. Given these conditions, we can calculate the joint Laplacian score of A by (3). A smaller joint Laplacian score of A means that the local structure can be maintained better by using this feature subset, or A is more important. On the basis of Definition 1, we can describe the novel criterion as follows.

Definition 2. *Given the feature set* F, *a feature subset* $A \subseteq F$, *and any feature* $f_k \in A$, *the joint local preserving ability of* f_k *under* A *is defined as*

$$L(f_k|A) = J\left(A - \{f_k\}\right) - J(A) \tag{4}$$

where $A - \{f_k\}$ *denotes that feature* f_k *is removed from* A, *and* $J(A - \{f_k\})$ *and* $J(A)$ *are the joint Laplacian scores of feature subsets* $(A - \{f_k\})$ *and* A, *respectively.*

Definition 2 implies that the joint local preserving ability of feature f_k is related to two joint Laplacian scores. One is $J(A)$ that is the same for all $f_k \in A$. The other is $J(A - \{f_k\})$, the joint Laplacian score of the candidate subset $A - \{f_k\}$. We discuss two cases: $L(f_k|A) > 0$ and $L(f_k|A) \le 0$.

- When $L(f_k|A) \le 0$, $J(A - \{f_k\})$ is smaller than or equal to $J(A)$. In other words, the joint Laplacian score does not change or becomes small when feature f_k is removed from A. Note that LS-like algorithms prefer features with small Laplacian scores. In this case, we much prefer the feature subset $A - \{f_k\}$ to the one A. That is, f_k is less important.
- When $L(f_k|A) > 0$, $J(A - \{f_k\})$ is greater than $J(A)$. In this case, it is unwise to remove feature f_k from A.

In summary, the greater the value $L(f_k|A)$ is, the stronger the joint local preserving ability of f_k under A is. Thus, what we need to do is to remove these features with a weak joint local preserving ability. We could follow the way of IterativeLS, or deleting the weakest feature in each iteration. However, it is time-consuming if just one bad feature is removed in an iteration.

One goal of FBILS is to speed up the iterative procedure, which can be implemented by using the following deletion rule:

$$\forall f_k \in A, f_k \text{ is delected from } A, if \ L(f_k|A) < 0 \tag{5}$$

The deletion rule allows us to remove all possible bad features from the current feature subset in one time. The feasibility and reasonability of this rule is discussed later in Subsect. 3.3.

3.2   Algorithm Description

The detail algorithm description of FBILS is shown in Algorithm 1. Quite simply, FBILS requires constructing a neighborhood graph and calculating joint Laplacian scores of feature subsets in each iteration, which is repeated until all features are ranked.

The inputs of FBILS include the given dataset and the parameter K that is required for the neighborhood graph. In the first step, the subset of remaining features A is initialized to be a complete set, or the whole feature set. $\bar{A}$ is the ordered set and $\bar{A} = \emptyset$. Step 3 computes the weight matrix $\mathbf{S}_A$ using the dataset X_A in the current iteration. Step 4 computes the joint Laplacian score of the feature subset A by (3). Steps 5–8 get the joint local preserving ability for all features in A. Step 10 finds out the unimportant features in A with $L(f_k|A) \le 0$ and forms a temporary feature subset B. If B is empty, then FBILS would come to an end. In this case, features in A would be ranked according to the value of $L(f_k|A)$. If B is non-empty, then features in B would be ranked according to their joint local preserving ability. Then the ranked features are inserted in the list $\bar{A}$. Note that the smaller the value $L(f_k|A)$ is, the later is the insertion of the corresponding feature f_k in the list. If B is non-empty, features in B should be removed from A. Repeat Steps 3–18 until A is empty.

Finally, Algorithm 1 greturns the ranking of all features. Assume that the number of features required by a given task is provided in advance, say r. There are two ways to get r features using FBILS. One way is that we can only pick up the r top features in $\bar{A}$ after we perform Algorithm 1. The other way is to change the termination condition in Step 2 as $|A| > r$ and return the remaining feature subset A instead of $\bar{A}$. In Step 20, if $|A| < r$, we would get the $r - |A|$ top features in $\bar{A}$ and put them into A.

3.3 Properties Analysis

This part concerns about the properties of FBILS. We give a lemma and a theorem, and do not prove them for limitation of space.

Lemma 1. *For a non-empty feature subset A, its joint Laplacian score is positive, or*

$$J(A) \geq 0. \tag{6}$$

Lemma 1 indicates that the joint Laplacian score of feature subsets are greater than zero. The following theorem is related to our deletion rule (5).

Algorithm 1. Fast Backward Iterative Laplacian Score (FBILS)

Require: Dataset X with u samples and n features, and neighborhood parameter K;
Ensure: The ordered set of features;
 1: Initialize $A = \{f_1, \cdots, f_n\}$ and the ordered set $\bar{A} = \emptyset$;
 2: **while** $A \neq \emptyset$ **do**
 3:     Calculate the weight matrix $\mathbf{S}_A$ by (1) using X_A with u samples and $|A|$ features;
 4:     Compute the joint Laplacian score of A, $J(A)$ by (3);
 5:     **for** each $f_k \in A$ **do**
 6:         Compute the joint Laplacian score of $A - \{f_k\}$, $J(A - \{f_k\})$ by (3);
 7:         Compute $L(f_k|A)$ by (4);
 8:     **end for**
 9:     Let a temporary feature subset $B = \emptyset$;
10:     Find all features with $L(f_k|A) \leq 0$ and add them to B, or $B = \{f_k \in A \mid L(f_k|A) <= 0\}$;
11:     **if** $B = \emptyset$ **then**
12:         Rank features in A according to the descending order of $L(f_k|A)$, $f_k \in A$;
13:         Insert the ranked features into the head of the list $\bar{A}$;
14:         **break**;
15:     **else**
16:         Rank features in B according to the descending order of $L(f_k|A)$, $f_k \in B$;
17:         Insert the ranked features into the head of the ordered set $\bar{A}$;
18:         Update the set A by removing these features in B, or $A = A - B$;
19:     **end if**
20: **end while**
21: **return** The ordered set $\bar{A}$.

Theorem 1. *Let A be the feature subset. For $f_k \in A$, given the joint Laplacian score $J(A)$, the joint local preserving ability $L(f_k|A)$ and the Laplacian score $J_{LS}(f_k)$, the following relationships hold true:*

- *If $L(f_k|A) > 0$, then $J_{LS}(f_k) < J(A)$;*
- *If $L(f_k|A) \leq 0$, then $J_{LS}(f_k) \geq J(A)$.*

Theorem 1 implies the relationship between the Laplacian score of feature f_k and the joint Laplacian score of the feature subset A containing f_k when $L(f_k|A) \leq 0$ or $L(f_k|A) > 0$. When $L(f_k|A) \leq 0$, the Laplacian score of f_k is greater than or equal to the joint Laplacian score of A. The deletion rule states that those features with $L(f_k|A) \leq 0$ would be removed from the current feature subset. In other words, we would delete all $f_k \in A$ with $J_{LS}(f_k) \geq J(A)$, where $J(A)$ is taken as a threshold to make us select more than one feature in one iteration. According to the view of both LS and IterationLS, the greater $J_{LS}(f_k)$ is, the less important feature f_k is. Thus, our deletion rule is reasonable because it consists with the view.

Now, we analyze the computational complexity of FBILS. Without loss in generality, let A be the remaining feature subset in the current iteration. Let $|A| = n'$. Similar to IterativeLS, the computational complexity of constructing $\mathbf{S}_A$ is $O(n'u^2)$. If we directly compute joint Laplacian scores, which has a computational complexity of $O(n'^2u^2)$, FBILS would be slow. Fortunately, we could speed up this procedure. According to the proof of Theorem 1, the calculation of $J(A - \{f_k\})$ can be reduced. For limitations of space, we do not discuss it any more. Thus, the computational complexity of computing joint Laplacian scores is reduced to be $O(n'u^2)$. Then the overall time complexity of FBILS is between $O(nu^2)$ and $O(n^2u^2)$, which is related to the iteration number. Assume that the iteration number is T, then the complexity is $O(Tnu^2)$.

4 Experimental Analysis

In order to verify the feasibility and effectiveness of FBILS, simulation experiments were carried out on several UCI datasets [7] and microarray gene expression datasets [11]. We compared FBILS with both LS and IterativeLS and used the nearest neighbor classifier to measure the discriminant ability of selected features.

All experiments were performed in MATLAB 2015b and run in a hardware environment with an Intel Core i5 CPU at 2.60 GHz and with 8 GB RAM.

4.1 UCI Dataset

We considered 8 UCI datasets here and compared FBILS with Variance, LS and IterativeLS algorithms. The related information of 8 UCI datasets, including Australian, Glass, Heart, Ionosphere, Segment, soybeanLarge, Vehicle and Wdbc, is shown in Table 1. For these UCI datasets, the original features were normalized to the interval $[0, 1]$. To validate the ability to select features, we added n noise features to the original data before normalization, where n is the number of features. In the ranked feature list, we only considered the first n features. We thought that a good method for feature selection should rank these noise features towards the end of the list.

In order to obtain more convincing comparison results and eliminate accidental errors, we used 10-fold cross-validation. That is to say, the original dataset was randomly divided into ten equal-sized subsets. Then 9 subsets were used as the training set and the rest one was used as the test set. The 10 subsets were used as test sets in turn, and then the average of 10 times was calculated as the final result of classification. In addition, all compared algorithms require the parameter K for constructing the neighborhood graph. Let it vary in the set $\{1, 2, \cdots, 9\}$.

Table 2 shows the highest average accuracy with the corresponding standard deviation and optimal feature number of all compared algorithms, where the best values among compared methods are in bold. We can see that FBILS is superior to Variance, LS and IterativeLS on all datasets. For example, FBILS achieves the accuracy 92.00% on the SoybeanLarge dataset, followed by IterativeLS 90.80%. In a nutshell, FBILS can effectively rank features and make good ones at the top of feature list. Table 3 shows the running time and iteration numbers of FBILS and IterativeLS on the UCI datasets. Owing to the small number of features in UCI datasets, the time required for LS and variance is very small. Thus, we did not list them here. It can be clearly seen that the number of iterations of FBILS is much smaller than IterativeLS, and the running time of FBILS is also less.

Table 1. Description of UCI datasets

No.	Dataset	#Sample	#Attribute	#Class
1	Australian	690	14	2
2	Glass	214	9	6
3	Heart	301	13	2
4	Ionosphere	351	34	2
5	Segment	2310	19	7
6	SoybeanLarge	250	35	14
7	Vehicle	846	18	4
8	Wdbc	569	30	2

4.2   Microarray Gene Datasets

In experiments, FBILS was applied to microarray gene datasets, including Leukemia [8], Novartis [6], St. Jude Leukemia (SJ-Leukemia) [16], Lungcancer [3] and the central nervous system (CNS) [12]. It is well-known that the number of features is much greater than the number of samples in the gene datasets. The

Table 2. Average accuracy and standard deviation of different methods on UCI datasets

Dataset	FBILS	Variance	LS	IterativeLS
Australian	**83.33** ± 5.57 (11)	81.91 ± 4.06 (10)	81.33 ± 4.48 (8)	81.77 ± 4.02(8)
Glass	**62.32** ± 14.41 (7)	42.59 ± 11.26 (7)	60.74 ± 11.60 (4)	51.73 ± 11.41 (9)
Heart	**78.52** ± 7.16 (12)	76.30 ± 9.59 (7)	77.78 ± 9.56 (15)	76.30 ± 9.11 (10)
Ionosphere	**93.19** ± 3.26 (8)	91.46 ± 4.22 (8)	92.63 ± 6.62 (9)	92.06 ± 3.42 (11)
Segment	**97.27** ± 1.08 (18)	**97.27** ± 1.14 (12)	97.14 ± 1.14 (17)	96.84 ± 1.65 (12)
SoybeanLarge	**92.00** ± 5.96 (28)	88.00 ± 5.96 (33)	90.40 ± 6.31 (31)	90.80 ± 4.24 (27)
Vehicle	**73.50** ± 4.90 (17)	69.85 ± 4.57(12)	73.38 ± 2.27 (16)	73.16 ± 2.80(12)
Wdbc	**96.47** ± 3.31 (30)	92.43 ± 1.51 (19)	96.30 ± 1.81 (29)	96.30 ± 2.06 (23)

*Numbers in parentheses are optimal feature numbers. The accuracy and standard deviation was showed by percentage values.

Table 3. Running time (sec.) and iteration numbers of FBILS and IterativeLS on UCI datasets

Dataset	FBILS	IterativeLS
Australian	**0.494** ± 0.078 (2.9)	1.967 ± 0.027 (28)
Glass	**0.068** ± 0.011 (3)	0.129 ± 0.009 (18)
Heart	**0.097** ± 0.023 (3.3)	0.277 ± 0.018 (26)
Ionosphere	**0.475** ± 0.025 (4.7)	1.556 ± 0.126 (68)
Segment	**7.677** ± 0.316 (5.1)	34.501 ± 0.273 (38)
Soybeanlarge	**0.331** ± 0.025 (7.9)	1.030 ± 0.037 (70)
Vehicle	**1.233** ± 0.116 (5)	4.448 ± 0.071 (36)
Wdbc	**1.000** ± 0.034 (6)	3.968 ± 0.461 (60)

*Numbers in parentheses are mean number of iterations.

Table 4. Description of microarray gene datasets

No.	Dataset	#Sample	#Attribute	#Class
1	Lungcancer	197	1000	4
2	Novartis	103	1000	4
2	SJ–Leukemia	248	985	6
3	Leukemia	38	999	3
4	CNS	42	989	5

gene expression datasets we used have been processed as described in [11]. Further biological details about these datasets can be found in the referenced papers. Most data were processed on the Human Genome U95 Affymetrix ©microarrays. The leukemia dataset was from the previous-generation Human Genome HU6800 Affymetrix ©microarray. The relevant information of these datasets is summaries in Table 4.

Here, we also compared FBILS with the unsupervised methods: Variance, LS and IterativeLS. In order to obtain convincing comparison results and eliminate accidental errors, as in the previous section, we used cross-validation. Because the number of samples in datasets is small, three-fold cross-validation was applied. In each trail, we randomly selected 2/3 of the samples as the training set, and the remaining 1/3 of samples as the test set. The experimental results were reported on the well-defined test sets. According to the statement in [14], we can know that we need 400 genes at most to complete the classification task of microarray gene data. Therefore, we analyze on the first 400 top features.

Figure 1 gives the classification accuracy vs. feature number on five microarray gene datasets. From Fig. 1, we can see that FBILS is obviously superior to other three methods on CNS, Lungcancer and Novartis datasets. In addition, FBILS can quickly achieve a better classification performance. We summarized the highest accuracy of compared methods in Table 5 according to Fig. 1, where bold numbers are the best results among compared methods. FBILS algorithm achieves better accuracies on all five gene datasets. On the Leukemia datasets, FBILS has the same accuracy as LS and IterativeLS. On the Lungcancer dataset, the accuracy of FBILS is 1.5% higher than LS. Table 6 shows the running time of the four methods on the gene dataset. The Variance and LS methods are fast without iteration. Two iterative methods, FBILS and IterativeLS take more time. However, it can be clearly seen that FBILS runs many times faster than IterativeLS.

Table 5. Average accuracy and standard deviation comparison on five microarray gene datasets

Dataset	FBILS	Variance	LS	IterativeLS
CNS	**87.63** ± 5.09 (40)	85.07 ± 14.71 (83)	86.81 ± 8.96 (138)	87.42 ± 15.16 (187)
Leukemia	**96.97** ± 5.25 (73)	93.94 ± 10.49 (144)	**96.97** ± 5.25 (59)	**96.97** ± 5.24 (358)
Lungcancer	**96.44** ± 2.29 (142)	94.93 ± 5.90(208)	94.94 ± 3.16 (292)	94.40 ± 7.11 (286)
Novartis	**100** ± 0.00 (37)	99.07 ± 0.78(359)	98.96 ± 2.24 (375)	99.05 ± 3.53 (199)
SJ-leukemia	**99.20** ± 0.70 (326)	98.01 ± 1.81 (230)	98.01 ± 1.81 (217)	98.01 ± 1.81 (142)

*Numbers in parentheses are optimal feature numbers. The accuracy and standard deviation was showed by percentage values.

Table 6. Running time (sec.) of different methods on five microarray gene datasets

Dataset	FBILS	Variance	LS	IterativeLS
CNS	0.083 ± 0.032	0.003 ± 0.006	0.030 ± 0.010	13.700 ± 0.430
Leukemia	0.073 ± 0.006	0.007 ± 0.012	0.033 ± 0.006	14.577 ± 0.520
Lungcancer	0.780 ± 0.144	0.003 ± 0.006	0.233 ± 0.006	122.940 ± 3.498
Novartis	0.243 ± 0.085	0.007 ± 0.012	0.100 ± 0.017	45.143 ± 0.326
SJ-leukemia	0.950 ± 0.128	0.007 ± 0.006	0.287 ± 0.006	154.700 ± 8.819

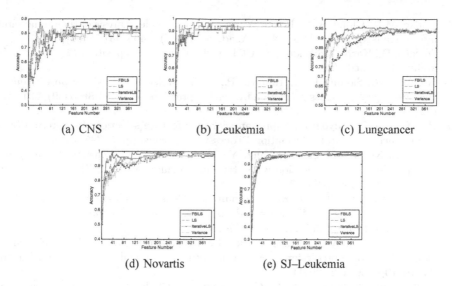

(a) CNS   (b) Leukemia   (c) Lungcancer

(d) Novartis   (e) SJ–Leukemia

Fig. 1. Accuracy vs. feature number on five gene datasets

5 Conclusion

This paper concentrates on unsupervised feature selection and proposes an algorithm called FBILS. FBILS aims to speed up the iterative process and maintains
the local manifold structure. Different from existing LS-like methods, FBILS
evaluates the joint locality preserving ability of features instead of Laplacian
score, and picks up more than one features in one time. On eight UCI and five
microarray gene datasets, a series of experiments were conducted for evaluating
the proposed method. FBILS retains the highest classification accuracy on most
datasets. From the running time of the UCI and gene dataset, we know that
FBILS is much faster than of IterativeLS.

References

1. Amiri, S.H., Jamzad, M.: Automatic image annotation using semi-supervised generative modeling. Pattern Recogn. **48**(1), 174–188 (2015)
2. Benabdeslem, K., Hindawi, M.: Constrained laplacian score for semi-supervised
   feature selection. In: Gunopulos, D., Hofmann, T., Malerba, D., Vazirgiannis, M.
   (eds.) ECML PKDD 2011. LNCS (LNAI), vol. 6911, pp. 204–218. Springer, Heidelberg (2011). https://doi.org/10.1007/978-3-642-23780-5_23
3. Bhattacharjee, A., Richards, W.G., Staunton, J., et al.: Classification of human
   lung carcinomas by mRNA expression profilingreveals distinct adenocarcinomas
   sub-classes. Proc. Natl. Acad. Sci. **98**(24), 13790–13795 (2001)
4. Bishop, C.M.: Neural Networks for Pattern Recognition. Oxford University Press,
   Oxford (1995)
5. Collins, R.T., Liu, Y., Leordeanu, M.: Online selection of discriminative tracking
   features. IEEE Trans. Pattern Anal. Mach. Intell. **27**(10), 1631–1643 (2005)

6. Cooke, M.P., Ching, K.A., Hakak, Y., et al.: Large-scale analysis of the human and mouse transcriptomes. Proc. Natl. Acad. Sci. **99**(7), 4465–447 (2002)

7. Dheeru, D., Karra Taniskidou, E.: UCI machine learning repository (2017). http:// archive.ics.uci.edu/ml

8. Golub, T.R., Slonim, D.K., Tamayo, P., et al.: Molecular classification of cancer: class discovery and class prediction by gene expression. Science **286**(5439), 531–537 (1999)

9. He, X., Cai, D., Niyog, P.: Laplacian score for feature selection. In: International Conference on Neural Information Processing Systems, pp. 507–541 (2005)

10. Luo, M., Nie, F., Chang, X., Yang, Y., Hauptmann, A.G., Zheng, Q.: Adaptive unsupervised feature selection with structure regularization. IEEE Trans. Neural Netw. Learn. Syst. **29**(4), 944–956 (2018)

11. Monti, S., Tamayo, P., Mesirov, J., Golub, T.: Consensus clustering: a resampling-based method for class discovery and visualization of gene expression microarray data. Mach. Learn. **52**(1–2), 91–118 (2003)

12. Pomeroy, S., Tamayo, P., Gaasenbeek, M., et al.: Gene expression-based classification and outcome prediction of central nervous system embryonal tumors. Nature **415**(6870), 436–442 (2001)

13. Sheikhpour, R., Sarram, M.A., Gharaghani, S., Chahooki, M.A.Z.: A survey on semi-supervised feature selection methods. Pattern Recogn. **64**, 141–158 (2017)

14. Shich, M.D., Yang, C.C.: Multiclass SVM-REF for product from feature selection. Expert Syst. Appl. **35**(1–2), 531–541 (2008)

15. Song, X., Zhang, J., Han, Y., Jiang, J.: Semi-supervised feature selection via hierarchical regression for web image classification. Multimedia Syst. **22**(1), 41–49 (2014). https://doi.org/10.1007/s00530-014-0390-0

16. Yeoh, E.J., Ross, M.E., Shurtleff, S.A., et al.: Classification, subtype discovery, and prediction of outcome in pediatricacute lymphoblastic leukemia by gene expression profiling. Cancer Cell **1**(2), 133–143 (2002)

17. Zhu, L., Miao, L., Zhang, D.: Iterative laplacian score for feature selection. In: Liu, C.-L., Zhang, C., Wang, L. (eds.) CCPR 2012. CCIS, vol. 321, pp. 80–87. Springer, Heidelberg (2012). https://doi.org/10.1007/978-3-642-33506-8_11

Improving Low-Resource Chinese Event Detection with Multi-task Learning

Meihan Tong[1,2], Bin Xu[1,2(✉)], Shuai Wang[3], Lei Hou[1,2], and Juaizi Li[1,2]

[1] Department of Computer Science and Technology, Beijing National Research Center for Information Science and Technology, Tsinghua University, Beijing 100084, China
tongmeihan@gmail.com, {xubin,houlei,lijuanzi}@tsinghua.edu.cn
[2] Knowledge Intelligence Research Center, Institute for Artificial Intelligence, Tsinghua University, Beijing 100084, China
[3] SLP Group, AI Technology Department, JOYY Inc, Beijing, China
wangshuai1@yy.com

Abstract. Chinese Event Detection (CED) aims to detect events from unstructured sentences. Due to the difficulty of labeling event detection datasets, previous approaches suffer from severe data sparsity problem. To address this issue, we propose a novel Lattice LSTM based multi-task learning model. On one hand, we utilize multi-granularity word information via Lattice LSTM to fully exploit existing datasets. On the other hand, we employ the multi-task learning mechanism to improve CED with datasets from other tasks. Specifically, we combine Name Entity Recognition (NER) and Mask Word Prediction (MWP) as two auxiliary tasks to learn both entity and general language information. Experiments show that our approach outperforms the six SOTA methods by 1.9% on ACE2005 benchmark. The source code is released on https://github.com/tongmeihan1995/MLL-chinese-event-detection .

Keywords: Chinese Event Detection · Multi-task learning · Lattice LSTM

1 Introduction

Chinese Event Detection (CED) aims to extract event triggers from sentences [3]. As illustrated in Fig. 1, CED needs to identify word "卸下" (resigned) in S_1 as the event trigger for *Resign* event. CED supports a large number of applications, and benefits various downstream tasks such as event knowledge graph building [16].

Mainstream methods can be divided into two categories: feature-based and neural-based models. Feature-based models [6,15] build lexical features and sentence features with NLP tools to detect event. Neural-based models adopt dynamic pooling CNN [2], attention enhanced Bi-LSTM [23] or Nugget Proposal Network [8] to automatically obtain useful features to improve CED.

Although previous methods have achieved great success, the performance of these supervised methods is largely limited by the amount of annotated data.

© Springer Nature Switzerland AG 2020
G. Li et al. (Eds.): KSEM 2020, LNAI 12274, pp. 421–433, 2020.
https://doi.org/10.1007/978-3-030-55130-8_37

Namely, labeled data sparsity is the bottleneck for CED [11]. Benchmark datasets for CED are small: 697 articles for ACE2005[1]. Due to the complexity of CED, it is time-consuming and labor-intensive to manually label more data. Some distant supervision methods [20] have been proposed to automatically augment datasets with existing knowledge bases. However, due to the ambiguity of event trigger (illustrated in Fig. 1), corpora labeled in this way are of low quality and insufficient coverage. In addition, compared with English, Chinese knowledge bases are much smaller [18], which limits the ability of distant supervision methods to address the data sparsity of CED.

Fig. 1. An instance of CED

We address the data sparsity issue from two aspects. On one hand, we delve into the characteristics of Chinese to utilize textual information in more detail. On the other hand, we introduce the idea of multi-task learning to leverage the annotated data from other tasks to improve CED. We illustrate the details as follows.

Chinese is a non-delimiter language, so Chinese words have different granularity and sub-word can also be a word. For instance, "英首相" (British Prime Minister) can be considered as a coarse-grained word, or can be disassembled as two fine-grained words "英" (British) and "首相" (Prime Minister). Previous methods totally ignore the fine-grained word information, which makes them unable to leverage the fine-grained words information to understand the semantics of coarse-grained words. As illustrated in Fig. 1), coarse-grained word "英首相" (British Prime Minister) is a rare word whose semantic is difficult to learn, but with the semantic support of the fined-grained word "首相" (Prime Minister) which means the head of the government, we have more confidence to say that "卸下" (resigned) triggers a *Resign* event instead of a *Transport* event. In our architecture, we employ Lattice LSTM to simultaneously capture word information of different granularity.

For the multi-task learning, we notice that named entity has a great impact on event classification [10]. In S_1, knowing the entity type of "卡梅伦" (Cameron) is a politician, we have more evidence to infer that "卸下" (resigned) triggers a *Resign* event. Fortunately, as a well-studied task, Named Entity Recognition (NER) has large-scale annotated corpora, making it convenient to learn entity

---

[1] https://catalog.ldc.upenn.edu/LDC2006T06.

information by employing NER as an auxiliary task. We also notice that general language information obtained from large-scale plain data is helpful to disambiguate the event trigger. Recently, by pre-training on large-scale corpora, many works [5] have achieved state-of-the-art performance on multiple information extraction tasks. Following these works, we design a variant of language model called Mask Word Prediction (MWP) to learn general language information in plain text.

In this paper, we propose a novel **Multi-Learning** enhanced **Lattice LSTM** model called MLL to address the data sparsity issue of CED. We formula CED as a sequence labeling task. Specifically, we exploit Lattice LSTM to encode multi-granularity word information by adding extra paths to traditional LSTM cell, and employ NER and MWP to leverage entity and general language information to eliminate event ambiguity. CED, NER and MWP share the parameters of feature extraction layer, and then we adopt three different matrices to map them to respective CRF layers for sequence labeling. Finally, we evaluate our model on benchmark ACE2005. Experiments show that our model outperforms six SOTA methods.

Our contributions can be summarized as: 1) To the best of our knowledge, we propose a novel MLL model for Chinese Event Detection, which leverages multi-granularity word information to eliminate event trigger ambiguity, and successfully utilize such information via a bi-directional Lattice LSTM. 2) Instead of only utilizing CED annotated corpus, we innovatively leverage multi-task learning to capture entity information from entity-rich corpus and general language information from plain text to improve CED. Experiment results show that the two auxiliary tasks NER and MWP complement each other by improving the precision and recall respectively. 3) We surpass six SOTA methods on benchmark ACE2005, which raises the F1-value by 1.9%.

2   Problem Definition

Our multi-task learning model involves three tasks: Chinese Event Detection (CED) and two auxiliary tasks, i.e., Named Entity Recognition (NER) and Mask Word Prediction (MWP). We first define the common symbols shared by the three tasks and then introduce the task-specific symbols.

Given a sentence $S = \langle x_1, x_2, \ldots, x_n \rangle$, where x_j stands for the j-th character and $x_{k:j}$ represents the word composed by character sequence $\langle x_k, x_{k+1}, \ldots, x_j \rangle$. For instance, in S1, x_7 refers to character "首" (prime) and $x_{7:8}$ refers to word "首相" (Prime Minister). Since Chinese words have various granularity, different words can end with the same character. We denote $X_j = \{x_{k_1:j}, x_{k_2:j}, \ldots, x_{k_l:j}\}$ as the collection of words ending with the j-th character x_j. $I_j = \{k_1, k_2, \ldots, k_l\}$ records the start position of these words. For example, in S1, both "英首相" (British Prime Minister) and "首相" (Prime Minister) are end with character "相" (Minister), so $X_8 = \{$英首相 (British Prime Minister), 首相 (Prime Minister)$\}$ and $I_8 = \{6, 7\}$. $X = \bigcup_{i=1}^{n} X_i$ represents all of the words detected from sentence S.

According to the definition of Automatic Content Extraction (ACE) [3], **Chinese Event Detection (CED)** aims to identify event trigger from sentences. Formally, given a sentence $S_c = \langle x_1, x_2, \ldots, x_n \rangle$ and word collection X_c, CED task aims to determine the event type y_j triggered by the j-th character x_j in S_c. Event type label $y_i \in \{Resign, Injure, \ldots, Die\}$ refers to the 33 predefined event classes in ACE system. $Y_c = \langle y_1, y_2, \ldots, y_n \rangle$ denotes the event trigger label sequence for S_c. **Named Entity Recognition (NER)** aims to identify whether the i-th character x_i in sentence S_n is a named entity (name of people, organization, places, etc.). For both CED and NER, we adopt BMES mechanism to label event trigger and named entity. **Mask Word Prediction (MWP)** is a whole-word masking task. At the input side, we randomly replace some characters with symbol "MASK". At the output side, we desire the model to predict the masked characters. Formally, given a sentence S_m that lacks the i-th word, MWP needs to find the original word x_i.

Essentially, given the sentence S_c, S_n, S_m and its annotated labels Y_c, Y_n, Y_m from CED, NER, MWP tasks respectively, our model estimates the sum of the probabilities of $P_c(Y_c|S_c)$, $P_n(Y_n|S_n)$ and $P_m(Y_m|S_m)$.

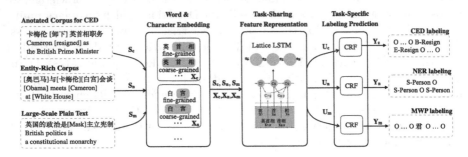

Fig. 2. The overall architecture of MLL. From left to right is Word and Character Embedding, Task-Sharing Semantic Representation and Task-Specific Label Prediction. All tasks share the first two layers, and the last layer is specific for each task.

3   Methodology

We propose a **M**ulti-task **L**earning enhanced **L**attice LSTM model, called MLL, to handle data sparsity issue in CED. In this section, we first introduce the architecture of MLL (Fig. 2), and then introduce each component in detail. MLL consists of three modules. **Word and Character Embedding** detects multi-granularity words from sentence (like "首相" (Prime Minister), "英首相" (British Prime Minister) in S1) and jointly transforms characters (like "英" (British), "首" (Prime), "相" (Minister)) and multi-granularity words into the same semantic embedding space. **Task-Sharing Semantic Representation** deploys bi-directional Lattice LSTM to simultaneously encode character sequences information and multi-granularity word sequences information. Comparing with traditional LSTM cell which only accepts a single input, the input to Lattice LSTM cell can be a collection of multi-granularity words that end with the same character. In S1, the input of the character "相" (Minister) not only comes from character "首" (Prime), but also from overlapping word "首相" (Prime Minister) and "英首相" (British Prime Minister). **Task-Specific Labeling Prediction** transforms the task-sharing features into task-specific features, and then deploys CRF to find the optimal labeling sequence for each task. The parameters of Word and Character Embedding and Task-Sharing Semantic Representation are shared by three tasks, while the parameters of Task-Specific Labeling Prediction are task-specific.

3.1 Word and Character Embedding

In this section, we generate word collection X from sentence S. We first construct a large word dictionary, and then look up multi-granularity words from sentences, including the overlapping ones.

To construct a large word dictionary, we first employ three word segmentation tools (NLPIR, THULAC and LTP) to split sentences in Gigaword corpus. In this way, we obtain different word sequences for the same sentence. Then we filter out low-frequency words. Since the word dictionary is large (containing 5.7k Chinese characters and 698.7k words), we exploit Trie Dictionary algorithm to improve query efficiency.

After detecting word list $X_j = \{x_{k_1:j}, x_{k_2:j}, \ldots, x_{k_l:j}\}$ for each character x_j, we represent characters and words with pre-trained embedding. We obtain the pre-trained embedding following [21], which adopts word2vec as model. $\mathbf{x_j}$ and $\mathbf{x_{k:j}}$ refer to the embedding representation for character x_j and word $x_{k:j}$.

3.2 Task-Sharing Feature Representation

Traditional LSTM cell can receive only one input, word embedding or character embedding. Therefore, previous LSTM-based models can only utilize word information from one specific segmentation. However, Lattice LSTM [22] can accept inputs of different lengths, which is especially useful for language without natural word separator like Chinese. With multi-input Lattice LSTM cell (shown in Fig. 3), our model is able to consider words with different granularity, which provide richer sentence semantics.

Formally, at the j-th step, the input of Lattice LSTM is grouped into two categories: the current character $\mathbf{x_j}$ and the word list $\mathbf{X_j} = \{\mathbf{x_{k_1:j}}, \mathbf{x_{k_2:j}}, \ldots, \mathbf{x_{k_l:j}}\}$, where l denotes the number of words ended with the j-th character. Noted that l varies according to the context. For instance in S1, $X_7 = \{$ "英首相" (British Prime Minister), "首相" (Prime Minister)$\}$ has a length of 2, $X_2 = \{$ "卡梅伦" (Cameron)$\}$ has a length of 1 and $X_0 = \{\}$ has a length of 0.

Character Representation. Standard LSTM is used to encode the current character x_j.

$$f_j = \sigma(W_a[x_j; h_{j-1}] + b)$$
$$i_j = \sigma(W_a[x_j; h_{j-1}] + b)$$
$$o_j = \sigma(W_a[x_j; h_{j-1}] + b) \tag{1}$$
$$\tilde{c}_j = \tanh(W_a[x_j; h_{j-1}] + b)$$

where f_j, i_j and o_j are the forget gate, input gate and output gate for character x_j, and h_{j-1} is the final hidden representation of previous character x_{j-1}. ";" represents the concatenation operation.

Word Representation. For each word $x_{k:j} \in X_j$, we obtain its cell state $c_{k:j}$ by deploying a variant of LSTM cell.

$$f_{k:j} = \sigma(W_b[x_{k:j} + b)$$
$$i_{k:j} = \sigma(W_b[x_{k:j}; h_{j-1}] + b)$$
$$\tilde{c}_{k:j} = \tanh(W_b[x_{k:j}; h_{j-1}] + b) \tag{2}$$
$$c_{k:j} = f_{k:j} \cdot c_{j-1} + i_{k:j} \cdot \tilde{c}_{k:j}$$

where c_{j-1} is the final cell representation of previous character x_{j-1}.

In the same way, we calculate cell state $c_{k:j}$ for each $x_{k:j} \in X_j$. Finally, we can obtain the collection of cell states $C_j = \{c_{k_1:j}, c_{k_2:j}, \ldots, c_{k_l:j}\}$.

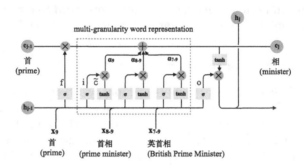

Fig. 3. The details of multi-granularity word information representation.

Hybrid Representation. As illustrated in Fig. 3, we aggregate character representation $\tilde{c}_j$ and word representation $C_j = \{c_{k_1:j}, c_{k_2:j}, \ldots c_{k_l:j}\}$ by self-attention mechanism. Attention weights are calculated by the following equations.

$$\alpha_j = \frac{\exp i_j}{\exp i_j + \sum_{k' \in I_j} \exp i_{k':j}}$$
$$\alpha_{k:j} = \frac{\exp i_{k:j}}{\exp i_j + \sum_{k' \in I_j} \exp i_{k':j}} \tag{3}$$

where $I_j = \{k_1, k_2, \ldots, k_l\}$ indicates the start index of word list X_j.

Finally, we can calculate the cell representation c_j and hidden representation h_j for the j-th character $\mathbf{x_j}$ by

$$c_j = \alpha_j \cdot \tilde{c}_j + \sum_{k \in I_j} \alpha_{k:j} \cdot c_{k:j}$$

$$h_j = o_j \cdot \tanh(c_j)$$

(4)

where o_j refers to the output gate representation of character $\mathbf{x_j}$ defined in Eq. 1.

Now, we can obtain the task-sharing feature H by concatenating hidden representation h_j. In order to capture both the forward and backward information, we adopt the Bi-directional Lattice LSTM model.

$$h_i = [\boldsymbol{h_j}; \boldsymbol{h_j}]$$

$$H = [h_1; h_2; \ldots; h_n]$$

(5)

3.3 Task-Specific Labeling Prediction

CED Output: For mainline task Chinese event detection, we adopt a fully connected layer to map the task-sharing feature H into the task-specific representation U_c by

$$U_c = W_c \cdot H + b_c$$

(6)

Then a CRF layer is used to predict the event trigger label based on task-specific feature $U_c = [u_1; u_2; \ldots; u_n]$. The probability of a label sequence $P_c(Y_c|S_c)$ is

$$P_c(Y_c|S_c) = \frac{\exp \sum_j^n (W_{y_j} u_j + b_{y_{j-1}:y_j})}{\sum_{Y_c \in E(Y_c)} \exp \sum_j^n (W_{y_j} u_j + b_{y_{j-1}:y_j})}$$

(7)

where S_c is the training sentence of Chinese event detection task, Y_c stands for the labels of the event type, $E(Y_c)$ is the full permutation of Y_c.

NER and MWP Output: Now, we transform task-sharing feature H to task-specific representations U_n and U_m for two auxiliary tasks NER and MWP. Analogous to the calculation formulas of U_c (defined in Eq. 6) and $P_c(Y_c|S_c)$ (defined in Eq. 7), we employ fully-connected layers and CRF layers to calculate U_n, $P_n(Y_n|S_n)$ for NER task and U_m, $P_m(Y_m|S_m)$ for MWP task respectively.

3.4 Joint Training

In this section, we introduce the details for training three tasks simultaneously. Instead of alternate training, we accumulate the loss from three tasks and jointly optimize the model parameters. In this way, the parameters from Task-Sharing Feature Representation will be updated with regard to all the three tasks, which helps the model to discover features that are beneficial for all the three tasks and prevent the model from over-fitting one specific task. The final loss function is:

$$
L(\theta) = -(\sum_{i=1}^{N_c} log(P_c(Y_{ci}|S_{ci})) + \sum_{i=1}^{N_n} log(P_n(Y_{ni}|S_{ni}))
$$
$$
+ \sum_{i=1}^{N_m} log(P_m(Y_{mi}|S_{mi})) + \frac{\lambda}{2}||\theta||^2)
$$

(8)

where N_c, N_n, N_m is the corpus length for CED, NER, MWP task respectively, θ is the parameter sets and λ is the weight of the L2 regulation.

4 Experiment

In this section, we evaluate MLL on widely-used benchmark ACE2005. We will first introduce the experiment settings, then demonstrate the effectiveness of our model by comparing with several baselines, and finally investigate whether the model successfully addresses the data sparsity problem.

4.1 Experimental Settings

Datasets. For our mainline task Chinese event detection task, we use ACE2005, which has 33 predefined event types respectively. Following [21], we split ACE2005 into training, validate and test sets with each having 569/64/64 articles. We utilize NLTK to separate articles into sentences. For auxiliary task NER, we adopt MSRA as benchmark dataset, which contains 39.0k Chinese annotated sentences. For auxiliary task MWP, we build the training corpus from Chinese Wikipedia[2]. We want masked words to be meaningful. Therefore, instead of random choice, we only mask words with outer links in Wikipedia. We filter the corpus w.r.t masked words frequency, and finally obtain 135.9k annotated sentences.

Experiment Detail. All of the three tasks share the same Chinese character and word dictionary. The dimension of character, word and LSTM hidden embedding are 50/50/200. Stochastic gradient descent (SGD) is used for optimization with learning rate as 1.5e-2, decay rate as 5e-2 and momentum as 0.9. The training is conducted on a TitanX GPU. The regularization weight λ and lattice dropout are set to 1e-8 and 0.5 respectively. For comparison, we use precision, recall and F1 score as evaluation metrics following [21].

Baselines. We compare our model with feature-based, word-based and character-based models. **Rich-C** [1] utilizes rich knowledge sources from character to discourse level to detect Chinese event from sentences. **DMCNN** [2] splits the max-pooling layer into two parts according to event trigger position.

---

[2] https://dumps.wikimedia.org.

Table 1. Overall Performance. The first model is the best feature-based model for ACE2005. C-BiLSTM and HNN are the best character/word-based models. NPN is the current state-of-the-art model. We directly cite the best experiment results from the original papers.

Model	Trigger identification			Trigger classification		
	P	R	F	P	R	F
Rich-C	58.9	68.1	63.2	58.9	68.1	63.2
DMCNN-C	60.1	61.6	60.9	57.1	58.5	57.8
C-BiLSTM	65.6	66.7	66.1	60.0	60.9	60.4
DMCNN-W	64.1	63.7	63.9	59.9	59.6	59.7
HNN	**74.2**	63.1	68.2	**77.1**	53.1	63.0
NPN	64.8	**73.8**	69.0	60.9	**69.3**	64.8
MLL (our)	**74.2**	69.9	**72.0**	68.7	64.8	**66.7**

The inputs to DMCNN-W and DMCNN-C are word sequences and character sequences respectively. **C-BiLSTM** [21] simultaneously considers n-gram information and sequence information at the character level. **HNN** [4] combines Bi-directional LSTM and convolutional neural networks to learn language-independent features for event detection task. **NPN** [8] proposes a nugget proposal network to deal with the word-trigger mismatch problem.

4.2 Overall Performance

Table 1 presents the overall results. We have the following observations. 1) MLL sets a new performance on benchmark ACE2005. Compared with baselines, MLL improves the F1 score by 1.9% on ACE2005 which proves the effectiveness of our model to handle labeled data sparsity issue of Chinese event detection. 2) MLL is consistently superior to feature-based models. Feature-based methods have pipeline error propaganda issue. Their accuracy is limited by the performance of word segmentation and feature extraction tools. 3) MLL significantly increases the recall comparing with character-based and word-based models, proving that jointly considering character, word, named entity and general language information can make MLL (our) more robust in semantic understanding. 4) MLL outperforms strong baseline NPN. NPN adopts two separate dynamic multi-pooling CNN to respectively encode character and word, and then integrates them via attention mechanism. We argue that our model combines character and word information at earlier stage, thus is more powerful in capturing sentences semantics of various granularity.

It would be interesting to see the improvement on each type and show which type benefits more from these auxiliary tasks. Compared with traditional LSTM (as shown in Fig. 4), MLL (ours) gains benefits on most of the event types in ACE2005. Among them, MLL achieves the most improvement on *Charge-Indict*, *Meet* and *Transfer-Owner*. We find that large proportion of the triggers in these

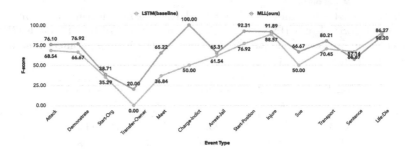

Fig. 4. Analysis from the event type: how and when MLL(ours) helps CED.

event types are unseen during training. We analyse the radio of unseen triggers in these event type, which shows that 38.7%/30.0%/30.0% of the triggers in *Transfer-Owner/Charge-Indict/Meet* are invisible during training. The discrepancy between training and test data leads to the inferior performance of the data-driven baseline. Our model surpasses the baseline by incorporating the entity information and general language information. Specifically, the entity information brings the co-occurrence and exclusion constraints between entity type and event type, which provides more insight for trigger disambiguation. The general language information prevents the model from overfitting the trigger word, and thus enhance unseen trigger detection.

4.3 Effectiveness of Multi-granularity Word Information

We remove the multi-task learning part of our model to analyze the effectiveness of Lattice LSTM. We compare our model with character-based model C-BiLSTM (character), word-based model C-BiLSTM (word) [21].

As illustrated in Fig. 5, Lattice LSTM outperforms character-based, word-based C-LSTM, which proves the effectiveness of multi-granularity word information in event detection. Specifically, 1) Lattice LSTM is superior to character-based model by significantly improving the precision (from 60.0% to 71.4%). Without multi-granularity word information, character-based model has limited ability to disambiguate the event trigger. For instance, without understanding that the word "死胡同" (dead end) means "没有路" (no roads), "死" (dead) can easily be mistaken for the trigger of a "Die" event. 2) Lattice LSTM outperforms word-based model by significantly improving the recall (from 54.2% to 57.7%). Word-based model regards coarse-grained word as basic labeling units, and thus cannot detect event trigger within a word. For instance, since the word "谋杀案" (homicide case) is fed into word-based model as a whole, word-based model cannot mark sub-word "谋杀" (homicide) as the trigger of "Kill" event.

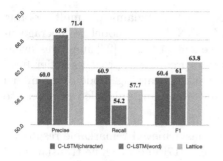

Methods	P	R	F
LSTM	69.9	50.6	58.7
LSTM+NER	71.3	52.6	60.5
LSTM+NER+MWP	69.0	**57.0**	**62.4**
lattice	71.4	57.7	63.8
lattice+NER	77.6	57.7	65.8
lattice+NER+MWP	68.7	**64.8**	**66.7**

Fig. 5. Multi-granularity word enhancement.

Fig. 6. Enhancement from auxiliary task.

4.4 Effectiveness of Auxiliary Task

We adopt the standard LSTM as task-sharing feature representation layer to prove that our multi-task learning mechanism is still effective without Lattice LSTM.

As illustrated in Fig. 6, NER auxiliary task consistently improves the precision by 1.4% and 6% for LSTM and Lattice LSTM respectively. We argue that with additional supervision from NER, our model can leverage entity information to eliminate the ambiguity of the event trigger. Considering the example of "卡梅伦黯然离开国会" (Cameron left Congress in dismay), knowing that "卡梅伦" (Cameron) is a politician, we will have more confidence that "离开" (left) is the event trigger for *Resign* rather than *Transport*. Besides, both LSTM + NER + MWP and lattice + NER + MWP are superior to LSTM + NER and lattice + NER by significantly increasing the recall (+1.9/+0.9), which proves the effectiveness of MWP task. A possible reason is that both NER and CED are human-labeled corpus, containing noise patterns that mislead the training. With the aid of MWP, the model is able to learn general language information from large-scale plain corpus, resulting in more robust in semantic understanding.

If we combine the results from Fig. 5 and Fig. 6, we have an interesting discovery: with much higher level semantic meaning captured by model (from character, word to named entity), the precision gradually increases (from 60.0, 69.8 to 77.6). This is consistent with human cognition: higher level of semantic knowledge improves the understanding of the text.

5 Related Work

Event Detection. Numerous methods have been adopted to handle Chinese event detection. We divide them into two categories: feature-based model and neural-based model. **Feature-based models** adopt manually constructed features to classify the event triggers. These features include lexical features [1] (such as n-gram sequence, part-of-speech tags, named entity), sentence-level features [14] (such as syntactic feature, semantic role labeling, event argument labeling) and document-level features [7]. **Neural-based models** exploit end-to-end neural networks to improve CED. DMCNN adopts a dynamically multi-pooling CNN [2,13] and others employ the hybrid CNN/LSTM model to leverage n-gram and sequential information to detect event [4,9,21]. [23] improves event detection via document information. However, previous approaches suffer from data sparsity issue. We address it via Multi-task Learning model. Our model MLL jointly exploits rich NER resources and abundant unlabeled resources to improve CED.

Multi-task Learning. Multi-task learning has achieved remarkable results on widely NLP tasks such as domain-specific NER [17] and machine translation of low-resource language [19]. [12] is the most relevant work, which adopts Word Sense Disambiguation(WSD) as auxiliary task to improve event detection. However, WSD has two drawbacks. Firstly, the training corpus of WSD is hard to obtain, while our auxiliary tasks NER and MWP either have a large number of existing annotation corpora or do not need manual annotation. Secondly, it does not employ language model task, so general language information cannot be utilized.

6 Conclusion

We propose a Lattice-LSTM based Multi-task Learning model to address the data sparsity issue of Chinese event detection. Specifically, we employ Lattice LSTM to capture multi-granularity word information to fully exploit exiting datasets, and leverage two auxiliary tasks NER and MWP to capture named entity and general language information to utilize datasets from other tasks. Experiment demonstrates that MLL outperforms traditional methods and achieves state-of-the-art performance on benchmark ACE2005.

Acknowledgments. This work is supported by the National Key Research and Development Program of China (2018YFB1005100 and 2018YFB1005101), NSFC key projects (U1736204, 61533018), and grants from Beijing Academy of Artificial Intelligence (BAAI2019ZD0502) and the Institute for Guo Qiang, Tsinghua University (2019GQB0003). It also got partial support from National Engineering Laboratory for Cyberlearning and Intelligent Technology, and Beijing Key Lab of Networked Multimedia.

References

1. Chen, C., Ng, V.: Joint modeling for Chinese event extraction with rich linguistic features. In: COLING, pp. 529–544 (2012)
2. Chen, Y., Xu, L., Liu, K., Zeng, D., Zhao, J.: Event extraction via dynamic multi-pooling convolutional neural networks. In: ACL, vol. 1, pp. 167–176 (2015)
3. Doddington, G.R., Mitchell, A., Przybocki, M.A., Ramshaw, L.A., Strassel, S.M., Weischedel, R.M.: The automatic content extraction (ACE) program-tasks, data, and evaluation. In: LREC (2004)

4. Feng, X., Qin, B., Liu, T.: A language-independent neural network for event detection. Sci. China Inf. Sci. **61**(9), 1–12 (2018). https://doi.org/10.1007/s11432-017-9359-x
5. Lee, J., et al.: Biobert: a pre-trained biomedical language representation model for biomedical text mining. Bioinformatics **36**(4), 1234–1240 (2020)
6. Li, Q., Ji, H., Huang, L.: Joint event extraction via structured prediction with global features. In: ACL (2013)
7. Liao, S., Grishman, R.: Using document level cross-event inference to improve event extraction. In: ACL, pp. 789–797 (2010)
8. Lin, H., Lu, Y., Han, X., Sun, L.: Nugget proposal networks for Chinese event detection. arXiv preprint arXiv:1805.00249 (2018)
9. Lin, H., Lu, Y., Han, X., Sun, L.: Cost-sensitive regularization for label confusion-aware event detection. arXiv preprint arXiv:1906.06003 (2019)
10. Liu, J., Chen, Y., Liu, K.: Exploiting the ground-truth: an adversarial imitation based knowledge distillation approach for event detection. In: AAAI, pp. 6754–6761 (2019)
11. Liu, S., Chen, Y., He, S., Liu, K., Zhao, J.: Leveraging framenet to improve automatic event detection. In: ACL, vol. 1, pp. 2134–2143 (2016)
12. Lu, W., Nguyen, T.H.: Similar but not the same: word sense disambiguation improves event detection via neural representation matching. In: EMNLP (2018)
13. Makarov, P., Clematide, S.: UZH at TAC KBP 2017: event nugget detection via joint learning with softmax-margin objective. In: TAC (2017)
14. McClosky, D., Surdeanu, M., Manning, C.D.: Event extraction as dependency parsing. In: ACL, pp. 1626–1635 (2011)
15. Miwa, M., Sætre, R., Kim, J.D., Tsujii, J.: Event extraction with complex event classification using rich features. J. Bioinform. Comput. Biol. **8**(01), 131–146 (2010)
16. Rospocher, M., et al.: Building event-centric knowledge graphs from news. J. Web Semant. **37**, 132–151 (2016)
17. Wang, X., et al.: Cross-type biomedical named entity recognition with deep multi-task learning. arXiv preprint arXiv:1801.09851 (2018)
18. Yang, T.H., Huang, H.H., Yen, A.Z., Chen, H.H.: Transfer of frames from English framenet to construct Chinese framenet: a bilingual corpus-based approach. In: LREC (2018)
19. Zaremoodi, P., Buntine, W., Haffari, G.: Adaptive knowledge sharing in multi-task learning: improving low-resource neural machine translation. In: ACL (2018)
20. Zeng, Y., et al.: Scale up event extraction learning via automatic training data generation. In: AAAI (2018)
21. Zeng, Y., Yang, H., Feng, Y., Wang, Z., Zhao, D.: A convolution BiLSTM neural network model for Chinese event extraction. In: Lin, C.-Y., Xue, N., Zhao, D., Huang, X., Feng, Y. (eds.) ICCPOL/NLPCC -2016. LNCS (LNAI), vol. 10102, pp. 275–287. Springer, Cham (2016). https://doi.org/10.1007/978-3-319-50496-4_23
22. Zhang, Y., Yang, J.: Chinese NER using lattice LSTM. arXiv preprint arXiv:1805.02023 (2018). https://arxiv.org/pdf/1805.02023
23. Zhao, Y., Jin, X., Wang, Y., Cheng, X.: Document embedding enhanced event detection with hierarchical and supervised attention. In: ACL, vol. 2, pp. 414–419 (2018)

Feature Selection Using Sparse Twin Support Vector Machine with Correntropy-Induced Loss

Xiaohan Zheng[1] , Li Zhang[1,2](✉) , and Leilei Yan[1]

[1] School of Computer Science and Technology, Joint International Research Laboratory of Machine Learning and Neuromorphic Computing, Soochow University, Jiangsu 215006, Suzhou, China
{20184227056,20184227032}@stu.suda.edu.cn, zhangliml@suda.edu.cn
[2] Provincial Key Laboratory for Computer Information Processing Technology, Soochow University, Jiangsu 215006, Suzhou, China

Abstract. Twin support vector machine (TSVM) has been widely applied to classification problems. But TSVM is sensitive to outliers and is not efficient enough to realize feature selection. To overcome the shortcomings of TSVM, we propose a novel sparse twin support vector machine with the correntropy-induced loss (C-STSVM), which is inspired by the robustness of the correntropy-induced loss and the sparsity of the ℓ_1-norm regularization. The objective function of C-STSVM includes the correntropy-induced loss that replaces the hinge loss, and the ℓ_1-norm regularization that can make the decision model sparse to realize feature selection. Experiments on real-world datasets with label noise and noise features demonstrate the effectiveness of C-STSVM in classification accuracy and confirm the above conclusion further.

Keywords: Twin support vector machine · Feature selection · Sparsity · Correntropy-induced loss

1 Introduction

Recently, feature selection has been a hot area of research to address the curse of dimensionality. Feature selection refers to select an optimal subset of original features, which retains valuable features and eliminates redundant features [14]. This procedure can reduce the complexity of processing data and improve the prediction performance [2]. Many methods about feature selection have been proposed, which can be broadly classified into three types: filter, wrapper, and embedded methods [6, 17, 29]. In this paper, we focus on feature selection using support vector machine (SVM) that is the most representative one of embedded methods.

This work was supported in part by the Natural Science Foundation of the Jiangsu Higher Education Institutions of China under Grant No. 19KJA550002, by the Six Talent Peak Project of Jiangsu Province of China under Grant No. XYDXX-054, by the Priority Academic Program Development of Jiangsu Higher Education Institutions, and by the Collaborative Innovation Center of Novel Software Technology and Industrialization.

© Springer Nature Switzerland AG 2020
G. Li et al. (Eds.): KSEM 2020, LNAI 12274, pp. 434–445, 2020.
https://doi.org/10.1007/978-3-030-55130-8_38

SVM, based on the principle of structural risk minimization and the theory of VC dimension, has been used to solve classification and regression problems and has a broad variety of application in real-word tasks [1,3,13,16,19,24,27]. For a binary classification problems, SVM aims at seeking a separating hyperplane to maximize the margin between positive and negative samples, which has excellent generalization performance [10]. Unfortunately, SVM has a high-computational complexity because it needs to solve an entire quadratic programming problem (QPP).

Lately, twin support vector machine (TSVM) has been proposed inspired by the idea of SVM [15]. Compared to SVM, TSVM attempts to find two hyperplanes by solving a pair of smaller QPPs for a binary classification task. Thus, TSVM works faster than SVM in theory and becomes a popular classifier. Many variants of TSVM have been proposed, such as twin bounded support vector machine (TBSVM) [20], least squares twin support vector machine (LSTSVM) [18], v-projection twin support vector machien (v-PTSVM) [9], locality preserving projection least squares twin support vector machine (LPPLSTSVM) [8], and new fuzzy twin support vector machine (NFTSVM) [7]. However, these methods do not have sparse representative models or cannot implement feature selection.

In order to improve the feature selection ability or the sparseness performance of TSVM-like methods, many scholars have proposed same improved methods. For example, ℓ_p-norm least square twin support vector machine (ℓ_p-LSTSVM) was proposed by Zhang et al. [30], which can realize feature selection by introducing an adaptive learning procedure with the ℓ_p-norm ($0 < p < 1$). Sparse non-parallel support vector machine (SNSVM) was proposed in [23]. By replacing the hinge loss with both the ϵ-insensitive quadratic loss and the soft margin quadratic loss, SNSVM has better sparseness performance than TSVM. Tanveer [22] proposed a new linear programming twin support vector machines (NLPTSVM) that uses the ℓ_1-norm regularization, distance and loss term, which causes the robustness and sparsity of NLPTSVM.

In real applications, data often contains noises or outliers, which would have a negative effect on the generalization performance of the learned model. To remedy it, Xu et al. [26] proposed a novel twin support vector machine (Pin-TSVM) inspired by the pinball loss. Pin-TSVM has favorable noise insensitivity, but it does not consider the sparseness of model. That is to say, the feature selection performance of Pin-TSVM is poor.

In this paper, we propose a novel sparse twin support vector machine with the correntropy-induced loss, called C-STSVM. The correntropy-induced loss function is a smooth, nonconvex and bounded loss, which was designed for both classification and regression tasks in [21]. For a classification task, the correntropy-induced loss measures the similarity between the prediction value and the true value from the perspective of correntropy. In addition, we know that the ℓ_1-norm regularization can induce a sparse model from [28] and [31]. C-STSVM is to minimize three terms: the distance term, the correntropy-induced loss term and the ℓ_1-norm regularization term. In doing so, we expect that C-STSVM have the ability to perform feature selection and the robustness to outliers. To find the solution to C-STSVM in the primal space, we design an alternating iterative method with the help of the half-quadratic (HQ) optimization. Experimental results verify the validity of the these theories.

The paper is organized as follows. Section 2 dwells on the proposed C-STSVM, including the introduction of the correntropy-induced loss and the description of the optimization problems and the solutions of C-STSVM. Section 3 discusses experimental results. The last section contains the conclusions.

2   Sparse Twin Support Vector Machine with Correntropy-Induced Loss

In this paper, we consider a binary classification problem and define a normal training set $X = \{(\mathbf{x}_1, y_1), \cdots, (\mathbf{x}_{n_1}, y_1), (\mathbf{x}_{n_1+1}, y_2), \cdots, (\mathbf{x}_{n_1+n_2}, y_2)\}$ where $y_1 = 1$ and $y_2 = -1$ are the positive and negative labels, respectively. Let $\mathbf{X}_1 = [\mathbf{x}_1, \cdots, \mathbf{x}_{n_1}]^T \in \mathbb{R}^{n_1 \times m}$ be the positive sample matrix, $\mathbf{X}_2 = [\mathbf{x}_{n_1+1}, \cdots, \mathbf{x}_{n_1+n_2}] \in \mathbb{R}^{n_2 \times m}$ be the negative sample matrix, $\mathbf{e}_1$ and $\mathbf{e}_2$ be vectors of all ones with appropriate size.

2.1   Correntropy-Induced Loss

For any sample $(\mathbf{x}_i, y_i)$ in the given training set X, the predicted value of $\mathbf{x}_i$ is defined as $f(\mathbf{x}_i)$. A loss function can be used to measure the difference between the predicted value $f(\mathbf{x}_i)$ and the true value y_i. Different loss functions would result in various learners.

Here, we introduce the correntropy-induced loss function. Correntropy is a nonlinear measure of the similarity between two random variables. Inspired by that, Singh [21] presented the correntropy-induced loss function that is to maximize the similarity between the predicted values and the true values for classification tasks. The correntropy-induced loss function has the form:

$$L_C(y_i f(\mathbf{x}_i)) = \beta \left[1 - \exp\left(-\frac{(1 - y_i f(\mathbf{x}_i))^2}{2\sigma^2} \right) \right] \tag{1}$$

where $\beta = [1 - \exp(-\frac{1}{2\sigma^2})]^{-1}$ and $\sigma > 0$ is the parameter. Note that $L_C(0) = 1$ when $y_i f(\mathbf{x}_i) = 0$.

As an mean square error in reproducing kernel Hilbert space, the correntropy-induced loss can further approximate a transition from like the ℓ_2-norm to like the ℓ_0-norm. The curve of the correntropy-induced loss is shown in Fig. 1. We can see that the correntropy-induced loss is smooth, non-convex and bounded. For its boundedness, the correntropy-induced loss is robust to outliers.

2.2   Optimization Problems

Similar to TSVM [15], the aim of C-STSBM is to seek two optimal hyperplanes: positive and negative ones, where the positive hyperplane is closer to the positive

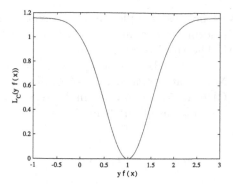

Fig. 1. Curve of correntropy-induced loss with $\sigma = 0.5$.

samples and as far as possible from the negative samples, and the same goes for the negative hyperplane. The two optimal hyperplanes are defined by the following discrimination functions:

$$\begin{cases} f_1(\mathbf{x}) = \mathbf{x}^T \mathbf{w}_1 + b_1 \\ f_2(\mathbf{x}) = \mathbf{x}^T \mathbf{w}_2 + b_2 \end{cases} \tag{2}$$

where $\mathbf{w}_1$ and $\mathbf{w}_2$ are the weight vectors for positive and negative classes, respectively, b_1 and b_2 are the thresholds for two classes, respectively. To represent these weight vectors and thresholds, we construct four non-negative vectors and four non-negative variables, which is $\boldsymbol{\beta}_+^*, \boldsymbol{\beta}_+, \boldsymbol{\beta}_-^*, \boldsymbol{\beta}_-, \gamma_+^*, \gamma_+, \gamma_-^*$ and γ_-, and let

$$\begin{cases} \mathbf{w}_1 = \boldsymbol{\beta}_+^* - \boldsymbol{\beta}_+ \ \ with \ \ \boldsymbol{\beta}_+^* \geq 0 \ and \ \boldsymbol{\beta}_+ \geq 0 \\ \mathbf{w}_2 = \boldsymbol{\beta}_-^* - \boldsymbol{\beta}_- \ \ with \ \ \boldsymbol{\beta}_-^* \geq 0 \ and \ \boldsymbol{\beta}_- \geq 0 \\ b_1 = \gamma_+^* - \gamma_+ \ \ with \ \ \gamma_+^* \geq 0 \ and \ \gamma_+ \geq 0 \\ b_2 = \gamma_-^* - \gamma_- \ \ with \ \ \gamma_-^* \geq 0 \ and \ \gamma_- \geq 0 \end{cases} \tag{3}$$

Then, (2) can be rewritten as

$$\begin{cases} f_1(\mathbf{x}) = \mathbf{x}^T (\boldsymbol{\beta}_+^* - \boldsymbol{\beta}_+) + (\gamma_+^* - \gamma_+) \\ f_2(\mathbf{x}) = \mathbf{x}^T (\boldsymbol{\beta}_-^* - \boldsymbol{\beta}_-) + (\gamma_-^* - \gamma_-) \end{cases} \tag{4}$$

C-STSVM can be described as the following pair of QPPs:

$$\begin{cases} \min_{\boldsymbol{\beta}_+^*, \boldsymbol{\beta}_+, \gamma_+^*, \gamma_+} & \frac{1}{2} \| \mathbf{X}_1 (\boldsymbol{\beta}_+^* - \boldsymbol{\beta}_+) + (\gamma_+^* - \gamma_+) \|_2^2 \\ & + C_1 \left(\|\boldsymbol{\beta}_+^*\|_1 + \|\boldsymbol{\beta}_+\|_1 + \gamma_+^* + \gamma_+ \right) + \frac{C_2}{n_2} \sum_{i=1}^{n_2} L_C(-f_1(\mathbf{x}_{2i})) \\ \min_{\boldsymbol{\beta}_-^*, \boldsymbol{\beta}_-, \gamma_-^*, \gamma_-} & \frac{1}{2} \| \mathbf{X}_2 (\boldsymbol{\beta}_-^{*T} - \boldsymbol{\beta}_-^T) + (\gamma_-^* - \gamma_-) \|_2^2 \\ & + C_3 \left(\|\boldsymbol{\beta}_-^*\|_1 + \|\boldsymbol{\beta}_-\|_1 + \gamma_-^* + \gamma_- \right) + \frac{C_4}{n_1} \sum_{i=1}^{n_1} L_C(f_2(\mathbf{x}_{1i})) \end{cases} \tag{5}$$

where $C_i > 0, i = 1, 2, 3, 4$ are parameters chosen a priori and $\| \cdot \|_i, \ i = 1, 2$ is the ℓ_i-norm. The first term of the first QPP in (5) is to minimize the distance between the outputs of positive samples and the hyperplane $f_1(\mathbf{x}) = 0$. The second term is the ℓ_1-norm regularization term that can reduce a sparseness solution of C-STSVM. The third term is to minimize the sum of correntropy-induced loss function, which makes the negative samples as far as from the positive hyperplane. Note that QPPs in (5) are similar to each other in form. Thus, for the second QPP in (5), we have a similar explanation.

We denote the first two terms of QPPs in (5) by

$$
\begin{cases}
H_1(\boldsymbol{\beta}_+^*, \boldsymbol{\beta}_+, \gamma_+^*, \gamma_+) \\
= \dfrac{1}{2}\|\mathbf{X}_1(\boldsymbol{\beta}_+^* - \boldsymbol{\beta}_+) + \mathbf{e}_1(\gamma_+^* - \gamma_+)\|_2^2 + C_1\left(\|\boldsymbol{\beta}_+^*\|_1 + \|\boldsymbol{\beta}_+\|_1 + \gamma_+^* + \gamma_+\right) \\
H_2(\boldsymbol{\beta}_-^*, \boldsymbol{\beta}_-, \gamma_-^*, \gamma_-) \\
= \dfrac{1}{2}\|\mathbf{X}_2(\boldsymbol{\beta}_-^{*T} - \boldsymbol{\beta}_-^T) + (\gamma_-^* - \gamma_-)\|_2^2 + C_3\left(\|\boldsymbol{\beta}_-^*\|_1 + \|\boldsymbol{\beta}_-\|_1 + \gamma_-^* + \gamma_-\right)
\end{cases}
\tag{6}
$$

The third term in the first QPP of (5) can be expressed as:

$$
\frac{C_2}{n_2}\sum_{i=1}^{n_2} L_c(-f_1(\mathbf{x}_{2i})) = \frac{C_2}{n_2}\sum_{i=1}^{n_2}\left[1 - \exp\left(-\frac{(1 + (\boldsymbol{\beta}_+^* - \boldsymbol{\beta}_+)^T\mathbf{x}_{2i} + (\gamma_+^* - \gamma_+))^2}{2\sigma^2}\right)\right]
\tag{7}
$$

Let

$$
L_1(\boldsymbol{\beta}_+^*, \boldsymbol{\beta}_+, \gamma_+^*, \gamma_+) = \frac{C_2}{n_2}\sum_{i=1}^{n_2}\exp\left(-\frac{(1 + (\boldsymbol{\beta}_+^* - \boldsymbol{\beta}_+)^T\mathbf{x}_{2i} + (\gamma_+^* - \gamma_+))^2}{2\sigma^2}\right)
$$

Then minimizing (7) is identical to maximizing $L_1(\boldsymbol{\beta}_+^*, \boldsymbol{\beta}_+, \gamma_+^*, \gamma_+)$.

Thus, we can represent the first QPP in (5) as

$$
\max_{\boldsymbol{\beta}_+^*, \boldsymbol{\beta}_+, \gamma_+^*, \gamma_+} \quad -H_1(\boldsymbol{\beta}_+^*, \boldsymbol{\beta}_+, \gamma_+^*, \gamma_+) + L_1(\boldsymbol{\beta}_+^*, \boldsymbol{\beta}_+, \gamma_+^*, \gamma_+)
\tag{8}
$$

Similarly, the second QPP in (5) can be rewritten as follows:

$$
\max_{\boldsymbol{\beta}_-^*, \boldsymbol{\beta}_-, \gamma_-^*, \gamma_-} \quad -H_2(\boldsymbol{\beta}_-^*, \boldsymbol{\beta}_-, \gamma_-^*, \gamma_-) + L_2(\boldsymbol{\beta}_-^*, \boldsymbol{\beta}_-, \gamma_-^*, \gamma_-)
\tag{9}
$$

where

$$
L_2(\boldsymbol{\beta}_-^*, \boldsymbol{\beta}_-, \gamma_-^*, \gamma_-) = \frac{C_4}{n_1}\sum_{i=1}^{n_1}\exp\left(-\frac{(1 - (\boldsymbol{\beta}_-^* - \boldsymbol{\beta}_-)^T\mathbf{x}_{1i} - (\gamma_-^* - \gamma_-))^2}{2\sigma^2}\right).
$$

2.3 Solutions

In this subsection, we turn the optimization problems (8) and (9) into two HQ optimization ones and use the alternating iterative method to find their solutions.

We first define two auxiliary variables $\mathbf{v} = [v_1, \cdots, v_{n_2}]^T$ and $\mathbf{v}' = [v_1', \cdots, v_{n_1}']^T$, where $v_i < 0$ and $v_i' < 0$, and then construct two convex functions

$$
\begin{cases}
G_1(\mathbf{v}) = \frac{C_2}{n_2}\sum_{i=1}^{n_2} g(v_i) = \frac{C_2}{n_2}\sum_{i=1}^{n_2}(-v_i \log(-v_i) + v_i) \\
G_2(\mathbf{v}') = \frac{C_4}{n_1}\sum_{i=1}^{n_1} g(v_i') = \frac{C_4}{n_1}\sum_{i=1}^{n_1}(-v_i' \log(-v_i') + v_i')
\end{cases}
\tag{10}
$$

Because each $g(v_i), i = 1, \cdots, n_2$ and $g(v_i'), i = 1, \cdots, n_1$ is independent of others, we analyze each $g(v_i)$ and $g(v_i')$ separately.

Based on [4,5], we can derive the conjugate function $g^*(u_i)$ and $g^*(u_i')$ of $g(v_i)$ and $g(v_i')$ respectively:

$$\begin{cases} g^*(u_i) = \sup_{v_i < 0}\{u_i v_i - g(v_i)\} \\ g^*(u_i) = \sup_{v_i' < 0}\{u_i' v_i' - g(v_i')\} \end{cases} \tag{11}$$

where v_i and v_i' are the optimization variables of the right hand of (11) respectively, and the supremums can be achieved at $v_i = -exp(-u_i)$ and $v_i' = -exp(-u_i')$ respectively. Substituting $v_i = -exp(-u_i)$ and $v_i' = -exp(-u_i')$ into (11), we have

$$\begin{cases} g^*(u_i) = \exp(-u_i) \\ g^*(u_i') = \exp(-u_i') \end{cases} \tag{12}$$

Hence, we have the conjugate function $G_1^*(\mathbf{u})$ and $G_2^*(\mathbf{u}')$ of $G_1(\mathbf{v})$ and $G_2(\mathbf{v}')$ respectively, their forms as

$$\begin{cases} G_1^*(\mathbf{u}) = \frac{C_2}{n_2}\sum_{i=1}^{n_2} \exp(-u_i) = \frac{C_2}{n_2}\sum_{i=1}^{n_2}\sup_{v_i<0}\{u_i v_i - g(v_i)\} \\ G_2^*(\mathbf{u}') = \frac{C_4}{n_1}\sum_{i=1}^{n_1} \exp(-u_i') = \frac{C_4}{n_1}\sum_{i=1}^{n_1}\sup_{v_i'<0}\{u_i' v_i' - g(v_i')\} \end{cases} \tag{13}$$

where $\sup\{\cdot\}$ represents the upper bounded of a variable.

Let $u_i = \frac{(1+(\boldsymbol{\beta}_+^* - \boldsymbol{\beta}_+)^T \mathbf{x}_{2i} + (\gamma_+^* - \gamma_+))^2}{2\sigma^2}$, $u_i' = \frac{(1-(\boldsymbol{\beta}_-^* - \boldsymbol{\beta}_-)^T \mathbf{x}_{1i} - (\gamma_+^* - \gamma_+))^2}{2\sigma^2}$, (13) can be described as

$$\begin{cases} G_1^*(\mathbf{u}) = \frac{C_2}{n_2}\sum_{i=1}^{n_2} \exp\left(-\frac{(1+(\boldsymbol{\beta}_+^* - \boldsymbol{\beta}_+)^T \mathbf{x}_{2i} + (\gamma_+^* - \gamma_+))^2}{2\sigma^2}\right) = \sup_{\mathbf{v}<0}\left\{\frac{C_2}{n_2}\sum_{i=1}^{n_2} h(v_i|\boldsymbol{\beta}_+^*, \boldsymbol{\beta}_+, \gamma_+^*, \gamma_+)\right\} \\ G_2^*(\mathbf{u}') = \frac{C_4}{n_1}\sum_{i=1}^{n_1} \exp\left(-\frac{(1-(\boldsymbol{\beta}_-^* - \boldsymbol{\beta}_-)^T \mathbf{x}_{1i} - (\gamma_+^* - \gamma_+))^2}{2\sigma^2}\right) = \sup_{\mathbf{v}'<0}\left\{\frac{C_4}{n_1}\sum_{i=1}^{n_1} h(v_i'|\boldsymbol{\beta}_-^*, \boldsymbol{\beta}_-, \gamma_-^*, \gamma_-)\right\} \end{cases} \tag{14}$$

where $h(v_i|\boldsymbol{\beta}_+^*, \boldsymbol{\beta}_+, \gamma_+^*, \gamma_+) = \left(v_i\frac{(1+(\boldsymbol{\beta}_+^* - \boldsymbol{\beta}_+)^T \mathbf{x}_{2i} + (\gamma_+^* - \gamma_+))^2}{2\sigma^2} - g(v_i)\right)$, $i = 1, \cdots, n_2$ and $h(v_i'|\boldsymbol{\beta}_-^*, \boldsymbol{\beta}_-, \gamma_-^*, \gamma_-) = \left(v_i'\frac{(1-(\boldsymbol{\beta}_-^* - \boldsymbol{\beta}_-)^T \mathbf{x}_{1i} - (\gamma_+^* - \gamma_+))^2}{2\sigma^2} - g(v_i')\right)$, $i = 1, \cdots, n_1$.

Obviously, we have

$$\begin{cases} L_1(\boldsymbol{\beta}_+^*, \boldsymbol{\beta}_+, \gamma_+^*, \gamma_+) = \sup_{\mathbf{v}<0}\left\{\frac{C_2}{n_2}\sum_{i=1}^{n_2} h(v_i|\boldsymbol{\beta}_+^*, \boldsymbol{\beta}_+, \gamma_+^*, \gamma_+)\right\} \\ L_2(\boldsymbol{\beta}_-^*, \boldsymbol{\beta}_-, \gamma_-^*, \gamma_-) = \sup_{\mathbf{v}'<0}\left\{\frac{C_4}{n_1}\sum_{i=1}^{n_1} h(v_i'|\boldsymbol{\beta}_-^*, \boldsymbol{\beta}_-, \gamma_-^*, \gamma_-)\right\} \end{cases} \tag{15}$$

where the supremums of $L_1(\boldsymbol{\beta}_+^*, \boldsymbol{\beta}_+, \gamma_+^*, \gamma_+)$ and $L_2(\boldsymbol{\beta}_-^*, \boldsymbol{\beta}_-, \gamma_-^*, \gamma_-)$ are achieved at

$$\begin{cases} v_i = -\exp\left(-\frac{(1+(\boldsymbol{\beta}_+^* - \boldsymbol{\beta}_+)^T \mathbf{x}_{2i} + (\gamma_+^* - \gamma_+))^2}{2\sigma^2}\right), i = 1, \cdots, n_2 \\ v_i' = -\exp\left(-\frac{(1-(\boldsymbol{\beta}_-^* - \boldsymbol{\beta}_-)^T \mathbf{x}_{1i} - (\gamma_-^* - \gamma_-))^2}{2\sigma^2}\right), i = 1, \cdots, n_1 \end{cases} \tag{16}$$

respectively. Thus, the optimization problems (8) and (9) with four variables can be turned to a HQ optimization problem with five variables:

$$
\begin{cases}
\max_{\beta_+^*,\beta_+,\gamma_+^*,\gamma_+,\mathbf{v}<0} & -H_1(\beta_+^*,\beta_+,\gamma_+^*,\gamma_+) + \frac{C_2}{n_2}\sum_{i=1}^{n_2} h(v_i|\beta_+^*,\beta_+,\gamma_+^*,\gamma_+) \\
\max_{\beta_-^*,\beta_-,\gamma_-^*,\gamma_-,\mathbf{v}'<0} & -H_2(\beta_-^*,\beta_-,\gamma_-^*,\gamma_-) + \frac{C_4}{n_1}\sum_{i=1}^{n_1} h(v_i'|\beta_-^*,\beta_-,\gamma_-^*,\gamma_-)
\end{cases}
\tag{17}
$$

From (17), we use the alternating iterative method to solve the optimization problem (17).

First, given $(\beta_+^*,\beta_+,\gamma_+^*,\gamma_+)$ and $(\beta_-^*,\beta_-,\gamma_-^*,\gamma_-)$ to optimize $\mathbf{v}$ and $\mathbf{v}'$ respectively. So that, (17) can be reduced to two independent functions with only respect to v_i or v_i':

$$
\begin{cases}
\max_{\mathbf{v}<0} & \frac{C_2}{n_2}\sum_{i=1}^{n_2} h(v_i|\beta_+^*,\beta_+,\gamma_+^*,\gamma_+) \\
\max_{\mathbf{v}'<0} & \frac{C_4}{n_1}\sum_{i=1}^{n_1} h(v_i|\beta_-^*,\beta_-,\gamma_-^*,\gamma_-)
\end{cases}
\tag{18}
$$

Second, given $\mathbf{v}$ and $\mathbf{v}'$ to optimize $(\beta_+^*,\beta_+,\gamma_+^*,\gamma_+)$ and $(\beta_-^*,\beta_-,\gamma_-^*,\gamma_-)$ respectively. The optimization problems (17) can be rewritten as:

$$
\begin{cases}
\max_{\beta_+^*,\beta_+,\gamma_+^*,\gamma_+} & -H_1(\beta_+^*,\beta_+,\gamma_+^*,\gamma_+) + \frac{C_2}{n_2}\sum_{i=1}^{n_2} \frac{v_i\left(1+(\beta_+^*-\beta_+)^T\mathbf{x}_{2i}+(\gamma_+^*-\gamma_+)\right)^2}{2\sigma^2} \\
\max_{\beta_-^*,\beta_-,\gamma_-^*,\gamma_-} & -H_2(\beta_-^*,\beta_-,\gamma_-^*,\gamma_-) + \frac{C_4}{n_1}\sum_{i=1}^{n_1} \frac{v_i'\left(1-(\beta_-^*-\beta_-)^T\mathbf{x}_{1i}+(\gamma_-^*-\gamma_-)\right)^2}{2\sigma^2}
\end{cases}
\tag{19}
$$

To solve the optimization problem (19) easily, we rewrite it as follows:

$$
\begin{cases}
\min_{\beta_+^*,\beta_+,\gamma_+^*,\gamma_+,\xi} & H_1(\beta_+^*,\beta_+,\gamma_+^*,\gamma_+) + C_2'\xi^T\Omega\xi \\
s.t. & -(\mathbf{X}_2(\beta_+^*-\beta_+)+(\gamma_+^*-\gamma_+)) = 1-\xi \\
\min_{\beta_-^*,\beta_-,\gamma_-^*,\gamma_-,\xi'} & H_2(\beta_-^*,\beta_-,\gamma_-^*,\gamma_-) + C_4'\xi'^T\Omega'\xi' \\
s.t. & (\mathbf{X}_1(\beta_-^*-\beta_-)+(\gamma_-^*-\gamma_-)) = 1-\xi'
\end{cases}
\tag{20}
$$

where $\xi = [\xi_1,\cdots,\xi_{n_2}]^T$ and $\xi' = [\xi_1',\cdots,\xi_{n_1}']^T$ are the slack variables, $\xi_i = 1+(\beta_+^*-\beta_+)^T\mathbf{x}_{2i}+(\gamma_+^*-\gamma_+)$, $i=1,\cdots,n_2$, $\xi_i' = 1-(\beta_-^*-\beta_-)^T\mathbf{x}_{1i}-(\gamma_-^*-\gamma_-)$, $i=1,\cdots,n_1$, $C_2' = C_2/(2n_2\sigma^2)$, $C_4' = C_4/(2n_1\sigma^2)$ and $\Omega = \frac{1}{n_2}diag(-\mathbf{v}) \in \mathbb{R}^{n_2\times n_2}$, $\Omega' = \frac{1}{n_1}diag(-\mathbf{v}') \in \mathbb{R}^{n_1\times n_1}$.

Further, let $\alpha_1 = \left[\beta_+^{*T},\beta_+^T,\gamma_+^*,\gamma_+\right]^T$, $\alpha_2 = \left[\beta_-^{*T},\beta_-^T,\gamma_-^*,\gamma_-\right]^T$, $\zeta_1 = \left[C_1\mathbf{1}_m^T,C_1\mathbf{1}_m^T,C_1,C_1\right]^T$, $\zeta_2 = \left[C_3\mathbf{1}_m^T,C_3\mathbf{1}_m^T,C_3,C_3\right]^T$, $\mathbf{M}_1 = [\mathbf{X}_2,-\mathbf{X}_2,1,-1]$, $\mathbf{M}_2 = [-\mathbf{X}_1,\mathbf{X}_1,-1,1]$, and

$$
\mathbf{Q}_1 = \begin{bmatrix}
\mathbf{X}_1^T\mathbf{X}_1 & -\mathbf{X}_1^T\mathbf{X}_1 & 0.5\mathbf{X}_1^T\mathbf{e}_1 & -0.5\mathbf{X}_1^T\mathbf{e}_1 \\
-\mathbf{X}_1^T\mathbf{X}_1 & \mathbf{X}_1^T\mathbf{X}_1 & -0.5\mathbf{X}_1^T\mathbf{e}_1 & 0.5\mathbf{X}_1^T\mathbf{e}_1 \\
0.5\mathbf{e}_1^T\mathbf{X}_1 & -0.5\mathbf{e}_1^T\mathbf{X}_1 & \mathbf{e}_1^T\mathbf{e}_1 & -\mathbf{e}_1^T\mathbf{e}_1 \\
-0.5\mathbf{e}_1^T\mathbf{X}_1 & 0.5\mathbf{e}_1^T\mathbf{X}_1 & -\mathbf{e}_1^T\mathbf{e}_1 & \mathbf{e}_1^T\mathbf{e}_1
\end{bmatrix}
$$

$$
\mathbf{Q}_2 = \begin{bmatrix}
\mathbf{X}_2^T\mathbf{X}_2 & -\mathbf{X}_2^T\mathbf{X}_2 & 0.5\mathbf{X}_2^T\mathbf{e}_2 & -0.5\mathbf{X}_2^T\mathbf{e}_2 \\
-\mathbf{X}_2^T\mathbf{X}_2 & \mathbf{X}_2^T\mathbf{X}_2 & -0.5\mathbf{X}_2^T\mathbf{e}_2 & 0.5\mathbf{X}_2^T\mathbf{e}_2 \\
0.5\mathbf{e}_2^T\mathbf{X}_2 & -0.5\mathbf{e}_2^T\mathbf{X}_2 & \mathbf{e}_2^T\mathbf{e}_2 & -\mathbf{e}_2^T\mathbf{e}_2 \\
-0.5\mathbf{e}_2^T\mathbf{X}_2 & 0.5\mathbf{e}_2^T\mathbf{X}_2 & -\mathbf{e}_2^T\mathbf{e}_2 & \mathbf{e}_2^T\mathbf{e}_2
\end{bmatrix}
$$

Then substitute the equality constraints into the objective function in (20), we can derive that

$$
\begin{cases}
\min_{\boldsymbol{\alpha}_1} \frac{1}{2}\boldsymbol{\alpha}_1^T \mathbf{H}_1 \boldsymbol{\alpha}_1 + \mathbf{f}_1^T \boldsymbol{\alpha}_1 \\
\min_{\boldsymbol{\alpha}_2} \frac{1}{2}\boldsymbol{\alpha}_2^T \mathbf{H}_2 \boldsymbol{\alpha}_2 + \mathbf{f}_2^T \boldsymbol{\alpha}_2
\end{cases}
\tag{21}
$$

where $\mathbf{H}_1 = 2\lambda_1' \mathbf{M}_1 \boldsymbol{\Omega} \mathbf{M}_1 + \mathbf{Q}_1$, $\mathbf{H}_2 = (2\lambda_2' \mathbf{M}_2^T \boldsymbol{\Omega}' \mathbf{M}_2 + \mathbf{G})$, $\mathbf{f}_1^T = 2C_2' 1^T \boldsymbol{\Omega} \mathbf{M}_1 + \boldsymbol{\zeta}_1^T$, and $\mathbf{f}_2^T = 2C_4' 1^T \boldsymbol{\Omega}' \mathbf{M}_2 + \boldsymbol{\zeta}_2^T$.

Once we have the values of $\boldsymbol{\alpha}_1$ and $\boldsymbol{\alpha}_2$, the hyperplanes $f_1(\mathbf{x}) = 0$ and $f_2(\mathbf{x}) = 0$ can be obtained. Hence, given a new sample $\mathbf{x}$, its class $i(i = 1, 2)$ is

$$
class\ i = \begin{cases}
+1, & |f_2(\mathbf{x})| - |f_1(\mathbf{x})| > 0 \\
-1, & otherwise
\end{cases}
\tag{22}
$$

3   Numerical Experiments

In this section, we carry out experiments to testify the validity of the novel algorithm C-STSVM. We compare C-STSVM with the state-of-the-art methods, including SVM [10], TSVM [15], TBSVM [20], KCC [25], Pin-TSVM [26], LSTSVM [18], ℓ_pLSTSVM [30], NLPTSVM [22] and SNSVM [23], where SVM, TSVM, and TBSVM are three traditional algorithms, KCC is a linear classifier with the correntropy-induced loss, Pin-TSVM and LSTSVM introduce the pinball loss and the square loss into TSVM, respectively, and the others were proposed for obtaining sparse models.

For all SVM-like methods, the linear kernel or the linear version is adopted. All experiments are implemented in MATLAB R2016a on Windows 10 running on a PC with a 3.0 GHz Intel Core and 8 GB of memory.

3.1   Data Description and Experimental Setting

We carry out experiments on nine UCI datasets [11]: Australian (690 samples and 14 features), Breast (288 samples and 9 features), German (1000 samples and 24 features), Heart (270 samples and 13 features), Pima (768 samples and 8 features), Sonar (208 samples and 60 features), Tic_tac_toe (958 samples and 9 features), Vote (435 samples and 16 features) and Wdbc (569 samples and 30 features).

The repeated double cross validation [12] was used to select parameters and give the final average result. The five-fold cross validation method is used in twice. First, each dataset is randomly divided into five parts, where one part is taken as the test set at a time and the remaining four parts are used as the calibration set. Next, each calibration set is randomly split into five parts, where we take four parts as training set and the rest part as the validation set. In this process, we train models on the training set using the regularization parameters in the range of $\{2^{-3}, \cdots, 2^3\}$ one by one, and apply the trained-model to the validation set to select optimal parameters. Then we train a model on the calibration set using the optimal parameters and apply the trained-model to the test set to obtain the result.

This process is repeated five times, and the average results of five trials are reported. In addition, the parameter σ in both KCC and C-STSVM is fixed at $\sigma = 0.5$ empirically.

3.2   Robustness to Outliers

In order to demonstrate the robustness to outlier of C-STSVM clearly, we test our method C-STSVM and the other comparison methods on the nine UCI datasets with 0% and 10% label noise.

Table 1. Accuracy obtained on original UCI datasets

Datasets	SVM	TSVM	TBSVM	KCC	Pin-TSVM	LSTSVM	ℓ_pLSTSVM	NLPTSVM	SNSVM	C-STSVM
Australian	86.96±2.98	87.54±2.32	87.54±3.10	86.96±3.22	78.93±19.39	86.81±2.24	87.68±3.10	87.10±2.51	87.69±3.91	**87.82±4.22**
Breast	68.23±3.02	72.48±4.71	68.20±3.78	70.04±5.22	63.48±19.49	67.93±5.67	72.58±4.89	67.90±6.71	72.58±2.72	**73.28±0.82**
German	77.30±1.30	76.90±1.29	77.40±1.47	77.80±2.02	75.10±1.52	77.30±1.89	77.30±1.64	76.30±0.97	77.40±2.07	**78.10±0.82**
Heart	**84.44±2.81**	83.33±2.93	83.70±2.75	81.85±4.01	81.85±7.22	82.59±1.01	82.59±2.11	83.70±2.03	82.59±2.11	82.59±3.36
Pima	76.56±1.95	76.43±1.43	76.83±3.05	76.31±2.59	72.66±5.89	75.52±1.92	76.56±2.18	75.29±2.61	76.82±2.35	**76.96±2.28**
Sonar	77.96±5.85	73.13±7.46	75.02±7.55	75.60±7.85	**80.41±10.65**	75.04±6.28	76.01±7.77	77.01±5.80	77.93±4.97	74.67±10.44
Tic_tac_toe	65.34±0.15	68.89±1.40	68.47±1.23	69.52±0.99	65.34±0.15	69.52±1.47	**69.62±1.17**	65.34±0.15	68.47±2.27	66.39±1.01
Vote	92.20±3.83	91.74±4.13	93.80±2.05	93.35±3.35	77.74±24.65	93.12±3.00	93.80±1.91	94.02±1.71	92.43±3.26	**94.26±1.78**
Wdbc	**98.25±1.07**	98.07±1.14	96.31±1.91	96.48±2.18	70.69±30.55	97.19±1.32	97.72±0.99	97.89±0.78	96.66±2.12	97.54±1.32

Table 1 shows the results for the original UCI datasets and the best classification accuracy is highlighted in bold. It can be seen that C-TSVM has satisfying results. The C-TSVM method obtains the highest classification accuracy on five out of nine datasets, while SVM obtain the best classification accuracy on Heart and Wdbc dataset, and ℓ_pLSTSVM and Pin-TSVM has the best performance on Tic_tac_toe and Sonar dataset respectively. Then, we corrupt the label of each calibration set. For each calibration set, the ratio of label noise is 10%. In this case, the average accuracy and standard deviation are presented on Table 2. As shown in Table 2, C-TSVM has the best performance on six out of nine datasets in terms of classification accuracy. TSVM and Pin-TSVM has the best accuracy on Australian and Sonar dataset respectively.

From the comparison between Table 1 and 2, we can state the following points: (1) The accuracy of C-STSVM has a small change by increasing the number of label noise and has the best classification accuracy on the most of datasets; (2) The accuracy of Pin-TSVM on German and Pima dataset have a substantial reduction by increasing the number of label noise; (3) SVM has the best accuracy on two out of original nine datasets. However, as the number of label noise increases, its classification advantage does not seem to be maintained. (4) The performances of TSVM, TBSVM, KCC, LSTSVM, ℓ_pLSTSVM, NLPTSVM and SNSVM method are average. In summary, C-STSVM has the better robustness and classification accuracy to outlier.

Table 2. Accuracy obtained on UCI datasets with 10% label noise

Datasets	SVM	TSVM	TBSVM	KCC	Pin-TSVM	LSTSVM	ℓ_pLSTSVM	NLPTSVM	SNSVM	C-STSVM
Australian	86.96±2.98	**88.12±2.48**	87.25±3.17	87.39±2.90	86.52±2.79	87.53±2.16	87.68±2.46	86.81±2.87	87.83±2.86	87.68±3.16
Breast	69.68±6.00	66.45±6.43	67.54±8.43	66.79±2.36	68.70±9.97	67.89±3.41	68.27±3.71	67.54±3.40	65.74±5.59	**73.27±1.77**
German	74.40±1.67	75.60±2.38	75.50±1.70	75.90±1.78	45.60±19.43	75.30±1.52	75.80±1.89	74.90±0.74	75.30±1.30	**76.60±1.95**
Heart	81.48±6.55	82.59±4.26	81.11±3.31	83.33±5.56	81.85±5.91	82.59±5.00	83.33±4.34	82.22±5.34	82.96±4.42	**84.81±4.79**
Pima	77.08±3.10	75.78±1.99	75.52±2.63	76.95±3.01	58.19±21.37	76.04±2.09	76.43±2.99	76.43±1.43	76.69±1.77	**77.21±2.16**
Sonar	74.67±9.86	75.01±5.47	70.74±8.34	74.07±7.81	**78.92±5.30**	74.08±4.93	75.00±5.00	77.43±6.15	74.11±7.97	76.95±8.42
Tic_tac_toe	63.99±1.46	63.26±1.61	62.74±1.39	64.20±1.54	59.29±4.54	63.16±1.23	63.78±0.81	64.99±1.97	64.09±0.77	**65.03±0.73**
Vote	93.12±3.00	93.57±2.74	92.88±3.15	93.12±3.00	93.12±0.30	91.50±4.77	93.34±3.35	93.12±0.30	92.89±3.35	**94.48±2.07**
Wdbc	92.08±1.90	94.91±2.65	91.74±2.71	91.39±2.90	92.45±1.76	93.31±1.85	93.85±2.90	95.26±2.91	91.75±3.13	**96.13±1.61**

3.3 The Ability of Feature Selection

We add 50 noise features to each UCI dataset to validate the ability of feature selection of these methods, where the last 50 features are noise features and the others are valid features. These noise features are drawn from the Gaussian distribution with 0 mean and 0.01 variance. Table 3 summarizes the results of numerical experiments. It is easy to see that the accuracy of C-STSVM is significantly better than other methods on most of datasets. Pin-TSVM and NLPTSVM algorithm has the best on Sonar and Wdbc dataset respectively, followed by C-STSVM. Compare Table 1 and Table 3, we can find that the increase in noise features does not negatively affect classification accuracy of C-STSVM, and it still maintains better classification performance than other methods.

In order to clearly show the feature selection performance, we calculated the ratio of sum of the absolute values of the w_1 (w_2) corresponding to the valid features to the sum of the absolute values of total w_1 (w_2) and Table 4 shows the results of some TSVM-like methods. In theory, the greater the contribution of a feature to the classification result, the greater the weight value corresponding to this feature. Hence, for a weight value obtained by training some algorithm, the greater the proportion of weight values corresponding to the valid features, the better the feature selection performance of this algorithm. In Table 4, the highest proportion is shown in bold figures. Table 4 depicts NLPTSVM and C-STSVM has the best feature selection performance on the most of datasets, followed by Pin-TSVM. But the accuracy of C-STSVM is higher than NLPTSVM and Pin-TSVM. The feature selection performance of TSVM, TBSVM, LSTSVM, ℓ_pLSTSVM, and SNSVM are unsatisfactory. The results in Table 3 and Table 4

Table 3. Accuracy obtained on UCI datasets with 50 noise features

Datasets	SVM	TSVM	TBSVM	KCC	Pin-TSVM	LSTSVM	ℓ_pLSTSVM	NLPTSVM	SNSVM	C-STSVM
Australian	86.81±3.36	86.95±3.55	87.39±4.16	87.83±3.52	87.40±4.36	87.39±4.42	87.54±4.03	87.39±1.51	87.24±3.50	**88.12±3.17**
Breast	70.41±5.57	70.05±5.59	71.16±5.81	69.68±4.23	70.04±6.10	69.67±2.70	70.75±2.77	67.50±6.19	70.04±3.75	**72.58±0.56**
German	77.10±1.34	74.90±1.47	75.10±1.52	75.40±1.52	75.20±1.52	75.60±1.98	75.60±1.75	74.60±2.33	76.60±2.22	**77.30±0.97**
Heart	84.81±3.04	82.59±3.61	81.48±5.07	81.48±6.14	82.96±4.01	80.74±3.84	80.37±3.10	81.85±4.22	83.70±3.04	**85.19±2.27**
Pima	76.31±2.40	72.92±1.70	73.57±1.31	75.52±2.79	73.18±1.73	72.91±2.32	73.05±1.57	75.39±2.16	75.78±2.55	**76.94±1.64**
Sonar	78.86±5.41	57.21±5.71	60.48±8.76	74.06±6.17	**78.47±7.15**	70.04±10.69	70.10±10.50	74.57±7.56	76.01±5.89	77.01±8.28
Tic_tac_toe	65.66±0.38	66.08±0.78	66.81±0.95	67.02±1.07	59.68±14.09	65.87±0.54	65.97±0.44	65.34±0.15	66.50±1.84	**68.06±1.90**
Vote	91.51±4.36	93.35±3.03	92.65±3.28	92.42±3.30	92.89±3.06	93.35±0.34	92.66±3.47	92.66±3.47	92.66±3.48	**94.95±1.29**
Wdbc	97.19±1.31	94.73±1.62	95.26±1.57	95.25±1.85	92.98±2.51	95.08±1.74	95.78±2.02	**97.54±1.32**	95.77±2.47	97.36±2.16

Table 4. The proportions of w_1 and w_2 corresponding to valid features

	TSVM		TBSVM		Pin-TSVM		LSTSVM		ℓ_pLSTSVM		NLPTSVM		SNSVM		C-STSVM	
	w_1	w_2	w_1	w_2	w_1	w_2	w_1	w_2	w_1	w_2	w_1	w_2	w_1	w_2	w_1	w_2
Australian	16.66%	41.36%	16.83%	29.49%	91.47%	91.28%	18.11%	18.11%	15.16%	33.62%	**100.00%**	**100.00%**	29.84%	25.55%	**100.00%**	**100.00%**
Breast	3.48%	5.44%	3.76%	5.32%	76.39%	76.34%	4.88%	4.61%	5.21%	5.53%	**100.00%**	**100.00%**	19.79%	18.89%	87.70%	90.21%
German	15.66%	18.02%	21.07%	18.21%	67.45%	52.51%	17.80%	19.04%	18.26%	19.21%	48.83%	43.28%	18.96%	20.72%	**87.70%**	**90.21%**
Heart	9.88%	8.82%	8.42%	8.84%	82.23%	**75.68%**	9.15%	10.63%	12.86%	10.45%	12.07%	19.52%	32.33%	38.53%	**100.00%**	46.16%
Pima	23.07%	22.24%	24.33%	22.20%	19.35%	20.78%	23.55%	23.81%	24.10%	23.10%	31.91%	**36.61%**	25.69%	24.83%	**80.90%**	20.98%
Sonar	57.03%	49.20%	60.23%	49.99%	65.50%	78.27%	53.75%	45.44%	48.97%	48.24%	95.71%	52.30%	55.65%	58.61%	**98.80%**	**88.32%**
Tic_tac_toe	4.30%	4.05%	4.32%	3.80%	74.52%	72.58%	4.25%	4.29%	3.83%	4.22%	0.00%	0.00%	4.43%	3.89%	**90.93%**	**87.23%**
Vote	17.92%	24.24%	17.21%	24.84%	48.86%	70.11%	22.81%	18.36%	16.57%	24.65%	**100.00%**	**100.00%**	44.47%	46.00%	**100.00%**	94.16%
Wdbc	80.16%	67.21%	88.29%	70.81%	95.88%	**95.91%**	82.71%	59.67%	65.33%	54.85%	94.75%	89.25%	55.43%	56.57%	**99.94%**	65.80%

undoubtedly prove that C-STSVM is significantly better than other compared methods in feature selection performance and the classification accuracy.

4   Conclusion

This paper proposes a novel sparse twin support vector machine with the correntropy-induced loss (C-STSVM). Because the correntropy-induced loss has favorable robustness to outliers and the ℓ_1-norm regularization can induce a sparse model, we expect that C-STSVM has a satisfactory robustness to outliers and a sparseness solution to implement feature selection. To validate the robustness of C-STSVM, we carry out experiments on nine UCI datasets with 0% and 10% label noise. To validate the performance of C-STSVM to select features, we conduct experiments on nine UCI datasets with 50 noise features. Experiments results confirm that C-STSVM is significantly better than other compared methods in robustness to outliers, feature selection performance, and classification accuracy.

Since C-STSVM achieves a good performance in the binary classification tasks, we plan to extend C-STSVM to regression estimation and multi-class classification tasks in future.

References

1. Adankon, M.M., Cheriet, M.: Model selection for LS-SVM: application to handwriting recognition. Pattern Recogn. **42**(12), 3264–3270 (2009)
2. Jain, A.K., Robert, P.W., Duin, J.M.: Statistical pattern recognition: a review. IEEE Trans. Pattern Anal. Mach. Intell. **22**(1), 4–37 (2000)
3. Arjunan, S.P., Kumar, D.K., Naik, G.R.: A machine learning based method for classification of fractal features of forearm sEMG using twin support vector machines. In: Proceedings of the IEEE Annual International Conference of the Engineering in Medicine and Biology Society, pp. 4821–4824 (2010)
4. Borwein, J., Lewis, A.: Convex Analysis and Nonlinear Optimization: Theory and Examples, 2nd edn. Springer, New York (2006). https://doi.org/10.1007/978-0-387-31256-9
5. Boyd, S.P., Vandenberghe, L.: Convex Optimization. Cambridge University, Cambridge (2004)
6. Chandrashekar, G., Sahin, F.: A survey on feature selection methods. Comput. Electr. Eng. **40**, 16–28 (2014)
7. Chen, S., Wu, X.: A new fuzzy twin support machine for pattern classification. Int. J. Mach. Learn. Cybernet. **9**, 1553–1564 (2018)
8. Chen, S., Wu, X., Xu, J.: locality preserving projection least squares twin support vector machine for pattern classification. Pattern Anal. Appl. **23**, 1–13 (2020)
9. Chen, W., Shao, Y., Li, C., Liu, M., Wang, Z., Deng, N.: v-projection twin support vector machine for pattern classification. Neurocomputing **376**, 10–24 (2020)
10. Cristianini, N., Shawe-Taylor, J.: An Introduction to Support Vector Machines and Other Kernel-Based Learning Methods. Cambridge University, Cambridge (2000)
11. Dheeru, D., Karra Taniskidou, E.: UCI machine learning repository (2017). http://archive.ics.uci.edu/ml

12. Filzmoser, P., Liebmann, B., Varmuza, K.: Repeated double cross validation. J. Chemom. **23**(4), 160–171 (2008)
13. Hua, X., Ding, S.: Locality preserving twin support vector machines. J. Comput. Res. Dev. **51**(3), 590–597 (2014)
14. Isabelle Guyon, A.E.: An introduction to variable and feature selection. J. Mach. Learn. Res. **3**, 1157–1182 (2003)
15. Jayadeva, Khemchandani, R., Chandra, S.: Twin support vector machine for pattern classification. IEEE Trans. Pattern Anal. Mach. Intell. **29**(5), 905–910 (2007)
16. Khan, N., Ksantini, R., Ahmad, I., Oufama, B.: A novel SVM+NDA model for classification with an application to face recognition. Pattern Recogn. **45**(1), 66–79 (2012)
17. Kohavi, R., John, G.H.: Wrappers for feature subset selection. Artif. Intell. **97**, 273–324 (1997). https://doi.org/10.1016/S0004-3702(97)00043-X
18. Kumar, M.A., Gopall, M.: Least squares twin support vector machine for pattern classification. Expert Syst. Appl. **36**, 7535–7543 (2009)
19. Liu, M., Dai, B., Xie, Y., Ya, Z.: Improved GMM-UBM/SVM for speaker verification. In: Proceedings of the IEEE International Conference on Acoustics, Speech, and Signal Processing, vol. 1, pp. 1925–1928 (2006)
20. Shao, Y., Zhang, C., Wang, X., Deng, N.: Improvements on twin support vector machine. IEEE Trans. Neural Netw. **22**(6), 962–968 (2011)
21. Singh, A., Principe, J.C.: A loss function for classification based on a robust similarity metric. In: Proceedings of IEEE International Joint Conference on Neural Network, pp. 1–6 (2010)
22. Tanveer, M.: Robust and sparse linear programming twin support vector machines. Cogn. Comput. **7**(1), 137–149 (2015)
23. Tian, Y., Ju, X., Qi, Z.: Efficient sparse nonparallel support vector machines for classification. Neural Comput. Appl. **24**(5), 1089–1099 (2013). https://doi.org/10.1007/s00521-012-1331-5
24. Wang, X., Wang, T., Bu, J.: Color image segmentation using pixel wise support vector machine classification. Pattern Recogn. **44**(4), 777–787 (2011)
25. Xu, G., Hu, B., Principe, J.C.: Robust C-loss kernel classifiers. IEEE Trans. Neural Netw. Learn. Syst. **29**(3), 510–522 (2018)
26. Xu, Y., Yang, Z., Pan, X.: A novel twin support-vector machine with pinball loss. IEEE Trans. Neural Netw. Learn. Syst. **28**(2), 359–370 (2017)
27. Xue, Z., Ming, D., Song, W., Wan, B., Jin, S.: Infrared gait recognition based on wavelet transform and support vector machine. Pattern Recogn. **43**(8), 2904–2910 (2010)
28. Zhang, L., Zhou, W.: On the sparseness of 1-norm support vector machines. Neural Netw. **23**(3), 373–385 (2010)
29. Zhang, X., Wu, G., Dong, Z., Curran, C.: Embedded feature-selection support vector machine for driving pattern recognition. J. Franklin Inst. **352**, 669–685 (2015)
30. Zhang, Z., Zhen, L., Deng, N., Tan, J.: Sparse least square twin support vector machine with adaptive norm. Appl. Intell. **41**(4), 1097–1107 (2014). https://doi.org/10.1007/s10489-014-0586-1
31. Zhu, J., Rosset, S., Hastie, T., Tibshirani, R.: 1-norm support vector machines. In: Proceedings of the 16th International Conference on Neural Information Processing Systems, vol. 16, no. 1, pp. 49–56 (2003)

Customized Decision Tree for Fast Multi-resolution Chart Patterns Classification

Qizhou Sun and Yain-Whar Si[✉]

Department of Computer and Information Science,
University of Macau Avenida da Universidade, Taipa, Macau, China
{yb87460,fstasp}@umac.mo

Abstract. Given the advancement in algorithmic trading, the needs for real-time monitoring of patterns and execution of trades in stock exchanges become increasing important for investors. However, real-time monitoring of patterns from a vast number of markets become inefficient when tens of thousands of time series are required to be processed. In order to alleviate these problems, we propose a novel approach called Multi-resolution Chart Patterns Classification (FMCPC) based on a decision tree. In the proposed approach, a Customized Decision Tree (CDT) is built for pattern matching. CDT starts with a detailed analysis of known patterns. CDT then forms generalizations of these examples by identifying commonalities for designing decision rules. To evaluate our approach, experiments are conducted on the real datasets containing 2,527,800 data points from 19,150 stocks across top 10 Exchanges in the world. Our results reveal that FMCPC with CDT can effectively identify the chart patterns within 1 min.

Keywords: Financial time series · Multi-resolution chart patterns · Customized decision tree

1 Introduction

According to statistics of World Federation of Exchange (WFE) [4], 70.2 trillion dollars worth of shares are traded on the world markets till March 2019. Financial chart patterns are one of the important trading signals adopted by investors for predicting market trends and making the trading decisions. Therefore, finding the most recent occurrence of patterns in a timely manner is crucial for stock market analysts. However, there are 250 market infrastructure providers and 48,000 listed companies and approximately 12 million data points (closing price) are generated in a year on WFE exchanges [4]. These markets are distributed around the world and their trading hours are different. Besides, some of the stocks could be listed on multiple markets simultaneously and existence of interdependency among markets is a subject of extensive investigation in finance area [11]. It also implies that stock markets can certainly affect each other and

© Springer Nature Switzerland AG 2020
G. Li et al. (Eds.): KSEM 2020, LNAI 12274, pp. 446–458, 2020.
https://doi.org/10.1007/978-3-030-55130-8_39

trading decisions should only be made after simultaneous observation of indicators from different stock markets. Therefore, it is crucial to find an effective way for fund managers and investors to constantly monitor stock market data for appearance of chart patterns in a highly efficient manner.

In addition, chart patterns classified from a time series can be *multi-resolution patterns* meaning that they could be formed with varying scales (length and height) and they can also be nested among each other. For example, two different sized and nested "Head and Shoulder" (Fig. 1(1) Top) can be identified from the blue input time series (Fig. 1(1) Bottom). In this diagram, a smaller "Head and Shoulder" is nested in a bigger "Head and Shoulder". Besides, smaller patterns could also be embedded in longer patterns of different types. For example, in Fig. 1(2), a "Head and Shoulder" is embedded in a "Triangle Symmetrical".

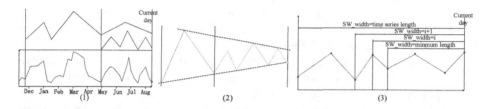

Fig. 1. (1) Two Head and Shoulders patterns segmented from the blue input time series. (2) a. Green line: Head and Shoulder pattern b. Green + blue line: Triangle Symmetrical pattern (3) A time series with different widths of the sliding window (SW) (Color figure online)

In order to find the most recent patterns, a common approach is to use a sliding window to extract subsequences from the input time series by gradually extending the left border while keeping the right border fixed to an anchor point (i.e. current day). This situation is shown in the right hand corner of Fig. 1(1). To extract the subsequences, the size of the sliding window is firstly set to the minimum length of the target pattern. Next, the left border is gradually extended until the window size is equal to the maximum monitoring period.

After the subsequences are extracted, they are segmented for pattern classification. These segmentation approaches include Perceptually Important Points (PIP) [2], PLA [8], PAA [9], and TP [16]. PIP is one of the most popular segmentation approaches adopted by the time series researchers [17]. During the segmentation step, the size of the subsequence is reduced/compressed to match the number of points from the target pattern. For example, in Fig. 1(1), the subsequence from May to *current day* depicted in blue line is reduced into 7 points for the small "Head and Shoulder" and the subsequence from December to *current day* is also reduced into 7 points for larger "Head and Shoulder". After the segmentation, the subsequence is ready to be processed by a suitable pattern matching method such as Rule-based (RB), Template-based (TB), Hybrid and Euclidean distance (ED) approaches. For the case of pattern discovery, segmentation and pattern matching can be performed in pair for the chart patterns reported in literature [1].

However, the above pattern matching is computationally expensive and does not scale up to the situation where tens of thousands of times series need to be scanned and classified on a daily or hourly basis. To alleviate this problem, we propose a novel approach called Customized Decision Tree (CDT) For Fast Multi-resolution Chart Patterns Classification (FMCPC).

In traditional rule-based (RB) classification approach, each rule defined for a chart pattern is examined one after another until all the rules have been processed. Rules checking are commonly performed in a sequential manner and in most of the cases, the same rules are examined repeatedly several times for reaching a conclusion. Therefore, this sequential approach to rule-based classification is highly inefficient. In order to alleviate this problem, we extract three attributes of the top trend and bottom trend lines of chart patterns and aggregate the common rules identified from [18] as the classification attributes to construct a customized decision tree for speeding up the pattern classification. The resulted tree (CDT) is able to effectively reduce the repeated comparison during the classification of 37 chart patterns.

In the experiments, we evaluated FMCPC on real datasets from 19,150 stocks across top 10 Exchanges. Experiment results show that FMCPC can find all charts patterns from 1 day to 1 year in length on a single stock in less than 1 second. It also can find the chart patterns of one to six months in length from 19,150 stocks within 1 min. The contributions of this paper are as follows:

1. A novel customized decision tree (CDT) built from the rules of 37 chart patterns is introduced.
2. A fast multi-resolution chart pattern classification (FMCPC) approach for efficient pattern classification is proposed.

The rest of this paper is organized as follows. Section 2 is the related work. In Sect. 3, we describe the trend line extraction and construction of customized decision tree. The proposed FMCPC algorithm is detailed in Sect. 4. The experiment results and performance are given in Sect. 5. Finally, we conclude our work in Sect. 6 with future work.

2   Related Work

In order to classify chart patterns, a time series should be first segmented to reduce the noise and less important data points. After that, the result is processed by a pattern matching algorithm. The existing segmentation algorithms include PIP, PLA, PAA, and TP. The pseudo code of PIP is detailed in Algorithm 1. The distance in Algorithm 1 includes Vertical Distance (VD), Euclidean Distance (ED) and Perpendicular Distance (PD) of [6].

In [16], Si and Yin proposed an approach called optimal binary search tree (OBST) by assigning a particular distance to the turning point (TP) and storing them into a heap according to their distance. In other word, OBST exploits the advantages of the heap data structure. In [7], Hu et al. proposed an approach called multi-resolution piecewise linear representation-important data point (MPLR-IDP). However, MPLR-IDP is essentially a variant of PIP by improving the specialized binary tree (SB-tree) proposed in [6].

As for pattern matching methods, Fu et al. compared template-based (TB) and rule-based (RB) approaches in [5]. Wan et al. [18] have formally defined a set of comprehensive rules for classification of 53 chart patterns. Based on the extensive analysis of rules given in [18], CDT proposed in this paper utilizes the attributes of top and bottom trend lines. It also aggregates most prominent rules shared among chart patterns as discriminants for achieving a fast classification result.

Algorithm 1: The pseudo code of PIP algorithm

Input: A time series T, n of PIP
Output: n PIPs
1 Initialize a point list S with the start point and end point of T;
2 **for** $S.length < n$ **do**
3   **foreach** *Two neighbored points* (P_{left}, P_{right}) *in* S **do**
4   **for** *Point P between* P_{left} *and* P_{right} *in* T **do**
5   **if** *P has maximum distance to* P_{left} *and* P_{right} **then**
6 | Insert P between P_{left} and P_{right} in S.
7   **end**
8   **end**
9   **end**
10 **end**

---

3 Customized Decision Tree

In "CBR decision tree for the stock selection" [19], Wang et al. use the technical indicators to build a decision tree for stock selection. Inspired by their work, in this paper we propose a Customized Decision Tree (CDT) for reducing the number of the comparisons for classification. CDT proposed in this paper utilizes top and bottom trend lines and the attributes aggregated from the common rules of chart patterns defined in [18]. In other words, CDT was built from the generalization of the formal definitions given in [18].

To construct CDT, we analyzed the formal specifications of the chart patterns compiled in [18]. From our analysis, we found that trend lines are one of the most common attributes used for identifying a pattern. Since trend lines are one of the most important features of chart patterns, they should be first extracted and identified prior to other attributes. Therefore, trend lines are adopted as one the most discriminating attributes in CDT for classification. The trend lines extraction is described in the following paragraphs.

Definition 1. *A Time Series is* $T = \{t_1, t_2, ..., t_i, ..., t_m\}$, *where* $1 \leq i \leq m, 2 \leq m$.

Definition 2. *A PIP list of T is* $S = \{s_1, s_2, ..., s_j, ..., s_n\}$, *where* $1 \leq j \leq n \leq m, 2 \leq n, s_j \in T, s_1 = t_1, s_n = t_m$.[1]

---
[1] j is determined when S is generated by Algorithm 1.

Definition 3. *Minor High (MH) is the salient point which is larger than its two nearest neighbors in S.*

Definition 4. *Minor Low (ML) is the salient point which is lower than its two nearest neighbors in S.*

Function 1. *Tindex(s) returns the corresponding data point index in T for the salient point (s) of S.*

Trend analysis is one of the important steps in technical analysis [1,12,13]. Basically, trends can be classified as up trend, down trend and horizontal. Their corresponding slopes are positive, negative and zero. In this paper, we focus on the top and bottom trends of the chart patterns. An example of top and bottom trends are shown in Fig. 2. In this pattern, top trend is formed by at least two MH points and bottom trend is formed by at least two ML points.

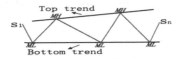

Fig. 2. Top trend and bottom trend

The least squares regression is adopted for calculating the a line $(y = \beta x + \alpha)$ based on MHs (or MLs) in Fig. 3. VD and ED of P to this line satisfies Eq. 1. If $\forall P \in MHs$ (or MLs), and $d_{ED} < d_{limit}$, this line can be considered as a trend line. d_{limit} is the distance of tolerance which can be defined by users. The slope of this trend line is β.

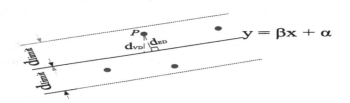

Fig. 3. Linear regression for the trend line

$$d_{ED} = \frac{d_{VD}}{\sqrt{1 + \beta^2}} \tag{1}$$

Moreover, patterns from "Head and Shoulders" category are frequently used as a reversal pattern for signaling a change in price direction. Trend line attribute can be extended to cover more chart patterns by combining basic trend lines.

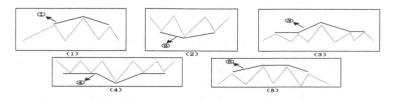

Fig. 4. Complex trend lines (1) up-down, (2) down-up, (3) horizontal-up-down-horizontal, (4) horizontal-down-up-horizontal and (5) up-horizontal-down

Figure 4 shows the complex trend lines which are from "Head and Shoulders" category.

In the above analysis of the trend line extraction, each trend line requires at least two MHs (or MLs). The lengths of MHs (or MLs) can also be used to classify other chart patterns when it is lower than 2 (See Table 1). As a result, three attributes are considered in the trend line extraction: type of trend line, slope and the lengths of MHs and MLs for chart pattern classification.

Rule Aggregation: Once the trends are extracted, they are used in rules for pattern classification. According to [18], lengths of MHs and MLs can be used to distinguish 37 chart patterns. The type of the trend lines can be used to distinguish 21 patterns. Slope can be used for classifying 6 patterns. From this observation, we list all the rules used for classification in Table 1. Columns in the table represent the attributes used for classification and each row represents pattern considered.

Based on Table 1, we have generated two decision trees based on C4.5 and CART algorithms. The resulted decision trees are depicted in Fig. 5[2] and 6 (see footnote 2). In these trees, we can observe that the maximum depth of the trees generated by C4.5 and CART are 7 and 15. This also highlight the fact that the

Fig. 5. The decision tree generated by C4.5.

Fig. 6. The decision tree generated by CART.

---

[2] Due to the space limitation, they cannot be enlarged. The main purpose of including these figures is to highlight the differences in tree structure.

Table 1. Attributes and rules of chart patterns

ID	MHL.length	S_FL	S_length	Fixed position	MHL_FS	Trend	Slope	MHL_parti_length	S_FS	SM_F	Patterns
1	MHs.length=0 MLs.length=1	$s_1 = s_n$	-	-	-	-	-	-	-	-	Rounding Bottoms
2	MHs.length=0 MLs.length=1	$s_1 < s_n$	-	-	-	-	-	-	-	-	Scallops Ascending
3	MHs.length=0 MLs.length=1	$s_1 > s_n$	-	-	-	-	-	-	-	-	Scallops Descending
4	MHs.length=1 MLs.length=0	$s_1 = s_n$	-	-	-	-	-	-	-	-	Rounding Tops
5	MHs.length=1 MLs.length=0	$s_0 < s_n$	-	-	-	-	-	-	-	-	Scallops Ascending and Inverted
6	MHs.length=1 MLs.length=0	$s_0 > s_n$	-	-	-	-	-	-	-	-	Scallops Descending and Inverted
7	MHs.length=1 MLs.length=2	-	S.length=5	-	-	-	-	-	-	-	Double Bottom Adam & Adam
8	MHs.length=1 MLs.length=2	-	S.length=7	MHs[1] is s_3 MLs[1] is s_2 MLs[2] is s_5	-	-	-	-	-	-	Double Bottom Adam & Eve
9	MHs.length=1 MLs.length=2	-	S.length=7	MHs[1] is s_4 MLs[1] is s_3 MLs[2] is s_6	MLs[1]=MLs[2]	-	-	-	-	-	Double Bottom Eve & Adam
10	MHs.length=1 MLs.length=2	-	S.length=7	MHs[1] is s_4 MLs[1] is s_3 MLs[2] is s_6	MLs[2]<MLs[3]	-	-	-	-	-	Cup with handle
11	MHs.length=1 MLs.length=2	-	S.length=9	-	-	-	-	-	-	-	Double Bottom Eve & Eve
12	MHs.length=2 MLs.length=1	-	S.length=5	-	-	-	-	-	-	-	Double Tops Adam & Adam
13	MHs.length=2 MLs.length=1	-	S.length=7	MLs[1] is s_3 MHs[1] is s_2 MHs[2] is s_6	-	-	-	-	-	-	Double Tops Adam & Eve
14	MHs.length=2 MLs.length=1	-	S.length=7	MLs[1] is s_4 MHs[1] is s_3 MHs[2] is s_6	MHs[0]=MHs[1]	-	-	-	-	-	Double Tops Eve & Adam
15	MHs.length=2 MLs.length=1	-	S.length=7	MLs[1] is s_4 MHs[1] is s_3 MHs[2] is s_6	MHs[1]>MHs[2]	-	-	-	-	-	Cup with handle Inverted
16	MHs.length=2 MLs.length=1	-	S.length=9	-	-	-	-	-	16	-	Double Tops Eve & Eve
17	MHs.length≥2 MLs.length≥2	-	-	-	-	MHs.trend=up MLs.trend=up	MHs.slope> MLs.slope	-	-	-	Broadening Wedges Ascending
18	MHs.length≥2 MLs.length≥2	-	-	-	-	MHs.trend=up MLs.trend=up	MHs.slope=MLs.slope	-	-	-	Three Rising Vallys
19	MHs.length≥2 MLs.length≥2	-	-	-	-	MHs.trend=up MLs.trend=up	MHs.slope< MLs.slope	-	-	-	Wedges,Rising
20	MHs.length≥2 MLs.length≥2	-	-	-	-	MHs.trend=up MLs.trend=down	-	-	-	s_1 >MHs[1]	Broading bottoms
21	MHs.length≥2 MLs.length≥2	-	-	-	-	MHs.trend=up MLs.trend=down	-	-	-	s_1 <MLs[1]	Broadening Tops
22	MHs.length≥2 MLs.length≥2	-	-	-	-	MHs.trend=down MLs.trend=up	-	-	-	-	Triangle Symetrical
23	MHs.length≥2 MLs.length≥2	-	-	-	-	MHs.trend=down MLs.trend=down	MHs.slope>MLs.slope	-	-	-	Broadening Wedges Descending
24	MHs.length≥2 MLs.length≥2	-	-	-	-	MHs.trend=down MLs.trend=down	MHs.slope<MLs.slope	-	-	-	Wedges,Falling
25	MHs.length≥2 MLs.length≥2	-	-	-	-	MHs.trend=down MLs.trend=down	MHs.slope=MLs.slope	-	-	-	Three Falling Peaks
26	MHs.length≥2 MLs.length≥2	-	-	-	-	MHs.trend=up MLs.trend=horizontal	-	-	-	-	Broadening Formation, Right-angled and Ascending
27	MHs.length≥2 MLs.length≥2	-	-	-	-	MHs.trend=down MLs.trend=horizontal	-	-	-	-	Triangles,Descending
28	MHs.length≥2 MLs.length≥2	-	-	-	-	MHs.trend=horizontal MLs.trend=down	-	-	-	-	Broadening Formation Right-angled and Descending
29	MHs.length≥2 MLs.length≥2	-	-	-	-	MHs.trend=horizontal MLs.trend=up	-	-	-	-	Triangle,Ascending
30	MHs.length≥2 MLs.length≥2	-	-	-	-	MHs.trend=horizontal MLs.trend=horizontal	-	MHs.length=MLs.length	$s_1>s_2$	-	Rectangle Bottoms
31	MHs.length≥2 MLs.length≥2	-	-	-	-	MHs.trend=horizontal MLs.trend=horizontal	-	MHs.length=MLs.length	$s_1<s_2$	-	Rectangle Tops
32	MHs.length≥2 MLs.length≥2	-	-	-	-	MHs.trend=horizontal MLs.trend=horizontal	-	MLs.length=3	$s_1>s_2$	-	Triple Bottoms
33	MHs.length≥2 MLs.length≥2	-	-	-	-	MHs.trend=horizontal MLs.trend=horizontal	-	MHs.length=3	$s_1<s_2$	-	Triple Tops
34	MHs.length≥2 MLs.length≥2	-	-	-	-	MHs.trend=up-down MLs.trend=horizontal	-	-	-	-	Head and Shoulder,Tops
35	MHs.length≥2 MLs.length≥2	-	-	-	-	MHs.trend=horizontal MLs.trend=down-up	-	-	-	-	Head and Shoulder,Bottoms
36	MHs.length≥2 MLs.length≥2	-	-	-	-	MLs.trend=horizontal-down-up-horizontal	-	-	-	-	Head and Shoulders Bottoms,Complex
37	MHs.length≥2 MLs.length≥2	-	-	-	-	MHs.trend=up-horizontal-down/ horizontal-up-down-horizontal	-	-	-	-	Head and Shoulders, Tops,Complex

generated trees are inefficient for classifying chart patterns from large volume of time series data.

In order to alleviate this problem, we aggregate the attributes from Table 1, as tree nodes to build a Customized Decision Tree (CDT) (See Fig. 8). The attributes of CDT include ID, length of MHs|MLs (MHL_length), comparison between first and last points of S (S_FL), length of S (S_length), Fixed position, comparison between first and second points of MHs|MLs (MHL_FS), Trend, Slope, first point comparison between S and MHs|MLs (SM_F), fixed length of MHs|MLs (MHL_fixed length), comparison between first and second points of S (S_FS), and Patterns. The main criteria in building the CDT is to choose the attributes which can differentiate the highest number of chart patterns and use these attributes in the highest levels of the tree. Fig. 8 shows the CDT with 4 levels for classifying 37 chart patterns. According to this CDT, minimum and maximum number of comparison required for classifying a pattern is 2 and 4.

4   Fast Multi-resolution Chart Pattern Classification

In this section, we describe how PIP segmentation and Customized Decision Tree (CDT) can be integrated for realizing fast multi-resolution chart pattern classification (FMCPC). The main process of FMCPC is depicted in Fig. 7. FMCPC first accepts a time series T with length m. Next, a subsequence is selected with

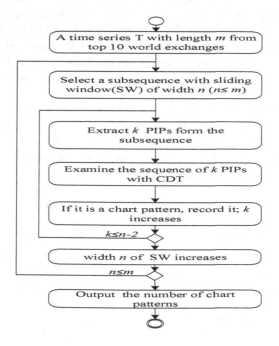

Fig. 7. Fast multi-resolution chart pattern classification

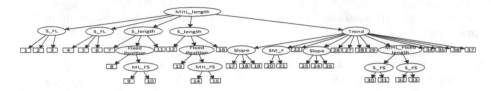

Fig. 8. CDT for classifying chart patterns. The eclipses are attributes and the number in rectangle represents the unique ID of the chart patterns. These IDs are listed in Table 1.

a sliding window of width n. Then, k PIPs are extracted from the classification process. The minimum numbers of PIPs required for each chart pattern are defined in [18]. The inner loop (See Fig. 7) examines all the PIP lists of different number ranging from minimum number to n. Finally, the resulted PIP list is passed to CDT for classification. If the given PIP list is correctly classified as a pattern, it will be recorded. The outer loop is used to repeat the processing of the whole time series T for the multi-resolution classification. Finally, FMCPC outputs all the classified chart patterns.

5   Experiments

In this section, we design two experiments to examine the efficiency of FMCPC and one experiment for the accuracy of CDT. All the experiments were conducted on the DELL Precision T7610 server with Intel (R) Xeon (R) CPU E5-2670 v2@2.50 GHz Memory 8G and Windows 7 Professional Service Pack 1. The program is written in Python and runs on Python 3.6.3 environment. The datasets used for the experiments are based on the top 10 world exchanges listed in [3].

The proposed FMCPC is evaluated on two cases: the efficiency in processing a single time series as well as in processing high volume of time series. We compare the proposed FMCPC with PIP+SPM (PIP segmentation with sequential rule-based pattern matching). To achieve fair comparison, we use the same set of rules for all pattern matching approaches considered in this paper.

Efficiency in Processing a Single Time Series: In this experiment, one-year stock data (from 2018-4-1 to 2019-3-31) of five companies (Apple (AAPL), Baidu (BIDU), Microsoft (MSFT), Cisco (CSCO) and Sina (SINA)) is used for the execution time comparison. In this experiment, 2 different approaches are evaluated: PIP+SPM and FMCPC. The width of the sliding window is incrementally increased by 1 day as illustrated in Fig. 1(3) to extract the subsequences from the time series. The total execution time of each approach and number of chart patterns (NCPs) found are listed in Table 2. From Table 2, we can observe that FMCPC is 4 times faster than PIP+SPM when the stocks of one year in length is tested for pattern classification. Besides, FMCPC is able to find the patterns in less than 1 s.

Table 2. Result statistics of five stocks on 2 approaches

Ticker	PIP + SPM	FMCPC	NCPs
AAPL	3.5568 s	**0.9048 s**	1016
BIDU	3.8844 s	**0.8112 s**	655
MSFT	3.9 s	**0.936 s**	423
CSCO	3.728 s	**0.948 s**	458
SINA	3.6345 s	**0.8736 s**	285

Table 3. Result statistics with SW of 1 and 6 months

Exchanges	NoS	Trading days	PIP+SPM$_1$	FMCPC$_1$	NCPs$_1$	PIP+SPM$_6$	FMCPC$_6$	NCPs$_6$
NYSE	2057	45,254	22.99 s	4.3 s	4765	25.54 s	7.45 s	4860
NASDAQ	3095	68090	27.32 s	5.13 s	6102	30.97 s	9.96 s	5859
London	1830	40,260	15.43 s	4.29 s	2863	17.53 s	5.05 s	3314
Tokyo	3822	84,084	33.69 s	6.27 s	7377	39.73 s	11.15 s	8001
Shanghai	1394	30,668	11.18 s	2.07 s	2576	14.18 s	4.38 s	3099
Hongkong	2115	46,530	17.19 s	2.82 s	3610	20.7 s	6.02 s	3778
Euronext	1009	22,198	8.58 s	1.7 s	1597	9.33 s	3.02 s	1901
Toronto	1297	28,534	10.7 s	2.4 s	2247	13.1 s	3.88 s	2918
Shenzhen	2027	44,594	17.44 s	3.35 s	4168	20.25 s	5.89 s	4409
Frankfurt	504	11,088	4.45 s	1.03 s	845	5.1 s	1.37 s	954
Total	19,150	421,300	168.97 s	33.36 s	36150	178.4 s	37.08 s	37,949

Efficiency in Processing High Volume Time Series Dataset: In this experiment, we compare the efficiency of the proposed approach (FMCPC) with the PIP+SPM approach. We collected the price data of 19,150 stocks which are listed in the top 10 exchanges of the world [3]. According to the analysis presented in [1], the average length of chart pattern is usually 1 and 6 months long. Based on this analysis, we conduct the multi-resolution chart pattern classification for patterns with 1 and 6 months in length. In the experiment, we set the current day to "2019-3-31". The experiment results for each sliding window width is summarized in Table 3. The columns of the tables represent the name of exchanges, number of stocks (NoS), trading days, execution time of PIP+SPM, execution time of FMCPC, and number of chart patterns (NCPs) found. From Table 3, we can observe that FMCPC can classify chart patterns of six months in length in 1 min from 19,150 time series.

The Accuracy of FMCPC: In this experiment, we compare FMCPC with Dynamic Time Warping (DTW) algorithm [10,14] in the chart pattern classification. DTW is the simplest and effective method in the single word voice recognition. We apply DTW to calculate the similarity between the reference chart pattern time series ($R = \{r_1, r_2, ..., r_k, ..., r_l\}$) and the target time series (T) through Eq. 2.

Table 4. Accuracy of FMCPC

Confusion matrix				FMCPC			DTW
				P	N	Acc	Acc
Head and Shoulders, Tops	T	42	6			94%	60%
	F	0	52				
Three Falling Peaks	T	73	3			95%	56%
	F	2	22				
Head and Shoulders, Bottoms	T	37	6			94%	61%
	F	0	57				
Triangles, Symmetrical	T	74	0			100%	45%
	F	0	26				
Broadening Formations, Right-Angled and Ascending	T	25	3			96%	67%
	F	1	71				
Broadening Wedges, Ascending	T	44	4			96%	41%
	F	0	52				
Triangles, Descending	T	33	0			100%	55%
	F	0	67				
Broadening Formations, right-angled and descending	T	27	5			95%	59%
	F	0	68				
Wedges, rising	T	42	0			100%	59%
	F	0	58				
Boradening Tops	T	50	1			99%	49%
	F	0	49				

$$D = \min \frac{\sum_{p=1}^{l} d(r_{k(p)}, t_{i(p)}) W_p}{\sum_{p=1}^{l} W_p} \tag{2}$$

$d(r_{k(p)}, t_{i(p)})$ is the locally shortest distance. W_p is the weight of each shortest distance. In this section, we set $W_p = 1$. Besides, the restrictions $(k(p)-k(p-1) \leq 1, i(p) - i(p-1) \leq 1)$ keep the monotonicity and continuity of DTW. Here, we select 10 chart patterns for evaluation. Since there is no benchmark datasets available for testing, we generate 100 synthetic patterns for each selected chart pattern. A time series can be represented as Eq. 3 in [15].

$$Y = WX + \varepsilon \tag{3}$$

where X is an non-stationary series. W is the coefficient matrix. ε is the Gaussian white noise. Y is the real time series. The process of synthetic patterns generation is as follows:

1. Design the templates of time series for each pattern.
2. Increase the lengths of pattern templates into 1, 2, ... ,10 months length.
3. Add the different Gaussian white noise to templates from the previous step.

Table 5. The execution time comparison between FMCPC and DTW (seconds)

Nos	44	66	88	110	132	154	176	198	220
FMCPC	0.1	0.17	0.25	0.31	0.51	0.4	0.51	0.52	0.7
DTW	1.02	2.54	4.15	6.24	8.64	11.84	15.44	19.04	23.28

For instance, we design the template of "Head and shoulders, Top" as [0.1, 0.6, 0.3, 0.9, 0.3, 0.6, 0.1]. Next, 15 points are added into this template series linearly in each segment to form a referenced chart pattern. Finally, 15 Gaussian white noises ($\mu = 0$, $\sigma = 1$) are added evenly. After these steps, each instance is labeled as true (T) or false (F) pattern by comparing with the chart pattern definitions given in [1]. In order to find a threshold (D) for DTW classification, we select 15% true time series and 15% false time series. Finally, the confusion matrix and the accuracy (Acc) of 10 chart patterns are given in Table 4. The execution speed of comparison is listed in Table 5 for classifying out 10 selected chart patterns. From these results, we can observe that the proposed FMCPC is superior than the DTW approach.

6 Conclusion

In this paper, we proposed a novel approach for fast classification of multi-resolution chart patterns. The advantages obtained from the proposed CDT are two folds. First, CDT allows transparent examination the rules used in the classification of patterns compared to other classification approaches such as artificial neural networks and support vector machines. Therefore, the CDT provides an explainable model for pattern classification. Second, since CDT was built with the aim of reducing fruitless comparisons in rule matching, it can be used for ultra-fast pattern matching applications from streaming time series in financial markets. This capability is crucial in real-time monitoring of huge volume of time series for pattern identification. To evaluate our approach, experiments are conducted on the real datasets containing 2,527,800 data points from 19,150 stocks across top 10 Exchanges in the world. In addition, 1000 synthetic time series are generated for evaluating the accuracy of FMCPC. Experiment results reveal that FMCPC achieves higher accuracy than DTW method. In addition, experiment results also reveal that FMCPC with CDT can efficiently and effectively identify the chart patterns when it is compared to PIP with rule-based pattern matching method. In addition, FMCPC is able to classify patterns with varying lengths (1 and 6 months) from time series of 10 stock exchanges within 1 min. As for the future work, we are planning to integrate the proposed model with trading strategies for testing in algorithmic trading platforms.

Acknowledgment. This research was funded by University of Macau (File No. MYRG2019-00136-FST).

References

1. Bulkowski, T.N.: Encyclopedia of Chart Patterns, 2nd edn. Wiley (2005)
2. Chung, F.L., Fu, T.C., Luk, R., Ng, V.: Flexible time series pattern matching based on perceptually important points. In: Proceedings of the Workshop on Learning from Temporal and Spatial Data, pp. 1–7 (2001)
3. Desjardins, J.: The 20 largest stock exchanges in the world (2017). https://www.visualcapitalist.com/20-largest-stock-exchanges-world/. Accessed 19 Apr 2019
4. The World Federation of Exchanges: Welcome to the future of markets (2019). https://www.world-exchanges.org/. Accessed 19 Apr 2019
5. Fu, T.C., Chung, F.L., Luk, R., Ng, C.M.: Stock time series pattern matching: template-based vs. rule-based approaches. Eng. Appl. Artif. Intell. **20**(3), 347–364 (2007)
6. Fu, T.C., Chung, F.L., Luk, R., Ng, C.M.: Representing financial time series based on data point importance. Eng. Appl. Artif. Intell. **21**(2), 277–300 (2008)
7. Hu, Y., Jiang, Z., Zhan, P., Zhang, Q., Ding, Y., Li, X.: A novel multi-resolution representation for streaming time series. Procedia Comput. Sci. **129**, 178–184 (2018)
8. Keogh, E., Chu, S., Hart, D., Pazzani, M.: An online algorithm for segmenting time series. In: Proceedings 2001 IEEE International Conference on Data Mining, pp. 289–296. IEEE (2001)
9. Keogh, E.J., Pazzani, M.J.: A simple dimensionality reduction technique for fast similarity search in large time series databases. In: Terano, T., Liu, H., Chen, A.L.P. (eds.) PAKDD 2000. LNCS (LNAI), vol. 1805, pp. 122–133. Springer, Heidelberg (2000). https://doi.org/10.1007/3-540-45571-X_14
10. Keogh, E.J., Pazzani, M.J.: Derivative dynamic time warping. In: Proceedings of the 2001 SIAM International Conference on Data Mining, pp. 1–11. SIAM (2001)
11. Khan, T.A.: Cointegration of international stock markets: an investigation of diversification opportunities. Undergrad. Econ. Rev. **8**(1), 7 (2011)
12. Kirkpatrick II, C.D., Dahlquist, J.A.: Technical Analysis: The Complete Resource for Financial Market Technicians. FT Press, Upper Saddle River (2010)
13. Meyers, T.: The Technical Analysis Course: Learn How to Forecast and Time the Market. McGraw Hill Professional, New York (2011)
14. Rakthanmanon, T., et al.: Searching and mining trillions of time series subsequences under dynamic time warping. In: Proceedings of the 18th ACM SIGKDD International Conference on Knowledge Discovery and Data Mining, pp. 262–270. ACM (2012)
15. Shumway, R.H., Stoffer, D.S.: Time Series Analysis and Its Applications: With R Examples. Springer, Heidelberg (2017). https://doi.org/10.1007/978-3-319-52452-8
16. Si, Y.W., Yin, J.: Obst-based segmentation approach to financial time series. Eng. Appl. Artif. Intell. **26**(10), 2581–2596 (2013)
17. Wan, Y., Gong, X., Si, Y.W.: Effect of segmentation on financial time series pattern matching. Appl. Soft Comput. **38**, 346–359 (2016)
18. Wan, Y., Si, Y.W.: A formal approach to chart patterns classification in financial time series. Inf. Sci. **411**, 151–175 (2017)
19. Wang, Y., Wang, Y.: A case-based reasoning-decision tree hybrid system for stock selection. World Acad. Sci. Eng. Technol. Int. J. Comput. Electr. Autom. Control Inf. Eng. **10**(6), 1223–1229 (2016)

Predicting User Influence
in the Propagation of Toxic Information

Shu Li[1,2,3,4], Yishuo Zhang[4,5], Penghui Jiang[1,2,3], Zhao Li[1,2,3],
Chengwei Zhang[1,2], and Qingyun Liu[1,2,3(✉)]

[1] Institute of Information Engineering, Chinese Academy of Sciences,
Beijing, China
liuqingyun@iie.ac.cn
[2] National Engineering Laboratory of Information Security Technologies,
Beijing, China
[3] School of Cyber Security, University of Chinese Academy of Sciences,
Beijing, China
[4] School of Information Technology, Deakin University,
Melbourne, VIC 3216, Australia
[5] Xinjiang Technical Institute of Physics and Chemistry,
Chinese Academy of Sciences, Urumqi, China

Abstract. With the advances of information technology, the Internet
has become an indispensable part of life. At the same time, toxic Information has become virulent and common on the Internet. Such information propagation can have a negative impact on individuals, organi-
sations and the society. Traditional approaches, such as detecting texts
and posts with toxic Information will eventually generate 'dark pools in
which the online propagation of toxic information will flourish. In this
study, we pay attention to influential users who evidently affect others
in the activities related to toxic information. A method of predicting
user influence was proposed. Compared to the existing literature, user
influence is assessed on the basis of users' text-based and behaviors-
based characteristics rather than the network structures only. Moreover,
whether the influential users have always been those with strong connections on the social networking site is also discussed. The effectiveness of
the proposed method is demonstrated in two real-world datasets.

Keywords: Toxic information · User influence · User propagation
ability

1 Introduction

The Internet provides the convenience of communication, transaction, entertainment for the Internet users. However, it also becomes the hotbed of toxic
information. There have been increasingly instances of toxic information propagation, such as hate speech, cyberbullying, pornography, violence crisis, racism,
anti-social behaviours and other cases online [5,16,17]. On the one hand, the

© Springer Nature Switzerland AG 2020
G. Li et al. (Eds.): KSEM 2020, LNAI 12274, pp. 459–470, 2020.
https://doi.org/10.1007/978-3-030-55130-8_40

characteristics of the network, such as the assumed anonymity and the freedom of speech, indulge the Internet users to freely disclose their ideologies or adopt an online aggression without considering the consequences of their behaviours. On the other hand, more seriously, some evil-minded people even utilized the Internet to disseminate violence, terrorism, or anti-social speech [1,25]. In 2009, [11] reported that terrorists used the Internet for fundraising, recruitment, and cyber attacks. In the annual report of Australian Institute of Criminology, an 18-year-old Australian teenager was lured into joining *ISIS* by their propaganda in the social networking site [3]. Importantly, such information have almost swept all mainstream social networking sites, including *Twitter* [19–21], *Instagram* [13], *Facebook* [22], *Whisper* [24], and *YouTube* [1]. This propagation of toxic information can mislead and provoke the hatred among innocent people, contribute to the crime against humanity, and further lead to social instability [1,25].

The wide propagation of toxic information on the Internet cannot be ignored because of its virulence to the public and society. Effective measures for countering toxic information have increasingly drawn attentions from the governments, companies, and researchers. In the spring of 2017, leading social networking sites including *Twitter*, *Facebook*, and *YouTube* were strongly criticised by the parliamentary committees in Germany and the UK for failing to take sufficient and effective measures against hate-speech. An Australian Harm Prevention Charity, Online Hate Prevention Institute (OHPI),[1] was created in 2012 as a dedicated institute for addressing online hate. Timothy Quinn and the Sentinel Project, a Canadian non-profit organization, co-founded the *Hatebase*[2], which is the world's largest structured repository of multilingual hate speech. *Hatebase* aims to reduce incidents of hate speech, to lessen the acceptability of hate speech, and to prevent violence which is predicated by hate speech.

Even though many research efforts have been devoted to building a healthy and clean virtual community for online users to obtain information or interact with others. The detection and prevention of toxic information propagation remains a significant and technically challenging task. Deleting and blocking posts, comments or other forms of content with toxic information has always been a feasible and preferred way to prevent their propagation on the Internet. However, the latest research published on *Nature* in 2019 studied online hate and predicted that traditional policing methods, such as deleting posts in an single platform is relatively coarse, can make matter worse, and will eventually generate global dark pools in which online propagation of toxic information will flourish [14]. [14] proposed that one strategy for combating online hate is to cultivate users as the "immune system" on the network. These users can announce the neutral speech, and then dilute the connection between users with different pole speech. Another strategy is to reduce the influence of toxic information on users by exposing contradictory words to them at the same time [14]. From two strategies of this latest research, we can see that users play more and more important roles and they deserve more concerns in the prevention of toxic

---

[1] https://ohpi.org.au/.
[2] https://hatebase.org/.

information propagation. Identifying these users and making full use of their roles and influence on information propagation is a promising method to defeat toxic information propagation. Therefore, in this study, instead of detecting whether textual content contains toxic information or not, we pay attention to those users who evidently affect others in the activities related to toxic information. These users are regarded as influential users in this study.

Identifying influential users on online social networking sites has been studied in existing literatures [2, 7]. However, to our knowledge, most previous studies are limited to network structures to analyse user influence, such as degree measure, core, betweenness, overlapped community bridges, and separated community bridges [31]. In this work, we firstly assess users influence based on users' text and behaviour characteristics, rather than the network structures. We then analyse the user propagation ability by visualizing community-based information on the social networking site.

The rest of the paper is organised as follows. Section 2 reviews related work on the detection of toxic information and the identification of influential user. Section 3 formalises the research problem. Section 4 presents the proposed methods for analysing the user influence and propagation ability. Section 5 demonstrates the effectiveness of the approach. Finally, Sect. 6 concludes our work.

2 Related Work

2.1 Toxic Information Detection

The Internet is flooded with a large number of user-generated data, and those user-generated data have been one of the important parts of the Internet content and drawn attentions as popular research topics [26, 28]. [27] used large-scale online reviews to explore dining preferences of tourists. [17] assessed self-report racism on the basis of routinely collected reviews available on tourism websites. Several researchers attempted to detect hate speech from the reviews on the social networking sites, including *Twitter* [20, 21, 30], *Instagram* [13], *Facebook* [15, 22], *Whisper* [24], and *YouTube* [1, 20]. Online reviews provide better insights and opportunities for text analysis task, also for toxic information detection.

Most of the previous studies on the detection of toxic information formulated it as a text classification task [8, 9, 29]. With the development of artificial intelligence, sophisticated models from machine Learning to deep learning, have been adopted to improve the performance of the detection of toxic information [4, 8, 9, 23, 29]. *SVM*, *Decision Trees*, and *Random Forests* are among the most popular algorithms in this context [12]. In this study, we not only used reviews-based, but also behaviours-base characteristics to represent users related online toxic information.

2.2 Influential User Identification

Influential users play significant roles in information propagation. Research has attempted to identify influential user along underlying social connection graphs, and analysed the propagation influence [10]. [31] verified the importance of highly-connected users by investigating the dissemination scale, the dissemination speed, and the controllability online social networks. They identified influential user according to core, betweenness, adn community bridges [31]. [6] measured social influence on the basis of indegree, retweets, and mentions, suggesting that network graph alone, such as indegree, revealed very little about user influence. Growing social networks and user generated content make it complicated and challenging to identify influential users. More factors, such as topic, time, social topology, play important roles in determining the user influence in the propagation of information. In this study, we attempted to assess influential users on the basis of users' text-based and behaviours-based characteristics rather than the network structures only.

3   Preliminaries

Toxic information is a collective concept, including hate speech, cyberbullying, abusive words, racism, terrorism, violent and pornographic content, anti-social behaviours and so forth [12]. In the propagation of toxic information, both user influence and user propagation ability are important. User influence is described as the ability of attracting attentions from other users. For example, users on the *Youtube* can leave their reviews after watching the video. The users can be regarded as influential users if their reviews get *likes* or *dislikes* exceeding certain thresholds. User propagation ability is to describe whether they have higher possibility to propagate their opinions to others. If users have strong social relationships or they are highly-connected users on the social networking site, their opinions are more likely to be distributed to the mass, which means that those users have a high propagation ability.

In our study, terrorism-related information and hate speech are demonstrated for research, and we predict user influence based on their reviews, behaviours and community attributes when they are involved in the propagation of toxic information. Let us formalise the research problem. Given the dataset D with U users and T texts in total, $D = (U, T)$. Let $U^M = \{u_1, u_2, ..., u_m\}$ denote a set of users, and for each user u_i, $T_i^R = \{t_1, t_2, ..., t_r\}$ is their text content. Suppose $L^P = \{l_1, l_2, ..., l_p\}$ represents the level of user influence, and then the research problem is to predict the $l_p \in L^P$ for each user u_i to represent their influence in the propagation of toxic information.

4   Method

Figure 1 presents the framework, and it consists of two components, user influence and user propagation ability. A new method, PUI, was proposed for predicting user influence in the propagation of toxic information context. Text-based and behaviours-based features are incorporated to represent users in PUI. Moreover, How to analyse user propagation ability is also described in this section.

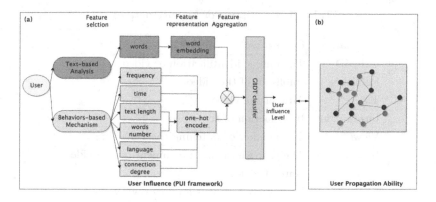

Fig. 1. The framework of user influence

4.1 Data Acquisition

Youtube[3] as a video sharing site and *Twitter*[4] as a microblog site are two representatives of today's social networking sites, where people around the world can obtain information and give their opinions. Therefore, in this study, experimental data are collected from these two platforms.

For *Youtube* dataset, we pay attention to videos related a terrorist organisation. *Youtube* provides video search function based on the specific hashtag. We selected hashtag `#hamas` to filter videos, then crawled descriptions of those videos, and further collected all reviews of each video. As such, we can obtain a comprehensive dataset related with one terrorist organisation as a reference for further analysis, marked as *Dataset(A)*. In this dataset, the description of one video consists of `VideoID`, `Publisher`, `Title`, `Introduction`, and `Views`. A typical review for each video includes `Reviewer`, `Date`, `Likes`, and `Review text` as listed in Table 1. Moreover, the subscription relationship between `Reviewer` and `Publisher` are collected.

Another dataset is about hate speech from *Twitter*. Waseem et al. collected online tweets over a period of two months and recruited experts to annotate these tweets as sexism, racism, both or neither [29]. The dataset was published as a list of $16,907$ tweet IDs and their labels[5]. By using the *Tweepy* library, we retrieved the tweets and also collected community-based information for the users who posts those tweets. Since some users have now been suspended, not all tweets in [29] were acquired. Table 2 presented the information of this dataset, expressed as *Dataset(B)*.

---

[3] https://www.youtube.com/.

[4] https://twitter.com/.

[5] https://github.com/ZeerakW/hatespeech/blob/master/NAACL_SRW_2016.csv.

Table 1. Description of *Youtube* dataset

Metadata	Description
VideoID	The unique identification of video
Publisher	Publisher of the video
Title	Title of the video
Introduction	Introduction of the video
Views	The number of watching the video
Reviewer	The user who make a comment on the video
Date	Date of review
Likes	The number of *likes* for this review
Review text	Content of the review

Table 2. Description of *Twitter* dataset

Metadata	Description
TweetID	The unique identification of the tweet
Tweet	Content of the tweet
Tweet label	The corresponding annotations of the tweet, racism, sexism or neither
Author	The user who post the tweet
AuthorID	The unique identification of the author

4.2   PUI: Predicting User Influence

User Text-Based Analysis. The Internet, as a global and virtual network, provides a setting within which people around the world can announce any speech or comment on any content online. Within a domain, users' standpoints are characterised by their speech or reviews, and users will show their more important and powerful influence if their opinions are accepted or even rejected by more people. Therefore, in this part, we pay attention to users' speech on one specific topic, and use text features to represent users.

Firstly, since text from the Internet can be noisy, we performed several steps to normalised these texts. Cleaning data was performed first, including expanding the abbreviations, deleting numbers and punctuations, removing stop words, converting characters to lower case, and replacing several special characters, such as "#" and "*". Moreover, we corrected some misspelled words; e.g., the word isreal was converted to israel. On the text topics, high-frequency terms are usually informative. Therefore, in this process, we also used the term frequency Cinverse document frequency (TF-IDF) technique to find the most frequently used terms and these terms were adopted to conduct word embedding.

Secondly, we represented the most frequently used terms as vectors by word embedding which is a method to convert the text data into numerical data, which means a word in the text space is map or embed to another numerical vector space. We selected $GloVe^6$ as the algorithm for word embedding. After that, each user was represented by aggregating the words of their texts, expressed as Eq. (1):

$$U_i^t = AVERAGE(\{W_j | j = 1, ..., N_i\}) \tag{1}$$

where N_i denotes the number of words of the texts for user U_i.

User Behaviours-Based Feature. When some users often view posts or watch videos on one specific topic or openly express their opinions frequently, they show their interest and activity on the information. This phenomenon can implicitly reflect or improve their influence to some extent. In this work, user behaviours features is based on such an assumption. In addition, the time attribute, the length of text, the number of notional words, the language, and the number of social connections are also featured as user characteristics. As such, the behaviours-based features for each user can be represented as Eq. (2):

$$U_i^b = (B_t \oplus B_l \oplus B_w \oplus B_g \oplus B_s \oplus B_f) \tag{2}$$

$$B_f = \{ \begin{matrix} R_{ratio}, for Dataset(A) \\ P_{prop}, for Dataset(B) \end{matrix} \} \tag{3}$$

where R_t, R_l, R_w, R_g and R_s are the vectors of the attribute of time, the text length, the words numbers, the language, and the social connection degree, respectively. R_{ratio} denotes the ratio of videos commented by the user to all videos collected, and P_{freq} indicates the proportion of containing hate speech terms defined by the public *Hatebase* repository for one user.

Predicting User Influence. For two different types of datasets, we selected review *likes* and frequency of publishing hate speech to represent user influence in *Dataset(A)* and *Dataset(B)*, respectively. We analysed the distribution of these two indicators and divided into two levels, influential users and normal users. We then trained a gradient boosted decision tree (GBDT) classifier to classify users based on users' text-based and behaviours-based features, represented as Eq. (4).

$$L_i = GBDT(U_i^t \oplus U_i^b) \tag{4}$$

4.3   Analysing User Propagation Ability

After identifying the influential users in the propagation of toxic information, the analysis of user propagation ability is also important. Users with high propagation ability can distribute toxic information to wider population and reach the receivers more swiftly, which means that those users deserve most of the attention when authorities want to curb information dissemination.

---

[6] https://nlp.stanford.edu/projects/glove/.

Intuitively, users with high propagation ability are those who are highly-connected in the online social network. However, in the propagation of toxic information, are the influential users always those with high propagation ability? In this study, we investigated and analysed user propagation ability with graph visualization platform *Gephi*[7].

5   Experiments and Analysis

5.1   Datasets

The first dataset employed in this study was collected from *Youtube* by adopting developed web-scraping and information extraction method as described in Sect. 4.1. We collected 583 videos related the "hamas" organisation in total, and ended up with 52, 558 reviews, as shown in Table 3. Among those reviews, not all of them got *likes*, and we ended up with 37, 596 *liked* reviews, accounting for about %71.53.

We also conducted the experiment on *Twitter* dataset as described in Sect. 4.1. 16, 907 tweet IDs and their corresponding annotations was released in this dataset. However, we only retrieved 9, 593 of the tweets and their authors because of accounts being suspended. We also collected social connections (e.g. followed) of those authors, and got a total of 1, 477 nodes and 3, 676 edges, as shown in Table 4.

Table 3. *Youtube* dataset

	Count
Videos	583
Reviews	52,558
Users	38,047

Table 4. *Twitter* dataset

	Count
Tweets	9,593
Authors	1,477
Followees	3,676

5.2   Predicting User Influence

To represent the influence of users, we classified users based on their text-based and behaviours-based features as described in Sect. 4.2.

For *Youtube* Dataset(A), we compared our method with three baseline methods, *Linear regression* (LR), *Support Vector Machine* (SVM), and a fully connected *neural network* (NN). The precision, recall, and F1 scores for all the methods are presented in Table 5, in which the best results are highlighted in bold. PUI performed better than other three classifiers.

---

[7] https://gephi.org/.

For the *Twitter* Dataset(B), we adopted the method LR+AUTH proposed by [18] as our baseline wherein they utilised both character n-grams and the author profiles to train the LR classifier and achieved better performance than the existing state-of-the-art. In Table 6, we compared LR+AUTH with our method, showing that our method outperformed LR+AUTH in recall and F1-score.

Although our method is more effective than other approaches in terms of the accuracy in predicting user influence, it performs differently on two datasets. After some analysis and comparison on the two datasets, we inferred that the distribution of users has an impact on the results.

Table 5. Performance on *Youtube* dataset

Method	Precision	Recall	F1
LR	59.07%	61.61%	60.08%
SVM	62.10%	67.08%	63.89%
NN	72.39%	71.23%	71.15%
PUI (Our method)	**79.22%**	**81.92%**	**80.54%**

Table 6. Performance on *Twitter* dataset

Method	Precision	Recall	F1
LR+AUTH	**87.57%**	87.66%	87.57%
PUI (Our method)	83.89%	**95.79%**	**89.69%**

5.3 The Effects of Behaviours-Based Features

In the proposed PUI method, user influence is predicted based on the text and behaviours features. In the section, we also tested the effect of the behaviours-based features. A variant of PUI was created by only considering text-based features. The results show that considering users' behaviours can gain the performance of predicting user influence, as shown in Table 7.

Table 7. The effects of behaviours-based feature on *Youtube* dataset

Method	Precision	Recall	F1
PUI	**79.22%**	**81.92%**	**80.54%**
PUI (no behaviours-based feature)	69.34%	72.16%	71.99%

5.4   User Propagation Ability Analysis

As there is no social relationship available in *Youtube*, we only analysed the user propagation ability in *Twitter* Dataset. Figure 2 gives the social graph, in which greener nodes represent users who are highly-connected in the online social network. In order to have a clearer comparison, the users with top three high level influence are also marked as red in this graph. We can see that users with strong connections in social networks may not always be the influential ones in the context of toxic information propagation. Moreover, we classified all users into two categories and utilised *Force Atlas* algorithm to layout the graph, as shown in Fig. 3. Red nodes represent users who have ever posted hate speech, and green ones indicate users without any hate speech. It can been seen that hate-users are more likely to be closely connected with other hate-users in the social networking site.

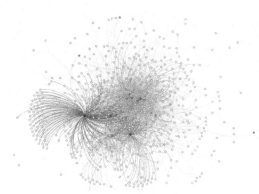
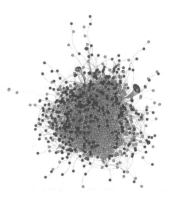

Fig. 2. The social graph of authors   **Fig. 3.** User classification (Color figure online)

6   Conclusions

Although the rise and advance of the Internet platforms provide great convenience in communication, transaction, entertainment for the Internet users, there have been increasingly substantial cases of negative and toxic information, such as hate speech, cyberbullying, pornography, violence crisis, racism, anti-social behaviours and other instances online. Unfortunately, detecting posts with toxic information has some difficulties in achieving more effective performance.

Inspired by the research [14] on *Nature* in 2019, this study focused on users rather than text detection in traditional methods. In this paper, we present PUI, a method that identifying influential users based on users' text and behaviours characteristics instead of the analysis on their network structures. Taking terrorism-related videos on *Youtube* and hate speech on *Twitter* as examples, the effectiveness of the proposed method is demonstrated. In addition, it is also discussed that the influential users have not always been those with strong connections on the social networking site, which is important in countering today's online propagation of toxic information.

Acknowledgment. This work is supported by Scientific Research Guiding Project (Grant No. Y9W0013401), Youth Innovation Promotion Association of CAS (Grant No. 1101CX0101) and Key Technical Talents Project of CAS (Grant No. Y8YY041101).

References

1. Alakrot, A., Murray, L., Nikolov, N.S.: Dataset construction for the detection of anti-social behaviour in online communication in Arabic. Procedia Comput. Sci. **142**, 174–181 (2018). https://doi.org/10.1016/j.procs.2018.10.473
2. Almgren, K., Lee, J.: A hybrid framework to predict influential users on social networks. In: The 10th International Conference on Digital Information Management, ICDIM 2015 (ICDIM), pp. 103–108 (2016). https://doi.org/10.1109/ICDIM.2015.7381864
3. Australian Informatics Olympiad Committee, Australian Government: Cybercrime in Focus. Annual Report (2015)
4. Badjatiya, P., Gupta, S., Gupta, M., Varma, V.: Deep learning for hate speech detection in tweets. In: Proceedings of the 26th International Conference on World Wide Web Companion, pp. 759–760. International World Wide Web Conferences Steering Committee (2017)
5. Pew Research Center: The Annual Cyberbullying Survey (2017). Online Harassment 2017
6. Cha, M., Haddadi, H., Benevenuto, F., Gummadi, K.P.: Measuring user influence in Twitter: the million follower fallacy. In: Fourth International AAAI Conference on Weblogs and Social Media (2010)
7. Chen, J., Deng, Y., Su, Z., Wang, S., Gao, C., Li, X.: Identifying multiple influential users based on the overlapping influence in multiplex networks. IEEE Access **7**, 156150–156159 (2019). https://doi.org/10.1109/ACCESS.2019.2949678
8. Corazza, M., et al.: Comparing different supervised approaches to hate speech detection. In: EVALITA 2018 (2018)
9. Davidson, T., Warmsley, D., Macy, M., Weber, I.: Automated hate speech detection and the problem of offensive language. In: Eleventh International AAAI Conference on Web and Social Media (2017)
10. Dey, K., Kaushik, S., Subramaniam, L.V.: Literature Survey on Interplay of Topics, Information Diffusion and Connections on Social Networks (2017). http://arxiv.org/abs/1706.00921
11. United Nations. Counter-Terrorism Implementation Task Force. Report of the Working Group on Countering the Use of the Internet for Terrorist Purposes (2009). http://www.un.orglterrorism/pdfs/wg6-internet_revI.pdf
12. Fortuna, P., Nunes, S.: A survey on automatic detection of hate speech in text. ACM Comput. Surv. **51**(4), 1–30 (2018). https://doi.org/10.1145/3232676. http://dl.acm.org/citation.cfm?doid=3236632.3232676
13. Hosseinmardi, H., Mattson, S.A., Rafiq, R.I., Han, R., Lv, Q., Mishra, S.: Detection of cyberbullying incidents on the Instagram social network. arXiv preprint arXiv:1503.03909 (2015)
14. Johnson, N., et al.: Hidden resilience and adaptive dynamics of the global online hate ecology. Nature **573**(7773), 261–265 (2019)
15. Kwan, G.C.E., Skoric, M.M.: Facebook bullying: an extension of battles in school. Comput. Hum. Behav. **29**(1), 16–25 (2013)

16. Label, D.T.: The Annual Cyberbullying Survey (2013). https://www.ditchthelabel.org/wp-content/uploads/2016/07/cyberbullying2013.pdf

17. Li, S., Li, G., Law, R., Paradies, Y.: Racism in tourism reviews. Tourism Manag. **80**, 104100 (2020)

18. Mishra, P., Del Tredici, M., Yannakoudakis, H., Shutova, E.: Author profiling for abuse detection. In: Proceedings of the 27th International Conference on Computational Linguistics, pp. 1088–1098 (2018). https://www.aclweb.org/anthology/C18-1093

19. Mondal, M., Silva, L.A., Correa, D., Benevenuto, F.: Characterizing usage of explicit hate expressions in social media. New Rev. Hypermedia Multimedia **24**(2), 110–130 (2018). https://doi.org/10.1080/13614568.2018.1489001

20. Mouheb, D., Ismail, R., Al Qaraghuli, S., Al Aghbari, Z., Kamel, I.: Detection of offensive messages in Arabic social media communications. In: Proceedings of the 2018 13th International Conference on Innovations in Information Technology, IIT 2018, pp. 24–29 (2019). https://doi.org/10.1109/INNOVATIONS.2018.8606030

21. Park, J.H., Fung, P.: One-step and Two-step Classification for Abusive Language Detection on Twitter (2017). http://arxiv.org/abs/1706.01206

22. Rodríguez, A., Argueta, C., Chen, Y.L.: Automatic detection of hate speech on Facebook using sentiment and emotion analysis. In: 2019 International Conference on Artificial Intelligence in Information and Communication (ICAIIC), pp. 169–174. IEEE (2019)

23. Salawu, S., He, Y., Lumsden, J.: Approaches to automated detection of cyberbullying: a survey. IEEE Trans. Affect. Comput. 1–25 (2017). https://doi.org/10.1109/TAFFC.2017.2761757

24. Silva, L., Mondal, M., Correa, D., Benevenuto, F., Weber, I.: Analyzing the targets of hate in online social media. In: Tenth International AAAI Conference on Web and Social Media (2016)

25. Subramani, S., Wang, H., Vu, H.Q., Li, G.: Domestic violence crisis identification from Facebook posts based on deep learning. IEEE Access **6**, 54075–54085 (2018)

26. Vu, H.Q., Li, G., Law, R., Zhang, Y.: Travel diaries analysis by sequential rule mining. J. Travel Res. **57**(3), 399–413 (2018)

27. Vu, H.Q., Li, G., Law, R., Zhang, Y.: Exploring tourist dining preferences based on restaurant reviews. J. Travel Res. **58**(1), 149–167 (2019)

28. Wang, X., Li, G., Jiang, G., Shi, Z.: Semantic trajectory-based event detection and event pattern mining. Knowl. Inf. Syst. **37**(2), 305–329 (2011). https://doi.org/10.1007/s10115-011-0471-8

29. Waseem, Z., Hovy, D.: Hateful Symbols or Hateful People? Predictive Features for Hate Speech Detection on Twitter, pp. 88–93 (2016). https://doi.org/10.18653/v1/n16-2013

30. Watanabe, H., Bouazizi, M., Ohtsuki, T.: Hate speech on Twitter: a pragmatic approach to collect hateful and offensive expressions and perform hate speech detection. IEEE Access **6**, 13825–13835 (2018). https://doi.org/10.1109/ACCESS.2018.2806394

31. Wen, S., Jiang, J., Xiang, Y., Yu, S., Zhou, W.: Are the popular users always important for information dissemination in online social networks? IEEE Netw. **28**(5), 64–67 (2014). https://doi.org/10.1109/MNET.2014.6915441

Extracting Distinctive Shapelets with Random Selection for Early Classification

Guiling Li[1,2(✉)] and Wenhe Yan[1]

[1] School of Computer Science, China University of Geosciences,
Wuhan 430074, China
guiling@cug.edu.cn, cugywh@163.com
[2] Hubei Key Laboratory of Intelligent Geo-Information Processing,
China University of Geosciences, Wuhan 430074, China

Abstract. Early classification on time series has attracted much attention in time-sensitive domains. The goal of early classification on time series is to achieve better classification accuracy, and meanwhile to make prediction as early as possible. Shapelets are local features of time series and have high discriminability. In shapelet-based early classification, due to the large number of shapelet candidates, it is challenging to discover shapelets more effectively. In this paper, we propose Early Random Selection Shapelet Classification on Time Series (EARSC). Firstly, we identify the representative time series for each class. Secondly, we extract shapelet candidates for the representative time series and then evaluate them to obtain prior knowledge. Thirdly, we design random selection strategy with prior knowledge to select the better shapelet and make early classification. Experimental results on 14 real datasets have shown the effectiveness of the proposed method.

Keywords: Time series · Early classification · Shapelets · Random selection

1 Introduction

Time series classification is one of the most important tasks in time series data mining. Especially in recent years, time series early classification has attracted much attention from the research community. Early classification can be used for some time-sensitive domains, such as medical early diagnosis, video surveillance, earthquake early-warning, network intrusion detection. For example, in early diagnosis for heart disease, abnormal ECG signals may indicate a specific heart disease that requires immediate treatment, so early diagnosis is very important in emergency health care area. An early classification model can make diagnosis when part of ECG time series data are obtained, ensuring that patients with heart disease receive early treatment.

The goal of early classification on time series is to make accurate prediction earlier, so there are two requirements for the early classification model of time

© Springer Nature Switzerland AG 2020
G. Li et al. (Eds.): KSEM 2020, LNAI 12274, pp. 471–484, 2020.
https://doi.org/10.1007/978-3-030-55130-8_41

series. On one hand, the model can make prediction as early as possible, but it needs to ensure that the prediction results are credible simultaneously. On the other hand, the accuracy of the model in classifying time series when obtaining part of the data should be comparable to that of the model with full data.

How to construct a good classifier for early classification on time series is challenging. For complete time series classification, we can extract specific features from full time series and then use the obtained features to construct a classifier. This method always aims to achieve high accuracy without taking into account the special requirements of early characteristics, so it is not suitable for early classification. Xing et al. [8,9] proposed ECTS (Early Classification on Time Series) method. ECTS is based on 1-nearest neighbor classification method, however, the method does not have interpretability. Ye et al. [4] proposed a new data mining primitive, called time series shapelets. Shapelets are the subsequences which could be the maximal representative of a class in time series, thus shapelets are suitable for time series classification. Xing et al. [10] proposed Early Distinctive Shapelet Classification (EDSC). EDSC constructed an early classification classifier by extracting shapelets for the first time, which made the classification results more accurate and interpretable.

The quality of shapelet determines the classification effect. Some methods first extract all shapelet candidates, then evaluate these candidates, and finally select the optimal shapelets. However, these methods generate a lot of redundant shapelets with more computation cost. In addition, the methods reduce the discriminability of shapelets affecting the classification accuracy.

In order to extract distinctive time series shapelets for early classification, we propose Early Random Selection Shapelet Classification on Time Series (EARSC), and conduct early classification experiments on real datasets to evaluate EARSC. The main contributions of this paper are summarized as follows.

(1) The representative time series for each class are first identified by distance calculation, then the shapelet candidates of the representative time series are evaluated to obtain the optimal shapelet length as prior knowledge.
(2) Random Selection with prior knowledge is proposed to extract shapelets, and then shapelets are selected for early classification.
(3) Experiments on real datasets demonstrate the effectiveness of the proposed method.

2   Related Work

The related work of this paper includes the following two research directions.

Shapelet Discovery. Ye et al. [4] first proposed the concept of shapelet. Shapelets are local features in time series which can maximally represent a class in time series. Therefore, shapelets have good interpretability. Ye et al. first adopted brute force search to find shapelets, and then designed two speedup techniques, i.e., early abandon and entropy pruning. However, the method is not efficient since it is based on brute force algorithm. Subsequently, some improved methods have been proposed. With dimensionality reduction technique such as symbolization, shapelets could be selected from symbolized sequences [18,23]. Subsequences containing key points could be considered as shapelet candidates, so shapelet candidates could be pruned with key points [5–7]. Mathematical optimization also was proposed to evaluate shapelet candidates [11,12]. Random selection strategy was also used for shapelet extraction [15].

Early Classification on Time Series. Early classification on time series is an important sub-problem in time series classification, and has attracted much attention of researchers recently [13,14]. Time series data have a temporary order increasing with time, thus early classification on time series is different from the traditional data classification. Xing et al. [8,9] proposed an early classification method ECTS (Early Classification on Time Series). ECTS is an 1-nearest neighbor classification method based on Minimum Prediction Length (MPL).

Some early classification models establish certain criteria to judge the credibility of classification result. Ghalwash et al. [3] proposed the Early Classification Model (ECM) algorithm by using HMM and SVM for early classification. Later, Mori et al. [20] proposed an early classification framework based on class discriminativeness and reliability of predictions (ECDIRE). Parrish et al. [22] proposed decision rules for incomplete data classification. Morid et al. [19] proposed an early classification method based on probabilistic classifiers and stopping rules. Multi-objective optimization methods [1,17,24] and deep learning [25–27] also have been proposed for early classification.

Due to good interpretability of shapelet, shapelet-based early classification methods have been proposed. Xing et al. [10] proposed an early classification algorithm Early Distinctive Shapelet Classification (EDSC). EDSC added early property to shapelets and selected shapelets according to the evaluation score. Ghalwash et al. [2] proposed the Modified EDSC with Uncertainty estimates (MEDSC-U). MEDSC-U added uncertainty parameters and evaluated the classification results through the uncertainty parameters. Karlsson et al. [16] proposed an early classification method based on random shapelet forest [15] and formulated two different decision tree traversal strategies, but the random shapelet could not explain the classification results well.

3 Preliminaries

In this section, we give some basic definitions used for shapelet discovery.

Definition 1. *Time Series. Time series is a set of real-valued variables with time order, represented by $T = \{t_1, t_2, t_3, \ldots, t_m\}$. Wherein, m is the length of time series T, and t_i is the data element at time stamp i of T.*

Definition 2. *Time Series Subsequence. Given a time series T with length m, a subsequence S is a sampling from contiguous positions with length l in time series T, represented by $S = \{t_p, \ldots, t_{p+l-1}\}$, $1 \leq p \leq m - l + 1$.*

Definition 3. *Distance Between Two Time Series. Given two time series T and R with the same length m, the distance between T and R is denoted by Dist(T,R). When using euclidean distance, Dist(T,R) can be calculated as:*

$$Dist(T, R) = \sqrt{\frac{1}{m} \sum_{i=1}^{m} (t_i - r_i)^2} \tag{1}$$

Definition 4. *Shapelet. Shapelet is a time series subsequence which is in some sense maximally representative of a class, and can be represented by a triple $f = (S, \delta, c)$, where S is a time series subsequence, δ is a distance threshold, and c is the class label. In this paper, we calculate the value of δ by using KDE method [10].*

Definition 5. *Best Match Distance (BMD). The best match distance between shapelet $f = (S, \delta, c)$ and time series T is defined as:*

$$BMD(f, T) = \min(Dist(S', S)), S' \in S_T^{|S|} \tag{2}$$

where, $S_T^{|S|}$ is a collection of subsequences with length $|S|$ in T.

Definition 6. *BMD-list. Given a shapelet f and a training dataset D containing r time series, BMD-list is a list of the BMDs between f and each time series in D, sorted in a non-descending order shown as follows:*

$$V_f = \{(d_1, c_1), (d_2, c_1), \ldots, (d_r, c_2)\} \tag{3}$$

where $d_i = BMD(f, T_i)$, $T_i \in D$, and $d_i \leq d_j$, for $i < j$, c_i is the class label of T_i.

Definition 7. *Earliest Match Length (EML). Given a shapelet $f = (S, \delta, c)$ and a time series T, EML is the minimal identifiable length of T, which means the shapelet f can classify time series T using its prefix from the beginning to the position EML(f, T).*

$$EML(f, T) = \min_{len(s) \leq i \leq len(T)} (dist(T[i - len(s) + 1, i], s) \leq \delta) \tag{4}$$

If f cannot classify time series T, we set $EML(f, T) = \infty$.

Definition 8. *Earliness. Earliness measures the early feature of shapelet f for early classification on a training dataset D.*

$$Earliness(f) = \frac{1}{\| D' \|} \sum_{T \in D} (1 - \frac{EML(f, T)}{length(T)}) \tag{5}$$

where D' represents a collection of time series that can be classified by shapelet f.

Definition 9. *Precision. Given a shapelet $f = (S, \delta, c)$ and a training dataset D, precision is defined as the proportion of time series in D that f could classify correctly.*

$$Precision(f) = \frac{|BMD(f,T) \leq \delta \wedge C(T) = c|}{|BMD(f,T) \leq \delta|}, T \in D \qquad (6)$$

Definition 10. *Recall. Given a shapelet $f = (S, \delta, c)$ and a training dataset D, recall is defined as the proportion of time series of class c in D that f could classify correctly.*

$$Recall(f) = \frac{|BMD(f,T) \leq \delta \wedge C(T) = c|}{|C(T) = c|}, T \in D \qquad (7)$$

4   Shapelet Discovery for Early Classification

Some shapelet-based time series classification algorithms search the feature space to discover the optimal shapelet, and then use the obtained shapelet for early classification. However, since any subsequence in time series could be a potential shapelet candidate, the feature space of shapelet candidates is huge. Moreover, some shapelet candidates are useless and redundant, so extracting and evaluating these shapelet candidates will cost a lot of time. If the useless and redundant shapelet candidates can be effectively filtered, the accuracy would be enhanced and the calculation cost can also be reduced.

The previous random method to discover shapelet is to randomly select a time series and then randomly select a subsequence with random starting point and length. The advantage of this method is its fast speed of extracting shapelet, but it needs to be combined with an ensemble method [16]. Otherwise, the classification is not accurate. Besides, the completely random results do not have interpretability. In this paper, we utilize the merit of random selection for shapelet. More importantly, we add prior knowledge for random shapelet selection without the requirement of ensemble, ensuring the reliability and accuracy of the classification results. In order to extract distinctive time series shapelets for early classification, we propose Early Random Selection Shapelet Classification on Time Series (EARSC). The workflow of EARSC is shown in Fig. 1.

Fig. 1. The workflow of EARSC

As shown in Fig. 1, the proposed method EARSC makes early classification as follows. Firstly, find representative time series for each class from the training set. Secondly, extract shapelet candidates in the representative time series and evaluate them, and then rank the lengths of shapelet candidates by quality score to obtain prior knowledge. Thirdly, utilize the prior information for random shapelet discovery to find the optimal shapelet. Finally, make early classification based on the selected shapelet.

4.1 Find Representative Time Series

In order to obtain prior knowledge for shapelet discovery, we first obtain the representative time series for each class in training set by distance calculation, and further extract shapelet candidates from these representative time series. There are three specific steps for finding representative time series.

Step 1 For each time series in training set D, calculate the average euclidean distance ave_{same} between this time series and all other time series with same class.

Step 2 For each time series in training set D, calculate the average euclidean distance ave_{nsame} between this time series and all time series with different classes.

Step 3 Calculate the $grade$ of each time series according to ave_{same} and ave_{nsame}. Higher $grade$ indicates that the time series can be a representative of the same class, the $grade$ is calculated by Eq. (8).

$$grade = \frac{ave_{nsame}}{ave_{same}} \tag{8}$$

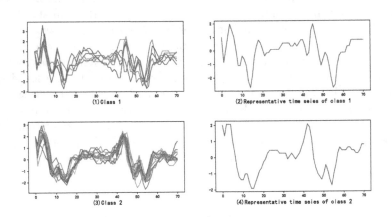

Fig. 2. Representative time series

Figure 2 shows the extracted representative time series for SonyAIBORobot-Surface1 dataset. Figure 2 (1) shows all time series of class 1 in time series training set, and Fig. 2 (2) shows a representative time series of class 1. Figure 2 (3) shows all time series of class 2 in training set, and Fig. 2 (4) shows the representative time series of class 2.

4.2   Find Best Length of Shapelet

We use the obtained representative time series for each class to explore useful information for shapelet extraction. Since any subsequence in time series can be regarded as a shapelet candidate, traditional method needs to extract continuous subsequences with each possible length. However, the quality of continuous subsequences varies. In random shapelet extraction, if a shapelet with better quality can be randomly selected with a higher probability, the accuracy of classification will be higher. In order to extract shapelet with better quality, we first extract and evaluate shapelet candidates in representative time series, and then sort the lengths of shapelet candidates by quality score defined as Eq. (10). The sorted array can provide prior knowledge for further random selection method to improve classification accuracy. The pseudo code of finding best length of shapelet is shown in Algorithm 1.

$$fscore = \frac{2 * Precision(f) * Recall(f)}{Precision(f) + Recall(f)} \tag{9}$$

$$score = \frac{2 * fscore * Earliness(f)}{fscore + Earliness(f)} \tag{10}$$

Algorithm 1. FindBestShapeletLength

Input: dataset of representative time series: D_{rts}
Output: shapelet lengths for each class of time series: $lenMatrix$
1: $lenMatrix \leftarrow \varnothing$
2: **for** *each time series T in D_{rts}* **do**
3:      *Initialize(lenArr)*
4:      **for** *each shapelet in T* **do**
5:          *CalculateScore(shapelet)*
6:          **if** *lenArr[shapelet.length] < shapelet.score* **then**
7:              *lenArr[shapelet.length] ← shapelet.score*
8:          **end if**
9:      **end for**
10:     *lenMatrix.add(lenArr)*
11: **end for**
12: **return** *lenMatrix*

In Algorithm 1, after obtaining the representative time series denoted by D_{rts}, line 3 initializes a length array *lenArr* with quality score for each class of time series. Lines 4–9 extract the shapelet candidates and then evaluate them. Lines 6–8 compare the length score of the current shapelet with the length score saved by *lenArr* and update *lenArr*. Line 10 stores *lenArr* into *lenMatrix*, and finally line 12 returns *lenMatrix* which stores the shapelet length array for each class.

4.3   Random Shapelet Discovery

In this paper, shapelets are discovered by randomly selecting starting point and length. Moreover, to ensure the reliability of random selection, we design two random rules as follows:

Rule 1   The length array obtained from representative time series for each class, is divided into two parts according to a certain ratio. Specifically, the first part of length array records the lengths of the better shapelets, and the second part of length array corresponds to the poor shapelets. Then the shapelet length is selected from these two parts with a certain probability.

Rule 2   The next randomly obtained shapelet cannot have overlapping part with the last shapelet. If there is overlap, perform random selection again by Rule 1.

Algorithm 2. RandomShapeletDiscovery

Input: training set: D, shapelet length array for each class: $lenMatrix$, selection parameter: k
Output: shapelet library:$library$
 1: $library \leftarrow \varnothing$
 2: **for** *each time series T in D* **do**
 3:     **for** $i = 1$ *to iterNum* **do**
 4:         $startPos \leftarrow RandomSelectStartPosition()$
 5:         $length \leftarrow RandomSelectLength(lenMatrix)$
 6:         **if** *satisfy Rules 1 and 2* **then**
 7:             $shapelet \leftarrow ExtractAndEstimateShapelet(T, startPos, length)$
 8:             $library.Add(shapelet)$
 9:         **end if**
10:     **end for**
11: **end for**
12: **return** $library$

The pseudo code of random shapelet discovery is shown in Algorithm 2. Wherein, $iterNum$ controls the number of randomly selected shapelets, and the value of $iterNum$ is the number of shapelets contained in each time series divided by selection parameter k. Lines 2–11 extract and evaluate shapelets from each time series. Lines 4–5 randomly select a starting point and length for shapelets. Line 6 judges whether the current random starting point and length are close to the previous random ones according to the rules. If not, lines 7–8 extract and evaluate the shapelet from the time series. Otherwise, reselect the starting point and length. Finally, the extracted shapelets are saved to the shapelet candidate set $library$ and line 12 returns $library$.

4.4   Shapelet Selection and Early Classification

When obtaining the shapelet candidate set *library*, shapelets need to be selected from *library* for early classification. We sort all the shapelets by quality score from highest to lowest.

In shapelet selection, we firstly select the shapelet $f = (S, \delta, c)$ with the highest quality score, and then mark all the samples covered by f in the training set. Wherein, the coverage condition is $dist(S, T) \leq \delta \wedge C(T) = c$, that is, the distance between the sample and f is less than δ and the class of sample and f is the same. Secondly, we select the shapelet $f' = (S', \delta', c')$ with the second highest quality score, continue to cover the unlabeled samples in the dataset, and then mark the samples covered by f'. Repeat this step until all samples are marked.

Early classification is to use part of the time series data to make prediction. First we obtain the selected shapelets and the minimum length l of selected shapelets. As the data elements of testing sample are read in continuously, use each shapelet to match the first l data elements of the sample. If the sample is matched successfully with one of the shapelets, return the class of the matching shapelet. If there is no match, continue to read the subsequent data elements of the sample and repeat the prediction process. If the classification cannot be performed at the end, the sample is unclassifiable.

5   Experiments

5.1   Experimental Setup

UCR time series dataset library [21] is a widely used standard time series classification archive. In order to verify the performance of the proposed method EARSC, we use 14 datasets of different categories from UCR time series datasets to conduct experiments. The information of 14 datasets is described in Table 1.

We select three baselines for comparison, 1NN-ED, ECTS [9] and EDSC [10], respectively. 1NN-ED is the nearest neighbor method based on Euclidean distance, applied for full time series classification. ECTS is an 1NN-based early classification method, which can handle the early classification of time series well. EDSC is an early classification algorithm based on shapelet, which has good earliness and interpretability.

We evaluate the effectiveness of early classification in terms of accuracy and earliness. The accuracy indicates the proportion of samples that can be correctly classified in testing set. Earliness refers to the average minimum prediction length of the time series for successful classification. The proposed method is implemented in JAVA language. The experimental results are obtained on the computer with Intel I5-7400, 3.0 GHz and memory 8G.

5.2   Results and Analysis

In the experiments, we set the minimum length of the shapelet $minL = 5$ and the maximum length of the shapelet $maxL = L/2$, where L is the length of

Table 1. Description of datasets

Datasets	Training	Testing	Length	Class
CBF	30	900	470	5
Coffee	28	28	286	2
DiatomSizeReduction	16	306	345	4
ECG200	100	100	96	2
FaceAll	560	1690	131	14
FaceFour	24	88	350	4
FacesUCR	200	2050	131	14
GunPoint	50	150	150	2
Lightning2	60	61	637	2
SonyAIBORobotSurface1	20	601	70	2
Symbols	25	995	398	6
SyntheticControl	300	300	60	6
Trace	100	100	275	4
TwoLeadECG	23	1139	82	2

a complete time series. In random shapelet discovery, the selection parameter k is set to 3 and the dividing ratio of length array for Rule 1 is 4:6. When $minL = 5, maxL = L/2$, the number of shapelets contained in a single time series is $sum = ((L - minL) + (L - maxL)) * (maxL - minL)/2$. So when k is equal to 3, we only need to extract $sum/3$ shapelet candidates. The parameter settings are the same for all the datasets, and the accuracy results are shown in Table 2.

It can be seen from Table 2, the average rank on accuracy of the proposed method EARSC is the best among the four methods. EARSC can perform best on accuracy for six datasets, however, both EDSC and 1NN-ED can only get the best accuracy on three datasets, while ECTS can only obtain the best accuracy on two datasets. Moreover, it is worth mentioning that, for the three datasets Coffee, SonyAIBORobotSurface1 and TwoLeadECG, the accuracy of EARSC is much improved compared to that of other methods. Especially for TwoLeadECG dataset, the accuracy of EARSC is 24.45% higher than ECTS, 9.45% higher than EDSC, and 22.45% higher than 1NN-ED. The reason is that, EARSC is a random selection shapelet-based early classification method. EARSC first finds representative time series and then finds the length of shapelets with better quality, so as to obtain prior knowledge for further random shapelet discovery. With prior knowledge, two random rules are designed for shapelet extraction. Therefore, the shapelets are discriminative and can improve the accuracy of classification.

In time series early classification, earliness is also an important evaluation criteria. Table 3 shows the earliness comparison of three methods in all 14 datasets.

Table 2. Accuracy values (%) for ECTS, EDSC, 1NN-ED and EARSC

Datasets	ECTS	EDSC	1NN-ED	EARSC
CBF	85	84	85	**87.55**
Coffee	75	75	75	**82.14**
DiatomSizeReduction	80	85	**93**	85.62
ECG200	89	85	88	**91**
FaceAll	**76**	66	71	75.98
FaceFour	**82**	75	78	76.14
FacesUCR	71	63	**77**	67.32
GunPoint	87	**94**	91	93.33
Lightning2	70	**80**	75	78.69
SonyAIBORobotSurface1	69	80	70	**85.69**
Symbols	81	51	**90**	51.46
SyntheticControl	88	89	88	**89.67**
Trace	74	**80**	76	76
TwoLeadECG	73	88	75	**97.45**
Average rank	2.928571	2.785714	2.464286	**1.821429**

Table 3. Earliness values (%) for ECTS, EDSC and EARSC

Datasets	ECTS	EDSC	EARSC
CBF	71.5	**31.85**	35.68
Coffee	83.94	54.23	**33.51**
DiatomSizeReduction	**14.88**	27.04	24.56
ECG200	60.11	**23.24**	28
FaceAll	63.85	**38.94**	45.97
FaceFour	72.26	47.98	**46.86**
FacesUCR	87.21	**51.58**	59.01
GunPoint	46.92	**45.58**	46.37
Lightning2	89.01	**55.14**	63.68
SonyAIBORobotSurface1	68.49	47.03	**44.02**
Symbols	**51.3**	60.25	67.15
SyntheticControl	87.88	**50.81**	50.89
Trace	50.72	**38.63**	51.60
TwoLeadECG	64.43	46.85	**45.90**
Average rank	2.642857	1.5	**1.857143**

It can be seen from Table 3 that, ECTS can only obtain the best earliness on two datasets, because ECTS is 1NN-based early classification. EDSC can obtain the best earliness on eight datasets and EARSC performs best on four datasets. The reason is that, when selecting shapelets with quality score, EDSC only considers precision and earliness, while EARSC considers precision, recall and earliness. From Tables 2 and 3, it is worth pointing out that, both earliness and accuracy of EARSC are higher than other methods for three datasets Coffee, SonyAIBORobotSurface1 and TwoLeadECG.

By evaluating accuracy and earliness simultaneously, we observe EARSC performs better than ECTS and EDSC. Moreover, with less data, EARSC can obtain the comparable accuracy with using the complete time series. It demonstrates that EARSC is suitable for early classification on time series.

6   Conclusions

In this paper, we propose Early Random Selection Shapelet Classification on Time Series (EARSC). Firstly, we identify the representative time series for each class. Secondly, we extract the shapelet candidates of the representative time series and make evaluation to obtain prior knowledge. Thirdly, we propose random selection with prior knowledge to discover shapelets, then select shapelets for early classification. Experiments on real datasets demonstrate the effectiveness of the proposed method. In future work, we will extend EARSC to multivariate time series early classification.

Acknowledgments. The authors would like to thank Prof. Eamonn Keogh and all the people who have contributed to the UCR time series classification archive for their selfless work. The work is supported by the National Natural Science Foundation of China (No. 61702468) and Open Research Project of The Hubei Key Laboratory of Intelligent Geo-Information Processing (No. KLIGIP-2018B03).

References

1. Ando, S., Suzuki, E.: Minimizing response time in time series classification. Knowl. Inf. Syst. **46**(2), 449–476 (2015). https://doi.org/10.1007/s10115-015-0826-7
2. Ghalwash, M.F., Radosavljevic, V., Obradovic, Z.: Utilizing temporal patterns for estimating uncertainty in interpretable early decision making. In: SIGKDD, New York City, USA, pp. 402–411. ACM (2014)
3. Ghalwash, M.F., Ramljak, D., Obradovic, Z.: Early classification of multivariate time series using a hybrid HMM/SVM model. In: 2012 IEEE International Conference on Bioinformatics and Biomedicine (BIBM), Philadelphia, USA, pp. 1–6. IEEE (2012)
4. Ye, L., Keogh, E.: Time series shapelets: a new primitive for data mining. In: SIGKDD, Paris, France, pp. 947–956. ACM (2009)
5. Zhang, Z., Zhang, H., Wen, Y., Yuan, X.: Accelerating time series shapelets discovery with key points. In: Li, F., Shim, K., Zheng, K., Liu, G. (eds.) APWeb 2016. LNCS, vol. 9932, pp. 330–342. Springer, Cham (2016). https://doi.org/10.1007/978-3-319-45817-5_26
6. Ji, C., Zhao, C., Li, P., et al.: A fast shapelet discovery algorithm based on important data points. Int. J. Web Serv. Res. **14**, 67–80 (2017)

7. Li, G., Yan, W., Wu, Z.: Discovering shapelets with key points in time series classification. Expert Syst. Appl. **132**, 76–86 (2019)
8. Xing, Z., Jian, P., Yu, P.S.: Early prediction on time series: a nearest neighbor approach. In: IJCAI, Pasadena, USA, pp. 1297–1302 (2009)
9. Xing, Z., Jian, P., Yu, P.S.: Early classification on time series. Knowl. Inf. Syst. **31**(1), 105–127 (2012)
10. Xing, Z., Pei, J., Yu, P.S., Wang, K.: Extracting interpretable features for early classification on time series. In: SDM, Mesa, USA, pp. 247–258. Omnipress (2011)
11. Grabocka, J., Schilling, N., Wistuba, M., et al.: Learning time-series shapelets. In: SIGKDD, , New York City, USA, pp. 392–401. ACM (2014)
12. Hou, L., Kwok, J., Zurada, J.: Efficient learning of timeseries shapelets. In: AAAI, Phoenix, USA, pp. 1209–1215. AAAI Press (2016)
13. Hsu, E.-Y., Liu, C.-L., Tseng, V.S.: Multivariate time series early classification with interpretability using deep learning and attention mechanism. In: Yang, Q., Zhou, Z.-H., Gong, Z., Zhang, M.-L., Huang, S.-J. (eds.) PAKDD 2019. LNCS (LNAI), vol. 11441, pp. 541–553. Springer, Cham (2019). https://doi.org/10.1007/978-3-030-16142-2_42
14. Gupta, A., Gupta, H., Biswas, B., Dutta, T.: An early classification approach for multivariate time series of on-vehicle sensors in transportation. IEEE Trans. Intell. Transp. Syst. 1–12 (2020). https://doi.org/10.1109/TITS.2019.2957325, https://ieeexplore.ieee.org/document/8949724
15. Karlsson, I., Papapetrou, P., Boström, H.: Forests of randomized shapelet trees. In: Gammerman, A., Vovk, V., Papadopoulos, H. (eds.) SLDS 2015. LNCS (LNAI), vol. 9047, pp. 126–136. Springer, Cham (2015). https://doi.org/10.1007/978-3-319-17091-6_8
16. Karlsson, I., Papapetrou, P., Boström, H.: Early random shapelet forest. In: Calders, T., Ceci, M., Malerba, D. (eds.) DS 2016. LNCS (LNAI), vol. 9956, pp. 261–276. Springer, Cham (2016). https://doi.org/10.1007/978-3-319-46307-0_17
17. Mori, U., Mendiburu, A., Miranda, I., Lozano, J.: Early classification of time series using multi-objective optimization techniques. Inf. Sci. **492**, 204–218 (2019)
18. He, Q., Dong, Z., Zhuang, F., Shang, T., Shi, Z.: Fast time series classification based on infrequent shapelets. In: 2012 11th International Conference on Machine Learning and Applications, Boca Raton, FL, pp. 215–219 (2012)
19. Mori, U., Mendiburu, A., Dasgupta, S., Lozano, J.: Early classification of time series by simultaneously optimizing the accuracy and earliness. IEEE Trans. Neural Netw. Learn. Syst. **29**(10), 4569–4578 (2018)
20. Mori, U., Mendiburu, A., Keogh, E., Lozano, J.A.: Reliable early classification of time series based on discriminating the classes over time. Data Min. Knowl. Disc. **31**(1), 233–263 (2016). https://doi.org/10.1007/s10618-016-0462-1
21. Dau, H., Keogh, E., Kamgar, K., et al.: The UCR Time Series Classification Archive (2019). https://www.cs.ucr.edu/~eamonn/time_series_data_2018/
22. Parrish, N., Anderson, H.S., Gupta, M.R., Hsiao, D.: Classifying with confidence from incomplete information. J. Mach. Learn. Res. **14**(1), 3561–3589 (2013)
23. Rakthanmanon, T., Keogh, E.: Fast shapelets: a scalable algorithm for discovering time series shapelets. In: SDM, Austin, USA, pp. 668–676 (2013)
24. Tavenard, R., Malinowski, S.: Cost-aware early classification of time series. In: Frasconi, P., Landwehr, N., Manco, G., Vreeken, J. (eds.) ECML PKDD 2016. LNCS (LNAI), vol. 9851, pp. 632–647. Springer, Cham (2016). https://doi.org/10.1007/978-3-319-46128-1_40
25. Wang, W., Chen, C., Wang, W., Rai, P., Carin, L.: Earliness-aware deep convolutional networks for early time series classification. CoRR abs/1611.04578 (2016)

26. Rußwurm, M., Lefèvre, S., Courty, N., Emonet, R., Körner, M., Tavenard, R.: End-to-end learning for early classification of time series. CoRR abs/1901.10681 (2019)
27. Hartvigsen, T., Sen, C., Kong, X., Rundensteiner, E.: Adaptive-halting policy network for early classification. In: KDD, New York, USA, pp. 101–110. ACM (2019)

Butterfly-Based Higher-Order Clustering on Bipartite Networks

Yi Zheng, Hongchao Qin, Jun Zheng[✉], Fusheng Jin, and Rong-Hua Li

Beijing Institute of Technology, Beijing, China
zhengjun_bit@163.com

Abstract. Higher-order clustering is a hot research topic which searches higher-order organization of networks at the level of small subgraphs (motifs). However, in bipartite networks, there are no higher-order structures such as triangles, quadrangles or cliques. In this paper, we study the problem of identifying clusters with motif of dense butterflies in bipartite networks. First, we propose a framework of higher-order clustering algorithm by optimizing motif conductance. Then, we prove that the problem can be transformed to computing the conductance of a weight graph constructed by butterflies, so it can be solved by eigenvalue decomposition techniques. Next, we analyse the computational complexity of the proposed algorithms and find that it is indeed efficient to cluster motif of butterflies in bipartite networks. Finally, numerous experiments prove the effectiveness, efficiency and scalability of the proposed algorithm.

Keywords: Bipartite network · Butterfly · Higher-order clustering

1   Introduction

With the development of artificial intelligence applications, more and more technologies related to knowledge graphs are applied in the fields of intelligent search, intelligent question answering, personalized recommendation, etc. In real world applications, bipartite graph is a common representation for knowledge graph. For example, a bipartite graph in Fig. 1 can represent the relationship between research topics and papers, where research topics form a vertex partition, papers form another vertex partition. Every paper has edges connection with the topic involved. In addition, the relationship between products and manufacturers, actors and movies can be expressed through a bipartite graph, like the bipartite network of research topics and papers. Therefore, the bipartite graphs are hyper-graphs that can represent the one-to-one, one-to-many and many-to-many relationships between entities [10], and play an irreplaceable role in network analysis.

Finding the dense interaction structure in the network can reveal the functions and associations between different entities [1,6,8,9,13,17]. High-order clustering can cluster the common structure *(motif)* which is the basic building block of complex networks [1,2,6,8–10,12,17]. The previous studies have focused on

G. Li et al. (Eds.): KSEM 2020, LNAI 12274, pp. 485–497, 2020.
https://doi.org/10.1007/978-3-030-55130-8_42

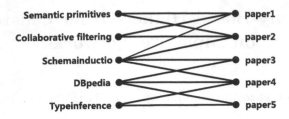

Fig. 1. Toy example of a bipartite graph

the problems of cluster cutting by edges and motifs. As the edge is not cohesive enough to show the clusters of the network, we often use a higher-order motif, like triangles, to build the clusters. Figure 1 is a bipartite graph which shows the relations of papers and research topics. However, it does not have any triangle ($\triangle$) in this bipartite graph, so it seems that higher-order clustering can not be conducted in this graph. Reconsider Fig. 1, we can observe topics "Semantic primitives" and "Collaborative filtering" are all linked to "paper 1" and "paper 2"; in contrary, "paper 1" and "paper 2" are all linked to topics "Semantic primitives" and "Collaborative filtering". The mentioned four nodes are densely connected in practice. Therefore, we can cluster bipartite graph by a common higher-order structure shaped like a butterfly ($\overline{\times}$).

In this work, the problem of clustering bipartite networks by motif of butterflies is addressed. Especially, our research have made the following major contributions.

- We introduce the concept of motif conductance, and propose the framework of higher-order clustering algorithm by optimizing motif conductance. We prove that the calculation of motif conductance of butterflies can be transformed to computing the conductance of a weight graph constructed by the butterfly. Therefore, we can solve the problem by matrix eigenvalue techniques.
- We analyse the computational complexity of the proposed algorithm. In addition, we compare the computational efficiency of higher-order clustering algorithm in ordinary and bipartite graphs, and conclude the reasons of why the algorithm performs faster in the bipartite graph.
- We conducted extensive experiments on five real-world application data sets. The results mean that our algorithm is highly efficient, efficient, and extensible.

The main content structure of this article is as follows: The second part introduces the problem model. The third part proposes the algorithm for finding higher-order clusters in bipartite graphs. The experimental results will be given in four parts. The fifth part introduces the relevant work of this article. The sixth part is the summary of the full text.

2   Preliminaries

G is an undirected and unweighted graph. In order to reveal the composition of the network, we defined network *motifs*.

Definition 1 (Motif). *Given graph $G = (V, E)$, a subgraph that appears frequently in G is called a motif M. For one node $v \in V$, M_v is a motif which contains v. The number of nodes in M is represented by $|M|$.*

One motif M is a triangle ($\triangle$) if $|M| = 3$ and the nodes in M are connected with each other in G. Similarly, one motif M is a butterfly ($\boxtimes$) if $|M| = 4$ and the nodes in M split into two groups of size 2 in which they are not linked inside each group but fully connected across the groups.

Definition 2 (Conductance). *There is a graph $G = (V, E)$ and a node subset $S \subseteq V$, the conductance of S, abbreviated as $\phi^{(G)}(S)$, is calculated as follows.*

$$\phi^{(G)}(S) = cut^{(G)}(S, \overline{S}) / min(vol^{(G)}(S), vol^{(G)}(\overline{S})). \tag{1}$$

Here, $\overline{S}$ and S make up the node set V, $cut^{(G)}(S, \overline{S})$ represents the number of edges cut dividing $\overline{S}$ and S, $vol^{(G)}(S)$ is the number of edges of the nodes in S.

Definition 3 (Conductance on Motif). *Given graph $G = (V, E)$ and motif M, the conductance of M is computed as follows.*

$$\phi_M^{(G)}(S) = cut_M^{(G)}(S, \overline{S}) / min(vol_M^{(G)}(S), vol_M^{(G)}(\overline{S})). \tag{2}$$

Here, $cut_M^{(G)}(S, \overline{S})$ is the number of motif examples cut, $vol_M^{(G)}(S)$ is the number of nodes in S in motif examples.

One motif instance is cut if the nodes in it are distributed in sets S and $\overline{S}$. We can convert the expression to the following form by a motif M:

$$cut_M^{(G)}(S, \overline{S}) = \sum_M I(\exists a, b \in M | a \in S, b \in \overline{S}), \tag{3}$$

$$vol_M^{(G)}(S) = \sum_M \sum_{i \in M} I(i \in S), \tag{4}$$

where $I(a)$ is a judgment function, if the result of statement a is true, the function value is 1; if the result of statement a is false, the function value is 0.

There is a graph G and one motif set M, one cluster S is a subgraph with dense M instances, where S should contain more examples of M, and avoid cutting the examples of M. Based on the motif conductance, the problem of this article is as follows:

Problem. When there is a bipartite graph G which is undirected and unweighted, and a motif M of butterfly ($\boxtimes$), we aim to look for the cluster S that minimizes the motif conductance ratio: $\phi_M^{(G)}(S) = cut_M^{(G)}(S, \overline{S}) / min(vol_M^{(G)}(S), vol_M^{(G)}(\overline{S}))$.

But finding cluster S with the smallest conductance is computationally infeasible [2], it is *NP-Hard*. Therefore, we extend the clustering method of eigenvalues and eigenvectors of Laplace matrix. Our method retains the characteristics of traditional spectral clustering: high computational efficiency and easy implementation. It is mathematically deduced that the obtained clustering is extremely close to the optimal [2].

3   Algorithms

3.1   Framework of Motif Conductance Optimization

In this section, we describe butterfly-based higher-order clustering algorithm on bipartite graph. The algorithm finds two sets S and $\overline{S}$, because $\phi_M^{(G)}(S) = \phi_M^{(G)}(\overline{S})$, then we use the smaller set of two as our final output target clustering in the bipartite graph. The algorithm is based on calculating the *motif adjacency matrix*. Next is its definition.

Definition 4 (Motif Adjacency Matrix). *There is a graph $G = (v, e)$ and motif M, the motif adjacency matrix W_M satisfies $(W_M)_{ij}$ = the number of times that nodes i and j participate in a motif simultaneously.*

Based on Definition 4, let G_M be the weighted graph which can induce W_M. Similar to Definition 3, the conductance of G_M can be calculated by Eq. (1) but $cut^{(G)}(S, \overline{S})$ is the sum of the weights of the cut edges dividing $\overline{S}$ and S. Furthermore, the *motif degree matrix* D_M satisfies $(D_M)_{ii} = \sum_{j=1}^{n}(W_M)_{ij}$. Therefore, the *motif laplacian matrix* can be calculated by $L_M = D_M - W_M$.

Theorem 1. *Given graph G and motif M. If $|M| = 3$, then the motif conductance equals the edge conductance on G_M [2]. If $|M| = 4$, then the motif conductance and the edge conductance on G_M are approximately equivalent.*

Proof. See Lemmas 1–4 and Theorem 2 in next subsection.

Based on the theorem above, Algorithm 1 efficiently finds the cluster S through the following steps: first of all, given the bipartite graph and motif M-*butterfly*, it can form the matrix W_M, D_M and the L_M. Then the eigenvector z and σ_i corresponding to the second minimum feature value of L can be calculated by the framework. The index value σ_i represents the index of the i_{th} largest value of z. Finally, by calculating the prefix sequence of σ_i, we can get the set S_r with the lowest conductance, where $S_r = \{\sigma_1, \ldots, \sigma_r\}$.

In the next part, we will show that if $|M| = 4$, $\phi^{(G_M)}(S_r) \approx \phi_M^{(G)}(S_r)$.

3.2   Motif Conductance with Butterflies $\approx$ Conductance of G_M

The following Lemma derives the relationship between the motif volume and the edge volume by the *motif adjacency matrix*.

Algorithm 1: Butterfly-based higher-order clustering algorithm on bipartite graph.

Input: (1)Undirected bipartite graph G=(V,E); (2)Motif butterfly M
Output: Butterfly-based cluster S
1 $W_M \leftarrow$ motif adjacency matrix of G;
2 $G_M \leftarrow$ a weight graph constructed by butterflies;
3 $D_M \leftarrow$ motif diagonal matrix of G;
4 $z \leftarrow$ a feature vector corresponding to the second smallest feature value of the Laplace matrix $L_M = D_M - W_M$;
5 $\sigma_i \leftarrow$ the index of z with i_{th} largest value;
6 $S \leftarrow min\phi^{(G_M)}(S_r)$, $S_r = \{\sigma_1, \ldots, \sigma_r\}$;
7 **return** $min(|S|, |\overline{S}|)$;

Lemma 1. *Given graph G and motif M with $|M| \geq 2$, and for any $S \subset V$, $vol_M^{(G)}(S) = \frac{1}{|M|-1} vol^{(G_M)}(S)$.*

Proof. Take one motif M as an example and $\{a_1, \ldots, a_{|M|}\}$ are nodes in motif. According to the Equation $S4$, the value of $(W_M)_{a_1,i}$ is all one, where $i = a_2, \ldots, a_{|M|}$. $(D_M)_{a_1,a_1}$ is equal to the addition of $\sum_i (W_M)_{a_1,i}$, where $i = a_2, \ldots, a_{|M|}$. Then the node a is added $|M| - 1$ times.

Next, we will be able to combine the different situations of cluster cutting with the quadratic form in the Lemma 2.

Lemma 2. *Let $y \in \{0,1\}^n$. Define x by $x_i = 1$ when $y_i = 1$ and $x_i = -1$ when $y_i = 0$. So for any graph laplacian matrix L, $4y^T Ly = x^T Lx$.*

Proof. $x^T Lx = \sum_{(i,j) \in E} w_{ij}(x_i - x_j)^2 = \sum_{(i,j) \in E} w_{ij}4(y_i - y_j)^2 = 4y^T Ly$

The following lemma gives the functional form of the four variables.

Lemma 3. *Let $x_a, x_b, x_c, x_d \in \{-1, 1\}$. Then the expression function that four variables are not exactly the same is*

$$8 \cdot I(x_a, x_b, x_c, x_d \text{ not all the same})$$
$$= (7 - x_a x_b - x_a x_c - x_a x_d - x_b x_c - x_b x_d - x_c x_d - x_a x_b x_c x_d). \quad (5)$$

We can further organize the above lemma into the following expression:

$$6 - x_a x_b - x_a x_c - x_a x_d - x_b x_c - x_b x_d - x_c x_d$$
$$= \begin{cases} 0 & x_a, x_b, x_c, x_d \text{ are the same} \\ 6 & three \text{ of } x_a, x_b, x_c, x_d \text{ are 1 or } -1 \\ 8 & two \text{ of } x_a, x_b, x_c, x_d \text{ are 1 or } -1. \end{cases} \quad (6)$$

Next, we can derive the relationship between the motif cut and the edge cut from this.

Lemma 4. *Given graph G and motif M with $|M| = 4$, and*

$$cut_M^{(G)}(S, \overline{S}) = \frac{1}{3} cut^{(G_M)}(S, \overline{S}) - \sum_{M_{a,b,c,d}} \frac{1}{3} \cdot I(exactly\ two\ of\ a, b, c, d\ in\ S), \forall S \subset V.$$

$$(7)$$

Proof. If the vector $x \in \{-1, 1\}^n$ is in set S that represents the cutting situation.

$$6 \cdot cut_M^{(G)}(S, \overline{S}) + \sum_{M_{a,b,c,d}} 2 \cdot I(exactly\ two\ of\ a, b, c, d\ in\ S)$$

$$= \sum_{M_{a,b,c,d}} 6 - x_a x_b - x_a x_c - x_a x_d - x_b x_c - x_b x_d - x_c x_d$$

$$= \sum_{M_{a,b,c,d}} \frac{3}{2}(x_a^2 + x_b^2 + x_c^2 + x_d^2) - (x_a x_b + x_a x_c + x_a x_d + x_b x_c + x_b x_d + x_c x_d)$$

$$= \frac{1}{2} x^T D_M x - \frac{1}{2} x^T W_M x = \frac{1}{2} x^T L_M x = 2 \cdot cut^{(G_M)}(S, \overline{S}).$$

Theorem 2. *Given graph G and motif M with $|M| = 4$, and $\forall S \subset V$,*

$$\phi_M^{(G)}(S) = \frac{cut_M^{(G)}(S, \overline{S})}{vol_M^{(G)}(S)} = \phi^{(G_M)}(S) - \frac{\sum_{M_{a,b,c,d}} I(exactly\ two\ of\ a, b, c, d\ in\ S)}{vol^{(G_M)}(S)}.$$

Proof. It can be easily proved by Lemmas 1 and 4.

Therefore, when the motif has four nodes, the motif conductance can be expressed by the conductance of constructed weighted graph and an additional formula. Next, we will analyze the performance of the additional formula in ordinary graphs and bipartite graphs.

Theorem 3. *Let G_1 be an ordinary graph, G_2 be a bipartite graph, where $V = R \cap L$ and $R \cup L = \emptyset$. Let $f(G) = \frac{\sum_{M_{a,b,c,d}} I(exactly\ two\ of\ a, b, c, d\ in\ S)}{vol^{(G_M)}(S)}$. Assuming that G_1 and G_2 have the same number of points and specify the same cutting sequence. Then for any $S \subset V$, in most cases, the value of $f(G_2)$ is less than the value of $f(G_1)$.*

Proof. Assuming that G_1 and G_2 are complete graphs with n nodes, the target cluster S contains nodes i and j, $f(G_1)$ and $f(G_2)$ are calculated respectively. For G_1, $f(G_1) = \frac{C_{n-2}^2 + 2 \cdot C_{n-2}^3}{3(2 \cdot C_{n-2}^2 + 2 \cdot C_{n-2}^3)} = \frac{2n-5}{6(n-1)}$. For G_2, (i) node i and j are in R, so $f(G_2) = \frac{C_{|L|}^2 + 2 \cdot C_{|R|-2}^1 \cdot C_{|L|}^1}{3(2 \cdot C_{|L|}^2 + 2 \cdot C_{|R|-2}^1 \cdot C_{|L|}^2)} = \frac{2|R|-3}{6|R|-6}$. If $|R| < \frac{n+2}{3}$, $f(G_2) < f(G_1)$. (ii) node i is in R and j is in L, $f(G_2) = \frac{C_{|R|-1}^1 \cdot C_{|L|-1}^1 + C_{|R|-1}^1 \cdot C_{|L|-1}^2 + C_{|R|-1}^2 \cdot C_{|L|-1}^1}{3(2 \cdot C_{|R|-1}^1 \cdot C_{|L|-1}^1 + C_{|R|-1}^1 \cdot C_{|L|-1}^2 + C_{|R|-1}^2 \cdot C_{|L|-1}^1)} = \frac{|R|+|L|-2}{3|R|+3|L|} = \frac{n-2}{3n}$. If $n > 4$, then $f(G_2) < f(G_1)$. Combined with the above, in most cases, the value of $f(G_2)$ is less than the value of $f(G_1)$.

Table 1. The values of $\phi^{(G_M)}(S)$ and $\phi_M^{(G)}(S)$ during calculation in Fig. 1

S	$cut_M^{(G)}(S,\overline{S})$	$min(vol)$	$\phi_M^{(G)}(S)$	$\frac{\sum 1}{vol^{(G_M)}(S)}$	$\phi^{(G_M)}(S)$
1	2	2	1	0	1
1, 2	3	4	0.7500	1/12	0.83
1, 2, 6	3	7	0.4286	2/21	0.5238
1, 2, 6, 7	2	10	0.2000	0	0.2000
1, 2, 6, 7, 3	1	7	0.1429	0	0.1429
1, 2, 6, 7, 3, 8	1	6	0.1667	1/42	0.1905
1, 2, 6, 7, 3, 8, 4	2	4	0.5000	0	0.5000
1, 2, 6, 7, 3, 8, 4, 9	1	2	0.5000	1/54	0.5185
1, 2, 6, 7, 3, 8, 4, 9, 5	1	1	1.0000	0	1.0000

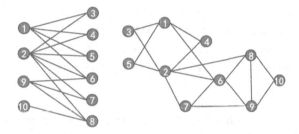

Fig. 2. The ordinary graph $G1$ and the bipartite graph $G2$

In summary, although $\phi_M^{(G)}(S)$ is not equal to $\phi^{(G_M)}(S)$, it can be observed from the Table 1 that the additional plenty is far less than the conductance on G_M, or even 0 in most cases. In practice, the additional formula has almost no effect on the result in large graph. In most cases, the value of the additional formula in bipartite graph is smaller than that in ordinary graph. Because of this, we can use results from G_M to cluster the bipartite graph.

3.3   Algorithm Analysis

The most common motifs in an ordinary graphs are small-sized cliques, such as triangles, but in the bipartite graph, we see that there are no clusters with more than two vertices, and there are no triangles. The most basic motif in the bipartite graph is the complete 2×2 biclique, also known as the butterfly. It is the smallest subgraph in the bipartite graph and the smallest unit of cohesion. It has been used to define basic metrics, such as clustering coefficients in the bipartite graph. It is the basic building block of bipartite networks, like a triangle in ordinary networks. So in this paper, we choose butterfly as the basic motif of bipartite graph.

Next we observe the computational efficiency in ordinary and bipartite graphs. There is an ordinary graph G_1 with butterflies in Fig. 2. First, to

$$W_M^1 \begin{bmatrix} 0&8&3&3&3&5&1&1&0&0 \\ 8&0&3&3&3&8&4&4&3&0 \\ 3&3&0&1&1&1&0&0&0&0 \\ 3&3&1&0&1&1&0&0&0&0 \\ 3&3&1&1&0&1&0&0&0&0 \\ 5&0&1&1&1&0&4&5&4&1 \\ 1&4&0&0&0&4&0&3&3&0 \\ 1&4&0&0&0&5&3&0&4&1 \\ 0&3&0&0&0&4&3&4&0&1 \\ 0&0&0&0&0&1&0&1&1&0 \end{bmatrix} \qquad W_M^2 \begin{bmatrix} 0&6&3&3&3&3&0&0&0&0 \\ 6&0&3&3&3&5&2&2&3&0 \\ 3&3&0&1&1&1&0&0&0&0 \\ 3&3&1&0&1&1&0&0&0&0 \\ 3&3&1&1&0&1&0&0&0&0 \\ 3&5&1&1&1&0&1&1&2&0 \\ 0&2&0&0&0&1&0&1&2&0 \\ 0&2&0&0&0&1&1&0&2&0 \\ 0&3&0&0&0&2&2&2&0&0 \\ 0&0&0&0&0&0&0&0&0&0 \end{bmatrix}$$

Fig. 3. Matrix W_M^1 for the graph G_1 and matrix W_M^2 for the graph G_2

Table 2. Information of datasets

Dataset	$|V| = n$	$|E| = m$	d_{max}	Diameter
email-Eu-core	1004	25571	546	201
ca-HepTh	9877	25998	79	17
ca-CondMat	23133	93497	132	14
facebook	22470	171002	398	152
git-web	37700	289003	1368	106

compare the algorithm applied in different graphs, we constructed a bipartite graph G_2 consistent with the topological structure of G_1. Then, we applied our algorithm to graph G_1 and G_2, and calculated the matrix W_M of the two graphs. As shown in Fig. 3, we can clearly see that the W_M^2 of the bipartite graph is sparser. Compared with the W_M^1, the sparse matrix W_M^2 can greatly save the calculation time and improve the computational efficiency during the calculation of eigenvectors.

Next, the computational complexity of the algorithm proposed in this paper will be analyzed. In general, the calculation of *motif adjacency matrix* W_M and the eigenvectors of *motif laplace matrix* L_M will affect the computational complexity of our algorithm. The number of edges is represented by m. In theory, using Nearly-Linear Time Solvers for Laplacian Systems [11], the computational complexity of the eigenvectors of *motif laplace matrix* is $O(m \, log^{O(1)} \, m)$. The calculation time to form W_M is equivalent to the time to identify all butterfly motifs in the bipartite graph and it has the complexity of $O(min(\sum_{u \in R}(d_u)^2, \sum_{v \in L}(d_v)^2))$ [10,14]. So, the final computational complexity of our algorithm is $O(m \, log^{O(1)} \, m + min(\sum_{u \in R}(d_u)^2, \sum_{v \in L}(d_v)^2))$.

In an ordinary graph and a bipartite graph, assuming that they have the same number of nodes n, the maximum number of butterflies in ordinary graph is C_n^4, and the maximum number of butterfly in bipartite graph is $C_{|R|}^2 \cdot C_{|L|}^2$. Although the magnitude of both is n^4, in practice, the number of butterflies in ordinary graph is several times or even tens of times in bipartite graph. Therefore, in the bipartite graph, the speed of generating the matrix W_M is much better than that in the ordinary graph. From this we can see that our framework applied in the bipartite graph, has a faster calculation speed than the ordinary graph, reduces the calculation time, and improves the efficiency of the optimization framework.

4   Experiments

The experimental results on datasets of different sizes will be presented in this part. Table 2 summarizes the basic information of different datasets. The data source comes from http://snap.stanford.edu/data/. Except for our proposed algorithm of *butterfly conductance cut (BCC)*, we consider two other cluster cutting methods: *the traditional conductance cut (TCC)* and the *K-means method (KMM)*. The *TCC* is a cluster method based on finding the minimum conductance by edges. The *KMM* is another widely used clustering algorithm. Each test result is to repeat the experiment more than 5 times, take the average value, and then record it.

Motif Conductance Results. As shown in Fig. 4, we compared the motif conductance of the clusters cut by three methods. It can be observed that the motif conductance of the cluster S in BCC is the lowest. As the dataset increases, the difference is much more obvious. The small conductance means that the cluster S contains dense butterflies. The result indicates that our framework can

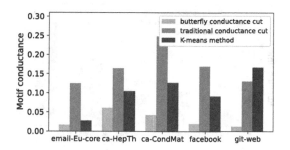

Fig. 4. Motif conductance of three cluster algorithms

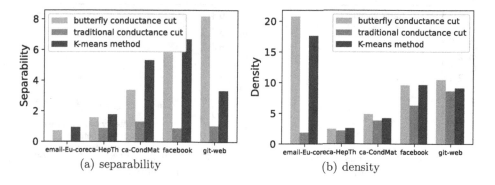

(a) separability

(b) density

Fig. 5. Separability and density of three cluster algorithms

avoid cutting butterfly instances and expose the cluster with dense butterfly instances in bipartite graphs.

Effectiveness Results. In order to better show the effect of clustering experiments, we have defined the following indicators:

Separability can intuitively show that a good cluster is basically separated from the rest of the graph, which indicates that the nodes in cluster S and the nodes in the rest of the graph have almost no connection edges. The ratio of the number of edges in S to the number of edges in $\overline{S}$ is Separability: $g(S) = \frac{m}{number\ of\ cut\ edges}$, the number of edges in S is denoted by m.

Density can intuitively establish a good cluster connection. It can indicate the closeness of the connection within the cluster. It measures the ratio of the number of edges between nodes and nodes in S: $g(S) = \frac{m}{n}$, where the number of edges in S is represented by m, and the number of nodes in S is represented by n.

Figure 5(a) shows the separability statistics of the three algorithms. It can be found from the data in Fig. 5(a) that the separability of the cluster cut by BCC is higher than TCC and KMM, and as the scale of the graph increases, the separability also increases gradually. This means that in the process of cutting, very few edges were cut, and a lot of edges were retained in the cluster S. Thus, the

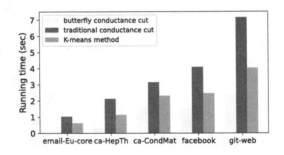

Fig. 6. Running time of three cluster algorithms

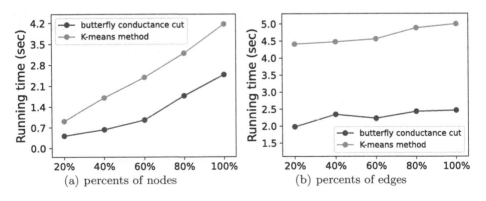

Fig. 7. Scalability testing on nodes and edges

structure of the original graphic was retained to the greatest extent. Similarly, Fig. 5(b) presents the data about the density of three algorithms. Therefore, the data suggest that the density of *BCC* is better than the density of *KMM*. This feature is more obvious in some big datasets like git-web. These experimental results support that our framework shows excellent performance in cluster cutting in bipartite graphs.

Efficiency Results. Figure 6 lists the running time of *BCC*, *TCC* and *KMM*. Obviously, the running time of *BCC* is much lower than the *TCC* and *KMM*. We can see that *BCC* needs about 11%–50% time of *TCC* and about 20%–60% of *KMM* on all the datasets. The results above indicate that the optimization framework in our work has better computational efficiency. The data suggest that our framework can efficiently cut dense clusters in bipartite graphs.

Scalability Results. We use the dataset git-web to analyze the scalability of algorithms. In the previous experiment, the running time of *TCC* is higher than that of the other two methods, so *TCC* is not considered in this experiment. In Fig. 7(a), input different percentage nodes of the graph, and record the running time of the algorithm *BCC* and *KMM* at different percentages. It is not difficult to see that the BCC's running time will grow steadily as the percentage of input nodes gradually increases.

The running time of *KMM* fluctuates greatly and is greater than that of *BCC*. In Fig. 7(b), input different percentage edges of the graph, and recorded the running time. It is not difficult to see that with the increase of the number of edges, the growth of *BCC* has a small fluctuation, while the running time of *KMM* has a slow increase. Through the above experiments, we analyzed that the running time of BCC mainly depends on the number of nodes in the input network, while the influence of the number of edges on the running time is small. Our algorithms can be shown to be more scalable than other algorithms and scalable when handling large networks.

5   Related Work

Some scholars have done a few research on clustering algorithms based on motifs. Jaewon Yang et al. [16] proposed a parameterless cluster recognition method by improving the spectral clustering algorithm. Hao Yin et al. [18] proposed a more comprehensive method of clustering at the edge of the network in a complex network. Jaewon Yang et al. [15] proposed a new paradigm for revealing the clustering of different modules in complex networks. Suraj Jain et al. [7] presented the approach of Sparse Grassmann Clustering (SGC) for cluster cutting. Hao Yin et al. [19] developed the Motif-based Approximate Personalized PageRank (MAPPR) algorithm to find clusters with minimal motif conductance. Previous studies have proposed many clustering methods, including the personalized PageRank method, graph diffusions, local spectral methods, modularity maximization, flow-based algorithms, and minimum degree maximization. The most relevant to our method are several clustering methods based on triangle motifs, but they all perform cluster cutting in ordinary graphs by changing different triangle motifs.

Higher-order structures (network motifs) are widely used in many fields, such as medicine, biology, transportation networks, social networks and so on. Hany Farid et al. [5] proposed a method to detect hidden messages using higher-order statistical models. Dermot M.F. Cooper et al. [4] studied the regulation of adenylate cyclase by studying its higher-order organization. Helmut Cölfen et al. [3] introduced the interaction between higher-order organizaitons that aggregate and crystallize in materials. More and more researchers are turning their attention to the problem of higher-order modes. Many existing researches are based on small-scale cliques such as triangular motifs, but some cliques with a larger structure are also crucial for the construction of complex networks, for example, four-node cliques in protein and protein interaction networks in medicine. However, clustering based on such high-order structure has not been well understood and studied.

6   Conclusions

In the work of this paper, we study how to identify higher-order clusters with motif of dense butterflies in bipartite networks. First, by optimizing the form of motif conductance, we propose the framework of higher-order clustering algorithm. Then, we prove that the problem can be transformed to computing the conductance of a weight graph constructed by the butterfly, so it can be solved by matrix eigenvalue decomposition techniques. Next, through code analysis, we summarize the computational complexity of the algorithm proposed in this article. A large number of experiments have proved the efficiency, robustness and scalability of the algorithm in our paper.

Acknowledgements. This research was supported by the National Key Research and Development Program of China (No. 2018YFB1004402).

References

1. Alon, U.: Network motifs: theory and experimental approaches. Nat. Rev. Genet. **8**(6), 450–461 (2007)
2. Benson, A.R., Gleich, D.F., Leskovec, J.: Higher-order organization of complex networks. Science **353**(6295), 163–166 (2016)
3. Colfen, H., Mann, S.: Higher-order organization by mesoscale self-assembly and transformation of hybrid nanostructures. Angew. Chem. **42**(21), 2350–2365 (2003)
4. Cooper, D.M., Crossthwaite, A.J.: Higher-order organization and regulation of adenylyl cyclases. Trends Pharmacol. Sci. **27**(8), 426–431 (2006)
5. Farid, H.: Detecting hidden messages using higher-order statistical models. In: IEEE IPIP, vol. 2, p. II (2002)
6. Hou, Y., Whang, J.J., Gleich, D.F., Dhillon, I.S.: Non-exhaustive, overlapping clustering via low-rank semidefinite programming, pp. 427–436 (2015)

7. Jain, S., Govindu, V.M.: Efficient higher-order clustering on the grassmann manifold. In: ICCV 2013, USA, pp. 3511–3518. IEEE (2013)
8. Mangan, S., Alon, U.: Structure and function of the feed-forward loop network motif. Proc. Natl. Acad. Sci. U.S.A. **100**(21), 11980–11985 (2003)
9. Milo, R., Shenorr, S.S., Itzkovitz, S., Kashtan, N., Chklovskii, D.B., Alon, U.: Network motifs: simple building blocks of complex networks. Science **298**(5594), 824–827 (2002)
10. Saneimehri, S., Sariyuce, A.E., Tirthapura, S.: Butterfly counting in bipartite networks, pp. 2150–2159 (2018)
11. Schaub, M.T., Trefois, M., Van Dooren, P., Delvenne, J.: Sparse matrix factorizations for fast linear solvers with application to laplacian systems. SIAM J. Matrix Anal. Appl. **38**(2), 505–529 (2017)
12. Seshadhri, C., Pinar, A., Kolda, T.G.: Wedge sampling for computing clustering coefficients and triangle counts on large graphs. Stat. Anal. Data Mining **7**(4), 294–307 (2014)
13. Smith, K.: On neighbourhood degree sequences of complex networks. Sci. Rep. **9**(1), 8340 (2019)
14. Wang, K., Lin, X., Qin, L., Zhang, W., Zhang, Y.: Vertex priority based butterfly counting for large-scale bipartite networks. Proc. VLDB Endow. **12**(10), 1139–1152 (2019). https://doi.org/10.14778/3339490.3339497
15. Yang, J., Leskovec, J.: Overlapping communities explain core-periphery organization of networks. Proc. IEEE **102**(12), 1892–1902 (2014)
16. Yang, J., Leskovec, J.: Defining and evaluating network communities based on ground-truth. Knowl. Inf. Syst. **42**(1), 181–213 (2013). https://doi.org/10.1007/s10115-013-0693-z
17. Yaveroglu, O.N., et al.: Revealing the hidden language of complex networks. Sci. Rep. **4**(1), 4547–4547 (2015)
18. Yin, H., Benson, A.R., Leskovec, J.: Higher-order clustering in networks. Phys. Rev. E **97**(5), 052306 (2018)
19. Yin, H., Benson, A.R., Leskovec, J., Gleich, D.F.: Local higher-order graph clustering. In: KDD 2017, pp. 555–564. ACM, New York (2017)

Learning Dynamic Pricing Rules
for Flight Tickets

Jian Cao$^{(\boxtimes)}$, Zeling Liu, and Yao Wu

Department of Computer Science and Engineering, Shanghai Jiaotong University,
Shanghai, People's Republic of China
{cao-jian,lzllzl2013051,wuyaoericyy}@sjtu.edu.cn

Abstract. It is possible and necessary to adjust the flight ticket prices
for each airlines dynamically in order to increase online travel agencies'
revenues. Unfortunately, the demands and the availability of flight tickets
change following very complex patterns so that it is very hard, if not
impossible, to adopt mathematical models to describe them and to derive
analytical solutions. We apply reinforcement learning approach to learn
dynamic pricing rules from a passenger simulator which can generate
passengers' responses according to flight tickets' prices. In order to make
passenger simulator more realistic, it adjusts it's inherent models based
on historical data and up-to-date data continuously. The experimental
results on a real-world data set show that our approach can learn dynamic
pricing rules efficiently.

Keywords: Dynamic pricing · Yield management · Reinforcement
learning · Simulation

1 Introduction

Dynamic pricing is a business strategy that adjusts the product price in a timely
fashion in order to allocate the right service, to the right customer, at the right
time [1]. Many previous researches focused on dynamic pricing problems by
modeling the customers' demand curves with variety kinds of mathematical formulas [2,3]. However, flight tickets are perishable products which have finite
time horizons and low marginal cost [4]. Especially, the demand curves of perishable products are non-stationary and, it's difficult to describe the demands
into mathematical functions as the demands of perishable products are affected
by huge amount of factors.

This paper presents a dynamic pricing rule learning method taking advantage
of reinforcement learning, which is a model-free approach [5]. It is not possible
to adjust the prices to observe passengers' responses in real environment, which
is a necessary for reinforcement learning. We implement a passenger simulator
which can predict whether a simulated passenger will buy a ticket or not in
terms of his preference. Comparing existing passenger simulator platform such
as PODS [6], our passenger simulator can adjust it's inherent models based on
historical and up-to-date data.

G. Li et al. (Eds.): KSEM 2020, LNAI 12274, pp. 498–505, 2020.
https://doi.org/10.1007/978-3-030-55130-8_43

2   Related Work

Many researchers have formulated dynamic pricing problem using different models [7–9]. Unfortunately, providing accurate analytical models to describe the changing demands and affecting factors is too difficult to make it to fit for the real situations. The choice behaviors are widely researched. For example, Cao J. et al. proposed models to describe passenger preferences to flights [10]. [11] studied the passengers' choice behaviors under flight delay. In addition, a set of passenger simulators have developed and the most famous one, PODS needs the order number as it's input [6].

Reinforcement Learning (RL), has emerged as a powerful tool for solving complex sequential decision-making problems. Q-learning [12] has been used in the flight ticket dynamic pricing problem [13], and other pricing problems such as cloud service [14]. However, using a look-up table enlarges the training space and as a result, Q-learning method converges rather slow. What's more, the possible sizes of action space and state space are limited. To solve these problems, Volodymyr Mnih et al. introduced an algorithm using a neural network instead of the lookup table in Q-learning [15]. Furthermore, in many problems, actions are made according to not only the current states but also the past states, thus the Markov decision problems turn into partially observed Markov decision problems. [16] proposed the method using recurrent network to solve these problems. We borrow the idea from [16] to implement DRQN model.

3   Passenger Simulator

Passenger simulator is an environment that provides feedback to the reinforcement learning module. Passengers can be divided into five types according to their attributes and the way they book tickets [17]. Based on the historical orders and other studies [10], the following three features are the most important ones in modeling a passenger's preference for a flight ticket:

1) Airlines: The airline preference is represented as:

$$A_u = [a_1, a_2, a_3, ..., a_n] \quad s.t. \quad 0 \leq a_i, \sum_{i=1}^{n} a_i = 1 \tag{1}$$

where n equals the number of different airlines appearing in the historical orders.

2) Class of Travel: Without loss of generality, in our study, class is divided into economy, business and commercial. Class preference is defined as:

$$C_u = [c_1, c_2, c_3, ..., c_m] \quad s.t. \quad 0 \leq v_i, \sum_{i=1}^{m} c_i = 1 \tag{2}$$

where m equals to the number of different classes appearing in the historical orders.

3) Price: For a given pair of departure and arrival cities, the price-sensitivity equals to:

$$P_{sen} = \frac{p_{high} - \bar{p}}{p_{high} - p_{low}} \tag{3}$$

where p_{high} and p_{low} represent the highest and the lowest price for the same flight, respectively. $\bar{p}$ denotes the average price of the tickets the passenger selected.

We use $Prf_u = <A_u, C_u, p_{senu}>$ to represent preference of passenger u. Similarly, a ticket can be represented by $Tck_t = <A_t, C_t, p_t>$. A_t is an n dimensional one-hot column vector indicating the airline information in this ticket. C_t is an m dimensional one-hot column vector indicating class information in this ticket. p_t is the price metrics.

The utility function for passenger u on a flight ticket t is denoted as U_u^t, which is defined as:

$$U_u^t = w_1(A_u \cdot A_t) + w_2(C_u \cdot C_t) + w_3 F(p_u, p_t)$$

$$F(x, y) = \begin{cases} 1 & x \geq y \\ \dfrac{2}{1+e^{\frac{2(y-x)}{x}}} & x < y \end{cases} \tag{4}$$

$$\sum_{i=1,2,3} w_i = 1$$

We define U^* as the threshold value that decides a passenger is satisfied or not. In other words, passenger u is satisfied with ticket t if and only if $U_u^t \geq U^*$.

The buying ratio is the proportion between the number of final orders and the number of passengers who want to book tickets. Suppose the buying ratio is $BR(qd, i)$, where qd represents the date to query tickets and i represents the number of days between query date and flight date. Suppose the buying ratio will become zero if flight date is later than the query date by MaxConv days. Therefore, the buying ratio can be estimated using a linear function:

$$BR(qd, i) = \alpha * \frac{MaxConv - i}{MaxConv}, \quad i = 0, 1, 2, ..., MaxConv \tag{5}$$

where α is a decay coefficient, which can be estimated based on actual orders.

Suppose d_2 and d_1 represents the dates that are two days and one day before the query date qd respectively. We use $Num_p(qd, i)$ to denote the number of passengers who are satisfied with the ticket and $Num_O(qd)$ represents the number of orders that are really placed on date qd. Since the decay coefficient is self-correlated, the decay coefficient on date qd can be estimated as follows:

$$\alpha_{qd} = \theta\alpha_{d_1} + (1 - \theta)\alpha_{d_2} \tag{6}$$

Where θ is self-correlation coefficient. And after we collect real daily orders whose number is $Num_O(qd)$, we can update the true value of decay coefficient on date qd since it satisfies:

$$Num_O(qd) = \sum_{i=0}^{MaxConv} Num_p(sd, i) \cdot (\alpha_{qd} \frac{MaxConv - i}{MaxConv}) \tag{7}$$

which implies:

$$\alpha_{qd} = \frac{Num_O(qd)}{\sum_{i=0}^{MaxConv} Num_p(sd, i) \cdot \frac{MaxConv-i}{MaxConv}} \tag{8}$$

The simulator will generate a group of passengers with quantity of $S_{qd,i}$, which represents the number of passengers who search for flights on date qd and flight date is i days later. Each passenger is generated according to passenger type distribution and his preference model generated randomly in terms of the historical data. The utility function is applied on each passengers to simulate the choice process. Each passenger chooses ticket t which maximizes U_u^t, if $U_u^t \geq U^*$, this passenger is called as a satisfied passenger. The number of satisfied passengers whose query date is qd and their flight date is i days later is denoted as $Num_p(sd, i), i = 0, 1, ...MaxConv$. Thus the predicted total order number on query date qd is:

$$Num_Oqd = \sum_{i=0}^{MaxConv} Num_p(sd, i) \cdot (\alpha_{qd} * \frac{MaxConv - i}{MaxConv}) \tag{9}$$

Finally, after we obtain real order number after date qd, we can update the decay coefficient according to Eq. 8.

4   A Reinforcement Learning Model for Learning Dynamic Pricing Rules

The flight ticket dynamic pricing decision process is characterized by following elements:

Decision Epoch: The OTA sells tickets over a fixed time horizon of T days. The days were indexed by t (t = 1, 2, ... T).

Agent: The OTA is the agent in our reinforcement learning problem.

State s: A state S_t at day t tells the agent what the environment is. The elements in state should be highly related to the pricing problems. The elements we are interested in are shown as follows:

Time: Time index t.

Order Quantity Prediction: The order quantity on day t with price adjust Δp is represented as:

$$Ord_{Pre}(t, \Delta p) \tag{10}$$

Specially, when no price adjustment is applied, $\Delta p = 0$.

Passenger Feedback: The real order number on query date t with price adjustment Δp is denoted as $Ord_{Real}(t, \Delta p)$. Passenger feedback on day $t - 1$ with price adjust Δp applied is formulated as:

$$UFB_{t-1,\Delta p} = Ord_{Real}(t - 1, \Delta p) - Ord_{Pre}(t - 1, \Delta p) \tag{11}$$

Thus the state on day t is represented as:

$$S_t = \;<t, UFB_{t-1,\Delta p}, Ord_{Pre}(t,0)> \tag{12}$$

where Δp represents the amount of price adjustment applied on day $t-1$.

Action Set A: Action set A is a list of real numbers, standing for how much the price adjustment will be. All the ticket prices will be changed by a_t on day t.

Reward: Let R^* be the revenue OTA can earn by selling one ticket at the original price. To simplify, we assume R^* to be a constant. The reward on day t by taking action a_t is:

$$r(t, a_t) = Ord_{Real}(t, a_t) * (R^* + a_t) \tag{13}$$

State Transition: State transition problem is rather clear in our model. At the end of day t, after we collect real order data from the market, the state changes into:

$$S_{t+1} = \;<t+1, UFB_{t,\Delta p}, Ord_{Pre}(t+1,0)> \tag{14}$$

In most real world environments, it's unlikely to provide full states to the agent. As a result, Markov assumption does not hold any more. The Markov decision problem that traditional DQN can solve falls into a partially observed Markov decision problem. Besides, flight ticket dynamic pricing problem can be also regarded as a time series problem since most states varies with time and are self-correlated to some extent. Recurrent network is a reliable solution to solve such problems, and as a result the Deep Recurrent Network algorithm, i.e., a LSTM network, is placed in DQN structure is proposed and applied in the pricing problem. All reinforcement learning elements discussed above remain the same except:

State: State model changes in DRQN model:

$$Recurrent S_t = \begin{bmatrix} S_{t-nsteps+1} \\ S_{t-nsteps+2} \\ \vdots \\ S_t \end{bmatrix} \tag{15}$$

Where $nsteps$ represents the time steps of LSTM structure used in the model.

The reward and state mentioned above are too large in amount, which undoubtedly leads to divergence in the reinforcement learning problem. These features are normalized as follows:

$$r^*(t, a_t) = \frac{Ord_{Real}(t, a_t) * (R^* + a_t)}{P} \tag{16}$$

where p represents the average ticket price calculated from historical data.

$$S_t^* = \;<\frac{t}{T}, \frac{UFB_{t-1,\Delta p}}{O_{max}}, \frac{Ord_{Pre}(t,0)}{O_{max}}> \tag{17}$$

where T represents total number of days mentioned in **Decision Epoch** and O_{max} represents the maximum order number in the past historical data.

5   Experiments

5.1   Dataset

We collected historical orders of flights from Beijing to Los Angeles from 2016 to 2018 from one online travel company. There are total 75238 orders in this period. Each record includes take off date, purchasing date, airline, seat class, and ticket price. In addition, the dataset also includes total search volumes, order numbers and available air ticket information in each day.

5.2   Passenger Choice Simulation

Feature distributions are calculated using the historical data and are provided to the simulator introduced above. ω_i and U^* are tuned using grid search method with step size 0.05. The training period starts from July 1st, 2017 till December 31st, 2017 with θ set to be 0.9 and $MaxConv$ set to be 180. For each set of parameters, a list of predicted order numbers is outputted by the simulator, which is denoted as Y_{ω,U^*}. At the same time, the real order number list in the same time period is defined as O. The optimal parameter values are chosen based on:

$$\omega_i, U^* = arg \max_{\omega_i, U^*} MAPE(Y_{\omega,U^*}, O)$$

$$MAPE(X, Y) = \sum_{i=1}^{|X|} |\frac{X_i - Y_i}{X_i}| * \frac{100}{|X|} \quad (18)$$

Furthermore, we test the performance of the simulator from Jan 1st, 2018 to May 1st, 2018 and the MAPE is 16.52%. The result shows the accuracy and robustness of our simulator.

5.3   Performance of DRQN Based Dynamic Pricing Strategy

Total time horizon length T is set to be 90. For reinforcement learning models, learning rate is set to be 0.01, batch size is set to be 128, memory size is set to be 400, LSTM time step is set to be 5 with 20 hidden units, and target network's parameters are updated every 300 steps. The algorithm is trained from Aug 1st, 2017 for 1000 episodes, and is tested from Dec 1st, 2017. R^* is chosen every 50 from 50 to 1000, and we repeat the experiments on different R^*.

We also compare DRQN based dynamic pricing strategy with the following approaches:

- Demand Modeling-based Pricing Strategy (DMPS): This strategy is based on the relation modeling between numbers of orders and prices. The price can be decided by maximizing the total revenues.
- Fixed Pricing Strategy (FPS): It adjusts a fixed amount of money to the original ticket price.

- Rule-based Pricing Strategy (RPS): It adds 5 yuan to the original price if yesterday's real order quantity is larger than the predicted one, vice versa.
- Q-Learning based Pricing Strategy (QLPS) [14]: It is a Q-Leaning based approach to learn how to adjust prices.

Table 1 lists the revenue gain percentages of different approaches based on multiple experiments. It lists the average values and the standard deviations. It also lists the average values which are ranked at the 25% and the 75% position.

Table 1. Performance comparisons of different pricing strategies

Pricing rule	Average (%)	Std (%)	25th quantile (%)	75th quantile (%)
DMPS	−8.43	5.79	−10.02	−4.66
FPS (−15)	−3.17	5.41	−6.46	0.11
FPS (−10)	10.04	4.83	6.93	11.89
FPS (−5)	−2.02	1.83	−3.03	−0.79
No rule applied	0	−	−	−
FPS (+5)	2.02	1.83	0.79	3.03
FPS (+10)	2.44	3.25	0.21	4.8
FPS (+15)	2.74	3.65	0.97	3.76
RPS	5.34	5.95	1.16	6.7
QLPS	−0.07	6.59	−0.08	0.23
DRQN	10.41	5.15	7.94	14.46

From the Table 1, it can be observed that DRQN achieves the best results comparing with other strategies. The performance of DRQN is also relative stable.

6   Conclusions

This paper proposed an approach which provides a real-time, interactive, robust dynamic pricing strategy using deep recurrent reinforcement learning algorithm (DRQN). This approach is trained by a passenger choice simulator. The simulator proposed in this paper shows a new methodology to simulate passenger actions and predict order quantity based on a passenger preference model, a buying ratio model and passenger utility functions. The simulator's performance is proofed by the experiments. The DRQN method takes advantages of traditional DQN method and recurrent networks and is able to solve the POMDPs.

References

1. Den Boer, A.V.: Dynamic pricing and learning: historical origins, current research, and new directions. Surv. Oper. Res. Manag. Sci. **20**(1), 1–18 (2015)

2. Wang, K., Zhang, A., Zhang, Y.: Key determinants of airline pricing and air travel demand in China and India: policy, ownership, and LCC competition. Transp. Policy **63**, 80–89 (2018)
3. Anjos, M.F., Cheng, R.C., Currie, C.S.: Maximizing revenue in the airline industry under one-way pricing. J. Oper. Res. Soc. **55**(5), 535–541 (2004)
4. Tekin, P., Erol, R.: A new dynamic pricing model for the effective sustainability of perishable product life cycle. Sustainability **9**(8), 1–22 (2017)
5. Van Hasselt, H., Guez, A., Silver, D.: Deep reinforcement learning with double q-learning. In: National Conference on Artificial Intelligence (2016)
6. Carrier, E.: Modeling airline passenger choice: passenger preference for schedule in the passenger origin-destination simulator (PODS). Doctoral dissertation, Massachusetts Institute of Technology (2003)
7. Gallego, G., Van Ryzin, G.: Optimal dynamic pricing of inventories with stochastic demand over finite horizons. Manag. Sci. **40**(8), 999–1020 (1994)
8. Bitran, G., Caldentey, R., Mondschein, S.: Coordinating clearance markdown sales of seasonal products in retail chains. Oper. Res. **46**(5), 609–624 (1998)
9. Lin, K.Y.: A sequential dynamic pricing model and its applications. Naval Res. Logistics (NRL) **51**(4), 501–521 (2004)
10. Cao, J., Yang, F., Xu, Y., Tan, Y., Xiao, Q.: Personalized flight recommendations via paired choice modeling (2017)
11. Jiang, H., Ren, X.: Model of passenger behavior choice under flight delay based on dynamic reference point. J. Air Transp. Manag. **75**, 51–60 (2019)
12. Sutton, R.S., Barto, A.G.: Reinforcement Learning An Introduction. MIT Press, Cambridge (1998)
13. Collins, A., Thomas, L.: Learning competitive dynamic airline pricing under different customer models. J. Revenue Pricing Manag. **12**(5), 416–430 (2013)
14. Ren, J., Pang, L., Cheng, Y.: Dynamic pricing scheme for IaaS cloud platform based on load balancing: a Q-learning approach. In: International Conference on Software Engineering (2017)
15. Mnih, V., et al.: Human-level control through deep reinforcement learning. Nature **518**(7540), 529 (2015)
16. Hausknecht, M., Stone, P.: Deep recurrent q-learning for partially observable MDPs. In: National Conference on Artificial Intelligence (2015)
17. Williams, K. R. (2017). Dynamic Airline Pricing and Seat Availability. Social Science Research Network

Author Index

Printed in the United States
By Bookmasters